Calabi-Yau Manifolds

A Bestiary for Physicists

2nd Ed.
2024

Családomnak

Calabi–Yau Manifolds

A Bestiary for Physicists

2nd Ed.
2024

Tristan Hübsch

Department of Physics and Astronomy
Howard University
USA

World Scientific

NEW JERSEY • LONDON • SINGAPORE • BEIJING • SHANGHAI • HONG KONG • TAIPEI • CHENNAI

Published by

World Scientific Publishing Europe Ltd.

57 Shelton Street, Covent Garden, London WC2H 9HE

Head office: 5 Toh Tuck Link, Singapore 596224

USA office: 27 Warren Street, Suite 401-402, Hackensack, NJ 07601

Library of Congress Cataloging-in-Publication Data
Names: Hübsch, Tristan, author.
Title: Calabi-yau manifolds : a bestiary for physicists / Tristan Hübsch,
Department of Physics and Astronomy, Howard University, USA.
Description: Second edition. | New Jersey : World Scientific, [2024] |
Includes bibliographical references and index.
Identifiers: LCCN 2024032191 | ISBN 9781800616424 (hardcover) |
ISBN 9781800616431 (ebook for institutions) | ISBN 9781800616448 (ebook for individuals)
Subjects: LCSH: Calabi-Yau manifolds. | Mathematical physics.
Classification: LCC QC20.7.M24 H83 2024 | DDC 530.15/607--dc23/eng/20241119
LC record available at https://lccn.loc.gov/2024032191

British Library Cataloguing-in-Publication Data
A catalogue record for this book is available from the British Library.

For any available supplementary material, please visit
https://www.worldscientific.com/worldscibooks/10.1142/Q0484#t=suppl

Desk Editors: Aanand Jayaraman/WSE-Gabriel Rawlinson

Typeset by Stallion Press
Email: enquiries@stallionpress.com

Preface, 2024 Edition

THE VERY FIRST SENTENCE of the original 1992 preface echoes presciently: active research and literature on Calabi–Yau models have grown tremendously over the past 32 years. The relevant online publication databases have by now consolidated, so that arXiv, inSPIRE-hep, Google Scholar and others have become standard resources via your favorite internet browser and search engine. The original edition of the *Bestiary* [272] was completed just before the arXiv repository and the World Wide Web sprouted.* More importantly, the subject itself has undergone a combinatorially accelerating development—in breadth, depth and inter-related complexity, and engendering paradigm shifts dubbed *string revolutions*: the second one, 1994–2003, the third one, 2003–, and some as yet unlabeled tectonic rumblings since around 2015–2017.

As of 2024, superstring applications include much more than just the smooth and perhaps mildly singular spaces that were tacitly taken for granted three decades ago. As originally indicated in § F.4 and now surveyed telegraphically in § F.8, the Lagrange multipliers in a *constrained nonlinear* σ*-model* (F.4.2) necessarily become propagating superfields, corresponding to the fiber-coordinates of the defining equation line bundles, giving rise to the *gauged linear* σ*-model* (GLSM) [241, 443]. This framework and its $(0,2)$-supersymmetric generalization [371, 459] quickly became the bedrock of the subsequent developments in the field, especially owing to their direct relation to (complex algebraic) toric geometry. This much broader

* Online encyclopedic resources such as Wolfram MathWorld, Wikipedia, Encyclopedia of Mathematics and nLab were also yet to emerge, in 1995, 2001, 2002 and 2008, respectively.

framework unquestionably showed [304, Fig. 19] that stringy compactifications necessarily include certain (stratified) pseudomanifolds in the "hybrid phases" of most GLSMs. Indeed, string theory inherently includes various so-called 'branes (0-, 1- and 2-branes are points, strings and membranes, respectively) [284, 283], so it should be unsurprising that string theory target spaces may themselves be stratified. Indeed, more rigorous examinations of the inclusion of 'branes introduced *derived categories* into the current vernacular [303]; see also [249]. Similarly, the study of *mirror symmetry* soon revealed relations between complex algebraic and symplectic geometry [468, 279], as well as the geometric Langlands program [421]. In fact, Calabi–Yau spaces (of various, unbounded dimensions) were found even in quantum field theory Feynman amplitude computations [339]!

The interested Reader has by now many more relevant sources even in book form: on string theory, from introductory [293, 251] to advanced [284, 275, 277, 249, 462], and even catering to mathematicians [262, 259, 290]; on toric geometry [260, 255, 270, 249, 257]; and a continuing sequence of collected works on the mirror "symmetry"/map [291]. It is then particularly gratifying to find that this modest volume seems to have weathered fairly well and remains usable through the substantial changes in methods, techniques and even modes of research.† This revised edition is thus perforce restricted to errata, an expanded bibliography and modest updating edits—ranging from brief 2024⌈*inline statements and footnotes*⌋ to short, highlighted passages, so-indicated for emphasis and perspective. These include amending my erstwhile focus on L^2-cohomology, such as in § 7.3, the closing of Chapter E and the discussion of $(1,1)^3$-couplings and normalizations in § 8.7.

Hindsight being perfect, I can but wistfully sigh at having mentioned the following only too briefly: (**1**) weighted projective spaces (§ 5.1), where the by now iconic "shield" plot of mirror pairs of hypersurfaces was found [353, 350]; (**2**) toric geometry (§ 4.3.2), which provides the richest known construction database of Calabi–Yau hypersurfaces [430, 431, 295]; and (**3**) having completely missed discussing the extension of the "standard" $\mathcal{T}_{\mathcal{M}} \hookrightarrow \mathcal{V}_{E_8}$ Ansatz (§ 0.2.3) to deformed and higher-rank

---

† Distance collaborations of course did exist even *then*: Refs. [148, 121] for example involved intense *bitnet* and (*inter-continental!*) "snail-mail" exchanges with co-authors whom I will have met in person only much later. However, the by now mundane and ubiquitous globe-spanning nearly instantaneous communications and *TB-size* data sharing were sheer science fiction in 1992.

sub-bundles [476], which results in the most astonishingly bountiful pool of constructions [365] (see the updating passages on p. 22). A just coverage of these and related topics would require orders of magnitude more space than an update to this text can possibly afford, and the interested Reader is referred to the literature. This revision is also greatly helped by Yang-Hui He's generous and kind Afterword, p. 371 and [402], providing an active researcher's independent perspective of the manifold developments since 1992, and especially on the rapidly increasing use of various forms of computer-aided methods within the last decade.

Whereas this revised edition may resemble a choir of echoes from what in this field is almost prehistory, it also curiously resonates with some novel developments and hopefully sets the stage for more: (**1**) Introduced in 2015, generalized complete intersections [296, 319, 386] unexpectedly include (both algebraic and non-algebraic) toric models [320, 321, 322]. This significantly complicates and expands the world-wide web of Calabi–Yau models discussed in Chapter D and [305, 430]. (**2**) Also in about 2015, the "metastring" framework [379] carefully retraced (in the underlying world-sheet quantum field theory) the left- and right-moving degrees of freedom down to the zero-modes, where it uncovered a symplectic, non-commutative and chiral doubling of the stringy target spacetime; see [325] for a scouting reconnoissance. (**3**) May the kind Reader indulge in one more rumbling: the gauge-neutral "E_6-singlets," corresponding to complex structure moduli for the *total space of* $\mathcal{T}_{\mathcal{M}}$—the bundle being embedded into one of the E_8 bundles to produce the E_6 gauge symmetry in the standard Ansatz. These chargeless fields interact with the charged fields (0.5.11), and so directly affect the dynamics of the Standard Model particles. There are also other gauge-neutral (or otherwise charged) objects with however less obvious phenomenological impact; see [479, 297] and [458, 436, 373, 330] for starters. Unlike the charged fields, for which the correspondence to $\mathcal{T}_{\mathcal{M}}$- and $\mathcal{T}^*_{\mathcal{M}}$-valued cohomology provides multiple rigorously well-established mathematical results (see Chapters 7–A and C), the E_6-singlets (Chapter B) are less well understood. Here, a syzygy of recent developments in computer-aided (and *-guided!*) computation, and in particular the physically relevant normalization of Yukawa couplings [345, 363], provide a promising new avenue for starting to deliver on the by now four-decade-old promise [160] of deriving the Standard Model from string theory.

— ❦ —

The opportunity to update the *Bestiary* [272] in its 32nd year is a wonderful pleasure, and I should like to thank Laurent Chaminade and Rosie Williamson at World Scientific for making this possible and for all the help with this project. The organization and content of the updates in this edition owe much to my continuing learning from all of my research collaborators and colleagues, Per Berglund most prominently in this field—but none of whom however could possibly be faulted for the omissions, misunderstandings and other mistakes that I am certain remain in no short supply. Over the past three decades, several colleagues have kindly reported errors in the original text and graciously suggested corrections, as collected in its errata web-page.[‡] Many thanks indeed to Vishnu Jejjala, Karol Palka, Andreas Schachner, Philippe Spindel and Nicholas Warner. While this 2024 edition includes those and many additional corrections and updates, all remaining mistakes (typographical and otherwise) are still completely mine.[§]

Tristan Hübsch
Alexandria, 2024

[‡] See online, at https://tristan.nfshost.com/ErrBeast.html.

[§] NISI ERRARE ESSET DIVINVM, NOS NON ESSEMVS; corrected from p. xi, with special thanks to Prof. M. Myerowitz Levine for historic and linguistic advice on this dusty adage.

Preface, 1992 Edition

Bestiary (bɛstɪərɪ) [ad. L. *bestiārius* 'a fighter with beasts in public spectacles,' and med. L. *bestiārium* a menagerie, also 'liber de bestiis compositus,' etc., ...]...
1871 *Sacristy* I. 7/1 The Bestiaries..are natural histories of animals treated so that the peculiarities of animals shall convey a wholesome moral.
— *The Oxford English Dictionary*

THE RAPIDLY INCREASING number of articles about Calabi–Yau spaces makes it very difficult, if not impossible, to fruitfully use and learn from the existing results without being forced to spend too much time merely in getting oriented in the literature. Yet, the physics application in compactifying (super)strings has attracted the attention of many a physicist, more often than not quite unprepared for the rather specialized techniques (and even jargon) with the use of which results can be obtained most readily. It is my sincerest hope that the present effort may provide the necessary ingredients for a bridge over the troubled waters that divide the existing results on Calabi–Yau spaces and the continually developing machinery of algebraic geometry from their vigorous physics application.

From the point of view of building superstring models with possibly realistic physics in 3+1-dimensional spacetime, Calabi–Yau spaces of complex 3-dimensions are the most interesting. In complex one and two dimensions, such spaces are unique up to diffeomorphisms: they are the torus (T^2) and the Kummer surface (K3), respectively. These have been encountered previously in various branches of physics and are relatively well known. On the other hand, the number of *known* topological types of Calabi–Yau 3-folds is well in the thousands and more and more new constructions are being uncovered daily.

Yau's proof [54] of Calabi's conjecture [83] guarantees the existence of a unique Ricci-flat Kähler metric (per choice of the Kähler class) on any compact Kähler space, precisely if the first Chern class vanishes. Regardless of dimension, all these are then called Calabi–Yau spaces. Motivated primarily by immediate physics application, however, we shall be interested in compact Calabi–Yau spaces almost exclusively of three complex dimensions. Accordingly, unless explicitly stated, by a Calabi–Yau space we shall understand the three-dimensional case.

Most of this volume is dedicated to methods of constructing and analyzing various Calabi–Yau 3-folds. To begin with, however, we first review some basics of the physics application in superstring models to note the properties of the Calabi–Yau 3-folds that have so far been recognized as physically interesting. The Reader should be aware that string theory with an 'internal' Calabi–Yau space is rapidly developing but rather incomplete and we necessarily rely on approximations. It is gratifying to learn, through particular examples and also general results, that the geometric analysis is quite robust and here we shall adhere to it. The last chapter is intended as a brief directory of some more "stringy" aspects of our subject.

— ❦ —

Algebraic curves were created by God, algebraic surfaces by the Devil *; it is then only human to create algebraic 3-folds. The world of Calabi–Yau 3-folds, even if only a very restricted set of algebraic 3-folds, is in its present stage of infancy most similar to a Universe still in its "first three minutes". New manifolds are being created every moment, joining the menagerie, some of them readily finding their place in this jigsaw puzzle while others seed entire new families, seemingly disconnected from the rest.

Our task therefore necessarily becomes taxonomy in part, guiding the interested Reader through the web of known Calabi–Yau manifolds and preparing the more adventurous Reader for a voyage into the jungle with many more beasts to be discovered. It should be obvious that this could not possibly have been intended as a definitive account on Calabi–Yau 3-folds; not even a compilation of all known facts and results. Rather, the focus

* This saying is attributed to Max Noether; see p. 272 of Ref. [37].

is primarily on techniques and methods which will hopefully have a long-lasting application.

The interested Reader undoubtedly has at least nodding acquaintance with the basics of differential and algebraic geometry which are by now fairly standard in theoretical physics. As a prerequisite we have in mind texts like the *Physics Report* by Eguchi, Gilkey and Hanson [19] and the last couple of chapters of *String Theory II* by Green, Schwarz and Witten [22], to name a few. With a little group theory, linear algebra and an open mind in one's arsenal, no problem should appear in mastering this bestiary.

— ❦ —

The recent literature on Calabi–Yau spaces is simply too abundant for the Bibliography included at the end to be complete in any sense; competing with the databases @BABBAGE.SISSA.IT, @CERNVM, @PUBLICATIONS.MATH.DUKE.EDU, @SLACVM, @XXX.LANL.GOV and others was not the intention. It has been compiled as a suggestion for possible further reading and references for aspects which could not be covered in sufficient detail. It necessarily reflects my personal preference and convenience and should by no means be taken as a value judgment. As regards the structure of the book and the selection and presentation of topics, these have grown out of my own learning and in this respect the innumerable discussions foremost with Paul Green but also with Philip Candelas, Michael Eastwood, Sylvester James Gates, Jr., Joseph Harris, Shin-Tung Yau and many others have taught me and helped a lot. Clearly however, they could not correct all faults of my design and misconceptions. In addition, I'm afraid, all the errors (typographical and otherwise) are entirely mine.† Comments and corrections are most welcome; I hope an opportunity will emerge to incorporate them. Special thanks to Ms. H.M. Ho of the World Scientific Pub. Co., for correcting more than a lion's share of typos and stylistic errors.

Also, I am indebted to all of my research collaborators and colleagues for sharing the joys of research, for countless explanations, discussions, arguments and disagreements—I know I have benefited from all of it and perhaps especially so from the repeated questions which have often helped me to understand my own thoughts better. A simple "thank you" is ... well, perhaps too simple to convey my gratitude, but in printed form, that

† ERRARE DIVINVM EST, ALITER NOS NON SIMVS.

will have to do. Special thanks to Donna D'Fini, for the cover photograph. Thank you all, very much indeed. The support by the DOE grant DE-FG02-88-ER25065 is gratefully acknowledged.

Tristan Hübsch

Cambridge and Austin, 1991
updated: Silver Spring, 1994

About the Author

Tristan Hübsch, born on December 22, 1958, in Novi Sad, Yugoslavia, completed his BSc in physics in 1981 at the Institute of Physics in the Faculty of Sciences of the University of Novi Sad, Yugoslavia. After two years of graduate study in the Department of Physics of the University of Zagreb, Yugoslavia, he transferred to the Department of Physics and Astronomy at the University of Maryland, College Park, in Maryland, USA, where he earned his PhD in physics in 1987.

After postdoctoral positions at the University of Texas (Austin, USA, 1987–1990) and Harvard University (Cambridge, USA, 1990–1992), he has been serving as a professor of physics in the Physics Department of Howard University in Washington, DC, USA, since fall 1992, where he received tenure in 1997 and full professorship in 2001. He has also been appointed as a research professor at the University of Central Florida (Orlando, Florida, USA) and the University of Maryland and as a visiting professor at the University of Novi Sad (Serbia) since 2007. Having participated in several spring and summer schools and workshops, often as an invited lecturer, he has given over 130 lectures around the world. His research has been supported by various granting agencies, including the US Department of Energy (1994–2013); his efforts in improving physics education were supported by the US Department of Education (MSEIP, 1996–2002). His professional publications include over 130 research articles and three books.

Contents

Chapter 0

Spiritus Movens

To begin, we briefly review supersymmetric compactification on Calabi–Yau spaces, having superstring models in mind. To avoid delving too deeply into the tangle of string theory and 2-dimensional superconformal field theories, we propose a deal: we begin with a pointillist sketch, shaded however by world-sheet instanton retouch. This will suffice to determine which characteristics of Calabi–Yau spaces are relevant for their immediate physics application and will provide a subject for most part of this volume. Towards the end, we return to more "stringy" aspects of Calabi–Yau models and include some more recent advances in this field.

0.1 A Pointillist Sketch

From the usual point of view, one begins with string theory in Minkowski spacetime of 9+1 dimensions (denoted M_{9+1}) and then attempts to replace M_{9+1} with $M_{3+1} \times \mathcal{M}$. The phenomenologically motivated $N=1$ local supersymmetry and various consistency requirements then imply that $\mathcal{M}$ is a Calabi–Yau space. This line of analysis was originally carried out by Candelas, Horowitz, Strominger and Witten [93] in the 'point-field limit,' i.e., restricting in the outset to the massless modes of the 9+1-dimensional string theory; they also assumed $\mathcal{M}$ to be smooth.

Instead of a system of (point-like) massless particles, for a string theory, one should consider a system of closed strings which propagate in some general 9+1-dimensional spacetime, X. As proper time of string propagation elapses, we obtain a one-parameter family of strings, the world-sheet, immersed in X. In this approach, it is natural to consider the so-called based loop space, $\Omega_* X$, and construct the Hilbert space from functions on Ω_* [127]. $\Omega_* X$ is the space of all closed paths which begin and end at some point $* \in X$. Notably, this space appears as the fiber of two very different *bundles*$^{(L)}$ over X:

$$\begin{array}{ccc} \Omega_* X \hookrightarrow \mathcal{L}X & & \Omega_* X \hookrightarrow \mathcal{P}_* X \\ \quad\ \downarrow\varpi & \text{and} & \quad\ \downarrow\pi \\ \quad\ X & & \quad\ X \end{array} . \tag{0.1.1}$$

In the first one, we simply sweep the base point, $*$, through X and obtain the so-called free loop space $\mathcal{L}X$; after all, there appears to exist no preferred point in our Universe, X. $\mathcal{L}X$ consists of all closed paths in X and is most naturally the (canonical) configuration space for a theory of closed strings propagating in X. 2024⌈*Since loops support both left- and right-moving modes, this picture naturally involves* oriented *loops and paths, as well as* oriented *loop spaces. As two oppositely oriented copies of any loop in the oriented loop space coincide precisely when the loop shrinks to a point—and those sweep out the base space —* oriented *loop spaces double-cover "ordinary" loop spaces, branching over (a section of) the base space. This description strongly resonates with the "metastring" re-examination of string theory [379] and its chiral doubling of spacetime.*⌋ In the second fibration, $\mathcal{P}_* X$ is the space of all paths $\widehat{*p}$ (open if $p \neq *$, and closed if $p = *$) in which $\Omega_* X$ is embedded as the closed paths. The projection π is given by the endpoints p ranging over all of X (see pp. 197–199 of Ref. [11]).

In general, a fibration $A \hookrightarrow B \twoheadrightarrow X$ allows one to determine the homotopy and (co)homology groups of any one of A, B, X in terms of the other two. Note however that each path $\widehat{*p}$ may be contracted to $* \in X$. That is, $\mathcal{P}_* X$ is contractible and so has trivial homotopy and (co)homology. As a result, using the right-hand-side fibration, the homotopy and (co)homology of $\Omega_* X$ are determined in terms of $\pi_\star(X)$, $H_\star(X)$ and $H^\star(X)$, respectively. Using then the left-hand-side fibration (0.1.1), the analogous data is also determined for $\mathcal{L}X$. For essentially flat spaces, this relation is straightforward and simple; depending on X, however, it may well be beyond our present means of calculation. Nevertheless, from the general framework of geometric quantization, we expect the Hilbert space of the complete string theory on X to be largely determined in a similar fashion. In other words, even though the fully fledged string theory is much more involved than the 'point-field' limit model, the spectrum of massless states and many physical observables can reliably be determined ignoring the stringiness.

— ❦ —

In the other, perhaps presently more popular approach to (super)strings, one considers 1+1-dimensional, (super)conformally invariant quantum field theories over the world-sheet (domain-space) [201]. A sector of each of these theories is given by a (flat) σ-model with the 3+1-dimensional Minkowski spacetime as the target space. The 'rest' of such a quantum field theory is then just about anything, as long as the complete quantum theory is consistent. It is the particular choice of the 'rest' of the theory which determines the physical observables in the effective, low-energy, 3+1-dimensional model. Consequently, two such models with identical 3+1-dimensional physics are considered equivalent even though they might appear as rather different 1+1-dimensional quantum field theories.

For our purposes, the 'rest' of the theory will always contain a 'Calabi–Yau sector' and a 'Yang–Mills sector.' The former is simply a non-linear σ-model, the target space of which is a Calabi–Yau space. The latter is more involved and gives rise to the Yang–Mills interactive fields in the effective model. In this approach, it is natural to consider $\Sigma\mathcal{M}$, the space of mappings $Z^\mu(\sigma)$ of the world-sheet Σ into the Calabi–Yau space $\mathcal{M}$, equipped with the (Euclidean) action functional

$$S_{\mathcal{M}} = \frac{1}{\pi\alpha'}\int_\Sigma \Big\{(B_{\mu\bar\nu} + iG_{\mu\bar\nu})(\bar\partial Z^\mu\,\partial Z^{\bar\nu}) \;-\; (B_{\mu\bar\nu} - iG_{\mu\bar\nu})(\partial Z^\mu\,\bar\partial Z^{\bar\nu})\Big\} \\ +\ \text{susy completion}\ldots \tag{0.1.2}$$

$S_{\mathcal{M}}$ is the integrated total 'energy' of Σ immersed in $\mathcal{M}$; here $\int_\Sigma$ is the reparametrization-invariant integration over the world-sheet Σ and $G_{\mu\bar\nu}(Z,\bar Z)$ is a Hermitian metric on $\mathcal{M}$. In practice, the σ-model is chosen to have full (2,2)-supersymmetry on the world-sheet, the background (vacuum) metric $G_{\mu\bar\nu}$ is Kähler and the background value of $B_{\mu\bar\nu}$ vanishes:

$$\begin{aligned} S_{\mathcal{M}} &= \int_\Sigma \mathrm{d}^2\sigma\,\mathrm{d}^2\varsigma\,\mathrm{d}^2\bar\varsigma\; K(\boldsymbol{Z},\bar{\boldsymbol{Z}})\ , \\ G_{\mu\bar\nu}(Z,\bar Z) &= \partial_\mu\,\partial_{\bar\nu}\,K(\boldsymbol{Z},\bar{\boldsymbol{Z}})\big|_{\varsigma=\bar\varsigma=0}\ , \end{aligned} \tag{0.1.3}$$

where $(\sigma^0,\sigma^1;\varsigma^\pm,\bar\varsigma^\pm)$ are coordinates on the $(2,2)$-super Riemann surface, $\boldsymbol{Z}^\mu$ are the coordinate (chiral) superfields and $\partial_\mu = \partial/\partial\boldsymbol{Z}^\mu$. Furthermore, the required conformal invariance of the σ-model implies that $G_{\mu\bar\nu}$ also must be Ricci-flat, to lowest order.

The 'point-field limit' results are obtained by restricting to the constant maps z^μ. The effects of non-constant maps are then treated as (σ-model) corrections. In particular, holomorphic maps which are not continuous deformations of the constant z^μ give rise to the so-called 'world-sheet instanton' corrections. In addition, choosing genus(Σ) $= 0$ corresponds to the string tree level, while the choice of a genus-n Riemann surface for Σ leads to the nth order string-corrections. Nevertheless, many of the results obtained in the lowest approximation and corrected for the world-sheet instanton effects are actually quite robust [240, 111, 113] in view of the required world-sheet and spacetime supersymmetry. In particular, this level of accuracy will perfectly suffice for our present purpose: to determine which geometrical characteristics of Calabi–Yau manifolds are physically relevant.

Associated to the action (0.1.2), the partition function is defined

$$\mathcal{Z}_{CY} = \int \mathsf{D}[\Sigma]\ \mathsf{D}[\boldsymbol{Z}]\ e^{-S_{\mathcal{M}}}\ , \tag{0.1.4}$$

where the integration over $\mathsf{D}[\Sigma]$ is taken to include a sum over genera and spin structures, integration over moduli and metrics of Riemann surfaces Σ; the action $S_{\mathcal{M}}$ is taken in the Euclidean formulation. Path-integration over $\mathsf{D}[\boldsymbol{Z}]$ ranges over the maps $Z^\mu\colon \Sigma \hookrightarrow \mathcal{M}$ and their superpartners $\zeta^\mu, \zeta^{\overline{\nu}}$. Including the sector pertaining to spacetime, M_{3+1}, and also the Yang–Mills sector,

$$\mathcal{Z}_{\text{tot.}} = \mathcal{Z}_{3+1\text{-spacetime}} \cdot \mathcal{Z}_{\text{CY}} \cdot \mathcal{Z}_{\text{YM}}\ ; \tag{0.1.5}$$

all of these sectors interact with each other. For example, the metric $G_{\mu\overline{\nu}}$ on the internal Calabi–Yau manifold depends on a number of parameters, which are spacetime dependent in general [158, 150]. Also, the Yang–Mills gauge fields are functions of both the 3+1-dimensional spacetime and the "internal" Calabi–Yau space. The complete target space of the total σ-model is then a combined space in which possibly different Calabi–Yau spaces $\mathcal{M}$ are fibered over M_{3+1} and the Yang–Mills sector is then fibered (varies, possibly discontinuously) over both.

Of course, this hierarchical structure of the target space of the total σ-model is a biased choice. From the point of view of the 1+1-dimensional field theory, each of these sectors—the spacetime M_{3+1}, the compactification space $\mathcal{M}$ and also the space of Yang–Mills degrees of freedom—is 'internal' and arises as spanned by the vibrations of the string. Generalizing, one should consider the space of all possible consistent world-sheet quantum field theories. Some of them will span the Minkowski spacetime, M_{3+1}, moreover with $N=1$ local supersymmetry and also possibly realistic gauge-interacting matter and so may come close to a theoretical model of the real world.

— ❦ —

Many such models have been constructed in a wide variety of quite dissimilar ways, in some of which the symmetries alone determine all physically relevant data and moreover at the fully fledged string level. In this respect, such constructions clearly outflank the geometrical σ-model description. Increasing evidence is however being accumulated indicating that these exactly soluble models correspond to certain very special (highly symmetric and possibly singular) Calabi–Yau spaces. Motivated by this, it has been conjectured [138, 30] that in fact all models with $N=1$ locally supersymmetric spacetime M_{3+1} also contain an 'internal' Calabi–Yau space. The converse, namely that every Calabi–Yau space offers a possibly realistic compactification, is beyond every physicist's doubt, although this has not been proven to every mathematician's satisfaction. Since there is a true *embarras des richesses* of Calabi–Yau spaces, we shall focus here on these but keep in mind that many other, possibly equivalent, constructions exist.

0.2 Why Calabi–Yau Spaces?

For ease of notation, we follow here the first attempt [93, 137] to make the heterotic string model [160] physically realistic.* We restrict to the massless modes of the $E_8 \times E_8$ heterotic string and replace the 9+1-dimensional Minkowski spacetime, M_{9+1}, with a product of a 3+1-dimensional maximally symmetric spacetime and some 'internal' compact manifold $\mathcal{M}$. For the resulting 3+1-dimensional effective model to possesses precisely $N=1$ local supersymmetry, consistency will require the 3+1-dimensional spacetime to be Minkowskian, M_{3+1}, and $\mathcal{M}$ to be Ricci-flat and Kähler if Riemannian.

0.2.1 $N=1$ supersymmetry

Recall that the supersymmetry parameter in the original heterotic string model is a spinor transforming as the $\mathbf{8}_S$ representation of the $SO(8)$ *helicity subgroup*$^{(L)}$ of the *Lorentz group*$^{(L)}$ $SO(1,9)$. Upon passing to the product $M_{3+1} \times \mathcal{M}$, the helicity subgroup is broken to $SO(2) \times SO(6)$ and the 8-component spinor gives rise to *four* 2-component spinors of $SO(2)$, corresponding to four independent supersymmetries in the 3+1-dimensional sense. One—and only one—linear combination of these is needed to

* ⌈*In fact, type IIB string theory and its F-theory cousin can also be used to produce physically realistic effective field theory models by compactifying the F-theory on an elliptically fibered Calabi–Yau 4-fold; see Refs. [437, 475] for reviews on the topic.*⌋

parametrize the desired $N=1$ (super)symmetry of the effective 'low energy' physics.

To this end, the background values of the fermionic fields are set to vanish (no preferred spinor in M_{3+1}). For consistency then, the supersymmetry transformations are required to maintain this Ansatz precisely 'along' one of the four component supersymmetries. Now, supersymmetry transformations act by

$$\delta_\varepsilon\langle \boldsymbol{B}\rangle = \varepsilon\cdot\langle \boldsymbol{F}\rangle~, \qquad \delta_\varepsilon\langle \boldsymbol{F}\rangle = \varepsilon\cdot\langle \boldsymbol{B}'\rangle~, \tag{0.2.1}$$

where the quantities $\langle \boldsymbol{B}\rangle$ and $\langle \boldsymbol{F}\rangle$ depend on the *background* fields and transform as tensorial (spinorial) representations of the Lorentz group. Since $\langle \boldsymbol{F}\rangle = 0$, we also need $\langle \boldsymbol{B}'\rangle = 0$, which then restricts the various background fields relating to the internal manifold $\mathcal{M}$ and thereby determines its geometry, i.e., the vacuum configuration. To stabilize this Ansatz, it turns out, one needs to modify the original action by adding certain curvature-squared terms; this in turn modifies the supersymmetry transformation rules ... Fortunately, these feedback corrections can be treated perturbatively [137].

Assuming maximal symmetry for the 3+1-dimensional spacetime,

$$R_{\mu\nu\rho\sigma} = \kappa\,(g_{\mu\rho}g_{\nu\sigma} - g_{\mu\sigma}g_{\nu\rho})~, \tag{0.2.2}$$

explicit calculation yields that $\kappa = 0$ and the 3+1-dimensional spacetime is Minkowskian order by order in the string tension, α'. Also, suffice it here just to state that the connection on $\mathcal{M}$ can consistently be chosen torsion-free so that we are left with Riemannian candidates for $\mathcal{M}$. Such manifolds are partially classified by their holonomy groups [8]. Since spinors in odd-dimensional spaces are (pseudo)real, if the internal space contains an odd-dimensional factor the resulting 3+1-dimensional model will necessarily have particles of both helicities. In particular, the left-handed neutrino would always be accompanied by its right-handed mirror image. It would be very hard, if not impossible, to resolve this violent contradiction with experiment in the present context; the possibilities at hand are then listed in Table 0.1.

Note that only for manifolds with $\mathfrak{su}(3)$ holonomy, precisely one spinor together with its CPT conjugate (in the 3+1-dimensional spacetime sense) is invariant under holonomy. Such a spinor, taken to be constant on $\mathcal{M}$, will also be covariantly constant on $\mathcal{M}$ (being invariant under holonomy, the parallel transport is trivial) and will guarantee simple ($N=1$) supersymmetry in M_{3+1}. We refrain from a detailed discussion of the other choices

Table 0.1: Possible holonomy algebras and corresponding transformation of spinors in Riemannian manifolds.

Dim.	Holonomy algebra	Tangent spinor
6	$\mathfrak{so}(6)$	$\mathbf{4} \oplus$ c.c.
	$\mathfrak{u}(3)$	$\mathbf{3}_{-2} \oplus \mathbf{1}_6 \oplus$ c.c.
	$\mathfrak{su}(3)$	$\mathbf{3} \oplus \mathbf{1} \oplus$ c.c.
4+2	$\mathfrak{so}(4) \times \mathfrak{so}(2)$	$(\mathbf{2}, \mathbf{1})_1 \oplus (\mathbf{1}, \mathbf{2})_{-1} \oplus$ c.c.
	$\mathfrak{su}(2) \times \mathfrak{so}(2)$	$\mathbf{2}_2 \oplus \mathbf{1}_{-1} \oplus \mathbf{1}_{-1} \oplus$ c.c.
	$[\mathfrak{su}(2) \times \mathfrak{u}(1)] \times \mathfrak{so}(2)$	$(\mathbf{2}, 0)_1 \oplus (\mathbf{1}, 3)_{-1} \oplus (\mathbf{1}, -3)_{-1} \oplus$ c.c.
2+2+2	$\mathfrak{so}(2) \times \mathfrak{so}(2) \times \mathfrak{so}(2)$	$(1,1,1) \oplus (-1,-1,1) \oplus (1,-1,-1)$ $\oplus (-1,1,-1) \oplus$ c.c.

above; suffice it here to refer to the existing literature [199] and merely say that no other of the above choices has been found consistent with the various constraints of physical reality.

In the above discussion, by 'holonomy group' we have meant the restricted *holonomy group* $^{(L)}$, which is in 1–1 correspondence with the *algebra* of the (full) holonomy group. If the manifold is multiply connected, the full and the restricted holonomy groups differ and this can have some interesting effects [193].

0.2.2 Ricci-flatness

In general, on a Kähler n-dimensional space, the holonomy algebra must be contained in $\mathfrak{u}(n)$. Also, the Riemann tensor simplifies:

$$R_{\mu\overline{\nu}\,\overline{\rho}}{}^{\overline{\sigma}} = \partial_\mu \Gamma_{\overline{\nu}\,\overline{\rho}}{}^{\overline{\sigma}} = R_{\mu\overline{\rho}\,\overline{\nu}}{}^{\overline{\sigma}}\,, \qquad \Gamma_{\overline{\nu}\,\overline{\rho}}{}^{\overline{\sigma}} = g^{\tau\overline{\sigma}}\,\partial_{\overline{\nu}}\, g_{\tau\overline{\rho}}\,, \tag{0.2.3}$$

and the Ricci tensor becomes:

$$R_{\mu\overline{\nu}} \stackrel{\text{def}}{=} R_{\mu\overline{\sigma}\,\overline{\nu}}{}^{\overline{\sigma}} = R_{\mu\overline{\nu}\,\overline{\sigma}}{}^{\overline{\sigma}}\,. \tag{0.2.4}$$

Considering the Riemann tensor as a $\mathfrak{u}(n)$-valued curvature (1,1)-form

$$\Theta_{\overline{\rho}}{}^{\overline{\sigma}} \stackrel{\text{def}}{=} \mathrm{d}z^\mu \wedge \mathrm{d}z^{\overline{\nu}} R_{\mu\overline{\nu}\,\overline{\rho}}{}^{\overline{\sigma}}\,, \tag{0.2.5}$$

we see, from Eq. (0.2.4), that the corresponding Ricci (1,1)-form

$$\Theta \stackrel{\text{def}}{=} \Theta_{\overline{\rho}}{}^{\overline{\rho}} \tag{0.2.6}$$

is the $\mathfrak{u}(1)$-valued trace of the Riemann (1,1)-form. Recall that $\mathfrak{u}(n) \approx \mathfrak{su}(n) \oplus \mathfrak{u}(1)$, where $\mathfrak{u}(1)$ is generated by the trace of the $\mathfrak{u}(n)$ generators.

For the holonomy on $\mathcal{M}$ to be $\mathfrak{su}(3)$ rather than the full $\mathfrak{u}(3)$, there must exist a metric for which the Ricci (1,1)-form and hence also the Ricci tensor vanishes. It would follow that this Ricci-flat metric should then be used in the Calabi–Yau σ-model action (0.1.2). However, quantum corrections will alter this choice. In general, the induced correction $\delta G_{\mu\overline{\nu}}$ is up to fourth-order in perturbations proportional to the Ricci tensor and so vanishes for our Ricci-flat choice of $G_{\mu\overline{\nu}}$. From the fourth order, $\delta G_{\mu\overline{\nu}}$ receives corrections that are proportional to powers of the Riemann tensor and are non-vanishing [161]. Nevertheless, the compactification Ansatz can be modified so that $\delta G_{\mu\overline{\nu}} = 0$ at any desired order in (the σ-model) perturbation theory [196]. Two important lessons have been learned in this process:

1. The physically relevant metric $G_{\mu\overline{\nu}}$ for the action (0.1.2) is not the Ricci-flat metric on $\mathcal{M}$ but can be obtained from it through a sequence of quantum corrections.
2. The cohomology class represented by the (1,1)-form

$$J \overset{\text{def}}{=} iG_{\mu\overline{\nu}}\, \mathrm{d}z^{\mu} \wedge \mathrm{d}z^{\overline{\nu}} \tag{0.2.7}$$

does not change in the quantum correction process. The fourth order correction is proportional to the double derivative of the Euler density and hence cohomologically trivial.

The above use of the Riemann and the Ricci (1,1)-form does not appear coordinate-independent. To settle this qualm, recall that the usual characteristic polynomial of an $n \times n$ matrix

$$\chi_{\lambda}[\boldsymbol{\Theta}] \overset{\text{def}}{=} \det[\boldsymbol{\Theta} - \lambda\mathbb{1}] = \sum_{k=0}^{n} (-)^{k}\, \lambda^{k}\, c_{n-k}[\boldsymbol{\Theta}] \tag{0.2.8}$$

and that, $c_k[\boldsymbol{\Theta}]$, the coefficients of λ^{n-k} are independent of the choice of basis. If in addition the components of $\Theta_i{}^j$ are closed (1,1)-forms, as is the case above, $c_k[\boldsymbol{\Theta}]$ are closed (k,k)-forms and their integrals are topologically invariant. The c_k are collected in the total *Chern class* $^{(L)}$

$$c[\boldsymbol{\Theta}] \overset{\text{def}}{=} \chi_{-1}\big[\tfrac{i}{2\pi}\boldsymbol{\Theta}\big] = \det\big[\mathbb{1} + \tfrac{i}{2\pi}\boldsymbol{\Theta}\big] = \mathbb{1} + c_1[\boldsymbol{\Theta}] + c_2[\boldsymbol{\Theta}] + \ldots \tag{0.2.9}$$

The above Ricci-flatness condition therefore translates into the vanishing of the first Chern class, $c_1[\boldsymbol{\Theta}] = c_1(\mathcal{M})$, which is coordinate-independent and in fact a topological characteristic. Indeed, Yau's proof of the Calabi conjecture guarantees that only the cohomology class of the Ricci two-form, rather than its explicit choice is the obstruction to finding a Ricci-flat Kähler metric.

0.2.3 The Yang–Mills bundle (standard Ansatz)

So far, we have seen the requirement of $N=1$ local supersymmetry in spacetime to restrict the compactification space $\mathcal{M}$ to be Ricci-flat. There are also further restrictions and they arise from requiring the effective model in 3+1-dimensional spacetime to be anomaly free. The discovery of Green and Schwarz [143], that anomaly freedom is guaranteed if the quantity

$$30 R_{[MN|K}{}^{L}\, R_{|PQ]L}{}^{K} \; - F^{\theta}_{[MN}\, F^{\theta}_{PQ]} \tag{0.2.10}$$

vanishes in vacuum, was really the trigger for most of the subsequent interest in (super)string compactification. Here $R_{MNP}{}^{Q}$ is the Riemann tensor on the 9+1-dimensional spacetime and F^{θ}_{MN} is the $E_8 \times E_8$ (or $Spin(32)/\mathbb{Z}_2$) Yang–Mills field strength. Of course, prior to compactification, the 9+1-dimensional spacetime is flat so that $\langle R_{MNP}{}^{Q}\rangle = 0$ and consequently $\langle F^{\theta}_{MN}\rangle = 0$ too. When compactifying the Riemann tensor develops a background value and so must also the Yang–Mills field strength.

The simplest non-trivial way to satisfy $30\langle R \wedge R\rangle = \langle F \wedge F\rangle$ is to set the background spin connection equal to the background Yang–Mills connection. Since the compactification space has $\mathfrak{su}(3)$ holonomy, one seeks an $SU(3)$ sub-bundle in the $E_8 \times E_8$ Yang–Mills bundle. Now, the fiber of one E_8 factor of the Yang–Mills bundle decomposes as

$$\mathbf{248} \;\to\; (\mathbf{78},\mathbf{1}) \oplus (\mathbf{27},\mathbf{3}) \oplus (\mathbf{27}^*,\mathbf{3}^*) \oplus (\mathbf{1},\mathbf{8}) \tag{0.2.11}$$

under the $E_6 \times SU(3)$ maximal subgroup. Group theoretically, $\mathbf{Tr}(F \wedge F)$ is proportional of the second Dynkin index and since only the $SU(3)$ part of the connection acquires a background value, the E_6 representations merely give rise to a multiplicity. We straightforwardly have

$$2 \cdot 27 \cdot I_2(\mathbf{3}) + 1 \cdot I_2(\mathbf{8}) \;=\; 2 \cdot 27 \cdot \tfrac{1}{2} + 1 \cdot 3 \;=\; 30\ , \tag{0.2.12}$$

while the Riemann tensor is already in the $\mathbf{3} \oplus \mathbf{3}^*$ vector representation, so that the analogous group-theoretical prefactor is $I_2(\mathbf{3} \oplus \mathbf{3}^*) = 2I_2(\mathbf{3}) = 1$. This exactly accounts for the extra factor of 30 in the expression (0.2.10).

In more geometrical terms, we find $\mathcal{V}$, an $SU(3)$ sub-bundle of the original $E_8 \times E_8$ Yang–Mills bundle, and identify it fiber by fiber with $\mathcal{T}_{\mathcal{M}}$, the holomorphic tangent bundle of $\mathcal{M}$. This is done in a way that at every point of $\mathcal{M}$, the holomorphic tangent space is identified with the representation-$\mathbf{3}$ fiber of $\mathcal{V}$.

— ❦ —

Let us mention that there are other possible ways to satisfy the vanishing of (0.2.10), where an $SU(n)$ sub-bundle of $E_8 \times E_8$ is non-vanishing on the background and $n > 3$. Some of these are constructed as holomorphic deformations of the standard Ansatz, but some involve quite elaborately constructed rank-n stable vector bundles over $\mathcal{M}$ [112].

0.3 Massless Modes

Having determined the background values of the supergravity and super Yang–Mills connections, consider now fluctuations of these connections around the chosen background. While a connection itself is not an exterior form (it transforms inhomogeneously), the fluctuations are. Each of the connections therefore gives rise to 9+1-dimensional spacetime exterior forms which are then to be decomposed according to the compactification Ansatz. This gives a list of fields among which the massless modes (harmonic forms) are to be found.

0.3.1 Lorentz symmetry breaking

Upon replacing M_{9+1} with a product (locally) of M_{3+1} and $\mathcal{M}$, only the $SO(1,3)$ subgroup of the original $SO(1,9)$ *Lorentz group*$^{(L)}$ provides still the usual Lorentz symmetry of the effective model. The various tensorial (spinorial) fields, such as the graviton (gravitino), ..., are rewritten as collections of fields that have definite transformation properties under $SO(1,3)$. Since $SO(1,3) \times SO(6)$ is a maximal subgroup of $SO(1,9)$, the representation of $SO(6)$ can be used to classify the modes. Furthermore, since we decided to compactify on manifolds of $SU(3)$ holonomy, the states will be classified by the $SO(1,3) \times SU(3)$ subgroup of $SO(9,1)$.

The task is also simplified if we consider the respective helicity subgroups instead of the full Lorentz groups, whence

$$SO(1,9) \longrightarrow SO(8) \supset SO(2) \times SO(6) \supset SO(2) \times SU(3)\ . \qquad (0.3.1)$$

The advantage here is that precisely the helicity groups (rather than the full Lorentz groups) describe physical degrees of freedom unambiguously.

Consider, for example, an 8-component Majorana–Weyl massless fermion Ψ in M_{9+1}. It satisfies the Dirac equation $\nabla\!\!\!/_{(9+1)}\,\Psi = 0$, which can be rewritten as

$$\nabla\!\!\!/_{(3+1)}\,\Psi(x,y) + \nabla\!\!\!/_{(6)}\,\Psi(x,y) = 0\ , \qquad (0.3.2)$$

with x and y coordinates of M_{3+1} and $\mathcal{M}$, respectively. The 'slash' implies contraction with suitably chosen Dirac matrices on M_{3+1} and $\mathcal{M}$, as indicated by the subscripts†. Using compactness of the 'internal' space $\mathcal{M}$, we expand

$$\Psi(x,y) = \sum_A \psi_A(x)\,\eta^A(y) \;, \tag{0.3.3}$$

where the η^A are analogous to the familiar spherical harmonics. Precisely the 0-modes $\mathring{\eta}^a$ obey

$$\nabla\!\!\!\!/_{(6)}\,\mathring{\psi}_a(x)\,\mathring{\eta}^a(y) = \mathring{\psi}_a(x)\big(\nabla\!\!\!\!/_{(6)}\,\mathring{\eta}^a(y)\big) = 0 \tag{0.3.4}$$

and therefore yield

$$\big(\nabla\!\!\!\!/_{(3+1)}\,\mathring{\psi}_a(x)\big)\,\mathring{\eta}^a(y) = 0 \;, \qquad \forall\,\mathring{\eta}^a(y) \;. \tag{0.3.5}$$

In such a decomposition, ψ_A transforms as a spinor of $SO(1,3)$, while η^A are spinors of $SO(6)$. By supersymmetry, however, both will be accompanied by bosonic superpartners, ϕ and φ respectively, that transform as tensors of the respective groups. Massless spinors will be accompanied by massless bosons:

$$\triangle_{(3+1)}\,\mathring{\phi}_a(x) = 0 \;, \qquad \triangle_{(6)}\,\mathring{\varphi}^a(y) = 0 \;. \tag{0.3.6}$$

The latter of these, for all the fields we shall have to consider, will turn out to be in 1–1 correspondence with certain harmonic forms on the 'internal' space $\mathcal{M}$. The correspondence of spinors to harmonic forms on a Calabi–Yau manifold can also be established by more direct means (see Ref. [93]).

The massless states of the heterotic string are those of the 9+1-dimensional spacetime, $N{=}1$ supergravity coupled to the supersymmetric $E_8 \times E_8$ Yang–Mills multiplet. We now discuss these two sectors in turn.

0.3.2 The supergravity multiplet

The supergravity multiplet may be represented by the following component fields:

$$g_{MN},\; B_{MN},\; \Phi,\quad \psi_M{}^A,\; \chi_A \;. \tag{0.3.7}$$

Here M, N are vectorial indices while A is spinorial; g_{MN} is symmetric while B_{MN} is skew-symmetric and $\psi_M{}^A$ satisfies

$$[\Gamma^M]_A{}^B\,\psi_M{}^A \;=\; 0 \;, \tag{0.3.8}$$

† Ref. [22], pp. 365–367, explains a subtlety in this procedure and how to deal with it.

where $[\Gamma^M]$ are Dirac matrices. In general, the bosonic fields are non-vanishing in vacuum and we consider their fluctuations to find massless fields. Alternatively, one may start from fluctuations of the Zehnbein ${E_M}^N$ and its superpartner ${E_M}^A$, which are tensor coefficients of a tangent bundle valued and spin bundle valued 1-form, respectively.

Upon compactification, we distinguish vectors (co)tangent to M_{3+1} and the 'internal' space $\mathcal{M}$:

$$\begin{array}{lllll} V^M \to & V^m & \oplus\ V^\mu & \oplus & V^{\overline{\mu}}\ , \\ \mathbf{8}_V & \to (\pm 2, \mathbf{1}) & \oplus\ (0, \mathbf{3}) & \oplus & (0, \mathbf{3^*})\ , \end{array} \tag{0.3.9}$$

where we have indicated the respective representations of $SO(8) \to SO(2)\times SU(3)$, using that. The Ψ^A spinors decompose,

$$\begin{array}{lllll} \Psi^A \to & \psi^\alpha\,\eta^a & \oplus\ \psi^\alpha\,\eta^0 & \oplus\ \psi^{\dot\alpha}\,\eta_a & \oplus\ \psi^{\dot\alpha}\,\eta_0\ , \\ \mathbf{8}_S & \to (+1, \mathbf{3}) & \oplus\ (+1, \mathbf{1}) & \oplus\ (-1, \mathbf{3^*}) & \oplus\ (-1, \mathbf{1})\ , \end{array} \tag{0.3.10}$$

while the Ψ_A spinors decompose in the conjugate way. In the formulae (0.3.9) and (0.3.10), we have already implied expansions as in Eq. (0.3.3). Thus, e.g., ψ^α is a two-component left-handed Weyl spinor in M_{3+1} spacetime while η^a is a commuting spinor on the Calabi–Yau space $\mathcal{M}$ and it is in 1–1 correspondence with tangent vectors [93].

Each component field of the supergravity multiplet is now decomposed according to Eq. (0.3.9), (0.3.10). For example,

$$\delta g_{MN} \to \begin{bmatrix} \delta g_{mn} & \delta g_{m\,\nu} & \delta g_{m\,\overline{\nu}} \\ \delta g_{\mu\,n} & \delta g_{\mu\nu} & \delta g_{\mu\overline{\nu}} \\ \delta g_{\overline{\mu}\,n} & \delta g_{\overline{\mu}\nu} & \delta g_{\overline{\mu}\,\overline{\nu}} \end{bmatrix}\ . \tag{0.3.11}$$

The component δg_{mn} expands into scalars on $\mathcal{M}$ so that the 0-mode of δg_{mn} (the M_{3+1} graviton) corresponds to the 1-dimensional $H^0(\mathcal{M}, \mathbb{C})$. Similarly, $i\delta g_{\mu\overline{\nu}}\,\mathrm{d}z^\mu \wedge \mathrm{d}z^{\overline{\nu}}$ is a $(1,1)$-form and the 0-modes correspond to elements of $H^1(\mathcal{M}, \mathcal{T}^*_{\mathcal{M}})$.

Having so decomposed the entire supergravity multiplet, we obtain several tensorial and spinorial quantities which combine into $N=1$ superfields. Among these, we are interested only in those that give rise to massless fields in 3+1-dimensional sense and the resulting list is given in the first part of Table 0.2.

0.3.3 The Yang–Mills multiplet

In compactifying, $E_8 \times E_8$ is broken to $E_6 \times SU(3) \times E_8$. Namely, once an $SU(3)$ sub-bundle of E_8 is identified with $\mathcal{T}_{\mathcal{M}}$ and the corresponding

Table 0.2: The list of massless superfields upon compactification.

Superfield	**Field content** $(+h.c.)$	**Cohomology**	**Number**[a]
$(2, 3/2)_\pm$	δg_{mn}, $(\psi_m{}^\alpha \eta^0)$	$H^0(\mathcal{M}, \mathbb{C})$	$h^{0,0} = 1$
$(1/2, 0)_+ + h.c.$	$\delta\Phi$, δB_{mn}, $(\chi^{\dot\alpha}\eta_0)$	$H^0(\mathcal{M}, \mathbb{C})$	$h^{0,0} = 1$
$(1/2, 0)_+ + h.c.$	$\delta g_{\mu\bar\nu}$, $\delta B_{\mu\bar\nu}$, $(\psi^{\dot\alpha}\eta_{\bar\mu\, a})$	$H^1(\mathcal{M}, \mathcal{T}^*_{\mathcal{M}})$	$h^{1,1} = b_2$
$(1/2, 0)_+ + h.c.$	$\delta g_{\mu\nu}$, $\delta g_{\bar\mu\bar\nu}$, $(\psi^\alpha \eta_{\bar\mu}{}^a)$	$H^1(\mathcal{M}, \mathcal{T}_{\mathcal{M}})$	$h^{2,1}$
$(1, 1/2)_\pm$	$\delta A_m{}^{(\mathbf{78},\mathbf{1})}$, $(\lambda^{\alpha\,(\mathbf{78})}\eta^{0\,(\mathbf{1})})$	$H^0(\mathcal{M}, \mathbb{C})$	$h^{0,0} = 1$
$(1/2, 0)_+ + h.c.$	$\delta A_{\bar\mu}{}^{(\mathbf{27},\mathbf{3})}$, $(\lambda^{\alpha\,(\mathbf{27})}\eta^{0\,(\mathbf{3})})$	$H^1(\mathcal{M}, \mathcal{T}^*_{\mathcal{M}})$	$h^{1,1} = b_2$
$(1/2, 0)_+ + h.c.$	$\delta A_{\bar\mu}{}^{(\mathbf{27}^*,\mathbf{3}^*)}$, $(\lambda^{\alpha\,(\mathbf{27}^*)}\eta^{0\,(\mathbf{3}^*)})$	$H^1(\mathcal{M}, \mathcal{T}_{\mathcal{M}})$	$h^{2,1}$
$(1/2, 0)_\pm$	$\delta A_{\bar\mu}{}^{(\mathbf{1},\mathbf{8})}$, $(\lambda^{\alpha\,(\mathbf{1})}\eta^{0\,(\mathbf{8})})$	$H^1(\mathcal{M}, \mathrm{End}\,\mathcal{T}_{\mathcal{M}})$	$h^1(\mathrm{End}\,\mathcal{T}_{\mathcal{M}})$
$(1, 1/2)_\pm$	$\delta A_m{}^{(\mathbf{248})}$, $(\lambda^{\alpha\,(\mathbf{248})}\eta^{0\,(\mathbf{1})})$	$H^0(\mathcal{M}, \mathbb{C})$	$h^{0,0} = 1$

[a]See text for the identifications.

connections are non-zero in vacuum, the physical gauge group will be the maximal subgroup of $E_8 \times E_8$ which commutes with this $SU(3)$. Eventually, further modification of the Ansatz will be needed to break E_6 to a physically more realistic subgroup thereof and we shall briefly discuss that below.

Combining the decomposition (0.2.11) for one factor in $E_8 \times E_8$ and (0.3.9)–(0.3.10) for both, a list of component fields is obtained. Expanding these on $\mathcal{M}$ and keeping the 0-modes, we complete[‡] the list of massless fields in the effective 3+1-dimensional model.

In Table 0.2, we have used the *Dolbeault Theorem* [(L)],

$$H^q(\mathcal{M}, \wedge^p \mathcal{T}^*_{\mathcal{M}} \otimes \mathcal{V}) = H^{p,q}(\mathcal{M}, \mathcal{V}) , \tag{0.3.12}$$

for any (sheaf of holomorphic germs of a) holomorphic vector bundle $\mathcal{V}$ over $\mathcal{M}$. In addition, since a Calabi–Yau manifold is equipped with a projectively unique (that is, defined up to an overall constant) covariantly constant holomorphic 3-form Ω, the mapping

$$\mathcal{T}_{\mathcal{M}} \xrightarrow{\cdot\,\Omega} \wedge^2 \mathcal{T}^*_{\mathcal{M}} \tag{0.3.13}$$

[‡] Explicit examples show that there occur additional 0-modes when $\mathcal{M}$ possesses special symmetries [138, 140, 76]; special treatment is required when $\mathcal{M}$ is singular [119, 170, 172].

is an isomorphism, and so is

$$H^q(\mathcal{M}, \mathcal{T}_{\mathcal{M}} \otimes \mathcal{V}) \xrightarrow{\cdot\,\Omega} H^q(\mathcal{M}, \wedge^2\mathcal{T}^*_{\mathcal{M}} \otimes \mathcal{V}) \, . \tag{0.3.14}$$

In particular, $H^q(\mathcal{M}, \mathcal{T}_{\mathcal{M}}) \approx H^{2,q}(\mathcal{M})$.

It is rather useful to note that the various cohomology groups occurring in Table 0.2 are closely related to the various structures on the Calabi–Yau space. It is standard that elements of $H^1(\mathcal{M}, \mathcal{T}_{\mathcal{M}})$ span local deformations of the complex structure of $\mathcal{M}$. Since the Kähler class itself belongs to $H^{1,1}(\mathcal{M})$, its variations are spanned by elements of $H^{1,1}(\mathcal{M})$. Finally, elements of $H^1(\mathcal{M}, \mathrm{End}\,\mathcal{T}_{\mathcal{M}})$ span local deformations of the complex structure of the holomorphic tangent bundle, $\mathcal{T}_{\mathcal{M}}$.

0.4 Holes in the Vacuum

Quite independently of the considerations so far, the massless spectrum can be changed in a very important way if the compact space $\mathcal{M}$ is multiply connected. Consider a closed loop Γ in $\mathcal{M}$ and the associated line integral

$$\Phi_\Gamma \stackrel{\text{def}}{=} \oint_\Gamma \Big(\mathrm{d}y^{\overline{\mu}} A_{\overline{\mu}}{}^{(\mathbf{78})}(y) + h.c. \Big) \, . \tag{0.4.1}$$

If Γ is not a contractible loop, it is not the boundary of a surface and this line-integral cannot be related to a surface-integral over the field strength. Thus, while the field strength vanishes in vacuum, the gauge-potential $A_m{}^{(\mathbf{78})}$ need not and the quantity Φ_Γ may acquire a non-vanishing vacuum expectation value (VEV).

This value necessarily breaks the E_6 symmetry since only those generators of E_6 which commute with Φ_Γ will remain gauge *symmetries.* So far, the situation is identical to the case described in pp. 70–72 of Ref. [48], with Φ_Γ playing the rôle of a Higgs multiplet. Indeed, using the Dynkin basis, each component of a representation of E_6 is labeled by a 6-plet of integers $[\,\rho_1, \ldots, \rho_6\,]$. Denoting the component of Φ_Γ which does not vanish in vacuum by $\langle\Phi_\Gamma\rangle = [\,a_1, \ldots, a_6\,]$, we can compute the mass-eigenvalue for each of the 78 gauge potentials and also for the $\mathbf{27}$–$\mathbf{27}^*$ component pairs [48]. Consider R, an arbitrary representation of E_6. Then

$$M_{[\,\rho^1,\ldots,\rho^6\,]} \;\propto\; (a_1\,\rho^1 + b\,\rho^2 + \ldots + a_6\,\rho^6) \tag{0.4.2}$$

is the mass-eigenvalue of the element of R, denoted by $[\,\rho^1, \ldots, \rho^6\,]$. Inserting the Dynkin labels for the elements of the gauge $\mathbf{78}$-plet or $\mathbf{27}$–$\mathbf{27}^*$pairs of matter we obtain their masses, all proportional to the compactification mass-scale which, essentially, is the inverse (average) size of $\mathcal{M}$.

It is then possible to determine the gauge symmetry spanned by the gauge potentials which remained massless or, alternatively, determine the vacuum configuration $[a_1, \ldots, a_6]$ for a certain subgroup of E_6 to remain massless. Thereafter, the eigenvalues of the

$$\big(\,\mathbf{27}^*\mathrm{s}\,\big|\,\langle\Phi_\Gamma\rangle\,\big|\,\mathbf{27}\mathrm{s}\,\big) \tag{0.4.3}$$

mass matrix are determined by contracting $[a_1, \ldots, a_6]$ with the weights in the **27**-representation. If say, $h^{2,1} > h^{1,1}$, we have $\frac{1}{2}|\chi_E|$ **27**'s and additional $h^{1,1}$ **27**–**27*** pairs. Among the $h^{1,1}$ pairs of 27-plets, those with vanishing mass-eigenvalue will remain massless in addition to $\frac{1}{2}|\chi_E| = |h^{2,1} - h^{1,1}|$ complete **27**s. More precisely, that is the net count; once E_6 is broken, the various components of the $\frac{1}{2}|\chi_E|$ "complete" **27**s need not have come from the same original **27**s.

If Φ_Γ were a usual Higgs multiplet, we could choose the vacuum configuration $[\,a_1, \ldots, a_6\,]$ freely. In the present case, however, there is an important restraint stemming from the fact that Γ is non-contractible but that, for compact projective spaces with $SU(3)$ holonomy rather than a subgroup thereof, multiple connectedness must be of finite order [23, 79] (see Theorem 4.1). So, the path-ordered exponential

$$U_\Gamma \;\stackrel{\text{def}}{=}\; \mathcal{P}\exp\left\{\oint_\Gamma \Big(\mathrm{d}x^m A_m{}^{(\mathbf{78})} + h.c.\Big)\right\} \tag{0.4.4}$$

is an element of the *finite* structure group, $U_\Gamma \in \pi_1(\mathcal{M})$, and $(U_\Gamma)^n = 1$ for some $n \in \mathbb{Z}$. Among other things, this implies that the coefficients $a_1, \ldots, a_6$ in the configuration of $\langle\Phi_\Gamma\rangle$ must be understood modulo n. In the mass formulae (0.4.2) and (0.4.3), this may create additional zeroes.

0.5 Yukawa Couplings

Once we have obtained the spectrum of massless fields for the effective 3+1-dimensional model, it is natural to inquire about their interactions. One part is certainly the gauge interactions with the gauge bosons left massless in the compactifications process and is simply through the standard minimal coupling, given the various charges as derived from the initial E_6.

However, the matter (super)fields Φ_i also interact directly and the corresponding terms must appear in the effective superpotential. As is well known, this basic property of supersymmetric theories provides an important restriction for the possible interactions, since the superpotential can depend only on *chiral*$^{(L)}$ *superfields*$^{(L)}$ (its Hermitian conjugate, the anti-superpotential, of course depends on anti-chiral superfields and in a

Hermitian conjugate way). Since the masses of the fields in the effective low-energy models are not directly related to the compactification scale, there can be no mass-term in the superpotential at the compactification scale and its general form is

$$W\Big|_{\text{at } M_C} = \sum_{n=3}^{\infty} \frac{\kappa_{i_1\cdots i_n}}{M_C^{(n-3)}} \prod_{k=1}^{n} \Phi_{i_k}(x;\theta) \ , \tag{0.5.1}$$

where M_C is the compactification mass-scale. The leading such terms come directly from the 9+1-dimensional theory and can be deduced by studying the 9+1-dimensional action, more precisely, the coupling [220]

$$\int \mathrm{d}^{10}X \ \sqrt{-g_{(10)}} \ \mathbf{Tr}\left\{\overline{\lambda}(X) \cdot \not{A}(X) \cdot \lambda(X)\right\} \ . \tag{0.5.2}$$

Decomposing the E_8 gauge vector and gaugino as described above, we obtain terms of the generic form

$$\int \mathrm{d}^4x \ \sqrt{-g_{(4)}} \ \left[\kappa_{ijk} \ \psi^i(x) \, \phi^k(x) \, \psi'^k(x)\right] \tag{0.5.3}$$

$$\kappa_{ijk} \ \stackrel{\text{def}}{=} \ \mathbf{Tr} \int_{\mathcal{M}} \eta_i(y) \cdot \varphi_j(y) \cdot \eta'_k(y) \ , \tag{0.5.4}$$

where the indices i, j, k enumerate the respective 0-modes. The quantity (0.5.4) is a so called Yukawa coupling and is manifestly of physics interest.

By E_6 gauge invariance, there are only four types of such Yukawa couplings:

1. The $(\mathbf{27}^*)^3$ couplings. The $\mathbf{27}^*$s accompany $\mathrm{d}z^\mu \, \mathrm{d}z^{\overline{\nu}} \, e_{(A)\,\mu\overline{\nu}}$'s, i.e., elements of $H^1_{\bar\partial}(\mathcal{M}, \mathcal{T}^*_{\mathcal{M}}) = H^{1,1}(\mathcal{M})$. Using this property, this Yukawa coupling can be evaluated as[§]

$$\begin{aligned} \mathring{\kappa}_{ABC} &\stackrel{\text{def}}{=} \int_{\mathcal{M}} e_{(A)} \wedge e_{(B)} \wedge e_{(C)} \ , \\ &= \int_{\mathcal{M}} \left(\Omega^{\overline{\mu}\,\overline{\nu}\,\overline{\rho}} \ e_{(A)\,\sigma\overline{\mu}} \, e_{(B)\,\tau\overline{\nu}} \, e_{(C)\,\lambda\overline{\rho}} \ \overline{\Omega}^{\sigma\tau\lambda}\right) . \end{aligned} \tag{0.5.5}$$

Since there are no harmonic (2,0)- or (0,2)-forms on a manifold with $SU(n)$ holonomy (for $n{\neq}2$), it follows that $H^{1,1}(\mathcal{M}) = H^2(\mathcal{M}) = H^2(\mathcal{M}, \mathbb{Z}) \approx H_4(\mathcal{M}, \mathbb{Z})$, where the last relation is an isomorphism since both H^2 and H_4 are dual to H^4 (see *duality*$^{(L)}$). This will

---

§ The little circle atop $\mathring{\kappa}_{ABC}$ reminds us that this expression will acquire quantum corrections.

make it possible to compute $\mathring{\kappa}_{ABC}$, in favorable cases, from intersections of 4-cycles in $\mathcal{M}$. More formally, this Yukawa coupling realizes the mapping

$$\int : \quad \wedge^3 H^{1,1}(\mathcal{M}) \ \to \ \mathbb{C} \ . \tag{0.5.6}$$

These Yukawa couplings are a *topological* invariant.

2. The $\mathbf{27}^3$ couplings. Since the αth $\mathbf{27}$-plet accompanies $\mathrm{d}z^{\overline{\mu}}\,\varphi^{\rho}_{(\alpha)\,\overline{\mu}}\,\partial_\rho$'s, i.e., the αth element of $H^1_{\overline{\partial}}(\mathcal{M},\mathcal{T}_{\mathcal{M}})$, this Yukawa coupling is

$$\begin{aligned}\kappa_{\alpha\beta\gamma} &\stackrel{\text{def}}{=} \int_{\mathcal{M}} \Omega \wedge \Big(\varphi_{(\alpha)} \wedge \varphi_{(\beta)} \wedge \varphi_{(\gamma)}\Big) \cdot \Omega \ , \\ &= \int_{\mathcal{M}} \Big(\Omega^{\overline{\mu}\,\overline{\nu}\,\overline{\rho}}\ \varphi_{(\alpha)\,\overline{\mu}}{}^{\sigma}\ \varphi_{(\beta)\,\overline{\nu}}{}^{\tau}\ \varphi_{(\gamma)\,\overline{\rho}}{}^{\lambda}\ \Omega_{\sigma\tau\lambda}\Big) \ .\end{aligned} \tag{0.5.7}$$

Here $\Omega = \Omega_{\lambda\mu\nu}\mathrm{d}z^\lambda \wedge \mathrm{d}z^\mu \wedge \mathrm{d}z^\nu$ is the covariantly constant holomorphic 3-form. More formally, this Yukawa coupling realizes the mapping

$$\int \Omega \wedge (\quad) \cdot \Omega : \quad \wedge^3 H^1_{\overline{\partial}}(\mathcal{M},\mathcal{T}_{\mathcal{M}}) \to \mathbb{C} \ , \tag{0.5.8}$$

which is coordinate independent because $c_1 = 0$ and so $\wedge^3\mathcal{T}_{\mathcal{M}} \cdot \Omega \approx \mathbb{C}$.

3. The $\mathbf{1}^3$ couplings. Since the matter $\mathbf{1}$s accompany $\mathrm{d}z^{\overline{\mu}}\boldsymbol{s}_{(k)\,\overline{\mu}}{}^{\rho}{}_{\sigma}$ $\mathrm{d}z^\sigma\,\partial_\rho$'s, i.e., elements of $H^1_{\overline{\partial}}(\mathcal{M},\mathrm{End}\,\mathcal{T}_{\mathcal{M}}) \approx H^1_{\overline{\partial}}(\mathcal{M},\mathcal{T}^*_{\mathcal{M}} \otimes \mathcal{T}_{\mathcal{M}})$, this Yukawa coupling is

$$\begin{aligned}\mathring{\kappa}_{ijk} &\stackrel{\text{def}}{=} \int_{\mathcal{M}} \Omega \wedge \mathrm{Tr}\Big(\boldsymbol{s}_{(i)} \wedge \boldsymbol{s}_{(j)} \wedge \boldsymbol{s}_{(k)}\Big) \ , \\ &= \int_{\mathcal{M}} \Big(\Omega^{\overline{\mu}\,\overline{\nu}\,\overline{\rho}}\ \boldsymbol{s}_{(i)\,\overline{\mu}}{}^{\rho}{}_{\sigma}\ \boldsymbol{s}_{(j)\,\overline{\nu}}{}^{\sigma}{}_{\tau}\ \boldsymbol{s}_{(k)\,\overline{\rho}}{}^{\tau}{}_{\rho}\Big) \ .\end{aligned} \tag{0.5.9}$$

More formally, this Yukawa coupling realizes the mapping

$$\int \Omega \wedge : \quad \wedge^3 H^1_{\overline{\partial}}(\mathcal{M},\mathrm{End}\,\mathcal{T}_{\mathcal{M}}) \to \mathbb{C} \ , \tag{0.5.10}$$

because $\mathrm{Tr}(\boldsymbol{\lambda}^3) \in \mathbb{C}$ is an invariant for any matrix $\boldsymbol{\lambda}$, with $\boldsymbol{\lambda}^3$ denoting the usual matrix multiplication.

4. The mixed couplings. Using the expressions as above, this Yukawa coupling is

$$\begin{aligned}\mathring{\kappa}_{\alpha jC} &\stackrel{\text{def}}{=} \int_{\mathcal{M}} \Omega \wedge \mathrm{Tr}\Big(\varphi_{(\alpha)} \wedge \boldsymbol{s}_{(j)} \wedge e_{(C)}\Big) \ , \\ &= \int_{\mathcal{M}} \Big(\Omega^{\overline{\mu}\,\overline{\nu}\,\overline{\rho}}\ \varphi_{(\alpha)\,\overline{\mu}}{}^{\sigma}\ \boldsymbol{s}_{(j)\,\overline{\nu}}{}^{\tau}{}_{\sigma}\ e_{(C)\,\tau\overline{\rho}}\Big) \ .\end{aligned} \tag{0.5.11}$$

More formally, this Yukawa coupling realizes the mapping

$$\int \Omega\wedge : \ H^1_{\bar\partial}(\mathcal{M},\mathcal{T}_{\mathcal{M}}) \wedge H^1_{\bar\partial}(\mathcal{M},\mathrm{End}\,\mathcal{T}_{\mathcal{M}}) \wedge H^1_{\bar\partial}(\mathcal{M},\mathcal{T}^*_{\mathcal{M}}) \to \mathbb{C}\ , \tag{0.5.12}$$

since End $\mathcal{T}_{\mathcal{M}}$ is the traceless part of $\mathcal{T}^*_{\mathcal{M}} \otimes \mathcal{T}_{\mathcal{M}}$, which of course naturally contracts with $\mathcal{T}_{\mathcal{M}}$ from the first factor and $\mathcal{T}^*_{\mathcal{M}}$ from the third factor.

Remark: Since e_A, φ_α, $\mathcal{S}_i$ and Ω are all $\bar\partial$-closed (annihilated by $\bar\partial$), it is easy to show by partial integration that the value of the integrals is unchanged under the shift such as $\mathcal{S} \to \mathcal{S} + \bar\partial\alpha$. The couplings (0.5.5), (0.5.7), (0.5.9) and (0.5.11) therefore depend only on the cohomology classes represented by the e_A, φ_α and $\mathcal{S}_i$. In particular: the $(\mathbf{27}^*)^3$ couplings (0.5.5) depend only on the topological (diffeomorphism) class; (0.5.7), (0.5.9) and (0.5.11) depend (via Ω) on the complex structure.

0.6 Instanton Retouch

The various supersymmetries which have been required have rather important consequences regarding the σ-model/string corrections.

0.6.1 A simple non-renormalization theorem

A relatively general non-renormalization theorem (selection rule) for the superpotential in the effective low-energy theory can be derived using the chirality of the superpotential and the abelian infinitesimal symmetries related to the field B_{MN} in (0.3.7).

In a supersymmetric compactification, there are no quantum corrections to the superpotential of the effective low-energy model to all finite orders in the σ-model and string perturbation.

To prove this (see Ref. [240]), recall first that the original heterotic string theory in 9+1-dimensional Minkowski spacetime contained the 2-form $B = B_{MN}\,\mathrm{d}X^M \mathrm{d}X^N$, which occurs only through its field-strength, $H = \mathrm{d}B$. Manifestly, this implies the symmetry $B \to B + \mathrm{d}\Lambda$ with any 1-form Λ. Upon compactification, this symmetry is inherited by the resulting model to all orders in the σ-model perturbation theory—as can be verified by explicit computation [232].

Consider now the components which are cotangent to the internal manifold, $B_{\mu\bar\nu}$. Since the superpotential does not contain derivative couplings, it can depend on $B_{\mu\overline{\nu}}$ only through its occurrence in a factor $e^{-S_{\mathcal{M}}}$, where

$$S_{\mathcal{M}} = \ldots + \left(\int_{\Sigma} \mathrm{d}^2\sigma \, \epsilon^{ab} B_{\mu\overline{\nu}} \, (\partial_a Z^{\mu}(\sigma))(\partial_b Z^{\overline{\nu}}(\sigma)) = \int_{\mathrm{im}(\Sigma\to\mathcal{M})} B \right), \tag{0.6.1}$$

which is non-vanishing only for non-trivial maps $Z^{\mu} : \Sigma \to \mathcal{M}$, i.e., for world-sheet instantons. Owing to supersymmetry, the components of $B_{\mu\overline{\nu}}$ occur as imaginary parts of the scalar components of a chiral superfield where the real parts are provided by corresponding components of the metric, $G_{\mu\overline{\nu}}$, on the internal space. Among the latter, the dilatational mode, r, corresponds to the overall size of the internal manifold: r^3 is the volume of the internal space in units of the string tension, and $r^{-1/2}$ is the coupling constant of the σ-model. Similarly, the component of the 2-form B which is (co)tangent to the Minkowski spacetime is paired up with the dilaton, Φ, into the complex scalar component field of a chiral superfield.

Since the superpotential is chiral, it must be a holomorphic function of $G + iB$. As it is independent of B to all orders in perturbation theory, it must also be independent of G and so also of the σ-model perturbation coupling constant, r, and of the string coupling constant, the dilaton. ☑

In general, superpotential terms may be corrected only through world-sheet instanton effects and stringy non-perturbative effects. This does modify the conclusions in the closing remark on p. 18: $(\mathbf{27^*})^3$ couplings do depend on the complex structure and $\mathbf{27}^3$ couplings do depend on the Kähler class; see Ref. [249].

0.6.2 World-sheet instantons

As the only possible source of quantum corrections to the superpotential terms (save for the string non-perturbative effects, which are beyond the present understanding), we now consider the effect of world-sheet non-perturbative effects.

Recall that, in the σ-model action (0.1.2), the field combinations $B+iG$ and $B - iG$ couple to different derivatives of $Z, \overline{Z}$. One argues that in contributions to the superpotential and *assuming* supersymmetry not to be broken by quantum corrections, only maps for which

$$\bar{\partial} Z^{\mu}(\sigma) = 0 \tag{0.6.2}$$

can contribute [111]. These are the holomorphic instantons, i.e., classical configurations of holomorphic Z^{μ}, inequivalent to the 'tree-level' Z^{μ} which is constant on Σ. In other words, they arise as homotopically non-trivial holomorphic maps $\Sigma \to \mathcal{M}$.

Remark: The *assumption* of unbroken supersymmetry can be relaxed if a non-perturbative criterion for validity of supersymmetry can be evaluated favorably. One such criterion is the Witten index, which equals the Euler characteristic of the (total) spacetime; if non-zero, supersymmetry will not be broken and the non-renormalization follows [237].

Now, the world-sheet itself being a Riemann surface, it is mapped into some complex curve in the internal space $\mathcal{M}$ under Z. Since the harmonic (1,1)-forms are dual to 2-cycles (which may be thought of as formal sums of certain 2-dimensional subspaces) in $\mathcal{M}$, holomorphic instantons will necessarily interact with (1,1)-forms and will affect any result related to (1,1)-forms. Odd-dimensional subspaces, on the other hand, can only be interacted with through their boundaries or the spaces that they themselves bound. Thus, results that depend only on odd-dimensional *cycles* in $\mathcal{M}$ cannot be affected by holomorphic instantons and are in fact, by the above perturbative non-renormalization theorem, exact up to stringy non-perturbative effects. Since there are no 1-cycles or 5-cycles on Calabi–Yau 3-folds, this result is restricted to any numerical characteristic of $H_3(\mathcal{M})$.

In particular, we have that *triple*-(2,1) *Yukawa couplings cannot be renormalized.* (This can also be verified by explicit computation; see Ref. [113].)

Moreover, the world-sheet is naturally mapped to 2-cycles, which are dual to (1,1)-cohomology. To any basis for the latter, there corresponds a basis for the 2-cycles, hence a 'natural' parametrization for the instantons. Many features of their effects can therefore be obtained without any actual evaluation [114], aside from the fact that the evaluation also can be performed, at least in principle. The $(1,1)^3$ Yukawa coupling (0.5.5) is therefore seen to be corrected as

$$\begin{aligned}\kappa_{ABC} \stackrel{\text{def}}{=} & \int_{\mathcal{M}} \omega_{(A)} \wedge \omega_{(B)} \wedge \omega_{(C)} \\ & + \sum_{L} e^{-S_{\mathcal{M}}[J;L]} \Big(\int_L \omega_{(A)}\Big)\Big(\int_L \omega_{(B)}\Big)\Big(\int_L \omega_{(C)}\Big)\,,\end{aligned} \tag{0.6.3}$$

where L's label the instanton configuration and one needs to sum (integrate) over all such possible configurations. J represents the choice of the Kähler class on $\mathcal{M}$ and $S_{\mathcal{M}}[J;L]$ is simply the integral over L of the pull-back of J.

Remark: The action, of course, contains the supersymmetric completion, but care has to be taken with 0-modes of the various fields, as those do not come in Fermi–Bose pairs. In view of this, it is usually assumed that (in most physically relevant calculations) the non-zero modes (which are Fermi–Bose paired) cancel one another's contribution and only the 0-modes (the Fermi–Bose un-paired ones) do contribute. This assumption permits us to restrict to the un-paired 0-modes, which are finite in number on smooth $\mathcal{M}$, but may diverge if $\mathcal{M}$ becomes singular.

0.7 VEVs for Free

Suppose that in a particular compactification, a component of some **27** or a corresponding component of some $\mathbf{27}^*$ have no self-interactions; more precisely, that their interaction terms vanish when all other fields are set to zero. In such a situation, the vacuum expectation values (VEVs) $\langle \mathbf{27} \rangle$ and $\langle \mathbf{27}^* \rangle$ are free parameters of the theory. They are also called "flat directions" in the field space.

However, $\langle \mathbf{27} \rangle, \langle \mathbf{27}^* \rangle \neq 0$ implies a breakdown of the E_6 Yang–Mills symmetry and if there is one such pair of fields, it is well known from spontaneous symmetry breaking analyses (see Ref. [48, 44]) that the resulting theory will have $SO(10)$ Yang–Mills invariance. With another pair of such fields $SU(5)$ may be achieved, but care has to be taken, since the second pair of fields has to have a free VEV *after* the first two fields are shifted by their VEVs. In fact, in order to obtain $E_6 \to SU(3)_C \times SU(2)_W \times U(1)_Y$ and the matter spectrum anywhere near acceptable, it is likely that both a non-contractible gauge flux (see § 0.4) and two pairs of $\langle \mathbf{27} \rangle$, $\langle \mathbf{27}^* \rangle$ have to be generated¶. With a non-abelian structure group $\pi_1(\mathcal{M})$, non-commuting gauge flux configurations are possible and a single pair $\langle \mathbf{27} \rangle$, $\langle \mathbf{27}^* \rangle$ (or even none at all) may suffice.

Similarly, some of the chargeless fields might acquire a free VEV. Now, giving a non-zero VEV to a modulus field accompanying (2,1)-forms or another one accompanying (1,1)-forms corresponds, respectively, to shifting the complex structure or the Kähler class of the internal space. Since elements of $H^1(\mathcal{M}, \mathrm{End}\, \mathcal{T}_{\mathcal{M}})$ span linear deformations of the holomorphic tangent bundle, $\mathcal{T}_{\mathcal{M}}$, we expect that turning on a VEV for a **1** will correspond to a deformation of this structure. We will return to this issue later.

¶ Clearly, once the Yukawa couplings are known, the analysis of the matter potential will predict whether or not some of the scalars develops a non-zero VEV or if some of the VEVs remain free.

0.8 The Wish List (Seven Veils)

Given a particular Calabi–Yau manifold, we need to supply the following information for characterizing the effective particle physics model:

1. $\frac{1}{2}\chi_E$: the net number of generations predicted; preferably, three.
2. $h^{2,1} = \dim H^1(\mathcal{M}, \wedge^2\mathcal{T}_\mathcal{M}) = \dim H^1(\mathcal{M}, \mathcal{T}_\mathcal{M})$ and $h^{1,1} = \dim H^1(\mathcal{M}, \mathcal{T}^*_\mathcal{M})$: both the number of generations and the number of mirror-generations (generations of mirror particles) independently, not just $\frac{1}{2}\chi_E = (h^{1,1} - h^{2,1})$. For phenomenological reasons, one needs at least $\min(h^{1,1}, h^{2,1}) = 2$, but not much bigger. (Generally, proliferation of particles can easily have devastating effects on the low-energy model.)
3. $\dim H^1(\mathcal{M}, \operatorname{End}\mathcal{T}_\mathcal{M})$: the number of light chargeless matter fields.
4. $\pi_1(\mathcal{M})$: through non-contractible gauge flux lines, it measures the ability to break E_6 to a more sensible Yang–Mills group. $\mathbb{Z}_2$ is "not enough," at least $\pi_1 = \mathbb{Z}_3$ is required for potentially successful phenomenology.
5. The $\langle \mathbf{27}^3 \rangle$, $\langle \mathbf{27}^{*3} \rangle$, $\langle \mathbf{1}^3 \rangle$, $\langle \mathbf{27} \cdot \mathbf{27}^* \cdot \mathbf{1} \rangle$ Yukawa (and also higher) couplings as given in Eq. (0.5.1).
6. Kinetic terms in the Lagrangian for the massless chiral fields (0.3.3)-paired with $\alpha, \beta \in H^q(\mathcal{M}, \mathcal{V})$ are normalized by the metric $N(\alpha, \beta) := \int_\mathcal{M} (\overline{*_\mathcal{V}\,\alpha}) \wedge \beta$—the eigenstates of which thus provide the physically relevant basis. In the underlying world-sheet quantum field theory, $N(\alpha, \beta)$ is by definition the Zamolodchikov metric for the scalar fields spanning the **27**s, **27***s and **1**s [120]. $N_{\mathbf{27}}(\alpha, \beta)$ is the well understood Weil–Petersson metric on the space of complex structures of $\mathcal{M}$. $N_{\mathbf{27}^*}(\alpha, \beta)$ is the analogous metric on the space of (complexified) Kähler classes, equal to the Weil–Petersson metric for the mirror of $\mathcal{M}$. The analogous $\mathcal{T}_\mathcal{M}$-complex structure moduli space metric, $N_{\mathbf{1}}(\alpha, \beta)$, however remains less well understood.
7. σ-model and string corrections to the results in 3., 5. and 6.

This is indeed a rather taxing schedule and in the explicitly analyzed cases only the first few tasks have been completed. Indeed, we are at a serious loss for ammunition and tactics for a *complete* assessment of the last two issues in the general case. In geometrical terms, the normalization matrices for the kinetic terms are components of a Zamolodchikov-type metric on the dual parameter space. This however does not provide a simple computational tool and exact results are known only in some simple cases.

Also, the above non-renormalization theorems do protect some of the results from σ-model/string corrections but, unfortunately, rather little is known outside the shielding scope of these theorems; see however Refs. [219, 89, 59, 60, 88].

2024 *The landscape of Calabi–Yau compactifications aiming for a minimally supersymmetric Standard Model (presumably containing the inhabitable portion of the general string landscape [374, 289]) has developed radically in the past three decades and is already described in book form [266]. The Hodge numbers (the 2nd veil, above) had been computed for all CICY 3-folds (see discussion on p. 51) in [148], and also for the physically more relevant free finite quotients of CICYs (the 4th veil, above) more recently [333, 364]; orbifolds with singularities in need of smoothing (by blowups or otherwise: see Chapters 4 and D) and* orientifolds *[355] are actively explored. Other significant developments in this regard include: (**1**) The Hull–Strominger (torsion) and $SU(3)$-structure extension of the standard Ansatz [409]; see, e.g., [252, 432, 299]. (**2**) Gauge bundle generalizations of the standard Ansatz, $(\mathcal{T}_{\mathcal{M}} \rightsquigarrow \mathcal{V}) \subset \mathcal{V}_{E_8}$ with rank $\mathcal{V}=3,4,5$ (see Refs. [476, 373, 449]) are estimated to harbor the most astonishing numbers of minimally supersymmetric Standard Models [365]: a mole ($\sim 10^{23}$) within CICYs, and $\sim 10^{723}$ in toric hypersurfaces; see also [370, 472, 400] and references therein. (**3**) Instead of the earlier-seeming sparsity [349, 334, 348] of passable approximations to the Standard Model, we are now facing a mind-boggling* embarrassment of riches*—for a systematic study of which there is no hope without artificial intelligence (AI) and other computer-aided, i.e., machine learning (ML) methods; see the updating close of § 4.1.3, the Afterword, p. 371, and, e.g., Refs. [398, 344, 315]. In fact, computational methods for systematic exploration of* physically normalized *Yukawa couplings throughout the above-mentioned pools of candidate Calabi–Yau compactifications of the $E_8 \times E_8$ heterotic string for 3-generation models with the "standard" embedding [345] and a "non-standard" embedding [363] are now finally becoming feasible—four decades after the exuberant promise [160].*

Finally, since just before the dawn of this millennium, we have known that our 3+1-dimensional spacetime is not Minkowski-flat but is rather well described asymptotically by the de Sitter geometry—which must break supersymmetry and so is much harder to describe; for a recent review, see [326]. Whether or not the Calabi–Yau compactifications as discussed herein (with a vanishing cosmological constant) can be uplifted *to accommodate the minuscule but* positive *cosmological constant is a highly contested question, subject to intense research. Suffice it here to mention that it* may be possible *to achieve this by, e.g., compactifying on barely and intricately smoothed conifolds (see § D.2) [314]. Searching for Calabi–Yau compactifications that may provide for a physically realistic de Sitter uplift again seems daunting (if not impossible) without substantial AI/ML help and is enabled by the recently (2022) discovered ability to machine-identify Calabi–Yau geometries with highly curved regions [315]. Alternatively, the de Sitter geometry of our 3+1-dimensional world may be successfully reproduced in "bubble" worlds inside a much larger 9+1-dimensional spacetime [411, 306, 324, 369]. Such 9+1-dimensional spacetimes might be described as "Wick-rotated"*

Calabi–Yau 5-folds—of which there are many more than 3-folds, so their exploration seems even more hopeless without AI help.

With best of hopes for future development in the areas pertaining to the last points (to which we will return), we concentrate on seeking and studying Calabi–Yau beasts for which the armament and spells provided herein suffice as far as stripping off the first four veils and getting a grip on the other three. The advanced Reader will hopefully enjoy learning some deeper and further secrets of the priesthood, for which the listed literature is recommended; those divulged herein reflect my own winding voyage and resulting opinions—may you learn from my errors.

— ❦ —

Several methods of creating Calabi–Yau manifolds have been employed in the past few decades. Roughly speaking, they come in two varieties:

In the first one, a Calabi–Yau manifold is realized as the space of solutions to a system of algebraic equations and is thereby embedded in some bigger and better understood space in which the equations are formulated; we may refer to such methods as 'algebraic'. In the second one, some space of complex three dimensions is being modified and possibly glued to another such construction; we may call these 'surgical,' stretching somewhat the usual meaning in mathematics.

Needless to say, the two methods are often combined to bring out the best in each. Algebraic techniques typically rely on polynomial algebra and are thus rather familiar. Algebraic and also some surgical methods yield access to numerical characteristics through straightforward albeit possibly tedious computations. On the other hand, certain techniques of surgery are on occasion indispensable but their effect on numerical characteristics is sometimes less well understood. We begin by describing the most often used algebraic techniques and discuss the various surgical techniques as suitable opportunities arise.

Given a system of K constraint equations $P^a(w) = 0$ in the embedding space $\mathcal{X}$, it is useful to consider the system of hypersurfaces X_a, each of which is defined as the zero-set of a single constraint. The space of common solutions is evidently $\mathcal{M} = \bigcap_a X_a \subset \mathcal{X}$. In an important class of such constructions, the hypersurfaces meet *transversely*. That is, at every point of $\mathcal{M}$, the gradients of $\{X_a\}$ span a non-degenerate system and $\mathrm{d}P^1 \wedge \cdots \wedge \mathrm{d}P^K$ is non-zero everywhere on $\mathcal{M}$. Such an $\mathcal{M}$ is a *complete intersection*$^{(L)}$ and has $\dim \mathcal{M} = \dim \mathcal{X} - K$. For a toy example, we may describe a circle as the complete intersection of two spheres (each defined

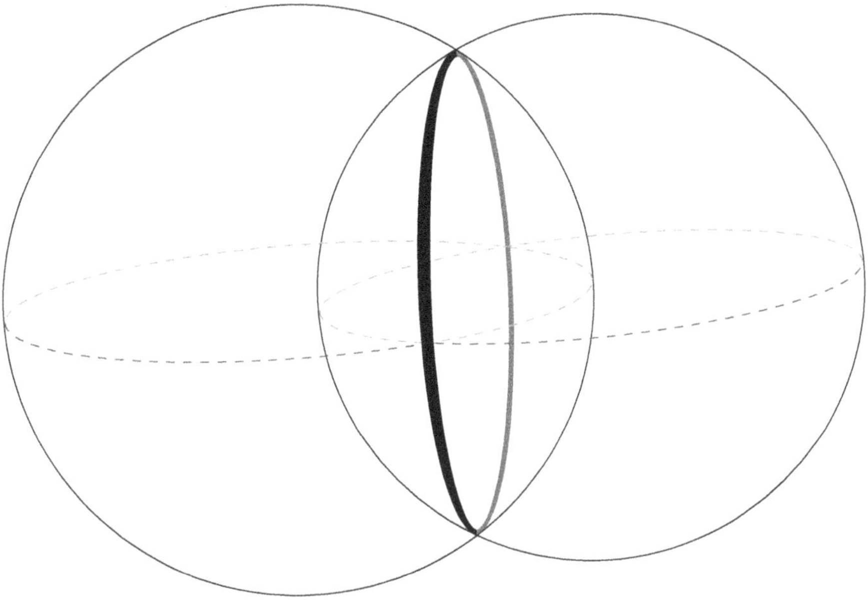

Figure 0.1: A circle as a complete intersection of two spheres.

by a quadratic equation) in $\mathbb{R}^3$, as in Figure 0.1. Both exceptional cases, a) when the spheres meet in a single point and b) when they coincide, fail our transversality test and their intersection is considered *singular*. (When fully disconnected, their intersection *void*.) Complete intersections are perhaps the easiest of the constructions and will serve to discuss many of the techniques. Subsequently, we shall also encounter other types of algebraic constructions, where manifestly $\dim \mathcal{M} > \dim \mathcal{X} - K$. I am not aware of a general classification of such constructions or of a particular type of geometry universally suitable for their analysis and will therefore only be able to present some examples ... But now, onward!

Part I
Constructions

Chapter 1

Complex Kindergarten

Note: Practically everything in this volume concerns complex spaces, fiber *bundles*$^{(L)}$ and *sheaves*$^{(L)}$ over them and other complex and *holomorphic*$^{(L)}$ derived structures. The same will be true of *cohomology groups*$^{(L)}$ also, unless otherwise is explicitly stated. Finally, as usual in algebraic geometry, one works over a given algebraically closed field $\mathbb{k}$ and all dimensions refer to that field. Owing to the usual physics application, by default, we have in mind $\mathbb{k} = \mathbb{C}$ throughout this volume.

In this chapter we construct Calabi–Yau 3-folds as intersections of hypersurfaces in a complex projective space, and study some of the properties of such 3-folds. We begin with a review of the properties of complex projective spaces.

1.1 Complex Manifolds and Holomorphic Vector Bundles

A manifold is smooth if local differentiability has an invariant meaning although we may need to choose a connection to define covariant derivatives. Similarly, on a complex manifold, local holomorphicity has an invariant meaning. Thus a holomorphic vector bundle is a smooth vector bundle for which one knows which local *sections*$^{(L)}$ are holomorphic: since the transition functions are required to be holomorphic, the holomorphicity of the section is independent of the choice of coordinate neighborhoods and corresponding coordinates.

Most of the time, we will be discussing algebraic varieties, defined as the zero-set of a system of polynomials in certain holomorphic variables x^a. By choosing the x^a to be coordinates on a complex space to begin with and using holomorphic polynomials, the algebraic variety is a complex space by construction.

A crucial difference between the smooth and holomorphic categories is detected by the number of *global* sections. A smooth vector bundle over a smooth manifold always has infinitely many linearly independent global sections. In contrast, a holomorphic vector bundle over a compact complex manifold has always a finite number of independent global holomorphic sections—sometimes only of the 0-section. In other words, the holomorphic category is much more rigid. For one thing, in the smooth category, there always exist "bump-functions" so that smooth manifolds can be cut and pasted at will. Surgery in the holomorphic category is always much more restricted and may well be impossible on occasion.

Another important difference is that in the holomorphic category it is necessary to distinguish between sub-bundles and quotient bundles. Given a holomorphic vector bundle $\mathcal{B}$ over a complex manifold $\mathcal{M}$ which has a holomorphic sub-bundle $\mathcal{A}$, we can always form the quotient bundle $\mathcal{B}/\mathcal{A}$, which is spanned by holomorphic vectors of $\mathcal{B}$ regarded equivalent if they differ by a vector of $\mathcal{A}$. However, $\mathcal{B}/\mathcal{A}$ is almost never a holomorphic sub-bundle of $\mathcal{B}$. One can certainly choose (in many ways) a Hermitian metric on $\mathcal{B}$ and define $\mathcal{C}$ to be the Hermitian complement of $\mathcal{A} \subset \mathcal{B} = \mathcal{A} \oplus \mathcal{C}$. However, $\mathcal{C}$ is almost never a holomorphic bundle over $\mathcal{M}$: using the metric to define it introduces antiholomorphic variables in a non-trivial way.

The simplest vector bundles over a compact complex manifold $\mathcal{M}$ have unit rank, are called holomorphic line bundles and are of particular importance for us. Their fiber is isomorphic to $\mathbb{C}^1$. Under the (fiber-wise) tensor product operation, these bundles form a group known as the Picard group of $\mathcal{M}$. Now, given any holomorphic vector bundle $\mathcal{B}$ over $\mathcal{M}$ of rank k (the fibers of $\mathcal{B}$ are all $\mathbb{C}^k$-like) and any holomorphic sub-bundle $\mathcal{A}$, one can always form the respective determinant bundles $\det \mathcal{B}$ and $\det \mathcal{A}$ and these satisfy the identity

$$\det(\mathcal{B}) \;=\; \det(\mathcal{A}) \;\otimes\; \det(\mathcal{B}/\mathcal{A}) \;. \tag{1.1.1}$$

All three determinant bundles are line bundles over X and are therefore easier to deal with than the original vector bundles.

Suppose $\mathcal{X}$ is a complex manifold, $\mathcal{L}$ a holomorphic line bundle over $\mathcal{X}$ and f a non-trivial global holomorphic section of $\mathcal{L}$. We can define a subspace $\mathcal{M} \subset \mathcal{X}$ to consist of all points $x \in \mathcal{X}$ for which $f(x) = 0$. Regarding f as a mapping that takes all points $x \in \mathcal{M} \subset \mathcal{X}$ to 0, we write $\mathcal{M} = f^{-1}(0)$ and call $\mathcal{M}$ a hypersurface in $\mathcal{X}$.

Reminder: The usual gradient of $f(x)$ provides a normal to the hypersurface $\mathcal{M}$ and has an invariant meaning. To see this, choose Γ_μ to be a local connection. Then $\nabla_\mu f = \partial_\mu f + \Gamma_\mu \cdot f = \partial_\mu f$ at $x \in \mathcal{M}$, since $\Gamma_\mu(x) \cdot f(x) = 0$ wherever $f(x) = 0$.

Now, $\mathcal{M}$ fails to be a complex sub*manifold* if f is singular, i.e., if there is at least one point where both f and all its gradients $\partial_\mu f$ vanish. One might worry if there exist non-singular choices of f. To that end, we have the following version of Bertini's theorem:

Theorem 1.1 (Bertini). *Let $\mathcal{X}$ denote a compact complex manifold and $\mathcal{L}$ a holomorphic line bundle over $\mathcal{X}$. Suppose that, at every point $x \in \mathcal{X}$, at least one section of $\mathcal{L}$ is non-zero. Then a generic (= almost every) section f of $\mathcal{L}$ defines a non-singular hypersurface* $\mathcal{M}_f \stackrel{\text{def}}{=} f^{-1}(0)$.*

To be precise, the qualification "generic" means "all except for a subset of strictly smaller dimension". Also, note that the submanifolds $\mathcal{M}_f$ so defined are parametrized by the choice of the section f.

For a complex manifold $\mathcal{M}$ of complex dimension n, we write $\mathcal{T}_\mathcal{M}$ for the holomorphic tangent bundle of $\mathcal{M}$ and $\mathcal{T}^*_\mathcal{M}$ for its *holomorphic* complex dual, the cotangent bundle (which is *not* its complex conjugate, $\overline{\mathcal{T}}_\mathcal{M}$!). $\mathcal{T}^*_\mathcal{M}$ is also identified with the bundle of holomorphic 1-forms, $\Omega^1_\mathcal{M}$. The bundle of holomorphic n-forms (also called holomorphic volume forms) is the determinant bundle of $\mathcal{T}^*_\mathcal{M}$,

$$\Omega^n_\mathcal{M} = \wedge^n \Omega^1_\mathcal{M} = \det \mathcal{T}^*_\mathcal{M} \stackrel{\text{def}}{=} \mathcal{K}_\mathcal{M} \ , \tag{1.1.2}$$

and is also called the canonical bundle.

If a compact complex manifold $\mathcal{M}$ admits a holomorphic volume form Ω which is non-singular and has no zeros, then Ω must be a constant over $\mathcal{M}$ and the canonical bundle must be trivial so that all possible holomorphic volume forms are merely constant complex multiples of each other

$$\Omega' = \lambda\, \Omega \ , \qquad \lambda \textit{ constant} \text{ on } \mathcal{M} \ . \tag{1.1.3}$$

This book is devoted, for the most part, to the construction and study of such 3-folds as three dimensional algebraic varieties and we will soon turn to the construction of simplest such spaces. First however, we find it useful to review a number of general properties of such spaces.

* However, as Philip Candelas likes to point out, most examples that you in fact write down turn out to be singular. The veracity of this statement is an increasing and swiftly saturating function of the number of variables involved.

1.2 Calabi–Yau 3-Fold Specials

Some rather important properties of Calabi–Yau 3-folds are described in terms of their *cohomology groups* $^{(L)}$. The Reader is probably most familiar with the De Rham cohomology groups, which are defined as the quotient spaces of closed modulo exact exterior forms. That is, elements of $H^r_{\rm DR}(\mathcal{M})$ are *classes* of exterior r-forms ω such that

$$\mathrm{d}\omega \;=\; 0\;, \qquad \omega \;\simeq\; \omega' \;=\; \omega + \mathrm{d}\alpha\;. \tag{1.2.1}$$

On a complex manifold, the exterior derivative may be decomposed into the holomorphic and the anti-holomorphic part, $\mathrm{d} = \partial + \bar{\partial}$; also, any exterior r-form may be decomposed into (p,q)-forms such that $p+q=r$. The $\bar{\partial}$- (also known as Dolbeault) cohomology groups $H^{p,q}_{\bar{\partial}}(\mathcal{M})$ are then defined in precise analogy with the De Rham cohomology. The *Hodge numbers* $^{(L)}$ $h^{p,q} \stackrel{\rm def}{=} \dim H^{p,q}_{\bar{\partial}}(\mathcal{M})$ are topological invariants. More generally, for a vector bundle $\mathcal{V}$ over $\mathcal{M}$, $H^q(\mathcal{M},\mathcal{V})$ denotes the cohomology group of $\mathcal{V}$ valued q-forms; $H^{p,q}_{\bar{\partial}}(\mathcal{M}) = H^q(\mathcal{M}, \wedge^p \mathcal{T}^*_{\mathcal{M}})$.

Now, given the nowhere vanishing and projectively unique (that is, unique up to an overall complex non-zero scaling factor) holomorphic 3-form Ω on our 3-fold $\mathcal{M}$, we have the isomorphism

$$\cdot\Omega \;:\; \mathcal{T}_{\mathcal{M}} \;\stackrel{\sim}{\longrightarrow}\; \wedge^2\,\mathcal{T}^*_{\mathcal{M}}\;, \tag{1.2.2}$$

and therefore also

$$\cdot\Omega \;:\; H^q(\mathcal{M},\mathcal{T}_{\mathcal{M}}) \;\stackrel{\sim}{\longrightarrow}\; H^{2,q}(\mathcal{M})\;. \tag{1.2.3}$$

In other words, to every holomorphic tangent vector v^μ, we may assign a projectively unique holomorphic 2-form $\omega_{\nu\rho} \stackrel{\rm def}{=} v^\mu\,\Omega_{\mu\nu\rho}$. The 3-form Ω is in fact the (trivial) section of the (trivial) canonical bundle $\mathcal{K}_{\mathcal{M}} \approx \mathbb{C}_{\mathcal{M}}$.

In view of this, the usual Serre duality formula

$$H^q(\mathcal{M},\mathcal{V}^*) \;\approx\; H^{3-q}(\mathcal{M},\mathcal{V}^*\otimes\mathcal{K}_{\mathcal{M}}) \tag{1.2.4}$$

simplifies in that "$\otimes\mathcal{K}_{\mathcal{M}}$" may be omitted. For example,

$$H^q(\mathcal{M},\mathcal{T}_{\mathcal{M}})^* \;\approx\; H^{3-q}(\mathcal{M},\mathcal{T}^*_{\mathcal{M}})\;, \tag{1.2.5}$$

which, together with Eq. (1.2.3), implies

$$H^{2,q}(\mathcal{M})^* \;\approx\; H^{3-q}(\mathcal{M},\mathcal{T}^*_{\mathcal{M}}) \;\equiv\; H^{1,3-q}(\mathcal{M})\;. \tag{1.2.6}$$

Furthermore, since $H^{3,0}(\mathcal{M}) \approx \mathbb{C}$, there is a new duality, implying

$$H^{p,0}(\mathcal{M})^\star \;\approx\; H^{3-p,0}(\mathcal{M})\;. \tag{1.2.7}$$

In a more systematic manner, we have the following relations (see also the Lexicon entries):

Hodge star duality is a property of smooth compact spaces and is guaranteed by the existence and uniqueness of the standard volume-form†. It implies

$$h^{p,q} \;=\; h^{n-q,n-p} \;. \tag{1.2.8}$$

Complex conjugation is a symmetry of Kähler manifolds. To every harmonic (p,q)-form, it associates a harmonic (q,p)-form, so that

$$h^{p,q} \;=\; h^{q,p} \;. \tag{1.2.9}$$

Holomorphic duality is a special feature of Calabi–Yau spaces and is guaranteed by the existence and uniqueness of the holomorphic volume-form, Ω (because $h^{3,0}=1$). Given a $(0,q)$-class α, there is a unique $(0,3{-}q)$-class β such that $\int_{\mathcal{M}} \alpha\wedge\beta\wedge\Omega = 1$. Thus

$$h^{0,q} \;=\; h^{0,3-q} \;, \qquad h^{p,0} \;=\; h^{3-p,0} \;. \tag{1.2.10}$$

Considering Calabi–Yau manifolds in three complex dimensions, so that $h^{p,q} \equiv 0$ if $p+q>6$, the use of the above relations reduces the number of independent Hodge numbers:

$$\begin{matrix} & & & 1 & & & \\ & & h^{3,2} & & h^{2,3} & & \\ & h^{3,1} & & h^{2,2} & & h^{1,3} & \\ 1 & & h^{2,1} & & h^{1,2} & & h^{0,3} \\ & h^{2,0} & & h^{1,1} & & h^{0,2} & \\ & & h^{1,0} & & h^{0,1} & & \\ & & & h^{0,0} & & & \end{matrix} \;\longrightarrow\; \begin{matrix} & & & 1 & & & \\ & & h^{0,1} & & h^{0,1} & & \\ & h^{0,1} & & h^{1,1} & & h^{0,1} & \\ 1 & & h^{2,1} & & h^{2,1} & & 1 \\ & h^{0,1} & & h^{1,1} & & h^{0,1} & \\ & & h^{0,1} & & h^{0,1} & & \\ & & & 1 & & & \end{matrix} \;. \tag{1.2.11}$$

Finally, to determine $h^{0,1}$, the following result may be used.

Theorem 1.2 (Bochner). *For any harmonic (De Rham) s-form ω, let*

$$F(\omega) \;:=\; R_m{}^n\,\omega_{[nr_2\cdots r_s]}\,\omega^{[mr_2\cdots r_s]} \;+\; \frac{s-1}{2}\;R_m{}^n{}_p{}^q\,\omega_{[nqr_3\cdots r_s]}\,\omega^{[mpr_3\cdots r_s]} \;. \tag{1.2.12}$$

Then, if $F(\omega)$ is positive semi-definite, ω is covariantly constant.

† It should perhaps be pointed out that all the manifolds we shall consider have a single connected piece and hence for complex 3-folds, $h^{3,3}=1$ counts the single volume-form.

Here, $R_m{}^n$ and $R_m{}^n{}_p{}^q$ are the Ricci and the Riemann tensors, respectively.

Choosing the Ricci-flat metric on the Calabi–Yau 3-fold, $R_m{}^n$ vanishes, so $F(\omega) \equiv 0$ for 1-forms. Therefore, for ω to be harmonic, it also has to be covariantly constant. But ω transforms as $\mathbf{3} \oplus \mathbf{3}^*$ under the $SU(3)$ holonomy and so transforms non-trivially under parallel transport and cannot be covariantly constant. Therefore, $b_1 = 0$ and so $h^{1,0} = h^{0,1} = 0$ and we have just derived the general form of the so-called Hodge diamond for Calabi–Yau 3-folds:

$$
\begin{array}{ccccccc}
 & & & 1 & & & \\
 & & 0 & & 0 & & \\
 & 0 & & h^{1,1} & & 0 & \\
1 & & h^{2,1} & & h^{2,1} & & 1 \\
 & 0 & & h^{1,1} & & 0 & \\
 & & 0 & & 0 & & \\
 & & & 1 & & &
\end{array} \quad . \tag{1.2.13}
$$

Remark: For Calabi–Yau spaces of two or more complex dimensions, the Hodge diamond has the same basic form. The "outer" rim of Hodge numbers is completely determined: 1's in the corners and 0's otherwise, while the "inside" Hodge numbers may vary. In two complex dimensions there is only the K3 surface with $h^{1,1} = 20$; in higher dimensions more and more variant Hodge numbers appear.

Part of our goal is to determine both $h^{1,1}$ and $h^{2,1}$ for given constructions. In fact, we know that in general

$$
\chi_E(\mathcal{M}) = \sum_{r=0}^{\dim \mathcal{M}} (-)^r b_r, \qquad b_r = \sum_{p=0}^{r} h^{p,r-p} \ . \tag{1.2.14}
$$

In view of our general result for the Hodge diamond of Calabi–Yau manifolds, it then follows that

$$
\chi_E(\mathcal{M}) \ = \ 2(h^{1,1} - h^{2,1}) \tag{1.2.15}
$$

for all Calabi–Yau 3-folds. If the Euler characteristic is easily obtained as in the subsequent constructions, it suffices to determine independently only one of the two variant Hodge numbers.

Remark: There is another curious property of Calabi–Yau 3-folds. It has been observed that Calabi–Yau manifolds tend to come in mirror pairs, $(\mathcal{M}, \mathcal{W})$ such that

$$
H^{2,1}(\mathcal{W}) \ \cong_Q \ H^{1,1}(\mathcal{M}) \ , \qquad H^{1,1}(\mathcal{W}) \ \cong_Q \ H^{2,1}(\mathcal{M}) \ . \tag{1.2.16}
$$

Here, "$\cong_Q$" denotes an equivalence of the ring structures generated, with the exact (quantum) Yukawa couplings as the ring structure constants. For a number of Calabi–Yau 3-folds, such a mirror-pair can be constructed explicitly. Related to this is the fact that general properties of superconformal field theories guarantee analogous mirror-models [117]. Often, both the original compactification model and its mirror-model have a geometric interpretation as compactification on a Calabi–Yau manifold and its respective mirror-pair. However, there do exist Calabi–Yau 3-folds for which no mirror-space has been found even though a compactification mirror-model (so far with no geometric interpretation) does exist.

1.3 The Quintic Hypersurface in $\mathbb{P}^4$

Perhaps the simplest and certainly the most popular Calabi–Yau manifold is the quintic hypersurface, i.e., the space of solutions to a quintic holomorphic constraint in $\mathbb{P}^4$. We shall denote the family of such hypersurfaces (parametrized by the choices of the quintic constraint) by $[4\|5]$. To see how the choice of the degree of the polynomial comes about, consider a general homogeneous, holomorphic polynomial of degree q in the homogeneous coordinates of $\mathbb{P}^n$:

$$f(z) \stackrel{\text{def}}{=} f_{i_1 \cdots i_q}\, z^{i_1} \cdots z^{i_q} \ . \tag{1.3.1}$$

(Summation over repeated indices is implied through $i_k = 0, \ldots, n$.)

Since a homogeneous polynomial can be interpreted as a section of an appropriate line bundle it follows from Bertini's theorem that for generic f, the hypersurface of solutions to $f(z) = 0$, in $\mathbb{P}^n$, is non-singular. Moreover, it is a Kähler manifold, inheriting this property from the embedding space $\mathbb{P}^n$. Of course it has dimension $n-1$.

A relatively simple non-singular example is provided by taking

$$F(z) = \sum_{i=0}^{n} (z^i)^q \ , \tag{1.3.2}$$

the so-called Fermat polynomial. Let $\mathfrak{F} \stackrel{\text{def}}{=} F^{-1}(0)$ be the hypersurface of solutions to $F(z) = 0$ in $\mathbb{P}^n$. Now, $\mathrm{d}F(z) = q \sum_{i=0}^{n} \mathrm{d}z^i (z^i)^{q-1}$ and the only solution to $\mathrm{d}F = 0$ is $z^i = 0$, $\forall i$, which is not in $\mathbb{P}^n$ by definition. So, since $F(z) = 0$ and $\mathrm{d}F(z) = 0$ have no common solution, the gradient to the hypersurface $\mathfrak{F}$ is everywhere well defined and the hypersurface is smooth.

For $\mathfrak{F}$ to be a Calabi–Yau manifold, it has to admit a metric for which the Ricci tensor vanishes. Now, on a Kähler manifold we may without loss of generality restrict to Hermitian metrics and Riemannian (torsion-free) connections. We may take, e.g., the Fubini–Study metric $\mathbf{g}_{(FS)}$ on the $\mathbb{P}^n$

and the induced metric on the hypersurface, i.e., the restriction of $\mathbf{g}_{(FS)}$. Straightforward but tedious calculation yields

$$\mathbf{R}\big|_{\mathfrak{F}} = [q-(n+1)]\,\mathbf{g}_{(FS)}\big|_{\mathfrak{F}}\,, \tag{1.3.3}$$

up to total derivatives. Here $\mathbf{R}|_{\mathfrak{F}}$ is the restriction to the hypersurface of the Ricci tensor computed form $\mathbf{g}_{(FS)}$. Note that the choice of precisely $q = n+1$, hence a quintic in $\mathbb{P}^4$, essentially turns $\mathbf{R}|_{\mathfrak{F}}$ into a total derivative which is sufficient to show that a (different) metric does exist for which the Ricci tensor actually vanishes. Indeed, Yau's proof of Calabi's conjecture guarantees that the cohomology class of the Ricci two-form rather than the two-form itself is the only obstruction to constructing a Ricci-flat Kähler metric.

For future application, it will be useful to reformulate the latter result in terms of the Riemann curvature two-form $\boldsymbol{\Theta}$

$$\Theta_{\kappa}{}^{\lambda} = \frac{i}{2\pi}\,\mathrm{d}z^{\mu}\wedge\mathrm{d}z^{\overline{\nu}}\,R_{\mu\overline{\nu}\kappa}{}^{\lambda} \tag{1.3.4}$$

and the cohomology classes of the coefficient forms in the characteristic polynomial of $\boldsymbol{\Theta}$. Also, underlying most of our analysis is the relation between $\mathcal{T}_{\mathcal{M}}$ and $\mathcal{T}_{X}$, the holomorphic tangent bundles of the hypersurface $\mathcal{M}$ and of the embedding space $\mathcal{X}$, respectively, and the holomorphic line bundle $\mathcal{E}$ over $\mathcal{X}$, which corresponds to $f(z)$. More precisely, polynomials of the same homogeneity as $f(z)$ above can be interpreted as sections of a holomorphic line bundle $\mathcal{E}$. In the case of quintics over $\mathbb{P}^4$, $\mathcal{E}$ is $\mathcal{O}_{\mathbb{P}^4}(5)$, the fifth tensor power of the *hyperplane bundle* [(L)] over $\mathbb{P}^4$. In general, a homogeneous polynomial of degree k in the homogeneous coordinates of $\mathbb{P}^n$ supplies a holomorphic section of $\mathcal{O}_{\mathbb{P}^n}(k)$.

1.3.1 The normal bundle: adjunction formulae

Consider $\mathcal{T}_{\mathcal{M}}$, the holomorphic tangent bundle of $\mathcal{M}$. It is manifestly a sub-bundle of $\mathcal{T}_{\mathbb{P}^n}$, the corresponding quotient bundle is called the *normal bundle* of $\mathcal{M} \hookrightarrow \mathbb{P}^n$,

$$\mathcal{N}_{\mathbb{P}^n/\mathcal{M}} \stackrel{\text{def}}{=} \{\,\mathcal{T}_{\mathbb{P}^n}|_{\mathcal{M}}\,/\,\mathcal{T}_{\mathcal{M}}\,\}\ . \tag{1.3.5}$$

Let us choose a connection Γ_i on $\mathcal{E} = \mathcal{O}_{\mathbb{P}^n}(k)$ and represent a tangent vector, $X\in\mathcal{T}_{\mathbb{P}^n}$, at a point of $\mathbb{P}^n$ by $X^i\nabla_i$. The covariant X-directional gradient,

$$\nabla_X f = X^i\,(f_{,i} + \Gamma_i f) \stackrel{\text{def}}{=} f_X, \tag{1.3.6}$$

associates to each tangent vector X a quintic, f_X, and so defines a complex linear mapping from the tangent bundle of $\mathbb{P}^n$ to $\mathcal{E}$. Moreover, if we restrict to points of $\mathcal{M} \subset \mathbb{P}^n$, this mapping is holomorphic and independent of the choice of Γ_i since $f=0$ and so $\Gamma_i \cdot f=0$ on $\mathcal{M}$. This map is specified by the differential $\mathrm{d}f \in \mathcal{T}^*_{\mathbb{P}^n} \otimes \mathcal{E}$: the contraction $(\mathcal{T}_{\mathbb{P}^n} \ni X^i \nabla_i) \cdot (\mathrm{d}x^j f_{,j}) = X^i \delta_i^j f_{,j} = (f_X \in \mathcal{E})$ reproduces (1.3.6).

Now, this map yields zero whenever X is tangential to $\mathcal{M}$ since f vanishes everywhere on $\mathcal{M}$, so that there is no gradient 'along' $\mathcal{M}$.‡ In other words, the map annihilates precisely the $\mathcal{T}_{\mathcal{M}}$ sub-bundle of $\mathcal{T}_{\mathbb{P}^n}$, so that $\ker(\nabla_X f) = \mathcal{T}_{\mathcal{M}}$. Moreover, near $\mathcal{M}$, f can be used as a local coordinate in $\mathcal{X}$ that is transversal to $\mathcal{M}$: it vanishes at $\mathcal{M}$ and is non-zero away from $\mathcal{M}$.‡ Rescaling suitably $X' = \lambda X$, for $\lambda \neq 0$, the map $\nabla_{\lambda X} f$ will cover all of this local range of f so that, locally, $\nabla_X f$ indeed maps *onto* all of the appropriate fiber of $\mathcal{E}$. In fact, the map is onto also globally, since it is holomorphic and $\operatorname{rank}\mathcal{E} = \operatorname{rank}\mathcal{T}_{\mathcal{X}} - \operatorname{rank}\mathcal{T}_{\mathcal{M}}$.

Adjunction formula 1: This is encoded in the short *exact sequence*[(L)]

$$0 \to \mathcal{T}_{\mathcal{M}} \xrightarrow{\;i\;} \mathcal{T}_{\mathbb{P}^n}|_{\mathcal{M}} \xrightarrow[(\nabla_X f)]{\mathrm{d}f} \mathcal{E}|_{\mathcal{M}} \to 0 \,, \tag{1.3.7}$$

and serves to establish the identification

$$\mathcal{E}|_{\mathcal{M}} = \{\, \mathcal{T}_{\mathbb{P}^n}|_{\mathcal{M}} \,/\, \mathcal{T}_{\mathcal{M}} \,\} \,. \tag{1.3.8}$$

From this it follows by Eq. (1.3.5) that

$$\mathcal{E}|_{\mathcal{M}} = \mathcal{N}_{\mathbb{P}^n/\mathcal{M}} \,. \tag{1.3.9}$$

We refer to Eq. (1.3.9) as the Adjunction formula 1 (see p. 146 of Ref. [23]).

Adjunction formula 2: Given the short exact sequence such as (1.3.7), using formula (1.1.1) we infer

$$\det \mathcal{T}_{\mathbb{P}^n}|_{\mathcal{M}} = \det \mathcal{T}_{\mathcal{M}} \otimes \det \mathcal{E}|_{\mathcal{M}} \,, \quad \text{i.e.,} \quad \mathcal{K}^*_{\mathbb{P}^n}|_{\mathcal{M}} = \mathcal{K}^*_{\mathcal{M}} \otimes \det \mathcal{E}|_{\mathcal{M}} \,, \tag{1.3.10}$$

using the definition of the canonical bundle (1.1.2). Since these are line bundles,

$$\mathcal{K}_{\mathcal{M}} = (\mathcal{K}_{\mathbb{P}^n} \otimes \det \mathcal{E})|_{\mathcal{M}} \,. \tag{1.3.11}$$

We refer to Eq. (1.3.11) as the Adjunction formula 2 (see p. 147 of Ref. [23]).

---

‡ Physicists no doubt recall that the gradient of the electrostatic scalar potential, Φ, is always directed *away* from the surface of a grounded perfect conductor, where $\Phi=0$ by definition.

1.3.2 Chern classes

We recall that for a complex vector bundle $\mathcal{V}$ over a complex manifold $\mathcal{M}$, we have the *Chern classes* $^{(L)}$

$$c_r(\mathcal{V}) \in H^{2r}(\mathcal{M}) \;, \qquad r = 1, \ldots, \min(\operatorname{rank} \mathcal{V}, \dim \mathcal{M}) \;, \tag{1.3.12}$$

and the total Chern class

$$c(\mathcal{V}) = 1 + \sum_r c_r(\mathcal{V}) \;. \tag{1.3.13}$$

These represent the cohomology classes of the coefficient forms which are found in the expansion of the characteristic polynomial

$$\det\left[\, \mathbb{1} + \lambda \mathbf{F} \,\right] = 1 + \lambda \mathrm{Tr}[\mathbf{F}] + \lambda^2 \big(\mathrm{Tr}[\mathbf{F} \wedge \mathbf{F}] - 2[\mathrm{Tr}\mathbf{F}]^2 \big) + \ldots \tag{1.3.14}$$

where $\mathbf{F}$ is a suitably normalized curvature two-form on $\mathcal{V}$. In the case $\mathcal{V} = \mathcal{T}_X$, i.e., when $\mathbf{F} = \boldsymbol{\Theta}$ is the (suitably normalized) Riemann two-form, we write $c_i(X)$ or simply c_i if no confusion can arise.

Chern classes $\{c_r(\mathcal{V})\}$ are respectively represented by closed (r,r)-forms on $\mathcal{M}$, with which we freely identify them in the sequel, and satisfy three important identities:

$$c_1(\mathcal{V} \otimes \mathcal{L}) = c_1(\mathcal{V}) + k\, c_1(\mathcal{L}) \tag{1.3.15}$$

for any rank-k vector bundle $\mathcal{V}$ and any line bundle $\mathcal{L}$.

$$c(\mathcal{V}) = c(\mathcal{V}') \wedge c(\mathcal{V}'') \tag{1.3.16}$$

whenever

$$0 \to \mathcal{V}' \to \mathcal{V} \to \mathcal{V}'' \to 0 \tag{1.3.17}$$

is a short exact sequence (this includes the case $\mathcal{V} = \mathcal{V}' \oplus \mathcal{V}''$). Both identities follow from the use of the determinant in the definition; compare also with Eq. (1.1.1). Also, if $\mathbf{F}$ is the curvature matrix of a connection in a complex vector bundle X the dual connection in $\mathcal{V}^*$ then has curvature matrix $-\mathbf{F}$ and so

$$c_r(\mathcal{V}^*) \;=\; (-)^r\, c_r(\mathcal{V}) \qquad \text{no summation on } r \;. \tag{1.3.18}$$

Remark: The Chern classes are invariant under smooth deformations, i.e., in the differentiable category. Hence, to compute them, we can ignore the requirement of holomorphicity and simplify the bundle in question.

In particular, we may *smoothly* deform a rank-k (holomorphic) vector bundle $\mathcal{V}$ into a k-fold direct sum of line bundles, $\oplus_{i=1}^{k}\mathcal{L}_i$. Then,

$$c(\mathcal{V}) = c(\bigoplus_{i=1}^{k} \mathcal{L}_i) = \prod_{i=1}^{k} c(\mathcal{L}_i) = \prod_{i=1}^{k} (1 + \ell_i) \ , \tag{1.3.19}$$

where ℓ_i is the first Chern class of $\mathcal{L}_i$ (being of rank-1, $\mathcal{L}_i$ have no higher Chern classes). This is called the *splitting*$^{(L)}$ *principle* and can be used for computing the coarsest properties of most vector bundles that we shall encounter.

Now, using the splitting principle $c(\mathcal{V}) = c(\bigoplus_i \mathcal{L}_i)$ and Eq. (1.3.15)–(1.3.18), it follows that

$$c_1(\det \mathcal{V}) = c_1(\mathcal{V}) \ . \tag{1.3.20}$$

In particular, if $c_1[\mathcal{T}_{\mathcal{M}}] = 0$ then also $c_1[\det \mathcal{T}_{\mathcal{M}}] = c_1[\det \mathcal{T}^*_{\mathcal{M}}] = 0$ and since the canonical bundle, $\mathcal{K}_{\mathcal{M}} = \det \mathcal{T}^*_{\mathcal{M}}$ is a line bundle, its total Chern class is $c[\mathcal{K}_{\mathcal{M}}] = 1$. In other words, $\mathcal{K}_{\mathcal{M}}$ is holomorphically trivial; $\mathcal{K}_{\mathcal{M}} \approx \mathbb{C}_{\mathcal{M}}$.

Using the splitting principle, the total Chern class of a complex projective space is obtained [23, 19]

$$c(\mathbb{P}^n) = (1 + J)^{n+1} \ , \tag{1.3.21}$$

where $J = c_1(\mathcal{O}_{\mathbb{P}^n}(1))$ and is the cohomology class of the curvature two-form of the hyperplane bundle over $\mathbb{P}^n$. With $\mathcal{E} = \mathcal{O}_{\mathbb{P}^n}(q)$, we have

$$c(\mathcal{E}) = 1 + qJ \tag{1.3.22}$$

by Eq. (1.3.15). Moreover, from the sequence (1.3.7) and (1.3.16),

$$c(\mathcal{M}) = c(\mathbb{P}^n)/c(\mathcal{E}) = \frac{(1+J)^{n+1}}{(1+qJ)} \ . \tag{1.3.23}$$

This somewhat symbolic expression acquires its full meaning when expanded in symbolic Taylor series where products of J are wedge-products so that the series terminates at the $J^{\dim \mathcal{M}}$ term, since all forms, i.e., cohomology classes in question are here restricted to $\mathcal{M}$.

The result (1.3.3) is now recovered in the case $r = 1$ of

$$c_r = \Big[\sum_{k=0}^{r} \tbinom{n+1}{k} (-q)^{r-k} \Big] J^r \ , \tag{1.3.24}$$

so that

$$c_1 = \Big[(n+1) - q) \Big] J \ . \tag{1.3.25}$$

To conclude, Ricci-flatness is equivalent to the vanishing of the first Chern class and so we choose $q = n+1$ for the degree-q hypersurface in $\mathbb{P}^n$ to be a Calabi–Yau manifold $\mathcal{M}$. Being interested in $\dim \mathcal{M} = 3$, the choice of a quintic hypersurface in $\mathbb{P}^4$ ensues.

In addition to recovering (1.3.3), Eq. (1.3.24) offers some more information. Since c_p are closed (p,p)-forms, it is quite obvious that $(c_1 \wedge c_1 \wedge c_1)$, $(c_1 \wedge c_2)$ and (c_3) are all closed (3,3)-forms and may be integrated over $\mathcal{M}$ to give the Chern numbers C_1, C_2 and C_3, respectively. By Stokes' formula, these numbers are independent of the choice of the particular differential forms c_r representing the Chern classes and are topological in nature. In particular, the Gauss–Bonnet formula III (see p. 307 of Ref. [19] or pp. 415–416 of Ref. [23]) tells that $C_3(\mathcal{M}) = \chi_{_E}(\mathcal{M})$. Having chosen $c_1(\mathcal{M}) = 0$, C_3 is in fact the only non-vanishing Chern number.

To evaluate C_3, we need to integrate over $\mathcal{M}$ and for that we invoke the following general result about integration over complex submanifolds:

Theorem 1.3. *Let $\mathcal{M}$ be a complex submanifold of $\mathcal{X}$, $\dim \mathcal{M} = n$, $\dim \mathcal{X} = n+K$, with normal bundle $\mathcal{N}$, $\operatorname{rank} \mathcal{N} = K$. Then there is a closed (K,K)-form μ on $\mathcal{X}$ whose restriction to $\mathcal{M}$ represents the top Chern class of the normal bundle, $c_K(\mathcal{N})$. Moreover, if ω is any closed (n,n)-form on $\mathcal{X}$, then*

$$\int_{\mathcal{M}} \omega = \int_{\mathcal{X}} \mu \wedge \omega \,. \tag{1.3.26}$$

That is, the closed form μ restricts integration from $\mathcal{X}$ to $\mathcal{M}$ like a delta function.

Remark: If $\mathcal{N}_{\mathcal{X}/\mathcal{M}}$ is the restriction to $\mathcal{M}$ of some bundle over $\mathcal{X}$, as in the present application, $\mathcal{N}_{\mathcal{X}/\mathcal{M}} = \mathcal{E}|_{\mathcal{M}}$, then the existence of μ is straightforward; $\mu = c_k[\mathcal{E}]$, for $k = \operatorname{rank} \mathcal{E}$, so $\mu = qJ$ for $\mathcal{E} = \mathcal{O}_{\mathbb{P}^n}(q)$. In general, μ represents the Poincaré dual of the homology class $[\mathcal{M}] \in H_{2m}(\mathcal{X})$.

We normalize (the first equality is known as the *Wirtinger theorem*, which however typically uses a $\frac{1}{n!}$-multiple of the integral [23, p. 31])

$$\operatorname{Vol}(\mathbb{P}^n) \;=\; \int_{\mathbb{P}^n} J^n \;=\; 1 \,, \tag{1.3.27}$$

and compute, for the quintic in $\mathbb{P}^4$ ($\chi_{_E} = C_3$ for a 3-fold)

$$\begin{aligned} \chi_{_E}(\mathcal{M}) &= \textstyle\int_{\mathcal{M}} c_3(\mathcal{M}) \,, \\ &= \textstyle\int_{\mathbb{P}^4} [5\,J] \wedge [-1 \cdot 125 + 5 \cdot 25 - 10 \cdot 5 + 10]\, J^3 \;=\; -\,200 \,. \end{aligned} \tag{1.3.28}$$

1.4 Four More Easy Pieces

Having introduced Chern classes, we now pay a visit to the remaining four complete intersection Calabi–Yau manifolds in a single $\mathbb{P}^n$. We now have that

$$\mathcal{E} = \bigoplus_{a=1}^{K} \mathcal{E}_a \ , \qquad q_a \stackrel{\text{def}}{=} \deg(\mathcal{E}_a) \ , \tag{1.4.1}$$

where each $\mathcal{E}_a$, $a = 1, \dots, K$, is the bundle corresponding to a degree-q_a constraint polynomial $f^a(z)$, $z \in \mathbb{P}^n$. The direct sum, $\mathcal{E}$, therefore corresponds to the system of constraint polynomials. Each hypersurface in $\mathbb{P}^n$ has dimension $n{-}1$, so the intersection of k hypersurfaces has dimension $n{-}k$. Upon using up all K constraints, we want to obtain for their intersection $\dim M = 3$, whence $K = n{-}3$.

We may use Bertini's theorem iteratively to conclude that, in this case also, the generic choice of constraints leads to a non-singular complex manifold. We simply order the constraints and interpret each succeeding constraint as a holomorphic section of an appropriate line bundle, restricted to the intersection of hypersurfaces defined by the previous constraints. The hypotheses of Bertini's theorem are satisfied at each stage.

However, to verify for a particular choice of constraints that the resulting intersection of hypersurfaces is smooth, it is necessary to check that there is no point at which all the constraints vanish together with the wedge product of their gradients

$$\mathrm{d}f^1(z) \wedge \cdots \wedge \mathrm{d}f^K(z) \ . \tag{1.4.2}$$

This wedge product is the holomorphic volume form of the normal bundle and for it to vanish, either some of the gradients vanish or some (linear combinations) of them become parallel; clearly, this would detect a singularity at the resulting intersection. As in the case of a single constraint, this condition is independent of the coordinates used to check it.

Although at the time we derived the Adjunction formula (1.3.9), $\mathcal{E}$ denoted a single line bundle, the argument is equally valid in this more general situation. Moreover, the basic results on Chern classes listed above lead easily to the generalization of the Chern polynomial formula (1.3.23),

$$c(\mathcal{M}) = \frac{(1+J)^{n+1}}{\prod_{a=1}^{K}(1+q_a J)} \ , \tag{1.4.3}$$

and we obtain straightforwardly

$$c_1(\mathcal{M}) = n + 1 - \sum_{a=1}^{K} q_a \ . \tag{1.4.4}$$

For Ricci-flatness, i.e., vanishing first Chern class, we demand

$$\sum_{a=1}^{K} q_a \;=\; n+1\;, \tag{1.4.5}$$

since J is a positive (1,1)-form. This yields precisely five solutions:

1. The quintic in $\mathbb{P}^4$, denoted $[4\|5]$;
2. The intersection of two cubics in $\mathbb{P}^5$, $[5\|3\,3]$;
3. The intersection of a quadric and a quartic in $\mathbb{P}^5$, $[5\|2\,4]$;
4. The intersection of two quadrics and a cubic in $\mathbb{P}^6$, $[6\|2\,2\,3]$;
5. The intersection of four quadrics in $\mathbb{P}^7$, $[7\|2\,2\,2\,2]$.

We have here adopted the somewhat cryptic 'bra-ket' notation of Ref. [168]: the left-most, 'bra' entry is the dimension of $\mathbb{P}^n$ and the remaining, 'ket' entries are the degrees of homogeneity of the constraint polynomials.

1.4.1 The Euler numbers

Expanding the generalized Chern polynomial formula, we obtain for these five manifolds

$$c(\mathcal{M}) = 1 + \tfrac{1}{2}\Big[\big(\sum_{a=1}^{K} q_a{}^2\big) - (n+1)\Big] J^2 - \tfrac{1}{3}\Big[\big(\sum_{a=1}^{K} q_a{}^3\big) - (n+1)\Big] J^3\;, \tag{1.4.6}$$

where we have used the $c_1 = 0$ requirement (1.4.5).

To evaluate the Euler characteristics, we begin by observing that

$$c_K(\mathcal{E}) \;=\; \prod_{a=1}^{K} \deg(\mathcal{E}_a\; J) = \Big(\prod_{a=1}^{K} \deg \mathcal{E}_a\Big)\, J^K\;, \tag{1.4.7}$$

and then invoke Eq. (1.3.26) to obtain

$$\int_{\mathcal{M}} (\;\;) \;=\; \int_{\mathbb{P}^n} \Big(\prod_{a=1}^{K} (q_a J)\Big) \wedge (\;\;)\;, \tag{1.4.8}$$

with which we pull the integration over the submanifold $\mathcal{M}$ back to the ambient space. Hence,

$$\begin{aligned}
\chi_{E}[4\|5] &= 5\,[\tfrac{1}{3}(5-125)] &&= -200\;,\\
\chi_{E}[5\|3\,3] &= 3{\cdot}3\,[\tfrac{1}{3}(6-27-27)] &&= -144\;,\\
\chi_{E}[5\|2\,4] &= 2{\cdot}4\,[\tfrac{1}{3}(6-8-64)] &&= -176\;,\\
\chi_{E}[6\|2\,2\,3] &= 2{\cdot}2{\cdot}3\,[\tfrac{1}{3}(7-8-8-27)] &&= -144\;,\\
\chi_{E}[7\|2\,2\,2\,2] &= 2{\cdot}2{\cdot}2{\cdot}2[\tfrac{1}{3}(8-8-8-8-8)] &&= -128\;.
\end{aligned} \tag{1.4.9}$$

1.5 Characteristic Classes

In § 1.3.2, we have seen an example of what are known as characteristic classes. For future reference, we now briefly review some other examples; more details are found in Ref. [29].

We recall that the total Chern class of a rank-k vector bundle $\mathcal{V}$ over a manifold $\mathcal{M}$ is

$$c(\mathcal{V}) \;=\; \det\left[\mathbb{1} + \mathbf{F}\right] , \tag{1.5.1}$$

where $\mathbf{F}$ is the (suitably normalized) curvature (1, 1)-form of $\mathcal{V}$. Using the splitting principle (1.3.19),

$$c(\mathcal{V}) \;=\; \sum_{r=1}^{k} c_r(\mathcal{V}) \;=\; \prod_{i=1}^{k}(1+\ell_i) , \tag{1.5.2}$$

where

$$c_r(\mathcal{V}) \;=\; \sum_{i<j<\cdots<l} \underbrace{\ell_i \wedge \ell_j \wedge \cdots \wedge \ell_l}_{r \text{ times}} \tag{1.5.3}$$

is the rth Chern class of $\mathcal{V}$ and $c_r(\mathcal{V}) \equiv 0$ for $r > \dim \mathcal{M}$. ℓ_i is the first Chern class of the ith line bundle $\mathcal{L}_i$ into a sum of which we have deformed $\mathcal{V}$ (even though $\mathcal{V} \neq \oplus_i \mathcal{L}_i$ holomorphically!).

Using the ℓ_i, one can define the *Chern character*$^{(L)}$ to be

$$\mathrm{ch}(\mathcal{V}) \;\stackrel{\text{def}}{=}\; \sum_{i=1}^{k} e^{\ell_i} . \tag{1.5.4}$$

It follows that

$$\mathrm{ch}(\mathcal{V} \oplus \mathcal{V}') = \mathrm{ch}(\mathcal{V}) + \mathrm{ch}(\mathcal{V}') , \tag{1.5.5}$$

$$\mathrm{ch}(\mathcal{V} \otimes \mathcal{V}') = \mathrm{ch}(\mathcal{V}) \wedge \mathrm{ch}(\mathcal{V}') . \tag{1.5.6}$$

In general,

$$\mathrm{ch}(\mathcal{V}) = k + \sum\nolimits_{r=1}^{k} \mathrm{ch}_r(\mathcal{V}) , \tag{1.5.7}$$

$$\begin{aligned} &= k + c_1 + \tfrac{1}{2}(c_1{}^2 - 2c_2) + \tfrac{1}{6}(c_1{}^3 - 3c_1c_2 + 3c_3) \\ &\quad + \tfrac{1}{24}\big(c_1{}^4 - 4c_1{}^2c_2 + 4c_1c_3 + 2c_2{}^2 - 4c_4\big) \\ &\quad + \tfrac{1}{120}\big(c_1{}^5 - 5c_1{}^3c_2 + 5c_1{}^2c_3 + 5c_1c_2{}^2 - 5c_1c_4 - 5c_2c_3 + 5c_5\big) \\ &\quad + \tfrac{1}{720}\big(c_1{}^6 - 6c_1{}^4c_2 + 6c_1{}^3c_3 + 9c_1{}^2c_2{}^2 - 6c_1{}^2c_4 - 12c_1c_2c_3 + 6c_1c_5 \\ &\qquad - 2c_2^3 + 3c_3{}^2 + 6c_2c_4 - 6c_6\big) + \ldots \end{aligned} \tag{1.5.8}$$

where $\mathrm{ch}_r(\mathcal{V}) = \frac{1}{r!}S_r(\mathcal{V})$. The Chern classes c_j and the symmetric polynomials S_r (see *invariant polynomials*$^{(L)}$) are related through the Newton formulae

$$\sum_{j=0}^{r}(-)^j\, c_j(\mathcal{V}) \wedge S_{r-j}(\mathcal{V}) \;=\; 0\;, \qquad S_0 = k\;. \tag{1.5.9}$$

Again using the ℓ_i, one can define the *Todd class* as

$$\mathrm{td}(\mathcal{V}) \;\stackrel{\text{def}}{=}\; \prod_{i=1}^{k} \frac{\ell_i}{1-e^{-\ell_i}}\;. \tag{1.5.10}$$

It follows that

$$\mathrm{td}(\mathcal{V} \oplus \mathcal{V}') = \mathrm{td}(\mathcal{V}) \;\wedge\; \mathrm{td}(\mathcal{V}')\;. \tag{1.5.11}$$

In general,

$$\mathrm{td}(\mathcal{V}) = 1 + \sum\nolimits_{r=1}^{\infty} \mathrm{td}_r(\mathcal{V})\;, \tag{1.5.12}$$

$$\begin{aligned} &= 1 + \tfrac{1}{2}c_1 + \tfrac{1}{12}\big(c_1^{\,2}{+}c_2\big) + \tfrac{1}{24}\big(c_1c_2\big) \\ &\quad - \tfrac{1}{720}\big(c_1^{\,4}{-}4c_1^{\,2}c_2{-}3c_2^{\,2}{-}c_1c_3{+}c_4\big) \\ &\quad - \tfrac{1}{1440}\big(c_1^{\,3}c_2{-}c_1^{\,2}c_3{-}3c_1c_2^{\,2}{+}c_1c_4\big) \\ &\quad + \tfrac{1}{60480}\big(2c_1^{\,6}{-}12c_1^{\,4}c_2{+}5c_1^{\,3}c_3{+}11c_1^{\,2}c_2^{\,2}{-}5c_1^{\,2}c_4{+}11c_1c_2c_3{-}2c_1c_5 \\ &\qquad + 10c_2^{\,3}{-}c_3^{\,2}{-}9c_2c_4{+}2c_6\big) + \dots \end{aligned} \tag{1.5.13}$$

Furthermore, for any almost complex n-fold, we have also the *Pontrjagin classes* $\mathrm{p}_r \in H^{4r}(\mathcal{M},\mathbb{Z})$ [29],

$$\mathrm{p} \;\stackrel{\text{def}}{=}\; \sum\nolimits_{r=0}^{\lfloor \frac{n}{2} \rfloor}(-)^r\, \mathrm{p}_r \;\stackrel{\text{def}}{=}\; \sum\nolimits_{i,j=0}^{n}(-)^i c_i \wedge c_j, \tag{1.5.14}$$

so that[§]

$$\mathrm{p}_1 = c_1^{\,2} - 2c_2\;, \quad \mathrm{p}_2 = c_2^{\,2} - 2(c_1c_3 - c_4)\;, \quad \mathrm{p}_3 = c_3^{\,2} - 2(c_2c_4 - c_1c_5 + c_6)\;, \tag{1.5.15}$$

and so on.

We will also find useful the Hirzebruch L-polynomial

$$L(\mathcal{V}) \;\stackrel{\text{def}}{=}\; \prod_{i=1}^{k} \frac{\ell_i}{\tanh(\ell_i)}\;, \tag{1.5.16}$$

[§] By a characteristic class of a manifold, we mean the characteristic class of its tangent bundle, which is customary to omit in notation if no confusion can arise.

for which the expression ($\mathcal{V} \to \mathcal{T}_{\mathcal{M}}$)

$$L = \tfrac{1}{3}\mathrm{p}_1 + \tfrac{1}{3^2\cdot 5}(7\mathrm{p}_2 - \mathrm{p}_1{}^2) + \tfrac{1}{3^3\cdot 5\cdot 7}(62\mathrm{p}_3 - 13\mathrm{p}_2\mathrm{p}_1 + 2\mathrm{p}_1{}^3) + \dots \tag{1.5.17}$$

will suffice for our purposes; for further details, see Ref. [29].

Finally, for completeness, we give the so-called A-roof polynomial,

$$\hat{A}(\mathcal{V}) \stackrel{\text{def}}{=} \prod_{i=1}^{k} \frac{\frac{1}{2}\ell_i}{\sinh(\frac{1}{2}\ell_i)}, \tag{1.5.18}$$

$$= -\tfrac{2}{3}\mathrm{p}_1 - \tfrac{2}{3^2\cdot 5}(4\mathrm{p}_2 - 7\mathrm{p}_1{}^2) - \tfrac{2}{3^3\cdot 5\cdot 7}(16\mathrm{p}_3 - 44\mathrm{p}_2\mathrm{p}_1 + 31\mathrm{p}_1{}^3) + \dots \tag{1.5.19}$$

where the second row specializes to $\mathcal{V} = \mathcal{T}_{\mathcal{M}}$. $\hat{A}$ is associated with the index of the Dirac operator. Notice that (1.5.14), (1.5.17) and (1.5.19) are all $(2k, 2k)$-forms, and so inherently contain less information than (1.5.2), (1.5.4) and (1.5.10).

1.6 The Lefschetz Hyperplane Theorem

In certain favorable cases, and certainly in the five examples above, $h^{1,1} = b_2$ can be computed using a generalization of the Lefschetz hyperplane theorem which requires only 'rather topological' data.

We quote here the Lefschetz hyperplane theorem in the general form as proven by R. Bott [10]:

Theorem 1.4 (Lefschetz–Bott). *Let $\mathcal{L}$ be a positive holomorphic line bundle over an $n{+}1$-dimensional complex compact manifold $\mathcal{X}$ and let λ be a holomorphic section of $\mathcal{L}$. Then:*

$$\begin{aligned} &\pi_q\big(\lambda^{-1}(0)\big) \approx \pi_q(\mathcal{X}) && 0 \le q < n = \dim \lambda^{-1}(0), \\ &\pi_n\big(\lambda^{-1}(0)\big) \xrightarrow{j} \pi_n(\mathcal{X}) && \textit{is onto/surjective.} \end{aligned} \tag{1.6.1}$$

In particular, we then have

Corollary 1.1.

$$\begin{aligned} &H_q\big(\lambda^{-1}(0), \mathbb{Z}\big) \approx H_q(\mathcal{X}, \mathbb{Z}) && 0 \le q < n = \dim \lambda^{-1}(0), \\ &H_n\big(\lambda^{-1}(0), \mathbb{Z}\big) \xrightarrow{j} H_n(\mathcal{X}, \mathbb{Z}) && \textit{is onto/surjective.} \end{aligned} \tag{1.6.2}$$

Moreover, if $b_n(\mathcal{X}) = b_n\big(\lambda^{-1}(0)\big)$ the map j is also an isomorphism. Therefore, the corresponding map, $H^n(\mathcal{X}, \mathbb{Z}) \xrightarrow{j} H^n\big(\lambda^{-1}(0), \mathbb{Z}\big)$, is into/injective.

Table 1.1: The five standard Calabi–Yau complete intersections in $\mathbb{P}^n$ and some of their numerical characteristics.

Complete intersection	**Denoted**	χ_E	$h^{1,1}$	$h^{2,1}$
a quintic in $\mathbb{P}^4$	$\mathbb{P}^4[5] = [4\|5]$	-200	1	101
a quartic and a quadric in $\mathbb{P}^5$	$\mathbb{P}^5[4\ 2] = [5\|4\ 2]$	-176	1	89
two cubics in $\mathbb{P}^5$	$\mathbb{P}^5[3\ 3] = [5\|3\ 3]$	-144	1	73
a cubic and two quadrics in $\mathbb{P}^6$	$\mathbb{P}^6[3\ 2\ 2] = [6\|3\ 2\ 2]$	-144	1	73
four quadrics in $\mathbb{P}^7$	$\mathbb{P}^7[2\ 2\ 2\ 2] = [7\|2\ 2\ 2\ 2]$	-128	1	65

The application to the five cases above is straightforward. In each case, we have one or more polynomials, corresponding to a line bundle or a direct sum of line bundles. Each of these line bundles is positive since it is a positive power of the hyperplane bundle, which itself is positive.

Now we proceed by iteration: Take any of the line bundles to begin with and consider the hypersurface in $\mathbb{P}^n$ defined as the zero-set of a polynomial corresponding to that line bundle. Since the kth line bundle is positive over $\mathbb{P}^n$, it is also positive over the subspace which is the complete intersection of the previous $k-1$ hypersurfaces and so the Lefschetz hyperplane theorem applies again and guarantees $H^{p,q}(\mathcal{M}_k, \mathbb{Z}) = H^{p,q}(\mathbb{P}^n, \mathbb{Z})$ for $p+q \neq (n-k)$. Here $\mathcal{M}_k$ denotes the complete intersection of the first k hypersurfaces; clearly, $\mathcal{M} = \mathcal{M}_{n-3}$. Upon including all $(n-3)$ constraints, we obtain

$$H^{p,q}(\mathcal{M}, \mathbb{Z}) = H^{p,q}(\mathbb{P}^n, \mathbb{Z}) , \qquad p+q \neq 3 . \tag{1.6.3}$$

This certainly agrees with our general form of the Hodge diamond and moreover tells that $h^{1,1} = 1$ for all five above cases. From the above relation (1.2.15) we have that

$$h^{2,1} \;=\; h^{1,1} - \tfrac{1}{2}\chi_E \;=\; 1 - \tfrac{1}{2}\chi_E \tag{1.6.4}$$

for the five Calabi–Yau manifolds as summarized in Table 1.1.

1.7 Onward into the Jungle!

We shall soon generalize the five simple constructions of this chapter in several different aspects. In the next chapter, we consider complete intersections in a product of several complex projective spaces rather than a single one. In later chapters, we shall consider complete intersections in (products of) other complex spaces but also constructions which rely on various surgical techniques.

In favorable cases of more general complete intersections, a simple iterative application of the Lefschetz hyperplane theorem will provide $b_2 = h^{1,1}$ and, through the independent computation of χ_E, the only other variant Hodge number, $h^{2,1}$. Unfortunately, more often than not—in the general case—this algorithm will fail to work and we will need to consider different techniques.

We shall then spend considerable time in describing the application of spectral sequences which yield a much more complete description of any particular cohomology group on any complete intersection in any product of so-called 'flag spaces' (of which complex projective spaces are the simplest cases). We shall also discuss the effect of dividing out group actions, i.e., passing to the quotient with respect to the holomorphic action of some symmetry group. All this builds up towards computing the certain overlap integrals which are, in corresponding compactification models, interpreted as couplings among the various light fields.

Finally, Calabi–Yau spaces in general may be endowed with various choices of the complex structure and Kähler class and the physics interpretation of such a model does depend on these choices. In the latter part of the book, we shall discuss the geometry of the space of complex structures and the complexified Kähler cone.

Chapter 2

Complete Intersections in Products of Projective Spaces

The five constructions described in the previous chapter may be thought of as toy models. In this chapter, we study a large number of examples—each defined as a complete intersection of hypersurfaces in a product of projective spaces.

2.1 Construction

Extending the ideas of the previous chapter, we now consider embedding Calabi–Yau manifolds into a general product

$$\mathcal{X} \stackrel{\text{def}}{=} \mathbb{P}_1^{n_1} \times \cdots \times \mathbb{P}_m^{n_m} . \tag{2.1.1}$$

One may consider generalizing this strategy further by including other spaces in the product. On the one hand, weighted (anisotropic, so to speak) complex projective spaces provide a rich family of constructions; already as transverse hypersurfaces in a single weighted $\mathbb{P}^4$, there are some 6000 Calabi–Yau 3-folds [96]. Including on the other hand Grassmannians or, even more so, general flag spaces generalizes the embedding space $\mathcal{X}$ in yet another way. Furthermore, embeddings other than complete intersections can also be discussed (see § 3.5).

We shall return to some of these options later on; for now, we focus on complete intersection Calabi–Yau 3-folds. Each of these is well characterized by a so-called *configuration* and we now turn to describe these.

2.1.1 Configurations

A *complete intersection manifold* is defined to be a manifold $\mathcal{M}$ embedded as complete intersection of hypersurfaces

$$\mathcal{M} \stackrel{\text{def}}{=} X^1 \cap \cdots \cap X^K \hookrightarrow \mathcal{X} \tag{2.1.2}$$

in the embedding space $\mathcal{X}$. Each hypersurface X^a is defined as the zero locus of a suitably chosen holomorphic polynomial:

$$X^a \;:\; \xi^a(z_{(1)}, \ldots, z_{(m)}) = 0, \quad a = 1, \ldots, K, \tag{2.1.3}$$

which is homogeneous of degree q_a^r with respect to the homogeneous coordinates $\vec{z}_{(r)}$ of $\mathbb{P}_r^{n_r}$.

Each homogeneous polynomial ξ^a may also be thought of as a holomorphic section of the line bundle

$$\mathcal{E}_a \stackrel{\text{def}}{=} \bigotimes_r \mathcal{O}_r(q_a^r) \, . \tag{2.1.4}$$

Here $\mathcal{O}_r(1)$ denotes the hyperplane (line) bundle over $\mathbb{P}_r^{n_r}$. In other words, the bundle of hyperplanes each of which is defined by a linear constraint $\lambda_\mu z_{(r)}^\mu = 0$, where $z_{(r)}^\mu$ are homogeneous coordinates on $\mathbb{P}_r^{n_r}$ and $\lambda_\mu \in \mathbb{C}$ parametrize the hyperplane.

— ❦ —

Definitions and notation: It will be convenient to define a *configuration matrix* as a pair $[\mathbf{n}\|\mathbf{q}]$ of an m-dimensional positive integer valued column vector n_r and an $m \times K$-dimensional non-negative integer valued matrix q_a^r. We shall denote it by

$$\left[\begin{array}{c\|ccc} n_1 & q_1^1 & \cdots & q_K^1 \\ \vdots & \vdots & \ddots & \vdots \\ n_m & q_1^m & \cdots & q_K^m \end{array}\right] . \tag{2.1.5}$$

A similar notation,

$$\begin{array}{c} \mathbb{P}^{n_1} \\ \vdots \\ \mathbb{P}^{n_m} \end{array} \begin{bmatrix} q_1^1 & \cdots & q_K^1 \\ \vdots & \ddots & \vdots \\ q_1^m & \cdots & q_K^m \end{bmatrix} , \tag{2.1.6}$$

is also used in the literature. While the embedding space is a product of only complex projective spaces, the former notation is sufficient and we adhere to it. Otherwise, we shall list the factors of $\mathcal{X}$ explicitly, akin to the latter variant. More generally, the *configuration* will be denoted as the pair $[\mathcal{X}\|\mathcal{E}]$, where $\mathcal{E}$ is the vector bundle over $\mathcal{X}$ the sections of which form the defining system of constraints for the embedded 3-fold $\mathcal{M}$.

A configuration matrix specifies the embedding space $\mathcal{X}$ and the degrees of homogeneity of the defining polynomials, but does not specify the polynomials otherwise. In particular, e.g., the configuration matrix of the Tian–Yau manifold [223],

$$\left[\begin{matrix}3\\3\end{matrix}\left\|\begin{matrix}3&0&1\\0&3&1\end{matrix}\right.\right] , \tag{2.1.7}$$

represents the system

$$f_{abc}x^a x^b x^c = 0 \ , \quad g_{\alpha\beta\gamma}y^\alpha y^\beta y^\gamma = 0 \ , \quad h_{a\alpha}x^a y^\alpha = 0 \ , \tag{2.1.8}$$

but does not specify the coefficients f_{abc}, $g_{\alpha\beta\gamma}$ and $h_{a\alpha}$. Such coefficients occur in every constraint and remain undetermined by the configuration matrix.

As we shall discuss below, for each configuration, almost every choice of the coefficients in the defining polynomials leads to a smooth manifold and the space spanned by the coefficients has a single, connected component. Two different sets of coefficients (i.e., points in this space) correspond to two complete intersections which are in general different as complex manifolds. The space of these coefficients is therefore a parameter space for the complete intersection manifolds.

By a *configuration*, we shall mean *the family of all complete intersections* (whether manifolds or singular) defined by a system of polynomial constraints which are represented by a configuration matrix and parametrized by the space of coefficients. This collection of varieties forms a deformation class. We represent the configuration by the configuration matrix and for a particular *member* of the configuration, a complete intersection space $\mathcal{M}$, we write

$$\mathcal{M} \in \left[\begin{matrix}n_1\\ \vdots\\ n_m\end{matrix}\left\|\begin{matrix}q_1^1 & \cdots & q_K^1\\ \vdots & \ddots & \vdots\\ q_1^m & \cdots & q_K^m\end{matrix}\right.\right] \tag{2.1.9}$$

and say that $\mathcal{M}$ *belongs* to the configuration. Clearly,

$$\dim\mathcal{M} = \dim\mathcal{X} - K \ , \qquad \dim\mathcal{X} = \sum_{r=1}^{m} n_r \ . \tag{2.1.10}$$

Consider a configuration where the $\mathbf{q}$ matrix is block-diagonal

$$[\mathbf{n}\|\mathbf{q}] = \begin{bmatrix} \mathbf{n}' \\ \mathbf{n}'' \end{bmatrix}\!\!\left\|\begin{matrix} \mathbf{q}' & \mathbf{0} \\ \mathbf{0} & \mathbf{q}'' \end{matrix}\right] . \tag{2.1.11}$$

It is rather obvious that $[\mathbf{n}\|\mathbf{q}]$ describes a class of product spaces of the form $\mathcal{M} = \mathcal{M}' \times \mathcal{M}''$ and where

$$\mathcal{M}' \in [\mathbf{n}'\|\mathbf{q}'] , \qquad \mathcal{M}'' \in [\mathbf{n}''\|\mathbf{q}''] . \tag{2.1.12}$$

A configuration $[\mathbf{n}\|\mathbf{q}]$ is called *decomposable* if it can be brought into a block-diagonal form by permuting its rows and the columns of $\mathbf{q}$; otherwise we call it *indecomposable.*

Finally, note that a linear constraint in $\mathbb{P}^n$ defines a $\mathbb{P}^{n-1} \subset \mathbb{P}^n$ and is therefore uninteresting. Consequently, we require the degree of a constraint polynomial to be at least two if it depends on one $\mathbb{P}^{n_r}_r$ only. This is easily seen to be equivalent to the condition

$$\sum_{r=1}^{m} q_a^r \geq 2 , \qquad a = 1, \ldots, K \tag{2.1.13}$$

and we require all configuration matrices to satisfy this.

Ricci-flatness: Since the Chern polynomial is determined by the dimensions of the $\mathbb{P}^n$ factors in $\mathcal{X}$ and the degrees of homogeneity of the constraints, we can write a general formula for each configuration. Straightforwardly, $\mathcal{T}_{\mathcal{M}}$ is a sub-bundle of $\mathcal{T}_{\mathcal{X}}$, the holomorphic tangent bundle of the embedding space and an analogue of the *Adjunction formula*$^{(L)}$ 2, (1.3.7), implies the exactness of

$$0 \to \mathcal{T}_{\mathcal{M}} \xrightarrow{\;i\;} \mathcal{T}_{\mathcal{X}}|_{\mathcal{M}} \xrightarrow{\;\phi\;} \mathcal{E}|_{\mathcal{M}} \to 0 . \tag{2.1.14}$$

From this short exact sequence, the determinant formula (1.3.16) yields

$$\begin{aligned} c[\,\mathcal{X}\,\|\,\mathcal{E}\,] &= c[\,\mathcal{X}\,]\big/c[\,\mathcal{E}\,] , \\ c[\,\mathbf{n}\,\|\,\mathbf{q}\,] &= \frac{\prod_{r=1}^{m}(1+J_r)^{n_r+1}}{\prod_{a=1}^{K}(1+\sum_{s=1}^{m} q_a^s J_s)} . \end{aligned} \tag{2.1.15}$$

On expansion, the first Chern class is

$$c_1[\,\mathbf{n}\,\|\,\mathbf{q}\,] = \sum_{r=1}^{m} \Big(n_r + 1 - \sum_{a=1}^{K} q_a^r\Big)\, J_r . \tag{2.1.16}$$

Since each J_r is a positive (1,1)-form over $\mathbb{P}_r^{n_r}$, the only way to make the first Chern class of the configuration to vanish is to set

$$\sum_{a=1}^{K} q_a^r = n_r + 1 , \qquad \forall r = 1, \ldots, m . \tag{2.1.17}$$

This ensures that each smooth member of the configuration $[\mathbf{n}\|\mathbf{q}]$ admits a Ricci-flat Kähler metric (this can be suitably generalized for singular members).

— ❦ —

Often, we shall need to consider slightly less restricted configurations, for which the

$$\sum_{a=1}^{K} q_a^r \leq n_r + 1 , \qquad r = 1, \ldots, m \tag{2.1.18}$$

condition holds instead. If the inequality is strict for at least one r, a member $\mathcal{M}$ of such a configuration has a non-trivial anti-canonical bundle, $\mathcal{K}_{\mathcal{M}}^*$, which is generated by global sections. Generalizing earlier results,

$$\mathcal{K}_{\mathcal{M}}^* = \wedge^3 \mathcal{T}_{\mathcal{M}} = \bigotimes_{r=1}^{m} \mathcal{O}_r(\textstyle\sum_{a=1}^{K} q_a^r) . \tag{2.1.19}$$

Configurations for which Eq. (2.1.18) holds, with the inequality strict for at least one r, represent complete intersections that do not admit Ricci-flat metrics. Rather, the Ricci tensor is non-negative: it is positive along the $\mathbb{P}_r^{n_r}$ for which the inequality in Eq. (2.1.18) is strict, and vanishes (modulo total derivatives) along the $\mathbb{P}_r^{n_r}$ for which Eq. (2.1.18) becomes an equality. In general, we will adhere to the taxonomy of configurations as presented in Table 2.1. To conform with existing nomenclature for two and three

Table 2.1: A partial taxonomy of configurations.

Configuration	**Condition on $[\mathbf{n}\|\mathbf{q}]$**
Calabi–Yau	$\sum_{a=1}^{K} q_a^r = n_r + 1$, all r
Almost Ample	$\sum_{a=1}^{K} q_a^r \leq n_r + 1$, all r ("$<$" for at least one r)
Ample	$\sum_{a=1}^{K} q_a^r < n_r + 1$, all r

dimensions, respectively, we may say "(almost) del Pezzo" and "(almost) Fano" instead of "(almost) ample". Clearly, with $n = \dim \mathcal{M}$,

$$H^0(\mathcal{M}, \mathcal{K}_{\mathcal{M}}) = H^{n,0}(\mathcal{M}) \approx \begin{cases} \mathbb{C}, & \mathcal{M} \text{ is Calabi–Yau,} \\ 0, & \mathcal{M} \text{ is (almost) ample.} \end{cases} \tag{2.1.20}$$

Precisely Calabi–Yau n-folds have a nowhere vanishing holomorphic $(n,0)$-form. ⌈*Calabi–Yau complete intersections (by default and unless otherwise qualified, in products of projective spaces) are by now routinely dubbed "CICY" [85].*⌋

Of course, there exist configurations for which $[n_r+1 - \sum_{a=1}^{K} q_a^r]$ is positive for some r, negative for some other r and possibly zero for some third r. As opposed to ample configurations, these might be called *limping*. On the other hand, if $[n_r+1 - \sum_{a=1}^{K} q_a^r] \leq 0$, the configuration might be called *almost scant* (*scant* if the inequality is strict). When the canonical bundle $\mathcal{K}_{\mathcal{M}}$ is positive and non-trivial, $\dim H^0(\mathcal{M}, \mathcal{K}_{\mathcal{M}}) > 1$, there are more than one $(n,0)$-forms and, in general, each of them vanishes somewhere on $\mathcal{M}$.

2.1.2 Diagrams

We shall find it useful to introduce a diagrammatic rendition of the configuration matrices representing the deformation classes of complete intersection embeddings.

For every $\mathbb{P}^n$ factor in $\mathcal{X}$ we introduce a hollow circle with $n+1$ legs. For every constraint we introduce a dot. We connect some (possibly all) of the free legs into the dots. The requirement (2.1.13) obviously translates into the rule that at least two legs be connected into every dot.

Decomposable configurations correspond to disconnected diagrams with more than one connected part while indecomposable configurations have diagrams that are fully connected. Unless stated otherwise, we consider only indecomposable configurations and therefore only connected diagrams.

The dimension of the members of the configuration represented by a diagram is $\ell - m - K$, where ℓ is the number of legs, m is the number of circles and K is the number of dots. We shall also refer to $(\ell - m - K)$ as the dimension of the diagram; it is always non-negative since $\ell \geq 2m$ and $\ell \geq 2K$.

A diagram represents a Calabi–Yau configuration (and we call it a Calabi–Yau diagram) precisely if there are no loose legs. If there is at least one loose leg, the diagram corresponds to an almost ample configuration (and we call it almost ample). If every hollow circle has at least one leg loose, the diagram corresponds to an ample configuration (and we call the diagram ample also).

It will also be useful to distinguish the following two special types of diagrams (configurations):

Decomposing dots: A diagram is called *one-dot-decomposable*, denoted 1DD, if it breaks in two upon the deletion of some dot.

Proposition 2.1. *The complement of a decomposing dot in a diagram of dimension n has two components which represent (possibly singular) varieties of positive dimensions, which add up to $n+1$.*

✎ Prove this Proposition. ❧

We index a decomposing dot by these dimensions and refer to it as an x-y-dot, where x and y are the dimensions of the two components of the complement of the decomposing dot. When the whole diagram corresponds to 3-folds, there can only be decomposing 1-3- or 2-2-dots.

In a 1DD diagram with a decomposing x-y-dot, the decomposing dot corresponds to the defining constraint of a hypersurface in the product of two spaces of dimensions x and y, respectively.

Decomposing legs: A diagram is called *one-leg-decomposable*, denoted 1LD, if it breaks in two upon the deletion of a leg. Clearly, every 1LD diagram is also 1DD, since the decomposing leg connects to a dot, the removal of which also disconnects the diagram.

The complement of a decomposing leg in a diagram representing an n-fold has two components containing, respectively, the dot and the hollow circle incident with the deleted leg. We shall call these respectively the dot-complement and the circle-complement of the leg in question. We write ℓ_d, m_d and K_d for the number of legs, circles and dots in the dot-complement and ℓ_c, m_c and K_c for the corresponding numbers in the circle-complement. We index the leg by $(\ell_d{-}m_d{-}K_d)$.

Proposition 2.2. *The index of a decomposing leg in a diagram representing an n-fold may be $0,\ldots,n{-}1$.*

✎ Prove this Proposition. ❧

Lemma 2.1. *Every Calabi–Yau diagram representing an n-fold with a decomposing $(n{-}1)$-leg is equivalent to another one with no decomposing*

$(n-1)$-leg which is obtained by replacing the circle-complement of the decomposing $(n-1)$-leg with a single circle and the decomposing leg with two legs. Diagrammatically:

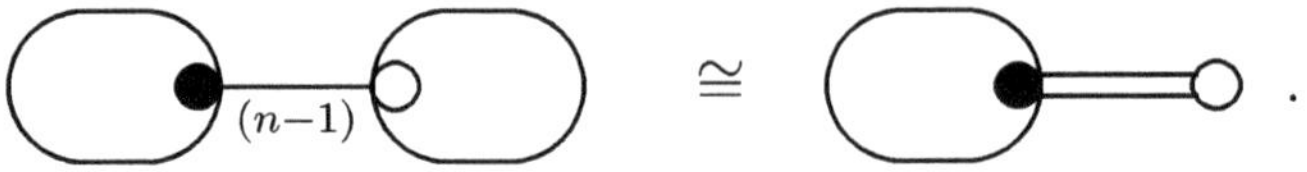

(2.1.21)

Proof. Consider the circle-complement of the decomposing $n-1$-leg together with the leg. This is one of the two components of the complement of the dot incident with the $n-1$-leg; call it F. By Proposition 2.1, F is one of the two connected components of the complement of the dot incident with the decomposing leg and represents (possibly singular) varieties of positive dimension d. We next have

$$\begin{aligned} n &= \ell - m - K \ , \\ &= (\ell_d - m_d - K_d) + 1 + (\ell_c - m_c - K_c) \ , \\ &= (n-1) + (\ell_F - m_F - K_F) \ , \end{aligned} \tag{2.1.22}$$

since $\ell_d - m_d - K_d = n - 1$ is used to index the leg. It follows that

$$d_F = \ell_F - m_F - K_F = 1 \ . \tag{2.1.23}$$

Since the complete diagram had no loose legs, F has precisely one—the decomposing one. So, F is an almost ample 1-fold and can only be a $\mathbb{P}^1$. ☑

Whenever possible, we use Lemma 2.1 to prune the diagrams and simplify the configurations. We also have the following easy corollary.

Corollary 2.1. *The dot-complement of a 0-leg can without loss of generality be reduced to*

(2.1.24)

Therefore

(2.1.25)

✎ Prove this Corollary. ❧

Remark: Note that, by swapping dots and circles, a Calabi–Yau diagram becomes another Calabi–Yau diagram. This corresponds to transposing the matrix $\mathbf{q}$ in $[\mathbf{n}\|\mathbf{q}]$ and changing $\mathbf{n}$ as appropriate for a Calabi–Yau configuration.

2.1.3 The Euler characteristic

Having obtained the total Chern class (2.1.15) for a generic configuration,

$$c[\mathbf{n}\|\mathbf{q}] = c_1^r J_r \;+\; c_2^{rs} J_r J_s \;+\; c_3^{rst} J_r J_s J_t \;, \tag{2.1.26}$$

we can list the three Chern classes:

$$c_1^r = \Big[n_r + 1 - \sum_{a=1}^{K} q_a^r \Big] \;=\; 0 \;, \tag{2.1.27}$$

$$c_2^{rs} = \tfrac{1}{2}\Big[-\delta^{rs}(n_r+1) + \sum_{a=1}^{K} q_a^r q_a^s \;+\; c_1^r c_1^s \Big] \;, \tag{2.1.28}$$

$$c_3^{rst} = \tfrac{1}{6}\Big[\,\delta^{rst}(n_r+1) - \sum_{a=1}^{K} q_a^r q_a^s q_a^t \;-\; c_2^{(rs} c_1^{t)} + c_1^r c_1^s c_1^t \Big] \;, \tag{2.1.29}$$

where the parenthesis around indices denotes symmetrization. Note the simplification which occurs for Calabi–Yau configurations.

The Euler characteristic is evaluated by finding the coefficient of the volume form of $\mathcal{X}$, $\bigwedge_{r=1}^{m} J_r{}^{n_r}$, in

$$\Big[\Big(\sum_{r,s,t=1}^{m} c_3^{rst} J_r J_s J_t \Big) \cdot \bigwedge_{a=1}^{K} \Big(\sum_{p=1}^{m} q_a^p J_p \Big) \Big] \;. \tag{2.1.30}$$

Imposing $c_1^r = 0$ for Calabi–Yau complete intersections, we obtain

$$\chi_E[\mathbf{n}\|\mathbf{q}] = \Big[\sum_{r,s,t=1}^{m} \tfrac{1}{3}\Big(\delta^{rst}(n_r+1) - \sum_{a=1}^{K} q_a^r q_a^s q_a^t \Big) J_r J_s J_t \cdot \bigwedge_{a=1}^{K} \Big(\sum_{p=1}^{m} q_a^p J_p \Big) \Big]_{\rm top} \;, \tag{2.1.31}$$

where the subscript "top" means the coefficient of $\bigwedge_{r=1}^{m} J_r{}^{n_r}$. This expression gets out of hand rather quickly for large configurations but is fortunately well suited for mechanized evaluation. Note that the $c_1^r = 0$ condition also implies that $(n_r+1) \le \sum_a (q_a^r)^3$, whence $\chi_E[\mathbf{n}\|\mathbf{q}] \le 0$.

The argument of §1.3.1 from the previous chapter can be repeated with only minor modifications:

$$\mathcal{N}_{\mathcal{X}/\mathcal{M}} \;=\; \mathcal{E} \;\stackrel{\rm def}{=}\; \bigoplus_{a=1}^{K} \mathcal{E}_a = \bigoplus_{a=1}^{K} \Big(\bigotimes_{r=1}^{m} \mathcal{O}_r(q_a^r) \Big) \tag{2.1.32}$$

is the normal bundle of the embedding $\xi\colon \mathcal{M} \hookrightarrow \mathcal{X}$ where $\xi \stackrel{\text{def}}{=} \oplus_a \xi_a$ is the collection of defining constraint polynomials and so a section of $\mathcal{E}$ and $\mathcal{M} = \xi^{-1}(0)$.

2.2 Bertini Variations

The method of constructing Calabi–Yau 3-folds described so far yields a very rich laboratory* and it is natural to inquire if these 3-folds are smooth.

2.2.1 A simple warm-up example

We wish to prove that there exist choices of defining polynomials $\{\xi^a\}$ in $\mathcal{X} = \prod_r \mathbb{P}_r^{n_r}$ with degrees prescribed by some configuration matrix (2.1.5) and such that the subspace $\mathcal{M} \hookrightarrow \mathcal{X}$ where $\xi^a = 0$, $a = 1, \ldots, K$, is smooth. This amounts to verifying that the subspace $\mathcal{M}$ has a system of linearly independent $(\dim \mathcal{X} - \dim \mathcal{M})$ gradients at every point. In other words, we require

$$\mathrm{d}^K \mathcal{N}_{\mathcal{X}/\mathcal{M}} \;=\; \bigwedge_{a=1}^{K} \mathrm{d}\xi^a(z) \neq 0 \;, \qquad \xi^a(z) = 0 \;, \quad z \in \mathcal{X} \;. \tag{2.2.1}$$

Take, e.g., the configuration $[4\|5]$ of quintics in $\mathbb{P}^4$. In view of Euler's theorem on homogeneous functions

$$z^a(\partial_a \xi) = 5\xi \;, \quad z^a \in \mathbb{P}^4 \;, \tag{2.2.2}$$

the five derivatives in $\mathrm{d}\xi = \mathrm{d}z^a(\partial_a \xi)$ are not independent of ξ. Requiring the derivatives $\mathrm{d}\xi$ to vanish where the defining polynomial ξ does imposes only four additional constraints which are independent for a general choice of ξ. This adds up to a system of five independent constraints in $\mathbb{P}^4$, is overdetermined and has no solution in general. Precisely this is the essence of the general argument; to gain familiarity with this, we shall briefly re-examine the present example and then formulate the general theorem.

— ❦ —

The most general homogeneous quintic in $\mathbb{P}^4$ can be written as

$$\xi(z; f) \;=\; f_{abcde}\, z^a\, z^b\, z^c\, z^d\, z^e \tag{2.2.3}$$

and has $\binom{4+5}{5} = 126$ coefficients f_{abcde} which must not vanish simultaneously. By homogeneity of $\xi(z; f)$, we may rescale all f_{abcde}'s by an

---

* Clearly, spaces of other dimensions or other types of canonical bundles are constructed as easily; see Table 2.1 on page 53 and subsequent discussion.

arbitrary complex number, so that $\{f_{abcde}\}$ span a $\mathbb{P}^{125}$. Furthermore, (projective) linear coordinate reparametrizations of the embedding space $\mathbb{P}^4$, $z^a \to \lambda^a{}_{a'} z^{a'}$, cannot change the locus of $\xi(z;f) = 0$. Therefore, we consider

$$f_{abcde} \simeq \lambda_a{}^{a'} \lambda_b{}^{b'} \lambda_c{}^{c'} \lambda_d{}^{d'} \lambda_e{}^{e'} f_{a'b'c'd'e'} , \qquad \lambda_a{}^b \in \mathrm{PGL}(5,\mathbb{C}) . \quad (2.2.4)$$

Thus, the effective parameter space is

$$\mathfrak{M} \stackrel{\text{def}}{=} \Big\{ \mathbb{P}\{f_{abcde}\} \Big/ \mathrm{PGL}(5,\mathbb{C}) \Big\} , \qquad (2.2.5)$$

since $\mathrm{PGL}(5,\mathbb{C})$ is the group of linear coordinate reparametrizations of the embedding space $\mathbb{P}^4$. Clearly, $\dim \mathfrak{M} = 101$.

More accurately, the action of the (projective) general linear group of linear transformations, $z^a \to \lambda^a{}_{a'} z^{a'}$, has been *lifted* in Eq. (2.2.4) to the parameter space $\{f_{abcde}\}$ and is manifestly non-linear. The structure of the quotient with respect to the equivalence (2.2.4) is therefore quite complicated and it is straightforward that the resulting orbit space is badly singular and rather difficult to study.

At any particular point, say $z = (1,0,0,0,0) \in \mathbb{P}^4$, the condition $\xi(z;f) = 0$ may now be viewed as a *linear* constraint on $\mathfrak{M}$, defining a hyperplane $\mathfrak{M}_z \subset \mathfrak{M}$ of *codimension*$^{(L)}$ one ($\dim \mathfrak{M} - \dim \mathfrak{M}_z = 1$). Similarly, the five constraints† $\partial_a \xi(z;f) = 0$ describe the intersection $\mathfrak{M}^\sharp_z$ of five hyperplanes, so $\dim \mathfrak{M} - \dim \mathfrak{M}^\sharp_z = 5$. Since $z^a \partial_a \xi(z;f) = 5\xi(z;f)$ by homogeneity, we know that $\mathfrak{M}^\sharp_z \subset \mathfrak{M}_z$.

Sweeping now z through all of $\mathbb{P}^4$, the hyperplane $\mathfrak{M}_z$ moves about in $\mathfrak{M}$ and in fact covers it all. To see this, note that for each $f_{abcde} \in \mathfrak{M}$, $\xi(z;f) = 0$ does have a solution in $\mathbb{P}^4$. Likewise, sweeping z through all of $\mathbb{P}^4$, the intersection of hyperplanes $\mathfrak{M}^\sharp_z$ sweeps out a four parameter family $\mathfrak{M}^\sharp \subset \mathfrak{M}$. Assuming that the five hyperplanes $\partial_a \xi(z;f) = 0$ were in general position, we have that

$$\dim \mathfrak{M}^\sharp \le \dim \mathfrak{M}^\sharp_z + \dim \mathbb{P}^4 \qquad (2.2.6)$$

and therefore

$$\dim \mathfrak{M} - \dim \mathfrak{M}^\sharp \ge \dim \mathfrak{M} - (\dim \mathfrak{M}^\sharp_z + \dim \mathbb{P}^4) = 5 - 4 = 1 . \quad (2.2.7)$$

Thus, $\mathfrak{M}^\sharp$ is of strictly smaller complex dimension than $\mathfrak{M}$.

† Recall that we need no connection for the gradients of ξ_a at the subspace, since $\xi_a = 0$ and $\nabla_a \xi = \partial_a \xi + \Gamma_a \xi = \partial_a \xi$.

For the varieties belonging to the above configuration, we can draw some important conclusions:

1. The space of smooth varieties $\mathfrak{M}-\mathfrak{M}^\sharp$ is dense and open. In other words, most choices of $f_{abcde} \in \mathfrak{M}$ correspond to smooth subspaces $\mathcal{M} \subset \mathcal{X}$.
2. Since $\dim_{\mathbb{R}} \mathfrak{M} - \dim_{\mathbb{R}} \mathfrak{M}^\sharp = 2$, for every point in $t^\sharp \in \mathfrak{M}^\sharp$, there exists at least one complex disk $\mathfrak{D} \subset \mathfrak{M}$ such that only $t^\sharp \in \mathfrak{D}$ corresponds to a singular $\mathcal{M} \subset \mathcal{X}$; all other points in $\mathfrak{D}$ correspond to smooth $\mathcal{M}$'s.
3. Any two smooth subspaces $\mathcal{M} \subset \mathcal{X}$ may be deformed into each other continuously, ranging over smooth subspaces only.
4. A configuration defines a deformation class of *embeddings* $\mathcal{M} \hookrightarrow \mathcal{X}$, only a set of positive *codimension*$^{(L)}$ of which consists of singular forms of $\mathcal{M}$.

Note that the deformation space of the abstract complex manifold $\mathcal{M}$ may well be larger than this deformation space of its embedding.

2.2.2 Every configuration contains smooth varieties

It should be clear that none of the above four conclusions depend on the particular choice of the configuration (and had nothing to do with requiring the first Chern class to vanish), wherefore one expects them to hold for any complete intersection.

Indeed, the following general theorem is easily generalized to fit our purpose.

Theorem 2.1 (Bertini). *A generic element of a linear system is smooth away from the base locus of the system.*

The base locus of the system consists of points in the embedding space where *all* the admissible (systems of) defining equations vanish. It must be noted that not infrequently, a system is smooth also at the base locus—however, that must be checked case by case.

Remark (base locus vs. transversality)**:** As the Reader may be unfamiliar with the notion of the base locus and since Bertini's theorem can be applied in many different variations, it is worth taking a closer look at it. Consider the case of a *generic* quintic in $\mathbb{P}^4$. Writing it in the form (2.2.3), it is easy to count 126 parameters, of which 101 are independent as we showed above. The base locus for this *generic* quintic consists of the set of points in $\mathbb{P}^4$

where the quintic vanishes for any choice of f_{abcde}. Equivalently, there are 101 independent quintic polynomials and the base locus is where they all vanish simultaneously, $z^a = 0$—which is not in $\mathbb{P}^4$. So the quintic (2.2.3) in $\mathbb{P}^4$ is smooth except (perhaps) for some special values of f_{abcde}.

The situation changes drastically if we require our quintic to have some special form, usually dictated by some symmetry. To see this, let us restrict to some specially symmetric quintics. Consider symmetries with the diagonal action

$$\pi_\omega^{[d_0,\dots,d_4]} \ : \ z^a \ \to \ \omega^{d_a} \, z^a , \quad a = 0, \dots, 4 \ . \tag{2.2.8}$$

Let us restrict to $\mathbb{Z}_5' \ltimes \mathbb{Z}_5{}^3$-symmetric quintics where the cyclic permutation symmetry $\mathbb{Z}_5'$ acts by swapping the coordinates z^a and $\mathbb{Z}_5{}^3$ is generated by $\pi_\alpha^{[1,0,0,0,4]}$, $\pi_\beta^{[0,1,0,0,4]}$ and $\pi_\gamma^{[0,0,1,0,4]}$ where α, β, γ are primitive fifth roots of 1. The only two admissible quintic polynomials under these restrictions are

$$F(z) \ \stackrel{\text{def}}{=} \ \sum_{a=1}^{5} (z^a)^5 \ , \qquad \Psi(z) \ \stackrel{\text{def}}{=} \ \prod_{a=0}^{4} z^a . \tag{2.2.9}$$

The base locus of $\mathbb{Z}_5' \ltimes \mathbb{Z}_5{}^3$-symmetric quintics is then the subspace of $\mathbb{P}^4$ where $F(z) = 0 = \Psi(z)$, which is a set of five quintic 2-folds intersecting in ten quintic curves which in turn meet in ten points[‡]. Bertini's theorem tells that the quintic hypersurface

$$F(z) \ - 5\psi \, \Psi(z) \ = \ 0 \tag{2.2.10}$$

may be singular only (1) at the base locus and arbitrary ψ or (2) away from the base locus but only for some special ψ. Explicit analysis of the system of gradients shows that such quintics are in fact smooth (transverse) at the base locus, for general ψ. The hypersurfaces (2.2.10) are singular only when $\psi^5 = 1$ or $\psi = \infty$.

Changing our symmetry requirements by including an R-symmetry such as $\pi_R^{[1,0,0,0,0]}$, only $F(z)$ becomes admissible. Alternatively, requiring a $\mathbb{Z}_5' \ltimes (\mathbb{Z}_2 \times \mathbb{Z}_5{}^3)$ symmetry, where the additional $\mathbb{Z}_2$ acts by partial reflection, $(z^1, z^2) \to (-z^1, -z^2)$, permits only $\Psi(z)$. In both cases with only one polynomial at hand, we cannot move away from the base locus and Bertini's theorem seems of no help. Explicit analysis of the system of gradients shows that $F(z) = 0$ is smooth everywhere, while $\Psi(z) = 0$ is (rather badly) singular. To be pedantic, write $F(z) \stackrel{\text{def}}{=} \sum_{a=1}^5 c_a (z^a)^5$ and note that the coefficients c_a may be absorbed by coordinate reparametrization—as long as $c_a \neq 0$. Allowing some of the c_a's to vanish reinterprets $F(z)$ as a linear combination of five *separate* $\mathcal{O}_{\mathbb{P}^4}(5)$-sections. The base locus of this *collection* is indeed the origin, $z^a = 0$, excluded from $\mathbb{P}^4$. Bertini's theorem then indeed guarantees that the generic $F(z)$, equivalent by coordinate reparametrization to the original definition in (2.2.9), is smooth away from the base locus, i.e., over the

[‡] Hint: $\Psi(z)$ vanishes where at least one $z^a = 0$ and so consists of five $\mathbb{P}^3$'s, intersecting in ten $\mathbb{P}^2$'s which meet in ten $\mathbb{P}^1$'s which in turn have five points in common.

entire $\mathbb{P}^4$. This need not always be as straightforward, and explicit verification of transversality may well be simpler.

— ❦ —

Since the set of choices of constraints leading to non-singular Calabi–Yau manifolds is an open connected subset of all possible choices of constraints for a given configuration, we have

Corollary 2.2. *All manifolds belonging to a given configuration are equivalent by deformation.*

By common abuse of language, we will say that a system of defining equations is smooth if it defines a smooth subspace. The above corollary then says that all choices of smooth defining equations corresponding to a single configuration matrix can be varied one into another continuously and moreover without ever forcing the defining equations to become singular.

2.3 Finiteness of the Family

We have seen that every configuration with $\dim \mathcal{X} - K = n$ and which satisfies Eq. (2.1.17) represents a (partial) deformation class of smooth Calabi–Yau n-folds, which are therefore topologically equivalent. The question of how many complete intersection Calabi–Yau 3-folds exist then reduces to finding out how many configuration matrices there are.

2.3.1 Infinitely many configurations

A simple algorithm for this task is as follows. It is clear that two configuration matrices $[\mathbf{n}\|\mathbf{q}]$ that differ by a permutation of rows of a permutation of the columns of $\mathbf{q}$ describe the same configuration. To obtain all possible configuration matrices for a particular choice of its size (m, the number of $\mathbb{P}^n$ factors in $\mathcal{X}$ and K, the number of constraints) one has to

1. List all possible positive-integer m-vectors $\mathbf{n}$ ($n_i \geq n_j$ if $i > j$), such that $\sum_{r=1}^{m} n_r = K + 3$.
2. For each $\mathbf{n}$, list all matrices $\mathbf{q}$ which satisfy Eq. (2.1.17) and (2.1.13).

The first listing is simple. The second is simplified if one regards the rth row of the matrix $\mathbf{q}$ as a K-digit number Q^r and order $\mathbf{q}$ so that $Q^r \geq Q^s$

if $r > s$. Starting with the smallest Q^r's satisfying Eq. (2.1.17), row by row and keeping the hierarchy, we find the next number which satisfies Eq. (2.1.17) by shifting

$$[\ldots, q_a^r, q_{a+1}^r, \ldots] \rightarrow [\ldots, (q_a^r+1), (q_{a+1}^r-1), \ldots] \,. \tag{2.3.1}$$

For any given $\mathbf{n}$, the iteration clearly terminates and the algorithm is very well suited for machine execution. Unfortunately, however, the complete algorithm would never terminate since the list of $\mathbf{n}$'s is unbounded, as we now show.

Consider for the moment only embeddings in products of m $\mathbb{P}^1$'s. We clearly need to impose $K = m - 3$ constraints. Let us start with

$$\left[\begin{array}{c||c} 1 & 2 \\ 1 & 2 \\ 1 & 2 \\ 1 & 2 \end{array}\right] \cong \tag{2.3.2}$$

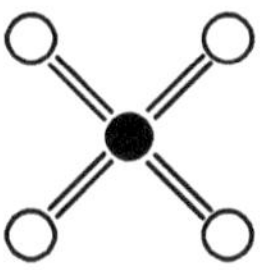

Add now a new $\mathbb{P}^1$ to the embedding space and replace the single constraint with two to obtain, say,

$$\left[\begin{array}{c||cc} 1 & 2 & 0 \\ 1 & 2 & 0 \\ 1 & 1 & 1 \\ 1 & 0 & 2 \\ 1 & 0 & 2 \end{array}\right] \cong \tag{2.3.3}$$

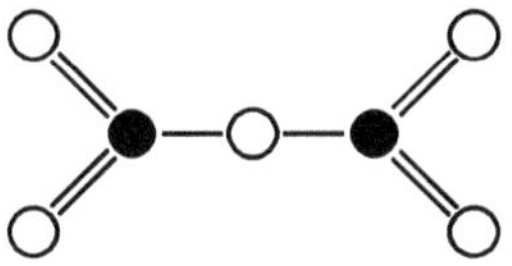

In Ref. [85] such a procedure was called 'splitting'. We can now split further, by adding one more $\mathbb{P}^1$ and replacing either of the constraints by two new constraints in the above fashion. This can clearly be repeated *ad infinitum.*

Having obtained such an 'augmented' configuration matrix however does not mean that the manifolds belonging to it are not equivalent up to deformation to the manifolds that belong to the original configuration.

2.3.2 Sufficiency of finitely many configurations

Indeed we now show that after a certain upper limit, no new configuration matrix corresponds to manifolds that are not equivalent by deformation to some already obtained.

Our proof was triggered by the observation that all configurations, after the upper limit, necessarily contain constraints that are effective only in a product of two $\mathbb{P}^1$'s and are bilinear. So, consider this sub-configuration,

$$\begin{bmatrix} 1 \\ 1 \end{bmatrix}\!\!\begin{Vmatrix} 1 \\ 1 \end{Vmatrix} \quad \equiv \tag{2.3.4}$$

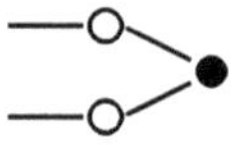

on its own. It defines a deformation class of complex 1-dimensional spaces. We compute

$$\begin{aligned} c\begin{bmatrix} 1 \\ 1 \end{bmatrix}\!\!\begin{Vmatrix} 1 \\ 1 \end{Vmatrix} &= \frac{(1+2J_x)(1+2J_y)}{(1+J_x+J_y)} = 1 + J_x + J_y \ , \\ \chi_E\begin{bmatrix} 1 \\ 1 \end{bmatrix}\!\!\begin{Vmatrix} 1 \\ 1 \end{Vmatrix} &= \int_{\mathbb{P}^1_x}\int_{\mathbb{P}^1_y} (J_x+J_y)\wedge(J_x+J_y) = 2 \ , \end{aligned} \tag{2.3.5}$$

and so it must be a $\mathbb{P}^1$. Therefore,

$$\begin{bmatrix} 1 \\ 1 \end{bmatrix}\!\!\begin{Vmatrix} 1 \\ 1 \end{Vmatrix} \equiv \mathbb{P}^1 \ , \qquad i.e., \tag{2.3.6}$$

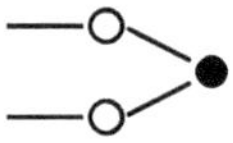

$\cong$

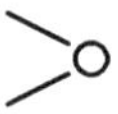

Any configuration (not necessarily Calabi–Yau) that contains this sub-configuration can be reduced by substituting a single $\mathbb{P}^1$:

$$\begin{bmatrix} \mathbf{n} \\ 1 \\ 1 \end{bmatrix}\!\!\begin{Vmatrix} 0 & \mathbf{q} \\ 1 & \mathbf{a} \\ 1 & \mathbf{b} \end{Vmatrix} \cong \begin{bmatrix} \mathbf{n} \\ 1 \end{bmatrix}\!\!\begin{Vmatrix} \mathbf{q} \\ \mathbf{a}{+}\mathbf{b} \end{Vmatrix} \ , \tag{2.3.7}$$

where **a** and **b** are some integer row-vectors. Because of this, we shall in addition to Eq. (2.1.13) require that constraints that that involve only two $\mathbb{P}^1$'s, say $\mathbb{P}^1_r$ and $\mathbb{P}^1_s$, satisfy a stronger condition

$$q^r_a + q^s_a \geq 3 \ , \qquad \forall a \ . \tag{2.3.8}$$

Theorem 2.2. *Finitely many configurations suffice to represent all topologically distinct complete intersection Calabi–Yau 3-folds.*

Proof. [145] To show that there are finitely many minimal configurations, let us consider

$$\mathcal{X} \stackrel{\text{def}}{=} \underbrace{\mathbb{P}^1 \times \cdots \times \mathbb{P}^1}_{s} \times \mathbb{P}^{n_1}_1 \times \cdots \times \mathbb{P}^{n_p}_p, \quad n_r \neq 1, \quad \forall r = 1, \ldots, p \,. \tag{2.3.9}$$

For future convenience we also define

$$\alpha = \sum_{r=1}^{p}(n_r - 1) \ , \qquad \Rightarrow \qquad p \leq \alpha \ , \tag{2.3.10}$$

where the inequality follows since $n_r \geq 2$.

The number of constraints necessary to obtain a variety of complex dimension three is

$$s + \Big(\sum_{r=1}^{p}\Big) - 3 = s + p + \alpha - 3 \tag{2.3.11}$$

and the total degree of all constraints is

$$2s + \Big(\sum_{r=1}^{p}(n_r + 1)\Big) = 2s + 2p + \alpha \ , \tag{2.3.12}$$

in order to satisfy Eq. (2.1.17). Since the constraints are at least of degree two, it follows that

$$2s + 2p + \alpha \ \geq \ 2(2 + p + \alpha - 3) \ , \qquad \Rightarrow \qquad \alpha \leq 6 \ , \tag{2.3.13}$$

and hence an upper limit

$$p \leq \alpha \leq 6 \ . \tag{2.3.14}$$

By the same argument that led to Eq. (2.3.8), we note that if a constraint is bilinear in two factors one of which is a $\mathbb{P}^1$, the other factor must be one of $\mathbb{P}_r^{n_r}$, $n_r \neq 1$. Let the number of such constraints be β. Note that β is also the total degree of these constraints in coordinates of the s $\mathbb{P}^1$'s and also in those of the p $\mathbb{P}_r^{n_r}$'s. For constraints of degree > 2, let γ denote the total degree (summed over constraints) in the coordinates of $\mathbb{P}^1$'s only. Then obviously, $\beta + \gamma = 2s$ to ensure Eq. (2.1.17) for the $\mathbb{P}^1$'s.

Since the bilinear constraints must involve a $\mathbb{P}_r^{n_r}$ also, $\beta \leq 2p + \alpha$. This follows since $2p + \alpha$ is the total degree of $\mathbb{P}_r^{n_r}$'s and we have, at most, one $\mathbb{P}^1$ for each $\mathbb{P}_r^{n_r}$. Furthermore, note that

$$\sum_{\text{all constr.}} (\text{degrees} - 2) = (2s + 2p + \alpha) - 2(s + p + \alpha - 3) = 6 - \alpha \tag{2.3.15}$$

is the number of constraints with degree > 2, if they are of degree precisely three.

On the other hand, for biggest γ, these constraints should involve only $\mathbb{P}^1$'s and so

$$\gamma \leq 3(6 - \alpha) \ . \tag{2.3.16}$$

Since $\beta + \gamma = 2s$, it follows that the number of $\mathbb{P}^1$'s is limited too,

$$s \leq 9 - (\alpha - p) \leq 9 \ , \tag{2.3.17}$$

where the last inequality follows from the relation (2.3.14).

Since both the numbers of $\mathbb{P}^1$'s and the number of $\mathbb{P}_r^{n_r}$'s ($n_r \neq 1$) is bounded from above by relations (2.3.14) and (2.3.17), respectively, the number of possible minimal configurations is finite. In fact, such lists have been assembled [58, 85, 131] and are contained within some 97,000 different configurations, with over 6000 distinct topological types of Calabi–Yau 3-folds [163]. ☑

2.3.3 Identities

The number of possibilities for minimal configurations are further reduced by identities like (2.3.6) and we now turn to discuss some of them. The basic form of the relations we seek is

$$\begin{bmatrix} \mathbf{n} \\ \mathbf{n}' \end{bmatrix}\!\!\begin{Vmatrix} \mathbf{P} & \mathbf{R} & \mathbf{0} \\ \mathbf{0} & \mathbf{R}' & \mathbf{Q}' \end{Vmatrix} \simeq \begin{bmatrix} \mathbf{n} \\ \mathbf{n}'' \end{bmatrix}\!\!\begin{Vmatrix} \mathbf{P} & \mathbf{R} & \mathbf{0} \\ \mathbf{0} & \mathbf{R}'' & \mathbf{Q}'' \end{Vmatrix} , \tag{2.3.18}$$

where $\mathbf{P} \ldots \mathbf{Q}''$ are integer matrices. The equivalence of the two configurations means that some of the manifolds belonging to one are equivalent up to deformation to the manifolds that belong to the other. If it can moreover be shown that the deformation class of manifolds belonging to the first configuration is isomorphic to the deformation class of manifolds belonging to the second, we will say that the configurations are *isomorphic* and write "$\approx$" instead of "$\simeq$". More often, however, one configuration will turn out to be a subset of the other (in the sense of the deformation classes they respectively define).

While there certainly may exist relations among pairs of configurations that do not fall in the above category, the relation (2.3.18) offers the advantage of studying the sufficient condition

$$\Big([\mathbf{n}' \| \mathbf{Q}'] \,,\, |\mathbf{R}'] \Big) \simeq \Big([\mathbf{n}'' \| \mathbf{Q}''] \,,\, |\mathbf{R}''] \Big) \,. \tag{2.3.19}$$

Here $[\mathbf{n}' \| \mathbf{Q}']$ and $[\mathbf{n}'' \| \mathbf{Q}'']$ ought to be equivalent configurations, corresponding to varieties of dimension N_Q. *In addition*, the bundles $|\mathbf{R}']$ and $|\mathbf{R}'']$ also have to be shown equivalent, *i.e.*, the polynomials over the manifolds belonging to the two respective configurations have to be equivalent too, since that is how the hypersurface in (2.3.18) is defined.

— ❦ —

Note that, for the two configurations in the relation (2.3.18) to be Calabi–Yau, the configurations in (2.3.19) must be almost ample. Moreover, let the matrices $\mathbf{R}$, $\mathbf{R}'$, $\mathbf{R}''$ have K_R columns. By a simple generalization of Proposition 2.1, it is obvious that if the configuration $[\mathbf{n} \| \mathbf{P}]$ corresponds to varieties of dimension N_P, we have that

$$N_P + N_Q - K_R = 3 \,. \tag{2.3.20}$$

Among these, the $N_Q = 1$, 2 and 3 cases are of particular interest, since (2.3.18) is then implied by relations among different embeddings of ample or almost ample varieties of complex dimension 1, 2 and 3 respectively and these have been classified (almost) completely. We now discuss these in turn.

1-fold identities: Complex 1-dimensional varieties, algebraic curves, are completely classified by their Euler characteristic. Moreover, for Eq. (2.1.17) to hold (or, for that matter, Eq. (2.1.18) for almost ample and ample 3-folds), the sub-configuration of algebraic curves occurring in relation (2.3.18) must be $\mathbb{P}^1$'s. We have just proven

Lemma 2.2. *If the sub-configurations*

$$[\mathbf{n}'\|\mathbf{Q}'] \qquad \textit{and} \qquad [\mathbf{n}''\|\mathbf{Q}''] \tag{2.3.21}$$

represent 1-dimensional varieties, then

$$\begin{bmatrix}\mathbf{n} \\ \mathbf{n}'\end{bmatrix}\!\!\left\|\begin{matrix}\mathbf{P} & \mathbf{R} & \mathbf{0} \\ \mathbf{0} & \mathbf{R}' & \mathbf{Q}'\end{matrix}\right]_{\textit{Calabi–Yau}} \simeq \begin{bmatrix}\mathbf{n} \\ \mathbf{n}''\end{bmatrix}\!\!\left\|\begin{matrix}\mathbf{P} & \mathbf{R} & \mathbf{0} \\ \mathbf{0} & \mathbf{R}'' & \mathbf{Q}''\end{matrix}\right]_{\textit{Calabi–Yau}} . \tag{2.3.22}$$

The earlier result, Lemma 2.1, is subsumed in scope by Lemma 2.2; the former is however quite easier to apply in practice. Also, according to Ref. [85],

$$[2\|2]\ , \qquad \begin{bmatrix}1 \\ 1\end{bmatrix}\!\!\left\|\begin{matrix}1 \\ 1\end{matrix}\right] \qquad \text{and} \qquad \begin{bmatrix}1 \\ 1\end{bmatrix}\!\!\left\|\begin{matrix}2 \\ 1\end{matrix}\right] \tag{2.3.23}$$

are all the different embeddings of $\mathbb{P}^1$'s that do occur in the computer listing complete intersection Calabi–Yau 3-folds. Their equivalence is, in fact, described already by Lemma 2.1.

2-fold identities: Very much like 1-folds, the topological types of almost ample (del Pezzo) 2-folds are classified by the Euler characteristic—except for $\chi_E = 4$, the so-called Hirzebruch surfaces. These latter 2-folds come in two homotopy (and therefore topological) types which need to be distinguished. As with 1-folds, one lists all possible almost ample 2-fold configurations and finds the smallest one for each Euler characteristic $\chi_E \neq 4$; for $\chi_E = 4$, the smallest configurations representing the two topologically distinct Hirzebruch surfaces are $\mathbf{F}_0 = \mathbb{P}^1 \times \mathbb{P}^1$ and $\mathbf{F}_1 = \left[\begin{smallmatrix}2 \\ 1\end{smallmatrix}\middle\|\begin{smallmatrix}1 \\ 1\end{smallmatrix}\right]$. The so obtained list of configurations then implies a number of identities among almost ample 2-fold configurations; see § 3.1 for more details on almost ample 2-folds, the distinction between the Hirzebruch surfaces and § 3.1.2 for a yet finer subtlety in these identities.

3-fold identities: As expected, the situation with almost ample 3-folds is more complicated. The ample (Fano) 3-folds were classified by Iskovskih (for $b_2 = 1$) [32] and Mori and Mukai (for $b_2 > 1$) [40], who classified 17 and 88 deformation classes, respectively. Semi-ample 3-folds which are not ample have not been classified yet. Listing all almost ample 3-fold configurations is comparable in number to listing all Calabi–Yau 3-fold configurations and has not been done either.

The Hodge numbers (see § 2.4.1 and Chapter A for techniques) provide at least a necessary condition to check conjectured identities. Perhaps not surprisingly, only some of those suspected from listing Calabi–Yau 3-fold configurations [85] turn out to be correct; unfortunately, little more has been worked out on this issue so far.

***n*-fold identities:** The situation with almost ample n-fold configurations is of course even less well understood. Suffice it here to state that the techniques for computation of the Hodge numbers provide at least some necessary criteria which can be used to verify any given conjectured identity.

2.4 $h^{1,1}$ in Favorable Configurations

Since the Calabi–Yau complete intersections we are studying are of the form $\mathcal{M} = \bigcap_{a=1}^{K} {\xi_a}^{-1}(0)$, we try to apply Lefschetz hyperplane theorem (see § 1.6) iteratively, including the factors ${\xi_a}^{-1}(0)$ one by one.

2.4.1 A simple iterating technique

Let us define

$$D_a \stackrel{\text{def}}{=} \{\, r \mid q_a^r > 0 \,\} \,. \tag{2.4.1}$$

We call a configuration *favorable* if the constraints can be ordered so that

$$\text{for} \quad k' < k, \qquad \text{either} \quad D_{k'} \cap D_k = \varnothing, \qquad \text{or} \quad D_{k'} \subseteq D_k \,. \tag{2.4.2}$$

Furthermore, let

$$\mathcal{X}_a \;=\; \prod_{\substack{r=1 \\ q_a^r > 0}}^{m} \mathbb{P}_r^{n_r}, \qquad \mathcal{M}_a \;=\; \bigcap_{b<a} {\xi_b}^{-1}(0) \,. \tag{2.4.3}$$

Note that, in the notation of Eq. (2.1.4), the line bundle $\mathcal{E}_a$ is positive over $\mathcal{X}_a$. However, ${\mathcal{E}_a}^{\iota_a}$ is not in general positive over $\mathcal{M}_a$; ι_a is the embedding $\mathcal{M}_a \hookrightarrow \mathcal{X}_a$. This implies certain limitations on the application of Lefschetz hyperplane theorem which will be shown below.

Theorem 2.3. *A Calabi–Yau manifold belonging to a favorable configuration*
(i) is simply connected;
(ii) has the $h^{1,1}$ at least equal to the number of $\mathbb{P}_r^{n_r}$ factors in the embedding space—if the configuration is also minimal.

> **Remark:** The minimality requirement, as needed in *(ii)*, is effectively achieved through the use of Lemma 2.1 and Corollary 2.1, or Lemma 2.2; see Ref. [148]. ⌈*Over the years, "favorable configuration" has been limited in the literature for when $H^{1,1}(\mathcal{M}\in[\mathcal{X}\|\mathcal{E}])$ is generated by the Kähler forms of the projective factors in $\mathcal{X}$, so $b_2=h^{1,1}(\mathcal{M})=h^{1,1}(\mathcal{X})$. However, it turns out that almost all Calabi–Yau complete intersections in products of projective spaces have a favorable embedding in this stricter sense [298], owing to the identities discussed in § 2.3.3.*⌋

The proof in Ref. [145] simply follows the ordering of constraints as above and applies the Lefschetz hyperplane theorem at every step.

2.4.2 Application and limitations

To clarify the technique and its limitations, we give some examples in detail.

1. Consider first $\mathcal{M} \in \left[\begin{smallmatrix}3\\2\end{smallmatrix}\middle\|\begin{smallmatrix}2&2\\2&1\end{smallmatrix}\right]$, for which $\chi_E = -120$. Both constraints depend non-trivially on homogeneous coordinates of both $\mathbb{P}^3$ and $\mathbb{P}^2$, i.e., are sections of line bundles that are positive over both factors in $\mathcal{X} = \mathbb{P}^3 \times \mathbb{P}^2$.
Firstly, using the Künneth formula

$$b_p(X \times Y) = \sum_{q=0}^{p} b_{p-q}(X) b_q(Y) \tag{2.4.4}$$

and the fact that $b_p(\mathbb{P}^n) = \begin{cases} 1 \text{ for } p \text{ even,} \\ 0 \text{ for } p \text{ odd,} \end{cases}$ we obtain

$$b_p(\mathcal{X}) = \begin{cases} 1,2,3,3,2,1, & \text{for } p = 0,2,4,6,8,10; \\ 0, & \text{for } p \text{ odd.} \end{cases} \tag{2.4.5}$$

Applying one constraint on the (5-dimensional) $\mathcal{X}$ one gets a (4-dimensional) submanifold for which b_p, $0 \le p < 4$, are the same as those for $\mathcal{X}$. Applying the other constraint yields $\mathcal{M}$ for which b_p equals those of $\mathcal{X}$ for $0 \le p < 3$ and so $b_0 = 1$, $b_1 = 0$ and $b_2 = h^{1,1} = 2$. Using $\chi_E = -120$, it follows that $h^{2,1} = 62$, and we have the complete Hodge diamond.

2. For our next example, we take a slightly more complicated configuration:

$$\mathcal{M} \in \begin{bmatrix} 2 \\ 2 \\ 1 \end{bmatrix}\!\!\begin{Vmatrix} 2 & 1 \\ 2 & 1 \\ 0 & 2 \end{Vmatrix} \quad \cong \tag{2.4.6}$$

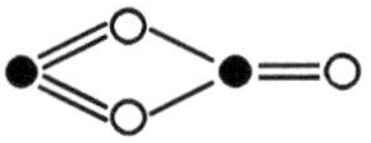

,

for which $\chi_{E} = -96$. Here the first constraint is independent of the coordinates of the third factor in $\mathcal{X}$, the $\mathbb{P}^1$, and the Lefschetz hyperplane theorem would not apply were we to consider the second constraint first. We therefore consider the sub-configuration

$$\begin{bmatrix} 2 \\ 2 \end{bmatrix}\!\!\begin{Vmatrix} 2 \\ 2 \end{Vmatrix} \quad \cong \tag{2.4.7}$$

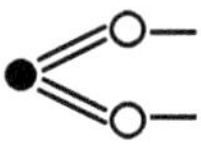

.

This defines a 3-dimensional hypersurface ($\mathcal{M}_1$) embedded in $\mathcal{X}_1 \stackrel{\text{def}}{=} \mathbb{P}^2 \times \mathbb{P}^2$. The Betti numbers of $\mathcal{M}_1$ are equal, by Lefschetz hyperplane theorem, to those of $\mathcal{X}_1$ for $p = 0, 1, 2$ and are $1, 0, 2$, respectively.

Next we impose the remaining constraint, which depends non-trivially on both $\mathcal{M}_1$ and $\mathbb{P}^1$ and embeds $\mathcal{M}$ in $\mathcal{M}_1 \times \mathbb{P}^1$ as a hypersurface. Using the Künneth formula,

$$b_p(\mathcal{M}_1 \times \mathbb{P}^1) = 1, 0, 3 \ , \quad p = 0, 1, 2 \ . \tag{2.4.8}$$

Since the constraint is non-trivial over the entire embedding space $\mathcal{M}_1 \times \mathbb{P}^1$, the Lefschetz hyperplane theorem applies again, equating the Betti numbers of $\mathcal{M}$ with those of $\mathcal{M}_1 \times \mathbb{P}^1$ up to, but not including the "middle dimension," i.e.,

$$b_p(\mathcal{M}) = 1, 0, 3 \ , \quad p = 0, 1, 2 \ . \tag{2.4.9}$$

Using $\chi_{E} = -96$, we have that $h^{2,1} = 51$.

3. Up to now, b_2 was equal to the number of factors in $\mathcal{X}$, but not so in

$$\mathcal{M} \in \begin{bmatrix} 3 \\ 1 \\ 1 \end{bmatrix}\!\!\begin{Vmatrix} 3 & 1 \\ 0 & 2 \\ 0 & 2 \end{Vmatrix} \quad \cong \tag{2.4.10}$$

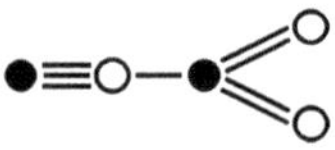

,

for which $\chi_{E} = -48$. Namely, the first constraint is trivial over the $\mathbb{P}^1 \times \mathbb{P}^1$ factor of the embedding space and we must begin with considering the sub-configuration $[3\|3]$. This is a deformation

class of cubic hypersurfaces ($\mathcal{M}_1$) in $\mathbb{P}^3$. The constraint being non-trivial, the Lefschetz hyperplane theorem applies and equates $b_p(\mathcal{M}_1)$ with $b_p(\mathbb{P}^3)$ but only up to the middle dimension, i.e., for $p = 0, 1$. As regards $b_2(\mathcal{M}_1)$, the Lefschetz hyperplane theorem only states that $b_2(\mathcal{M}_1) \geq b_2(\mathbb{P}^3)$.
In such an occasion, we use

$$\begin{aligned} \chi_E(\mathcal{M}_1) = \int_{[3\|3]} \frac{(1+\eta)^4}{(1+3\eta)} &= \int_{\mathbb{P}^3} (3\eta) \wedge (1 + \eta + 3\eta^2) \,, \\ = 3 \cdot 3 \int_{\mathbb{P}^3} \eta^3 = 9 \,. \end{aligned} \tag{2.4.11}$$

Since $\chi_E(\mathcal{M}_1) = \sum_{p=0}^{2} (-1)^p b_p(\mathcal{M}_1) = 9$, $b_{\dim \mathcal{M}_1 - p}(\mathcal{M}_1) = b_p(\mathcal{M}_1)$, it follows that $b_2(\mathcal{M}_1) = 7$.
Now we are ready to consider the second constraint acting in $\mathcal{M}_1 \times \mathbb{P}^1 \times \mathbb{P}^1$. By Künneth,

$$b_p(\mathcal{M}_1 \times \mathbb{P}^1 \times \mathbb{P}^1) = 1, 0, 9 \,, \quad p = 0, 1, 2 \,. \tag{2.4.12}$$

Being non-trivial in both $\mathcal{M}_1$ and $\mathbb{P}^1 \times \mathbb{P}^1$, the Lefschetz hyperplane theorem applies and equates $b_p(\mathcal{M})$ with $b_p(\mathcal{M}_1 \times \mathbb{P}^1 \times \mathbb{P}^1)$ up to $p = 2$ and in particular, $b_2(\mathcal{M}) = 9$. Since $\chi_E(\mathcal{M}) = -48$, $h^{2,1}(\mathcal{M}) = 33$.

4. Finally, we give an example where theorem 2.3 does not guarantee the applicability of the Lefschetz hyperplane theorem:

$$\mathcal{M} \in \begin{bmatrix} 2 \\ 2 \\ 1 \end{bmatrix}\!\!\begin{Vmatrix} 0 & 3 \\ 2 & 1 \\ 2 & 0 \end{Vmatrix} \quad \cong \tag{2.4.13}$$

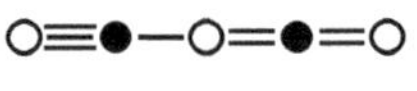

for which $\chi_E = -36$. The correct result here is $b_2 = 9$ and we will give all the details of this computation later.
Our present technique is not guaranteed to apply since the configuration is not favorable, i.e., the constraints cannot be ordered so as to satisfy Eq. (2.4.2).
If one starts with the first constraint, being non-trivial only over the latter two factors of the embedding space, we consider the sub-configuration $\left[\begin{smallmatrix} 2 \\ 1 \end{smallmatrix}\middle\|\begin{smallmatrix} 2 \\ 2 \end{smallmatrix}\right]$. This describes complex surfaces ($\mathcal{M}_1$) embedded in $\mathcal{X}_1 = \mathbb{P}^2 \times \mathbb{P}^1$. By Künneth,

$$b_p(\mathcal{X}_1) = 1, 0, 2 \,, \quad p = 0, 1, 2 \,. \tag{2.4.14}$$

the Lefschetz hyperplane theorem applies and equates $b_p(\mathcal{M}_1)$ with $b_p(\mathcal{X}_1)$ but only up to the middle dimension, i.e., for $p = 0, 1$. We again compute $\chi_E(\mathcal{M}_1) = 10$ using the Chern polynomial and thus have

$$\chi_E(\mathcal{M}_1) = \sum_r (-1)^r\, b_r(\mathcal{M}_1) = 10\ , \qquad \Longrightarrow \qquad b_2(\mathcal{M}_1) = 8\ . \tag{2.4.15}$$

Now, the second configuration is not non-trivial over the entire $\mathbb{P}^2 \times \mathcal{M}_1$, as it is trivial over the $\mathbb{P}^1$ factor of the embedding space of $\mathcal{M}_1$. Nevertheless, we shall show in § 3.1 that the line bundle—of which this constraint is a section—is positive, even though not manifestly so. By Künneth,

$$b_p(\mathbb{P}^2 \times \mathcal{M}_1) = 1, 0, 9\ , \quad p = 0, 1, 2\ . \tag{2.4.16}$$

Taking for granted at the moment that the Lefschetz hyperplane theorem applies, we have the Betti numbers of $\mathbb{P}^2 \times \mathcal{M}_1$ equated to those of $\mathcal{M}$ up though b_2:

$$b_2(\mathcal{M}) = 1, 0, 9\ , \quad p = 0, 1, 2\ . \tag{2.4.17}$$

As a cautionary remark, suffice it to note here that were we to employ the constraints in the reversed order and ignore the question of applicability, the wrong result $b_2(\mathcal{M}) = 3$ would ensue.

2.4.3 Redundancy of inherited (2,2)-forms

We have seen two examples in which b_2 of the complete intersection threefold is greater than that of the embedding space. In fact, for these, the generators of the 2-dimensional homology of the submanifold were inherited from the embedding space (cf. Corollary 1.1). Generally, additional cycles (here, 2-cycles) of the submanifold, which have not been inherited from the embedding manifold are usually called “vanishing cycles”. It appears natural to inquire if it is possible that b_2 of the members of a configuration could be less than that of the embedding space.

The above possibility may now be rephrased as the question whether there exist configurations in which some of the (independent) 2-cycles from the embedding manifold do not descent as independent cycles on the submanifold. Note that the appearance of additional, vanishing cycles may mask this effect so that b_2 of the complete intersection submanifold is nevertheless greater or equal to that of the embedding manifold. All configurations for which not all of the 2-dimensional homology of the embedding manifold descends to the complete intersection Calabi–Yau submanifold can easily be modified (see Lemma 2.1) into equivalent configurations where there is no such defect [148].

2.5 Embedding Characteristics

In addition to the Hodge numbers, there exist certain quantities which characterize the embedding of the Calabi–Yau manifold in $\mathcal{X} = \mathbb{P}_1^{n_1} \times \cdots \times \mathbb{P}_m^{n_m}$ and which can be computed with relative ease [95].

Consider a compact complex n-fold $\mathcal{M}$. Recall that the Euler characteristic is the index of the De Rham exterior derivative

$$\chi_E = \sum_r (-)^r \dim H^r_{\mathrm{DR}}(\mathcal{M}) = \int_{\mathcal{M}} c_n, \tag{2.5.1}$$

where the second equality is the statement of the Gauss–Bonnet formula. Similarly, on complex manifolds, there is the index of the $\bar{\partial}$-derivative (acting on the $(0,q)$-cohomology):

$$\chi^h = \sum_q (-)^q \dim H^{(0,q)}_{\bar{\partial}}(\mathcal{M}) = \int_{\mathcal{M}} \mathrm{td}_n, \tag{2.5.2}$$

also called the *holomorphic Euler characteristic* or the *Todd genus*. It is closely related to the *arithmetic genus*, $p_a \stackrel{\text{def}}{=} (-)^n(\chi^h - 1)$; i.e., $\chi^h = 1 + (-)^n p_a$. Also, there exists the *Hirzebruch signature*

$$\tau_H = \sum_{p,q} (-)^q \dim H^{(p,q)}_{\bar{\partial}}(\mathcal{M}) = \int_{\mathcal{M}} L_n, \tag{2.5.3}$$

and the index of the Dirac operator, which is the *A-roof genus*

$$\iota_{\mathrm{D}} = \int_{\mathcal{M}} \hat{A}_n. \tag{2.5.4}$$

It is not hard to see that the last three quantities vanish for a Calabi–Yau 3-fold and thus appear to be of little use. However, given a vector bundle over $\mathcal{M}$, we may consider the related "twisted indices"

$$\chi^h(\mathcal{V}) \stackrel{\text{def}}{=} \int_{\mathcal{M}} \mathrm{td}(\mathcal{M}) \wedge \mathrm{ch}(\mathcal{V}), \tag{2.5.5}$$

$$\tau_H(\mathcal{V}) \stackrel{\text{def}}{=} \int_{\mathcal{M}} L(\mathcal{M}) \wedge \mathrm{ch}^2(\mathcal{V}), \tag{2.5.6}$$

$$\iota_{\mathrm{D}}(\mathcal{V}) \stackrel{\text{def}}{=} \int_{\mathcal{M}} \hat{A}(\mathcal{M}) \wedge \mathrm{ch}^2(\mathcal{V}). \tag{2.5.7}$$

Here, $\mathrm{ch}^2(\mathcal{V})$ denotes the Chern character calculated from the double curvature 2-form, $2\mathbf{F}$, of $\mathcal{V}$; see § 1.5.

In general, no suitable vector bundle $\mathcal{V}$ intrinsic to our Calabi–Yau complete intersection 3-fold $\mathcal{M}$ is readily available for computation of the above indices. On the other hand, there are simple candidate vector bundles over the embedding space $\mathcal{X}$, which naturally restrict to vector bundles over $\mathcal{M} \subset \mathcal{X}$. These are of the general form

$$\mathcal{V}(\boldsymbol{\ell}\|\mathbf{k}) \stackrel{\text{def}}{=} \bigoplus_{r,a} (\mathcal{T}_r)^{\ell_r} \otimes \mathcal{O}_r(k_r^a)\Big|_{\mathcal{M}}, \qquad (2.5.8)$$

where $\mathcal{T}_r$ is the holomorphic tangent bundle of $\mathbb{P}_r^{n_r}$ and $\mathcal{O}_r(k_r^a)$ is the line bundle corresponding to degree-k_r^a polynomials over $\mathbb{P}_r^{n_r}$. Since all these are restrictions of bundles over the embedding space $\mathcal{X}$ to the $\mathbb{P}^n$ factors, the computation of these "twisted indices" is a straightforward generalization of the computation of the Euler characteristic.

Such "twisted indices" characterize the embedding $\mathcal{M} \hookrightarrow \mathcal{X}$ rather than the abstract manifold $\mathcal{M}$. Nevertheless, they may indeed be useful on a number of occasions. For example, in Ref. [58, 95], the divisibility properties of certain such "twisted indices" were used to determine which configurations define Calabi–Yau 3-folds that admit a freely acting symmetry D such that the quotient with respect to D would have $\chi_E = -6$. 2024⌈*Much more refined invariants characterizing an embedding are provided by (2.5.8)-valued cohomology groups; while tremendously tedious to compute, these are by now becoming accessible owing to both computer-aided techniques and some novel rigorous results [362]; see also § 7.*⌋

— ❦ —

The large collection of constructions of Calabi–Yau 3-folds as complete intersections in complex projective spaces (CICYs) appears to be merely the tip of an iceberg. It also appears that it is a fairly representative collection, in the sense that it exhibits phenomena which are quite generic for Calabi–Yau 3-folds and which we shall study in the sequel. At the same time, the analysis of such 3-folds is always straightforward—given the powerful tools of spectral sequences which we shall describe later—and often quite easy so that even more elementary and better known techniques suffice. 2024⌈ *Already by 1994, this pool of construction was vastly generalized to complete intersections of almost/semi-ample divisors in toric Fano varieties [309], and computational techniques to analyze them were developed over the next two decades. We will return to these in §§ 5.2 and F.8.*⌋

Chapter 3

Some More General Embeddings

In this chapter, we consider some simple generalizations of the complete intersection constructions in Chapter 2, still avoiding singular models.

3.1 Intersections in Products of Surfaces

One way to generalize the constructions of Chapter 2 is of course to consider complete intersections in products of spaces other than $\mathbb{P}^n$'s. Relatively simple additional factors in embedding spaces are provided by algebraic surfaces.

To formulate our complete intersections as in products of complex projective spaces, we require of these surfaces to have a non-trivial anti-canonical bundle with at least one non-vanishing section at every point of the surface. These sections will be used to construct the system of defining constraints. It turns out that all such surfaces in fact can be represented as complete intersections in products of complex projective spaces, so that this attempt does not enrich our bestiary by truly new examples. Nevertheless, the exercise proves useful as this alternative point of view will provide some new means of computation in Chapter 8.

3.1.1 Almost del Pezzo surfaces

An *almost del Pezzo surface*$^{(L)}$, $\mathcal{S}$, has a non-trivial anti-canonical bundle, $\mathcal{K}^*_{\mathcal{S}}$, which has at least one non-zero section at any point of $\mathcal{S}$. We require this with the forethought of constructing smooth 3-folds as complete intersections in products which include almost del Pezzo surfaces. Namely, if

$\mathcal{K}_{\mathcal{S}}^{*}$ is non-trivial and has at least one non-zero section at every point, it has to have at least two sections. This then essentially allows a suitable generalization of Bertini's theorem which will guarantee the existence of smooth complete intersections in products of such surfaces.

The alert Reader will have noticed that, on a surface $\mathcal{S}$, each section of course vanishes along a curve and that two curves in a surface generically do meet at an isolated number of points. However, a limited number of such points can be *blown up* $^{(L)}$ (a) without making the anti-canonical bundle $\mathcal{K}_{\mathcal{S}}^{*}$ non-positive and (b) leaving at every point one section non-zero. This is seen as follows. Let ϕ and ϕ' be two sections the zero-sets of which are two curves, C and C', respectively. Let $C \cap C' = \{p_i\} \subset \mathcal{S}$, which is where ϕ and ϕ' both vanish. In blowing these points up, we replace each of them with an *exceptional divisor* E_i of codimension 1. Now the intersection points $C \cap E_i$ and $C' \cap E_i$ are determined by the angle at which C and C' passed through p_i before it was blown up, so, unless (the zero-sets of) two sections are tangential at p_i, they will not vanish simultaneously after p_i is blown up. If they were tangential at p_i, two blowups would have been needed to disconnect them and so on.

Since the anti-canonical bundle is nontrivial, so is the canonical one. However, since the anti-canonical bundle has holomorphic sections its dual cannot have, $H^0(\mathcal{S}, \mathcal{K}_{\mathcal{S}}) = H^{2,0}(\mathcal{S}) = 0$. Almost del Pezzo surfaces are also *regular*, i.e., they have no harmonic (1,0)-forms and the Hodge diamond is

$$\begin{matrix} & & 1 & & \\ & 0 & & 0 & \\ 0 & & h^{1,1} & & 0 \\ & 0 & & 0 & \\ & & 1 & & \end{matrix} \qquad \Rightarrow \qquad h^{1,1} = \chi_{_E} - 2 \ . \tag{3.1.1}$$

Using the two different expressions for both the Euler characteristic (2.5.1) and the Hirzebruch signature (2.5.3), we obtain:

$$\begin{aligned} 2 + h^{1,1}(\mathcal{S}) &= \chi_{_E}(\mathcal{S}) = C_2(\mathcal{S}) \ , \\ 2 - h^{1,1}(\mathcal{S}) &= \tau_{_H}(\mathcal{S}) = \tfrac{1}{3}\big(C_1^2(\mathcal{S}) - 2C_2(\mathcal{S})\big) \ , \end{aligned} \tag{3.1.2}$$

with C_1^2 and C_2 denoting the Chern numbers. Therefore

$$C_1^2(\mathcal{S}) + C_2(\mathcal{S}) = 12 \ . \tag{3.1.3}$$

This also follows from the above Hodge diamond, using Eq. (1.5.13) and (2.5.2) and is known as an example of the *Noether formula* $^{(L)}$ [23]

$$C_1^2(\mathcal{S}) + C_2(\mathcal{S}) = 12\, \chi^h(\mathcal{S}) \ . \tag{3.1.4}$$

We note also the relation to the canonical bundle, $\mathcal{K}_{\mathcal{S}}$:

$$C_1^2(\mathcal{S}) = \mathcal{K}_{\mathcal{S}} \cdot \mathcal{K}_{\mathcal{S}}. \tag{3.1.5}$$

The dot-product occurring here is called the intersection number and is defined in general as follows. Given two line bundles $\mathcal{L}$ and $\mathcal{L}'$ over a surface $\mathcal{S}$, we define

$$\mathcal{L} \cdot \mathcal{L}' \stackrel{\text{def}}{=} \int_{\mathcal{S}} c_1[\mathcal{L}] \wedge c_1[\mathcal{L}'] \ . \tag{3.1.6}$$

Eq. (3.1.5) is now easily seen to follow from Eq. (1.3.20) and the above definition (3.1.6).

— ❦ —

We know that $\mathcal{K}_{\mathcal{S}}^*$ must have at least two (non-trivial, global and holomorphic) sections. Consequently, the m-fold tensor power of $\mathcal{K}_{\mathcal{S}}^*$ is also non-trivial and has (global, holomorphic) sections. Then ${\mathcal{K}_{\mathcal{S}}}^{\otimes m}$ is also non-trivial but cannot have sections, so all *plurigenera*$^{(L)}$ vanish:

$$\mathfrak{P}_m(\mathcal{S}) \stackrel{\text{def}}{=} H^0(\mathcal{S}, {\mathcal{K}_{\mathcal{S}}}^{\otimes m}) = 0 \ , \qquad m \geq 0 \ . \tag{3.1.7}$$

From this result and regularity ($h^{1,0} = 0$) it follows by the Castelnuovo–Enriques theorem [23] that $\mathcal{S}$ is *rational*$^{(L)}$. Thereupon, in view of the general classification theorems of complex surfaces, we know that $\mathcal{S}$ can only be one of the following:

1. $\mathbb{P}^2$ with finitely many (ρ, possibly $\rho = 0$) points blown up;
2. a Hirzebruch (rational, ruled) surface $\mathbf{F}_n$.

Remark: The almost del Pezzo surfaces of $\chi_E = 4$ are known as Hirzebruch surfaces $\mathbf{F}_n$. They are distinguished from each other by the exceptional divisor (with self-intersection equal to $-n$) which they contain. Note also that the Euler characteristic of a ρ-fold blowup of $\mathbb{P}^2$ is

$$\chi_E(\widetilde{\mathbb{P}^2}_{p_1 \cdots p_\rho}) = \chi_E(\mathbb{P}^2) + \rho = 3 + \rho. \tag{3.1.8}$$

— ❦ —

Restricting the range of n in the above list is rather easy:

Proposition 3.1. *No almost del Pezzo surface has a line of self-intersection less than* -2.

Proof. Let $\mathcal{S}$ be an almost del Pezzo surface with a line $L(\approx \mathbb{P}^1) \subset \mathcal{S}$ with self-intersection $-n < 0$. We know that the holomorphic tangent bundle of a line $L \approx \mathbb{P}^1$ is $\mathcal{T}_L = \mathcal{O}_L(2)$. So, $\mathcal{K}_L^* = \det \mathcal{T}_L = \mathcal{T}_L = \mathcal{O}_L(2)$ since $\operatorname{rank} \mathcal{T}_L = 1$. On the other hand, the normal bundle of $L \subset \mathcal{S}$ must be $\mathcal{N}_{\mathcal{S}/L} = \mathcal{O}_L(-n)$ for the self-intersection of L to be $-n$. But then

$$\mathcal{K}_{\mathcal{S}}^*|_L = \det \mathcal{T}_{\mathcal{S}}|_L \;=\; \mathcal{K}_L^* \otimes \mathcal{N}_{\mathcal{S}/L} \;=\; \mathcal{O}_L(2-n) \;, \tag{3.1.9}$$

and could not have sections if $n > 2$. ☑

We now turn to restricting the number of blowups in $\mathcal{S} = \widetilde{\mathbb{P}^2}_{p_1 \cdots p_\rho}$.

Proposition 3.2. *Let $\widetilde{\mathcal{S}}_p$ be the blowup of $\mathcal{S}$ at $p \in \mathcal{S}$. The sections of the anti-canonical bundle of $\widetilde{\mathcal{S}}_p$ are precisely the sections of $\mathcal{K}_{\mathcal{S}}^*$ which vanish at p.*

Proof. Let (z_1, z_2) be local coordinates on $\mathcal{S}$ at p and let $\zeta_1 \stackrel{\text{def}}{=} z_1$, $\zeta_2 \stackrel{\text{def}}{=} z_2/z_1$ whenever $z_1 \neq 0$. In the limit where $\zeta_2 =$*const.* while $\zeta_1 \to 0$, ζ_2 parametrizes the $\mathbb{P}^1$ into which p is blown up. Now, the left-hand-side of

$$\mathrm{d}z_1 \wedge \mathrm{d}z_2 \;=\; \zeta_1(\mathrm{d}\zeta_1 \wedge \mathrm{d}\zeta_2) \tag{3.1.10}$$

is a section of $\mathcal{K}_{\mathcal{S}}$, while the right-hand-side is a section of $\mathcal{K}_{\widetilde{\mathcal{S}}_p}$.

Away from the point which is blown up, the surfaces $\mathcal{S} \smallsetminus p$ and $\widetilde{\mathcal{S}} \smallsetminus \mathbb{P}^1$ are the same. Then, for a section ϕ of $\mathcal{K}^*_{\mathcal{S}\smallsetminus p} = \mathcal{K}^*_{\widetilde{\mathcal{S}}_p \smallsetminus \mathbb{P}^1}$,

$$\langle\, \phi \,|\, \mathrm{d}z_1 \wedge \mathrm{d}z_2 \,\rangle \;=\; \zeta_1 \,\langle\, \phi \,|\, \mathrm{d}\zeta_1 \wedge \mathrm{d}\zeta_2 \,\rangle \;. \tag{3.1.11}$$

So, if ϕ extends over $\widetilde{\mathcal{S}}_p$, we have

$$\lim_{\substack{\zeta_1 \to 0 \\ \zeta_2 \text{ fixed}}} \langle\, \phi \,|\, \mathrm{d}z_1 \wedge \mathrm{d}z_2 \,\rangle \;=\; 0 \;, \tag{3.1.12}$$

independently of ζ_2. Therefore:

$$\lim_{(z_1,z_2)\to(0,0)} \langle\, \phi \,|\, \mathrm{d}z_1 \wedge \mathrm{d}z_2 \,\rangle = 0, \tag{3.1.13}$$

and ϕ extends to $\mathcal{S}$, taking the value 0 at p.

Conversely, if ϕ is a section of the anti-canonical bundle of $\mathcal{S}$, vanishing at p, we can extend ϕ over $\widetilde{\mathcal{S}}_p$ by setting:

$$\langle\, \phi \,|\, \mathrm{d}\zeta_1 \wedge \mathrm{d}\zeta_2 \rangle|_{\text{at } (0,\zeta_2)} \;=\; \lim_{\substack{z_1 \to 0 \\ (z_2/z_1) \text{ fixed}}} \frac{1}{z_1} \langle\, \phi \,|\, \mathrm{d}z_1 \wedge \mathrm{d}z_2 \rangle \;. \tag{3.1.14}$$

☑

Proposition 3.3. *No almost del Pezzo surface has* $\chi_{_E} = 11$.

Proof. If $\mathcal{S}$ were a del Pezzo surface with $\chi_{_E} = 11$, it would have to be $\widetilde{\mathbb{P}^2}_{p_1\cdots p_8}$, an 8-fold blowup of $\mathbb{P}^2$. The anti-canonical bundle of a blowup of $\mathbb{P}^2$ is induced from the anti-canonical bundle of $\mathbb{P}^2$, which is $\mathcal{O}_{\mathbb{P}^2}(3)$ and has 10 independent sections. Since sections of $\mathcal{K}^*_{\mathcal{S}}$ are (the extensions of) those of $\mathcal{K}^*_{\mathbb{P}^2}$ which vanish at $p_1, \ldots, p_8$, we need to count the number of cubic polynomials over $\mathbb{P}^2$ which simultaneously vanish at 8 (generically chosen) points. It is not hard to verify that only two of ten cubics will vanish at all given 8 points and that they will also vanish at a ninth point. If this ninth point is not blown up, all sections of $\mathcal{K}^*_{\mathcal{S}}$ vanish there and $\mathcal{S}$ is not almost del Pezzo. If it is blown up, $\chi_{_E}(\mathcal{S}) = 12$. ☑

It is now not hard to see that there can be no almost del Pezzo surface $\mathcal{S}$ with $\chi_{_E} > 12$ either; essentially, when $\chi_{_E} > 12$, $\mathcal{K}^*_{\mathcal{S}}$ becomes non-positive and has no sections. Complete details of this and the previous proofs are found in Ref. [147].

3.1.2 Jumping deformations

Deformations are usually thought of as continuous variations of the complex structure. That this is not necessarily quite the complete story is demonstrated by the following simple example. Besides serving to illustrate what might be called jumping deformations, our simple example here presents a rather generic construction of models which incorporate such deformations.

The almost ample 2-fold configuration $\left[\begin{smallmatrix}3\\1\end{smallmatrix}\middle\|\begin{smallmatrix}1&1\\1&1\end{smallmatrix}\right]$ corresponds to a system of defining equations

$$\mathcal{S}_\varepsilon \in \begin{bmatrix}3\\1\end{bmatrix}\!\!\begin{Vmatrix}1&1\\1&1\end{Vmatrix} \quad : \quad \begin{cases} z_0\,w_0 & +\ z_1\,w_1 = 0\,, \\ z_2\,w_0 & +\ [\sum_{i=0}^{2} a_i z_i + \varepsilon z_3]\,w_1 = 0\,, \end{cases} \tag{3.1.15}$$

where we have, without loss of generality, chosen (z_0, z_1, z_2, z_3) and (w_0, w_1) for local homogeneous coordinates on $\mathbb{P}^3 \times \mathbb{P}^1$. The surface $\mathcal{S}_\varepsilon$ of common solution to the two Eq. (3.1.15) in $\mathbb{P}^3 \times \mathbb{P}^1$ is non-singular except in three special cases: (1) $a_1 = 0$, (2) $a_2 = 0$ and (3) $a_0 = -a_1 a_2 \neq 0$. So, except for these special cases, $\mathcal{S}_\varepsilon$ is smooth, has $\chi_{_E} = 4$ and so must be one of the Hirzebruch surfaces: $\mathbf{F}_n$, $n = 0, 1, 2$.

- When $\varepsilon \neq 0$, the equation

$$\Delta_{\varepsilon\neq0}(z_i) \stackrel{\text{def}}{=} \det \begin{bmatrix} z_0 & z_1 \\ z_2 & (\sum_{i=0}^{2} a_i z_i) + \varepsilon z_3 \end{bmatrix}_{\varepsilon\neq0} = 0 \tag{3.1.16}$$

embeds $\mathcal{S}_{\varepsilon\neq0}$ in $\mathbb{P}^3$ as a *smooth* quadric surface since $\mathrm{d}\Delta_{\varepsilon\neq0}\neq0$ over $\mathbb{P}^3$ and therefore $\mathcal{S}_{\varepsilon\neq0}\approx\mathbb{P}^1\times\mathbb{P}^1=\mathbf{F}_0$; this is an example of *splitting*$^{(L)}$.

• However, for the special value $\varepsilon=0$, both $\Delta_{\varepsilon=0}$ and $\mathrm{d}\Delta_{\varepsilon=0}$ vanish at the point $(0,0,0,1)\in\mathbb{P}^3$, so that the surface of solutions to $\Delta_{\varepsilon=0}=0$ in $\mathbb{P}^3$ is singular at $(0,0,0,1)$ although the surface of common solution to Eq. (3.1.15), $\mathcal{S}_{\varepsilon=0}\subset\mathbb{P}^3\times\mathbb{P}^1$ is smooth. In fact, over $(0,0,0,1)\in\mathbb{P}^3$, the surface $\mathcal{S}_{\varepsilon=0}$ contains an entire copy of $\mathbb{P}^1_w$ since with $z_0,z_1,z_2,\varepsilon=0$, both w_0,w_1 remain undetermined. Therefore, $\mathcal{S}_{\varepsilon=0}$ is the blowup of the singular space $z_0(\sum_i a_iz_i)=z_1z_2$ at the singularity $z_0,z_1,z_2=0$.

The above construction makes it manifest that $\mathcal{S}_{\varepsilon\neq0}=\mathbf{F}_0=\mathbb{P}^1\times\mathbb{P}^1$ is a deformation (parametrized by ε) of $\mathcal{S}_{\varepsilon=0}$, whence they must be of the same homotopy type. Then it must be that $\mathcal{S}_{\varepsilon=0}=\mathbf{F}_2$. Note however that $\mathcal{S}_\varepsilon$ is *rigid* for both $\varepsilon\neq0$ and $\varepsilon=0$. In other words, all $\mathcal{S}_\varepsilon$, $\varepsilon\neq0$ and arbitrary a_i, are equivalent up to linear reparametrizations. Similarly, all $\mathcal{S}_{\varepsilon=0}$ are equivalent. However, $\mathcal{S}_{\varepsilon=0}$ contains a line of self-intersection -2, while $\mathcal{S}_{\varepsilon\neq0}$ does not. It follows that the moduli space of the surfaces $\mathcal{S}_\varepsilon$ belonging to $\left[\begin{smallmatrix}3\\1\end{smallmatrix}\middle\|\begin{smallmatrix}1&1\\1&1\end{smallmatrix}\right]$ consists of two infinitesimally close but distinct points—one for $\mathbf{F}_0$ and one for $\mathbf{F}_2$.

2024 *This jumping deformation can in fact be more simply described by a more careful analysis of the explicit 1-parameter deformation family of smooth hypersurfaces:*

$$\{\,p_\epsilon(x,y)\stackrel{\text{def}}{=}x_0y_0{}^2+x_1y_1{}^2+\epsilon\,x_2y_0y_1=0\}\in\begin{bmatrix}2\\1\end{bmatrix}\!\!\begin{Vmatrix}1\\2\end{Vmatrix}\,. \tag{3.1.17}$$

With $(y_0,y_1)\in\mathbb{P}^1$, the gradient, $\partial p_\epsilon=(y_0{}^2,y_1{}^2,\ldots)$, cannot vanish for any $\epsilon\in\mathbb{C}$. The central, $\epsilon=0$ case was shown by F. Hirzebruch [269] to be the rational ruled surface $\mathbf{F}_2$, while the $\epsilon\neq0$ cases are all copies of $\mathbf{F}_0=\mathbb{P}^1\times\mathbb{P}^1$ [321]. Proving this distinction became straightforward by generalizing a construction pioneered in Ref. [296], wherein certain degrees, q^r_a, in a configuration (2.1.5) are allowed to be negative. The so-defined "generalized complete intersections" ("gCIs") vastly generalize the CICYs of Chapter 2 to "gCICYs" [296, 319, 386, 320], which contain (even within single deformation families) diverse MPCP-resolved toric hypersurfaces [321] from the much larger database [430, 431, 295], and so connect by explicit deformations (3.1.17) as well as the kind of singularizations discussed in Chapter D.

Similar situations are also found for $\chi_E>4$. For example,

$$\widetilde{\mathbb{P}}^2_{p_1,p_2,p_3}\in\begin{bmatrix}1\\1\\1\end{bmatrix}\!\!\begin{Vmatrix}1\\1\\1\end{Vmatrix} \tag{3.1.18}$$

precisely if the three points are not collinear [147]. To include also the non-generic case when the three points *are* collinear, one embeds

$$\widetilde{\mathbb{P}}^2_{p_1,p_2,p_3} \in \begin{bmatrix} 2 & \| & 1 & 1 & 1 \\ 1 & \| & 1 & 0 & 0 \\ 1 & \| & 0 & 1 & 0 \\ 1 & \| & 0 & 0 & 1 \end{bmatrix} . \tag{3.1.19}$$

A simple analysis of the defining equations verifies that the latter configuration contains surfaces each of which is a blowup of a singular surface contained in the former configuration. As in the previous, $\chi_E = 4$ case, however, the surfaces contained in both configurations are indeed deformations of this exceptional surface and they both have the same homotopy type.

— ❦ —

Up to the above subtlety in *discrete* changes in complex structure, the list of almost del Pezzo surfaces is exhausted by $\mathbb{P}^2$ and the members of the following nine almost ample 2-fold configurations:

$$\begin{array}{ccc} \mathbf{F}_0 = \begin{bmatrix} 1 \\ 1 \end{bmatrix}_4 , & \mathbf{F}_1 = \begin{bmatrix} 2 & \| & 1 \\ 1 & \| & 1 \end{bmatrix}_4 , & \begin{bmatrix} 2 & \| & 1 & 1 \\ 1 & \| & 1 & 0 \\ 1 & \| & 0 & 1 \end{bmatrix}_5 , \\ \begin{bmatrix} 1 & \| & 1 \\ 1 & \| & 1 \\ 1 & \| & 1 \end{bmatrix}_6 , & \begin{bmatrix} 2 & \| & 2 \\ 1 & \| & 1 \end{bmatrix}_7 , & [4\|2\;2]_8 , \\ [3\|3]_9 , & \begin{bmatrix} 2 & \| & 2 \\ 1 & \| & 2 \end{bmatrix}_{10} , & \begin{bmatrix} 2 & \| & 3 \\ 1 & \| & 1 \end{bmatrix}_{12} , \end{array} \tag{3.1.20}$$

where the subscripts denote the Euler characteristics. ⌊*Ref. [337] finds that each blowup of $\mathbb{P}^2$ is found in a (3.1.19)-like configuration with a $\mathbb{P}^1$ factor for each blowup; we will revisit this in § 8.6.*⌋

It is perhaps less than obvious that $\left[\begin{smallmatrix} 2 & \| & 2 \\ 1 & \| & 2 \end{smallmatrix}\right]$ represents double covers of $\mathbb{P}^2$ branched over a smooth quartic curve. As this type of analysis will be recurring, let us take a closer look at it. Denoting the homogeneous coordinates $x_i \in \mathbb{P}^2$ and $z_0, z_1 \in \mathbb{P}^1$, we write

$$\mathcal{S} \in \begin{bmatrix} 2 & \| & 2 \\ 1 & \| & 2 \end{bmatrix} \quad : \qquad Q_{00}(x)\, {z_0}^2 + 2Q_{01}(x)\, z_0\, z_1 + Q_{11}(x)\, {z_1}^2 \;=\; 0\,. \tag{3.1.21}$$

Now, for a generic point of $x \in \mathbb{P}^2$, this is a generic quadratic equation on $\mathbb{P}^1$, the solution of which is two points. So, almost every point in $\mathbb{P}^2$ is doubled in $\mathcal{S}$. Of course, when the discriminant $\Delta \stackrel{\text{def}}{=} ({Q_{01}}^2 - Q_{00}Q_{11})$ vanishes, the quadratic equation degenerates into a perfect square and the two solutions in $\mathbb{P}^1$ coincide. This happens at the quartic curve $\mathcal{Q} \subset \mathbb{P}^2$, defined by

$\Delta(x) = 0$. So, $\mathcal{S}$ is in fact a double cover of $\mathbb{P}^2$, branched however over the quartic $\mathcal{Q} \subset \mathbb{P}^2$. Finally, we note that $\mathcal{Q}$ is smooth for a generic choice of $Q_{ij}(x)$ because its singularities occur precisely where $\mathrm{d}\Delta(x) = 0$, which is easily seen to require all three quadrics $Q_{ij}(x)$ to vanish. However, these are three independent constraints in $\mathbb{P}^2$ and so cannot have a common solution in general. It turns out that every smooth quartic in $\mathbb{P}^2$ can also be represented as the discriminant of some quadric [230], and so the configuration (3.1.21) includes all double covers of $\mathbb{P}^2$ branched over smooth quartics.

3.1.3 Calabi–Yau hypersurfaces in $\mathcal{S} \times \mathcal{S}'$

Using these ten surfaces, we can now form arbitrary products and then consider Calabi–Yau complete intersections of hypersurfaces. Since all of these are covered as complete intersections in complex projective spaces, we forego a complete analysis here. Suffice it here merely to review the family of Calabi–Yau 3-folds obtained as hypersurfaces in products of some two almost del Pezzo surfaces [147].

One naïvely expects $\frac{10\cdot 11}{2} = 55$ different Calabi–Yau 3-fold hypersurfaces, one for each $\mathcal{S} \times \mathcal{S}'$ pair. This however is not quite true. Note first of all that the almost del Pezzo surfaces with $3 \le \chi_E \le 10$ are in fact ample; only the $\chi_E = 12$ surface is not. We here state

Theorem 3.1. *Let $\mathcal{S}$ and $\mathcal{S}'$ be two del Pezzo surfaces (that is, almost del Pezzo surfaces with $\chi_E = 3, \ldots, 10$). Then there are (a) $\frac{9\cdot 10}{2} = 45$ distinct Calabi–Yau 3-folds, with however (b) $\frac{8\cdot 9}{2} = 36$ distinct Hodge numbers:*

$$\chi_E \begin{bmatrix} \mathcal{S} & \| & \mathcal{K}^*_{\mathcal{S}} \\ \mathcal{S}' & \| & \mathcal{K}^*_{\mathcal{S}'} \end{bmatrix} = -2[12 - \chi_E(\mathcal{S})][12 - \chi_E(\mathcal{S}')] \,, \tag{3.1.22}$$

$$h^{1,1} \begin{bmatrix} \mathcal{S} & \| & \mathcal{K}^*_{\mathcal{S}} \\ \mathcal{S}' & \| & \mathcal{K}^*_{\mathcal{S}'} \end{bmatrix} = h^{1,1}(\mathcal{S}) + h^{1,1}(\mathcal{S}') \,. \tag{3.1.23}$$

Proof. (a) This is almost clear by construction. We only need to show that replacing $\mathbf{F}_0$ with $\mathbf{F}_1$ in the embedding spaces yields a different Calabi–Yau 3-fold; we defer this until Chapter 8. The statement about the Euler characteristic in part (b) follows by direct computation. For the statement about the Hodge numbers, we need the following Lemma.

Lemma 3.1. *The anti-canonical bundle of a del Pezzo surface is positive.*

Proof. By our earlier result, del Pezzo surfaces are the almost del Pezzo surfaces with $\chi_E = 3, \ldots, 10$. The statement is straightforward for the cases

$\chi_E = 3, \ldots, 9$ and is easily seen as follows. The anti-canonical bundle for all surfaces represented in (3.1.20) is obtained as the restriction to $\mathcal{S}$ of

$$\bigotimes_{r=1}^{m} \mathcal{O}_r(\hat{q}_r) , \qquad \hat{q}_r \stackrel{\text{def}}{=} n_r + 1 - \sum_{a=1}^{K} q_r^a . \tag{3.1.24}$$

Except for the $\chi_E = 10$ case, $\hat{q}_r > 0$ for all factors, $r = 1, \ldots, m$ and so $\mathcal{K}_{\mathcal{S}}^*$ is positive.

Consider now the case of $\left[\begin{smallmatrix}2\\1\end{smallmatrix}\middle\|\begin{smallmatrix}2\\2\end{smallmatrix}\right]$, that is, the family of double covers of $\mathbb{P}^2$ branched over a smooth quartic curve in $\mathbb{P}^2$. $\mathcal{K}_{\mathcal{S}}^*$ is then induced from $\mathcal{K}_{\mathbb{P}^2}^* = \mathcal{O}_{\mathbb{P}^2}(3)$ by the covering map. Since the hyperplane bundle is positive, the Chern class of $\mathcal{K}_{\mathcal{S}}^*$ has a representative, $\boldsymbol{\Theta}$, which is positive except at the branching curve where it is positive semi-definite and positive along the curve. Let $\mathfrak{d}$ be a smooth function on $\mathcal{S}$ which, near the branching set is the square of the distance from the branching set in some smooth metric. Then $\partial\bar{\partial}\,\mathfrak{d}$ is exact and positive semi-definite along the branching set and positive in directions transverse to it. For sufficiently small $\varepsilon > 0$, $(\boldsymbol{\Theta} - i\varepsilon\partial\bar{\partial}\,\mathfrak{d})$ is a positive representative of the first Chern class of $\mathcal{K}_{\mathcal{S}}^*$.

☑ (Lemma)

Having proved that $\mathcal{K}_{\mathcal{S}}^*$ is positive for all del Pezzo surfaces, the Lefschetz hyperplane theorem (see § 1.6) may be used to derive the statement (3.1.23) about the Hodge numbers $h^{1,1}$. Since $h^{2,1} = h^{1,1} - \frac{1}{2}\chi_E$ for a Calabi–Yau 3-fold, the more general statement *b)* follows. ☑

— ❦ —

We now remain with the Calabi–Yau hypersurfaces embedded in $\mathcal{S} \times \mathcal{Z}$, where $\mathcal{S}$ is any almost del Pezzo surface and $\mathcal{Z}$ is one with $\chi_E = 12$. For these, we need the following

Lemma 3.2. *For every almost ample surface* $\mathcal{Z}$ *with* $\chi_E = 12$*:* $\left[\begin{smallmatrix}\mathcal{Z}\\ \mathbb{P}^1\end{smallmatrix}\middle\|\begin{smallmatrix}\mathcal{K}_{\mathcal{Z}}^*\\ \mathcal{O}(1)\end{smallmatrix}\right] = \mathcal{Z}$.

Proof. Let $f : \mathcal{Z} \to \mathbb{P}^1$ be a map to an auxiliary $\mathbb{P}^1{}_{\mathcal{Z}}$. The anti-canonical bundle of $\mathcal{Z}$ has only two sections and so is induced from $\mathcal{O}_{\mathbb{P}^1_{\mathcal{Z}}}(1)$ by the map f and we write $f : \mathcal{K}_{\mathcal{Z}}^* \xrightarrow{\sim} \mathcal{O}_{\mathcal{Z}}(1)$. Then we have

$$\begin{array}{ccc} \mathcal{Z} & \times & \mathbb{P}^1 \\ \downarrow{\scriptstyle f} & & \downarrow{\scriptstyle \mathrm{Id}} \\ \mathbb{P}^1{}_{\mathcal{Z}} & \times & \mathbb{P}^1 \end{array} \tag{3.1.25}$$

Write the product map as $\mathfrak{f} = (f, \mathrm{Id})$, and $\mathcal{O}_{\mathcal{Z}}(1)$ and $\mathcal{O}(1)$ for the hyperplane bundles of $\mathcal{Z}$ and $\mathbb{P}^1$ respectively. Pulling back to $\mathcal{Z} \times \mathbb{P}^1$, we have that $\mathfrak{f}^*(\mathcal{O}_{\mathcal{Z}}(1), \mathcal{O}(1)) = \mathcal{K}^*_{\mathcal{Z}} \otimes \mathcal{O}(1)$. So, given a section ϕ of $\mathcal{O}_{\mathcal{Z}}(1) \otimes \mathcal{O}(1)$, $\varphi \stackrel{\text{def}}{=} \mathfrak{f}^*(\phi)$, is the pull-back section of $\mathcal{K}^*_{\mathcal{Z}} \otimes \mathcal{O}(1)$, so that the hypersurface defined as* $\varphi^{-1}(0)$ is the same as $\mathfrak{f}^{-1}(\phi^{-1}(0))$. Without loss of generality, $\mathbb{P}^1{}_{\mathcal{Z}}$ and $\mathbb{P}^1$ can be identified, so that $\phi^{-1}(0) = \mathrm{diag}(\mathbb{P}^1{}_{\mathcal{Z}} \otimes \mathbb{P}^1)$, whereby:

$$\mathfrak{f}^{-1}(\phi^{-1}(0)) \;=\; \{\,(x,y) \;:\; f(x) = y\,\} \;=\; \{(x, f(x))\,\} \;\approx\; \mathcal{Z}\,, \qquad (3.1.26)$$

with x denoting the points of $\mathcal{Z}$ and y the points of $\mathbb{P}^1$. ☑

Corollary 3.1. *For any almost del Pezzo surface $\mathcal{S}$ and $\mathcal{Z}$ one with $\chi_E(\mathcal{Z}) = 12$, there is an almost del Pezzo surface $\mathcal{Z}'$ with $\chi_E = 12$ such that:*

$$\begin{bmatrix} \mathcal{S} \\ \mathcal{Z} \end{bmatrix}\!\!\left\|\begin{matrix} \mathcal{K}^*_{\mathcal{S}} \\ \mathcal{K}^*_{\mathcal{Z}} \end{matrix}\right] = \begin{bmatrix} \mathcal{Z}' \\ \mathcal{Z} \end{bmatrix}\!\!\left\|\begin{matrix} \mathcal{K}^*_{\mathcal{Z}'} \\ \mathcal{K}^*_{\mathcal{Z}} \end{matrix}\right] \;. \qquad (3.1.27)$$

✎ Prove this. ❧

Since the $\chi_E = 12$ almost del Pezzo surfaces are not positive, the Lefschetz hyperplane theorem is not applicable to the Calabi–Yau manifolds considered in Corollary 3.1, which are all represented by the configuration

$$\left[\begin{matrix} 2 \\ 2 \\ 1 \end{matrix}\right\|\left.\begin{matrix} 3 & 0 \\ 0 & 3 \\ 1 & 1 \end{matrix}\right] \qquad (3.1.28)$$

the smooth members of which have $\chi_E = 0$. As we will discuss in more detail in Chapter 6, this can be viewed as a family of fibered spaces, where $\mathbb{P}^1$ is the base and the fibers are products of (possibly singular) tori; it follows that $h^{1,1} = 19$ for smooth such Calabi–Yau 3-folds [212].

We shall later return to Calabi–Yau hypersurfaces in products of surfaces to exemplify some handy computational techniques.

3.2 Complete Intersections in Products Involving Almost Fano 3-Folds

A further generalization of possible embedding spaces is provided by including (products of) almost Fano 3-folds. Unlike almost del Pezzo surfaces, many almost Fano 3-folds are not known to be represented as a complete intersection in a product of complex projective spaces. It is then possible that

* Recall that $\varphi^{-1}(0)$ denotes the hypersurface defined by $\varphi = 0$.

this addition to the set of considered embedding spaces provides genuinely new Calabi–Yau 3-folds.

3.2.1 Almost Fano 3-folds—an introduction

Similarly to the treatment in § 3.1, we define an almost Fano 3-fold to be an algebraic 3-fold $\mathfrak{F}$, which has a non-trivial anti-canonical bundle $\mathcal{K}^*_{\mathfrak{F}}$ with at least one non-zero section at every point of $\mathfrak{F}$, so $\mathcal{K}^*_{\mathfrak{F}}$ is *basepoint-free*$^{(L)}$. 2024⌈*Elsewhere in the literature, the term almost Fano*$^{(L)}$ *is used with a different definition; see the almost ample*$^{(L)}$ *Lexicon entry for disambiguation. Also, note the conceptual distinction between basepoint-free and transverse: the latter requires a particular* $\mathcal{K}^*_{\mathfrak{F}}$*-section,* f*, to have a non-zero normal,* $\mathrm{d}f$*, while the former requires at least one non-zero section at every point of* $\mathfrak{F}$*; see Remark on p. 58.*⌋

The Hodge diamond of an almost Fano 3-folds is†

$$\begin{array}{ccccccc} &&&1&&& \\ &&0&&0&& \\ &0&&h^{1,1}&&0& \\ 0&&h^{2,1}&&h^{2,1}&&0 \\ &0&&h^{1,1}&&0& \\ &&0&&0&& \\ &&&1&&& \end{array} \qquad \Longrightarrow \qquad \chi_E = 2(h^{1,1}+1-h^{2,1})\ . \tag{3.2.1}$$

We have used *Hodge star duality*$^{(L)}$ and complex conjugation. We see that there are two variant Hodge numbers, rather similarly to Calabi–Yau spaces. As with almost del Pezzo surfaces, the plurigenera vanish $\mathfrak{P}_m \stackrel{\text{def}}{=} H^0(\mathfrak{F}, \mathcal{K}_{\mathfrak{F}}{}^{\otimes m}) = 0$ for all $m > 0$. Now all three Chern numbers $C_1{}^3$, $C_1 \cdot C_2$ and $C_3 = \chi_E$ are non-vanishing. However, we have the following simple result.

Lemma 3.3. $C_1 \cdot C_2 = \int_{\mathfrak{F}} (c_1 \wedge c_2) = 24$ *for all almost Fano 3-folds* $\mathfrak{F}$.

Proof. This follows from the above Hodge diamond 3.2.1, using Eq. (1.5.13) and (2.5.2); to illustrate a useful idea however, another proof is included.

We embed a model of the K3 surface in $\mathfrak{F}$ as a hypersurface, using a section of the anti-canonical bundle $\mathcal{K}^*_{\mathfrak{F}}$; then $\mathcal{N}_{\mathfrak{F}/\mathrm{K3}} = \mathcal{K}^*_{\mathfrak{F}}$. Writing

---

† For Fano 3-folds, the Kodaira–Nakano Theorem 3.3 implies (3.2.1). For non-Fano *almost* Fano 3-folds, the defining requirement (start of § 3.2.1) implies that $[\mathcal{K}^*]$ is *basepoint-free*$^{(L)}$ and so also *nef*$^{(L)}$. If $d(\mathcal{K}^*) = C_1{}^3 \neq 0$ (see below), $\mathcal{K}^*$ is also *big*$^{(L)}$, and the Kawamata–Viehweg vanishing theorem [278, II] implies (3.2.1); see however § 3.2.4 for a further twist.

$c(\mathfrak{F}) = 1 + \mathfrak{f}_1 + \mathfrak{f}_2 + \mathfrak{f}_3$ and using Eq. (1.3.20), we know that $c_1[\mathcal{N}_{\mathfrak{F}/\mathrm{K3}}] = \mathfrak{f}_1$. Thus,

$$c(\mathrm{K3}) \;=\; \frac{c(\mathfrak{F})}{c(\mathcal{N}_{\mathfrak{F}/\mathrm{K3}})} \;=\; 1 + \mathfrak{f}_2 \;. \tag{3.2.2}$$

Therefore

$$24 = \chi_E(\mathrm{K3}) = \int_{\mathrm{K3}} \mathfrak{f}_2 = \int_{\mathfrak{F}} c_1[\mathcal{N}_{\mathfrak{F}/\mathrm{K3}}] \wedge \mathfrak{f}_2 = \int_{\mathfrak{F}} (\mathfrak{f}_1 \wedge \mathfrak{f}_2) \;. \quad ☑ \tag{3.2.3}$$

In addition, there is another numerical characteristic of complex varieties, called the *index*. It is the maximal integer $r > 0$, for which there exists a line bundle $\mathcal{L}$ over our almost Fano 3-fold $\mathfrak{F}$ such that $\mathcal{L}^{\otimes r} \approx \mathcal{K}^*_{\mathfrak{F}}$. In analogy, we define the *multiindex* $\vec{r} = (r_1, \ldots, r_k)$ where r_a are maximal integers for which $\mathcal{K}^*_{\mathfrak{F}} \approx \bigotimes_{a=1}^k \mathcal{L}_a^{\otimes r_a}$ and $\mathcal{L}_a$ are independent line bundles. For example, $r(\mathbb{P}^n) = n{+}1$ because $\mathcal{K}_{\mathbb{P}^n} \approx \mathcal{O}_{\mathbb{P}^n}(n{+}1)$ and $\vec{r}(\mathbb{P}^m \times \mathbb{P}^n) = (m{+}1, n{+}1)$ since $\mathcal{K}_{\mathbb{P}^m \times \mathbb{P}^m} \approx \mathcal{O}_{\mathbb{P}^n}(m{+}1) \otimes \mathcal{O}_{\mathbb{P}^n}(n{+}1)$. Note however that $r(\mathbb{P}^m \times \mathbb{P}^n) = \gcd(m{+}1, n{+}1)$. Given these "root" line bundles $\mathcal{L}$, i.e., $\mathcal{L}_a$ used to define the index r and the multiindex $\vec{r}$, we can also define the *degrees*

$$d \overset{\text{def}}{=} [\mathcal{L}]^3 \qquad\qquad \overset{\text{def}}{=} \int_{\mathfrak{F}} c_1(\mathcal{L})^3 \;, \tag{3.2.4}$$

$$d_{abc} \overset{\text{def}}{=} [\mathcal{L}_a] \cdot [\mathcal{L}_b] \cdot [\mathcal{L}_c] \overset{\text{def}}{=} \int_{\mathfrak{F}} c_1(\mathcal{L}_a) \wedge c_1(\mathcal{L}_b) \wedge c_1(\mathcal{L}_c) \;. \tag{3.2.5}$$

Unfortunately, all these characteristics, ${C_1}^3, h^{1,1}, h^{2,1}$ and in addition $r, \vec{r}, d, d_{abc}$, are not enough to label all the Fano 3-folds unambiguously (see Ref. [32] for the 18 Fano 3-folds with $h^{1,1}{=}1$ and Ref. [40] for the 87 ones with $2 \leq h^{1,1} \leq 10$). Allowing also non-ample almost ample 3-folds of course only complicates the classification; this remains an open subject. 2021⌈*Recent developments in surveying line bundles suitable for constructing Calabi–Yau 3-folds with rank$\geq$3 sub-bundles of E_8 (see § 0.2.3) also involve machine learning and provide much more information [336, 338, 302].*⌋

3.2.2 Some general theorems

A little thought reveals that the tricks we had used to derive Eq. (3.1.3), and here Lemma 3.3, indicate a more general framework. Indeed, we quote from Ref. [164], p. 149, without proof:

Theorem 3.2 (Riemann–Roch–Hirzebruch). *Let $\mathcal{X}$ be an algebraic manifold of dimension n and let $\mathcal{L}$ be a holomorphic line bundle over $\mathcal{X}$, with $c_1[\mathcal{L}] = \lambda$. The cohomology groups $H^q(\mathcal{X}, \mathcal{L})$ are finite dimensional and the (generalized) Euler–Poincaré characteristic is*

$$\sum_{q=0}^{n}(-)^q H^q(\mathcal{X},\mathcal{L}) \stackrel{\text{def}}{=} \chi_{_E}(\mathcal{X},\mathcal{L}) = \int_X \mathrm{ch}(\mathcal{L}) \wedge \mathrm{td}(\mathcal{X}) \stackrel{\text{def}}{=} \int_X \left[e^{\lambda} \prod_{i=1}^{n} \frac{\ell_i}{1-e^{-\ell_i}}\right] , \tag{3.2.6}$$

where

$$c(\mathcal{X}) = 1 + c_1 + c_2 + \cdots + c_n \stackrel{\text{def}}{=} \prod_{i=1}^{n}(1+\ell_i) . \tag{1.3.19$'$}$$

Note that integration over $\mathcal{X}$ picks out terms which are of degree n in $\{\lambda, \ell_i\}$. This is the celebrated Riemann–Roch theorem, generalized by Hirzebruch to manifolds $\mathcal{X}$ of arbitrary dimension.

Since the relation between ℓ_i and c_r is somewhat involved, we also quote (Ref. [164], p. 5) the expressions for cases $\dim \mathcal{X} = 1$, 2 and 3, respectively, as indicated by the subscript of $\mathcal{X}$:

$$\chi_{_E}(\mathcal{X}_{(1)},\mathcal{L}) = \int_X \left(\lambda + \tfrac{1}{2}c_1\right) , \tag{3.2.7}$$

$$\chi_{_E}(\mathcal{X}_{(2)},\mathcal{L}) = \int_X \left(\tfrac{1}{2}(\lambda^2 + \lambda c_1) + \tfrac{1}{12}({c_1}^2 + c_2)\right) , \tag{3.2.8}$$

$$\chi_{_E}(\mathcal{X}_{(3)},\mathcal{L}) = \int_X \left(\tfrac{1}{6}\lambda^3 + \tfrac{1}{4}\lambda^2 c_1 + \tfrac{1}{12}\lambda({c_1}^2 + c_2) + \tfrac{1}{24}c_1 c_2\right) . \tag{3.2.9}$$

Armed with this theorem, the Reader should have no difficulty recovering Eq. (3.1.3) and Lemma 3.3, by choosing $\mathcal{L} = \mathcal{O}_X$, the trivial holomorphic line bundle. Then, quite straightforwardly, $\chi_{_E}(\mathcal{X},\mathcal{O}) \equiv \chi^h(\mathcal{X}) = \dim H^0(\mathcal{X}) = 1$ for connected almost ample $\mathcal{X}$.

Another very useful result is the *Kodaira–Nakano vanishing theorem*, which we quote (Ref. [23], p. 154) without proof:

Theorem 3.3 (Kodaira–Nakano). *If $\mathcal{L}$ is a positive holomorphic line bundle over the compact complex algebraic manifold $\mathcal{X}$ of dimension n, then*

$$H^q(\mathcal{X}, \mathcal{L} \otimes \wedge^p \mathcal{T}^*_X) = 0 \quad \text{for } p+q > n. \tag{3.2.10}$$

A useful fact to recall when using this theorem is that $\wedge^n \mathcal{T}^*_{\mathcal{X}} \stackrel{\text{def}}{=} \mathcal{K}_{\mathcal{X}}$ is the canonical line bundle. In particular:

$$\mathcal{L}^* > 0 \implies H^q(\mathcal{X}, \mathcal{L}) = 0 \;, \qquad q \neq n; \tag{3.2.11}$$

$$\mathcal{L} \otimes \mathcal{K}^*_{\mathcal{X}} > 0 \implies H^q(\mathcal{X}, \mathcal{L}) = 0 \;, \qquad q \neq 0 \;. \tag{3.2.12}$$

✎ Derive/prove this. ❧

It is now straightforward to compute the number of sections of the mth power of the anti-canonical bundle (anti-plurigenera), when the anti-canonical bundle is positive:

$$\dim \mathcal{S} = 2 \;:\; \mathfrak{P}^m(\mathcal{S}) = \tfrac{1}{2} m(m+1) C_1{}^2 + 1 \;, \tag{3.2.13}$$

$$\dim \mathfrak{F} = 3 \;:\; \mathfrak{P}^m(\mathfrak{F}) = \tfrac{1}{12} m(m+1)(2m+1) C_1{}^3 + (2m+1) \;. \tag{3.2.14}$$

Note that $m(m+1) \in 2\mathbb{Z}$ and $m(m+1)(2m+1) \in 6\mathbb{Z}$, so that the obvious integrality of $\mathfrak{P}^m$ enforces $C_1{}^3(\mathfrak{F}) \in 2\mathbb{Z}$. This leads to

Definition 3.1. *The integer* $\mathrm{g} \stackrel{\text{def}}{=} 1 - \frac{1}{2} C_1{}^n$ *is called the genus of an n-fold.*

(This agrees with $n = 1$, $\mathrm{g} = 1 - \frac{1}{2}\chi_{_E}$ for Riemann surfaces.)

Now, sections of $\mathcal{K}^*_{\mathcal{X}}$ may be used to construct defining polynomials. While $C_1{}^n{>}0$, $\mathcal{K}^*_{\mathcal{X}}$ is positive, the Kodaira–Nakano theorem guarantees that $H^q(\mathcal{S}, \mathcal{K}^*_{\mathcal{S}})$ is non-zero only for $q = 0$, and $\chi_{_E}(\mathcal{X}_{(n)}, \mathcal{K}^*_{\mathcal{S}}) = \mathfrak{P}^1(\mathcal{X}_{(n)})$. However, when $C_1{}^n{=}0$, $\mathcal{K}^*_{\mathcal{X}}$ is not positive, the Kodaira–Nakano theorem does not apply, $H^q(\mathcal{S}, \mathcal{K}^*_{\mathcal{S}})$ is non-zero not only for $q{=}0$, so $\chi_{_E}(\mathcal{X}_{(n)}, \mathcal{K}^*_{\mathcal{S}}) \neq \mathfrak{P}^1(\mathcal{X}_{(n)})$. For the almost del Pezzo surface with $C_1{}^2{=}0$, we have $\mathfrak{P}^1(\mathcal{S}) = 1$, but there does exists another section; see Propositions 3.2 and 3.3. For the almost Fano 3-folds with $C_1{}^3{=}0$, $\mathfrak{P}^1(\mathfrak{F}) = 3$, but $\dim H^0(\mathcal{S}, \mathcal{K}^*_{\mathcal{S}})$ is only 2-dimensional! The $\mathcal{K}^*_{\mathcal{X}}$-valued cohomology will be calculated in § 9.6; suffice it here to observe that for all almost-but non-ample complete intersections of the type (3.2.23), $\left[\begin{smallmatrix} X \\ \mathbb{P}^1 \end{smallmatrix} \middle\| \begin{smallmatrix} c_1(X) \\ 1 \end{smallmatrix}\right]$, the anti-canonical bundle is (the pull-back of) $\mathcal{O}_{\mathbb{P}^1}(1)$, and so has precisely two sections. In any case, note that we do not really need to know a polynomial representation in terms of some suitable coordinates; many computations can be completed at a rather formal level.

3.2.3 Fano 3-folds

With almost ample *surfaces*, there was only one addition to the list of ample surfaces and all of them were represented by complete intersections

Table 3.1: Fano 3-fold complete intersections in $\mathbb{P}^n$.

$C_1{}^3$	χ_E	$b_{2,1}$	g	r	d	**Description**
64	4	0	33	4	1	$\mathbb{P}^3$
54	4	0	28	3	2	[4‖2]: a quadric in $\mathbb{P}^4$—also the double cover of $\mathbb{P}^3$, branched over $\mathbb{P}^1 \times \mathbb{P}^1 \subset \mathbb{P}^3$
24	−6	5	13	2	3	[4‖3]: a cubic in $\mathbb{P}^4$—also the triple cover of $\mathbb{P}^3$, branched over a smooth cubic in $\mathbb{P}^3$
32	0	2	17	2	4	[5‖2 2]: the intersection of two quadrics in $\mathbb{P}^4$
4	−56	30	3	1	4	[4‖4]: a quartic in $\mathbb{P}^4$—also the quadruple cover of $\mathbb{P}^3$, branched over K3 $\subset \mathbb{P}^3$
6	−36	20	4	1	6	[5‖3 2]: the intersection of a quadric and a cubic in $\mathbb{P}^5$
8	−24	14	5	1	8	[6‖2 2 2]: the intersection of three quadrics in $\mathbb{P}^6$

in products of complex projective spaces (3.1.20). With almost Fano 3-folds, there is more news.

Standard complete intersection Fano 3-folds: For one thing, as mentioned by Iskovskih, seven of the 18 Fano 3-folds with $b_2 = 1$ can be represented as complete intersections in a $\mathbb{P}^n$ and are here given in Table 3.1. Of course, complete intersections in products of more than one $\mathbb{P}^n$ would have $b_2 > 1$. So, with 11 Fano 3-folds with $b_2 = 1$ which are *not* complete intersections in a projective space, their inclusion as factors in the embedding space already enriches our menagerie.

Fano 3-fold branched coverings: A complete description of all 105 Fano 3-folds would take us far afield. Let us just note that some of them are easily described as multiple covers of $\mathbb{P}^3$, branched over certain surfaces in $\mathbb{P}^3$. This type of construction will be utilized in § 5.5 to produce Calabi–Yau 3-folds. Here we recall the discussion on page 81, shifted now into the context of 3-folds.

Consider the simple equation

$$z^k + B(x) = 0 , \qquad B(x) \stackrel{\text{def}}{=} \sum_{i=0}^{n} (x^i)^{kl} , \tag{3.2.15}$$

where $x \in \mathbb{P}^n$; $B(x)$ may be deformed into any other degree-kl polynomial, of course. While $B(x) \neq 0$, $z{=}-\sqrt[k]{B(x)}$ are k distinct solutions per point $x \in \mathbb{P}^n$; at $B(x) = 0$, $z{=}0$ is the only solution for z. The space of solutions thereby becomes a k-fold cover of $\mathbb{P}^n$, branched over the hypersurface $\mathcal{B} \subset \mathbb{P}^n$, defined by $B(x)=0$.

Theorem 3.4. *Let $\varpi : \overline{\mathcal{X}_{\mathcal{B}}}^{(k)} \to \mathcal{X}$ denote the k-fold cover of $\mathcal{X}$, branched over $\mathcal{B} \subset \mathcal{X}$, where $\mathcal{B}$ is defined as the zero-set of a section of a line bundle $\mathcal{L}^{\otimes k}$. Then*

$$\chi_{E}(\overline{\mathcal{X}_{\mathcal{B}}}^{(k)}) = k\chi_{E}(\mathcal{X}) - (k-1)\chi_{E}(\mathcal{B}) \ , \tag{3.2.16}$$

$$\mathcal{K}^{*}_{\overline{\mathcal{X}_{\mathcal{B}}}^{(k)}} = \varpi^{*}(\mathcal{K}^{*}_{\mathcal{X}} \otimes \mathcal{L}^{*\otimes(k-1)}) \ . \tag{3.2.17}$$

Sketch of proof: The first result is clear; $\overline{\mathcal{X}_{\mathcal{B}}}^{(k)}$ is constructed by (a) carving $\mathcal{B}$ out of $\mathcal{X}$, (b) copying $(\mathcal{X}-\mathcal{B})$ k times and then (c) gluing a single copy of $\mathcal{B}$ back. Since the Euler characteristic is additive under surgery‡, we have

$$\chi_{E}(\overline{\mathcal{X}_{\mathcal{B}}}^{(k)}) \ = \ k[\chi_{E}(\mathcal{X}) - \chi_{E}(\mathcal{B})] + \chi_{E}(\mathcal{B}) \ . \tag{3.2.18}$$

For a proof of the second statement, the Reader is referred to Ref. [41], p. 109, or Ref. [31], Theorem 5.5 on p. 202, which give the "ramification formula"

$$\mathcal{K}_{\overline{\mathcal{X}_{\mathcal{B}}}^{(k)}} \ = \ \varpi^{*}(\mathcal{K}_{\mathcal{X}}) \otimes \mathcal{L}(\mathcal{B}_{\varpi}) \ , \tag{3.2.19}$$

where $\mathcal{L}(\mathcal{B}_{\varpi})$ is the line bundle corresponding to the *branching locus*, $\mathcal{B}_{\varpi}$. Here we need the following:

Definition 3.2. *The* branching locus *of the covering map ϖ, denoted $\mathcal{B}_{\varpi}$, is the divisor obtained from $\mathcal{B} \subset \mathcal{X}$ by counting the points of $\mathcal{B}$ with multiplicity $k-1$.*

Since the multiplicity of $\varpi(p)$, $\mu[\varpi(p)]$, has an invariant meaning for all $p \in \overline{\mathcal{X}_{\mathcal{B}}}^{(k)}$, we count the points with multiplicity $(\mu[\varpi(p)]-1)$: only those in $\mathcal{B}$ remain, and with multiplicity $k-1$, as defined for $\mathcal{B}_{\varpi}$. But then $\mathcal{B}_{\varpi}$ corresponds to $\mathcal{L}(\mathcal{B}_{\varpi}) = \mathcal{L}^{\otimes(k-1)}$—rather than $\mathcal{L}^{\otimes k}$—and the second statement follows. ☑

Using this theorem, it is not hard to construct various multiple branched covers of the Fano 3-folds, keeping the anti-canonical bundle under control. We will use this to construct Calabi–Yau 3-folds as suitably branched multiple covers, but can clearly also construct many Fano and many almost (but not) Fano 3-folds. In fact, the proof of Lemma 3.1 is related to Theorem 3.4; the $(1,1)$-form $\partial\bar{\partial}\mathfrak{d}$ corresponds to $\mathcal{L}(\mathcal{B}_{\varpi})$.

‡ This can be shown, for example, by triangulation.

Looking back at Eq. (3.2.15), we see that for $l = 1$, z can be made into the $(n+2)^{th}$ homogeneous coordinate of a $\mathbb{P}^{n+1}$, in addition to the $n+1$ x's. Thus, these branched multiple covers straightforwardly become identified with simple hypersurfaces in $\mathbb{P}^{n+1}$. Now, for $l > 1$, the z and the x's scale differently, so they cannot span a simple $\mathbb{P}^{n+1}$. However, they can span a *weighted complex projective space*, $\mathbb{P}^{n+1}_{(1,\cdots,1,l)}$ (see Chapter 5), where

$$(x^0 : \ldots : x^n : z) \;\cong\; (\lambda x^0 : \ldots : \lambda x^n : \lambda^l z)\;. \tag{3.2.20}$$

Eq. (3.2.15) is invariant with respect to such a scaling transformation. So, for $k = 2$ and $l = 2, 3$, respectively, we obtain

$$\left[\mathbb{P}^4_{(1:1:1:1:2)} \,\big\|\, 4\right] \;\cong\; \overline{\mathbb{P}^3}^{(2)}_{[3\|4]}\;, \qquad \left[\mathbb{P}^4_{(1:1:1:1:3)} \,\big\|\, 6\right] \;\cong\; \overline{\mathbb{P}^3}^{(2)}_{[3\|6]}\;, \tag{3.2.21}$$

double covers of $\mathbb{P}^3$, branched over a quartic and over a sextic surface in $\mathbb{P}^3$, with $\mathcal{K}^* = \mathcal{O}_{\mathbb{P}^3}(2)$ and $= \mathcal{O}_{\mathbb{P}^3}(1)$, by Theorem 3.4.

Other Fano 3-folds: In addition, there are Fano 3-folds which are a little more complicated to describe. Suffice it here to mention three Fano 3-folds of $h^{1,1} = 1$ the description of which involves Grassmannians. Just like one uses the components of a complex $(n+1)$-vector for the homogeneous coordinates of $\mathbb{P}^n$, k complex n-vectors $[\vec{z}_i]$ are needed to provide coordinates for the Grassmannian $G_{(k,n)}$ of k-planes through the origin of $\mathbb{C}^n$. Recall that $\dim G_{(k,n)} = k(n-k)$, which is easily seen by noting that a coordinate patch is chosen, up to a permutation of columns, as

$$\begin{bmatrix} \vec{z}_1 \\ \vec{z}_2 \\ \vdots \\ \vec{z}_k \end{bmatrix} \longrightarrow \begin{bmatrix} 1 & 0 & 0 & \cdots & 0 & \zeta_{1,k+1} & \cdots & \zeta_{1,n} \\ 0 & 1 & 0 & \cdots & 0 & \zeta_{2,k+1} & \cdots & \zeta_{2,n} \\ \vdots & \vdots & \vdots & \ddots & \vdots & \vdots & \ddots & \vdots \\ 0 & 0 & 0 & \cdots & 1 & \zeta_{k,k+1} & \cdots & \zeta_{k,n} \end{bmatrix}, \tag{3.2.22}$$

leaving $k(n-k)$ local coordinates. Now, regarding the $\binom{n}{k}$ quantities $Z_{[j_1,\cdots,j_k]} \stackrel{\text{def}}{=} \epsilon^{i_1\cdots i_k} z_{i_1,j_1}\cdots z_{i_k,j_k}$ as homogeneous coordinates of a $\mathbb{P}^{\binom{n}{k}-1}$ defines the *Plücker embedding* $G_{(k,n)} \stackrel{\mathrm{P}}{\hookrightarrow} \mathbb{P}^{\binom{n}{k}-1}$.

So, for example, $G_{(2,5)} \stackrel{\mathrm{P}}{\hookrightarrow} \mathbb{P}^9$ embeds this 6-dimensional Grassmannian in $\mathbb{P}^9$. Intersecting it with three hyperplanes in $\mathbb{P}^9$ yields a 3-fold which is ample [32]. Similarly, Fano 3-folds are also obtained by intersecting $G_{(2,5)} \stackrel{\mathrm{P}}{\hookrightarrow} \mathbb{P}^9$, two hyperplanes and a quadric hypersurface in $\mathbb{P}^9$ or intersecting $G_{(2,6)} \stackrel{\mathrm{P}}{\hookrightarrow} \mathbb{P}^{14}$ with five hyperplanes in $\mathbb{P}^{14}$. By construction, the hyperplane bundle on the Plücker $\mathbb{P}^{\binom{n}{k}-1}$ equals the determinant of the

Table 3.2: Fano 3-fold complete intersections in Grassmannians.

$C_1{}^3$	χ_E	$b_{2,1}$	$\mathfrak{g}$	r	d	**Description**
40	4	0	21	2	5	The intersection of three hyperplanes and the Grassmannian $G_{(2,5)} \subset \mathbb{P}^9$; also $[\,G_{(2,5)}\|3\mathcal{L}\,]$
10	−6	5	6	1	10	The intersection of two hyperplanes, a quadric hypersurface and the Grassmannian $G_{(2,5)} \subset \mathbb{P}^9$; also $[\,G_{(2,5)}\|2\mathcal{L} \oplus \mathcal{L}^{\otimes 2}\,]$
14	−16	10	8	1	14	The intersection of five hyperplanes and the Grassmannian $G_{(2,5)} \subset \mathbb{P}^9$, also $[\,G_{(2,6)}\|5\mathcal{L}\,]$

Denote $\mathcal{L} \stackrel{\text{def}}{=} \det[\mathcal{S}^*]$, see Eq. (3.3.16); also, $2\mathcal{L} \stackrel{\text{def}}{=} \mathcal{L}\oplus\mathcal{L}$, $3\mathcal{L} \stackrel{\text{def}}{=} \mathcal{L}\oplus\mathcal{L}\oplus\mathcal{L}$, etc.

dual of the rank-k *universal sub-bundle*[(L)]—on Grassmannians, the analogue of $\mathcal{O}_{\mathbb{P}^n}(1)$; see Eq. (3.3.16). Therefore, these three Fano 3-folds can be described as complete intersections in Grassmannians, as presented in Table 3.2. We will return to these below.

Complementing these simple constructions, Ref. [32] and also Ref. [40] present several rather more involved cases, the description of which is beyond our present scope; [2024]⌈*see [342] for a database of Fano 3-folds.*⌋

3.2.4 Almost but not Fano 3-folds

On the other hand, from any almost Fano 3-fold, we can easily construct another one which surely is not Fano, that is, it is non-ample. We now describe this. Let $\mathfrak{F}$ be an almost Fano 3-fold with $c(\mathfrak{F}) = 1 + \mathfrak{f}_1 + \mathfrak{f}_2 + \mathfrak{f}_3$. Consider then $\mathfrak{F} \times \mathbb{P}^1$ and define a hypersurface using a section of $\mathcal{K}^*_{\mathfrak{F}} \otimes \mathcal{O}_{\mathbb{P}^1}(1)$. Stretching the configuration notation in an obvious way, we have

$$\begin{aligned} c(\widetilde{\mathfrak{F}}) = c\begin{bmatrix} \mathfrak{F} \\ \mathbb{P}^1 \end{bmatrix}\!\!\begin{Vmatrix} \mathfrak{f}_1 \\ J \end{Vmatrix} &= \frac{(1+\mathfrak{f}_1+\mathfrak{f}_2+\mathfrak{f}_3)(1+2J)}{(1+\mathfrak{f}_1+J)} \\ &= 1 + [J] + [\mathfrak{f}_2 + \mathfrak{f}_1 J] + [(\mathfrak{f}_3 - \mathfrak{f}_1\mathfrak{f}_2) + (\mathfrak{f}_2 - \mathfrak{f}_1{}^2)J]\ , \end{aligned} \tag{3.2.23}$$

where J represents $\mathcal{O}_{\mathbb{P}^1}(1)$. Straightforwardly,

1. $C_1{}^3(\widetilde{\mathfrak{F}}) = 0$;
2. $\chi_E(\widetilde{\mathfrak{F}}) = \chi_E(\mathfrak{F}) - C_1{}^3(\mathfrak{F})$;
3. $h^{1,1}(\widetilde{\mathfrak{F}}) = h^{1,1}(\mathfrak{F}) + 1$ if $\mathfrak{F}$ is in fact ample.

The condition in the last result is related to the fact (2.3.6) and says that simple iteration of this construction yields nothing new.

The class of almost Fano 3-folds we have just defined has a rather simple and more intrinsic geometric description. Consider the defining equation for $\widetilde{\mathfrak{F}}$:

$$K_0(z)x^0 + K_1(z)x^1 = 0\ , \qquad z \in \mathfrak{F}\ , \quad x \in \mathbb{P}^1\ ; \tag{3.2.24}$$

the $K_i(z)$ are sections of $\mathcal{K}^*_{\mathfrak{F}}$, as is dictated by $\mathfrak{f}_1 = c_1(\mathfrak{F})$. For a generic point $z \in \mathfrak{F}$, the above equation is simply a linear equation in two variables, x^0 and x^1, and has a single solution. At the special curve $C \subset \mathfrak{F}$, where $K_0(z) = 0 = K_1(z)$, the variables $x \in \mathbb{P}^1$ are not constrained at all. Thus, $\widetilde{\mathfrak{F}}$ is isomorphic to $\mathfrak{F}$ except along $C \subset \mathfrak{F}$, where each point of C is replaced by a copy of $\mathbb{P}^1$. We say that $\widetilde{\mathfrak{F}}_C$ is a *blowup*$^{(L)}$ of $\mathfrak{F}$ along $C \subset \mathfrak{F}$.

So, there are also non-ample almost Fano 3-folds extending the lists of Iskovskih and Mori and Mukai. A simple example is provided by

$$\widetilde{\mathbb{P}^3} \in \begin{bmatrix} 3 \\ 1 \end{bmatrix}\!\!\begin{Vmatrix} 4 \\ 1 \end{Vmatrix}\ , \qquad \begin{aligned} \chi_E(\widetilde{\mathbb{P}^3}) &= -60\ , & h^{1,1}(\widetilde{\mathbb{P}^3}) &= 2\ , \\ {C_1}^3(\widetilde{\mathbb{P}^3}) &= 0\ , & h^{2,1}(\widetilde{\mathbb{P}^3}) &= 33\ . \end{aligned} \tag{3.2.25}$$

✎ Prove that this is a blowup of $\mathbb{P}^3$ along a curve of genus $g = 34$. [Hint: the center of the blowup is a curve which is an intersection of two quartic surfaces (why?). Also, remember that for curves, $\chi_E = 2(2-g)$.] ❧

2024 *The anticanonical bundle of $\widetilde{\mathbb{P}^3}$ in (3.2.25) is non-positive, $\mathcal{K}^*_{\widetilde{\mathbb{P}^3}} \sim \left|\begin{smallmatrix}0\\1\end{smallmatrix}\right]$, but is induced from $\mathcal{O}_{\mathbb{P}^1}(1)$ and has two sections that cannot vanish simultaneously. Since $c_1(\widetilde{\mathbb{P}^3}) = J_{\mathbb{P}^1}$, its degree (3.2.4) is $\mathcal{d}(\widetilde{\mathbb{P}^3}) = {C_1}^3 = 0$. Thus, neither the Kodaira–Nakano nor the Kawamata–Viehweg theorems apply, and we cannot guarantee the zeros in the Hodge diamond (3.2.1). Nevertheless, explicit computation verifies this (see § 9.5), raising the question of whether having a basepoint-free $\mathcal{K}^*$ (with at least one non-zero section everywhere) alone guarantees (3.2.1).*

A classification of almost Fano 3-folds will have to be rather more involved than that of the ample subset. For example, consider the collection of hypersurfaces

$$\mathfrak{F} \in \begin{bmatrix} 3 \\ 1 \end{bmatrix}\!\!\begin{Vmatrix} k \\ d \end{Vmatrix}\ , \qquad \begin{aligned} \chi_E(\mathfrak{F}) &= 2[(6-4k+k^2) + 2d(1-k)^3]\ , \\ {C_1}^3(\mathfrak{F}) &= 2(4-k)^2(2d+3k-2dk)\ ; \end{aligned} \tag{3.2.26}$$

also, $h^{1,1}(\mathfrak{F}) = 2$ and $h^{2,1}(\mathfrak{F}) = (1-k)(3-3k+k^2+2d(1-k)^2)$. For $k = 1$, $\chi_E = 6$ and ${C_1}^3 = 54$ regardless of the value of d. Clearly, the spaces with various different $d \in \mathbb{Z}_{\geq 0}$ are not the same: the index $r_{d=1} = 1$ while $r_{d=2} = 3$. Also, for $d = 1$, $\mathfrak{F}$ is a blowup of $\mathbb{P}^3$ along a line $\mathbb{P}^1 \subset \mathbb{P}^3$.

However, for $d = 2$, $\mathfrak{F}$ is a double cover of $\mathbb{P}^3$, branched over a quadric surface which is singular at a point and where $\mathfrak{F}$ contains a $\mathbb{P}^1$ in place of the singularity. Of course, only the hypersurfaces with $d \leq 2$ and $k \leq 4$ are almost ample (Calabi–Yau if $d = 2$ and $k = 4$).

✎ Prove this. ❧

3.2.5 Calabi–Yau from almost Fano 3-folds (I)

One can of course consider Calabi–Yau complete intersections in spaces which contain almost Fano 3-fold factors and the simplest such constructions are simple hypersurfaces in the product $\mathfrak{F} \times \mathbb{P}^1$.

Writing again $c(\mathfrak{F}) = 1 + \mathfrak{f}_1 + \mathfrak{f}_2 + \mathfrak{f}_3$, we consider

$$\begin{aligned} c(\mathcal{M}) = c\begin{bmatrix} \mathfrak{F} \\ \mathbb{P}^1 \end{bmatrix}\!\!\begin{Vmatrix} \mathfrak{f}_1 \\ 2J \end{Vmatrix} &= \frac{(1 + \mathfrak{f}_1 + \mathfrak{f}_2 + \mathfrak{f}_3)(1 + 2J)}{(1 + \mathfrak{f}_1 + 2J)} \\ &= 1 + [\mathfrak{f}_2 + 2\mathfrak{f}_1 J] + [(\mathfrak{f}_3 - \mathfrak{f}_1\mathfrak{f}_2) - 2{\mathfrak{f}_1}^2 J] \ . \end{aligned} \tag{3.2.27}$$

It follows that

1. $\chi_E(\mathcal{M}) = 2\chi_E(\mathfrak{F}) - 2{C_1}^3(\mathfrak{F}) - 48$;
2. $h^{1,1}(\mathcal{M}) = h^{1,1}(\mathfrak{F}) + 1$ if $\mathfrak{F}$ is in fact ample.

Referring to the fact (2.3.6), the condition in the last result is sufficient, not necessary; there may be cases in which it could be relaxed.

Using these results and the data in Tables 3.1 and 3.2, a short list of Calabi–Yau manifolds of the type (3.2.27) are obtained. The three Calabi–Yau manifolds obtained in this manner from the Fano 3-folds in Table 3.2 are not related to complete intersections in products of complex projective spaces in any simple way; we find $\chi_E = -120$, -80 and -108 and $h^{1,1} = 2$ for all three.

— ❦ —

Just as the classification of almost del Pezzo surfaces may serve to derive identities between various almost ample 2-fold configurations[§], a future classification of almost Fano 3-folds can be used to derive identities between almost ample 3-fold configurations. Such identities can then be used to derive identities between various complete intersection constructions of Calabi–Yau 3-folds (and clearly, other varieties of the Reader's fancy) much the same as was done in proving Lemma 2.2.

[§] We now, of course, extend the meaning of a configuration to include products of complex projective spaces, almost del Pezzo surfaces and also almost Fano 3-folds.

Such identities are quite useful when looking for new Calabi–Yau 3-folds. For example, the class of non-ample almost Fano 3-folds $\widetilde{\mathfrak{F}}$ defined in Eq. (3.2.23) will not yield any new Calabi–Yau 3-folds. To see this, note firstly that the anti-canonical bundle of $\widetilde{\mathfrak{F}}$ is completely induced from $\mathbb{P}^1$ and is essentially $\mathcal{O}_{\mathbb{P}^1}(1)$. In the diagrammatic language of Chapter 2, we have that

$\widetilde{\mathfrak{F}} \approx$ 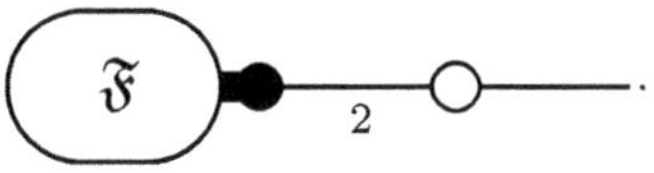 (3.2.28)

Therefore, thought of as a subdiagram of a Calabi–Yau diagram, $\widetilde{\mathfrak{F}}$ perfectly fits our Lemma 2.1. We can therefore conclude that

$$\left[\begin{array}{c||cc} \mathfrak{F} & 0 & \mathfrak{f}_1 \\ \mathbb{P}^1 & J & J \\ \text{anything} & & 0 \end{array}\right]_{\text{Calabi–Yau}} \cong \left[\begin{array}{c||c} \mathfrak{F} & \mathfrak{f}_1 \\ \mathbb{P}^1 & 2J \end{array}\right] . \tag{3.2.29}$$

In other words, the (deformation class of) Calabi–Yau 3-folds obtained as a complete intersection in however complicated embedding space with $\widetilde{\mathfrak{F}}$ as one of the factors, can just as well be obtained as a simple hypersurface in $\mathfrak{F} \times \mathbb{P}^1$!

While we have just made the class of non-ample almost Fano 3-folds $\widetilde{\mathfrak{F}}$ obsolete for constructing *new* Calabi–Yau spaces, there are many other non-ample almost Fano 3-folds the inclusion of which in considered embedding spaces may yield essentially new Calabi–Yau 3-folds. For instance, in the notation of § 3.2.3, the list of *splits* $^{(L)}$

$$\widetilde{\mathfrak{F}}^{(n)} \in \left[\begin{array}{c||ccc} \mathfrak{F} & \mathfrak{a}_1 & \cdots & \mathfrak{a}_n \\ \mathbb{P}^n & J & \cdots & J \end{array}\right] \qquad \sum_{i=1}^{n} \mathfrak{a}_i = \mathfrak{f}_1 \ , \quad n > 1 \ , \tag{3.2.30}$$

provides non-ample almost Fano 3-folds for which the Calabi–Yau hypersurfaces in $\widetilde{\mathfrak{F}}^{(n)} \times \mathbb{P}^1$ are not as trivially related to those in $\mathfrak{F} \times \mathbb{P}^1$ as when $n = 1$.

3.3 Kähler C-Spaces

In Chapter 2, we studied complete intersections in products of complex projective spaces. It appears natural to generalize this to products of arbitrary simply connected compact homogeneous spaces, which were classified and called C-spaces by Wang [231]. Baston and Eastwood [6] give a construction of Kähler (indeed, projective) such spaces which will be quite more familiar to most physicists, generalizing the notion of (complex) projective spaces, Grassmannians, ... into *generalized flag spaces*; we will generally denote them by $\mathbb{F}$. After some general definitions and remarks about such spaces, we will consider here only the simple case of single (Calabi–Yau) hypersurface and restrict moreover to the case of $b_2\{\mathbb{F}\} = 1$, as in Ref. [180].

3.3.1 Kähler C-spaces as flag spaces

General theory [231] says that every C-space $\mathbb{F}$, that is, simply connected compact homogeneous complex space, is homeomorphic to a torus-bundle over a product of certain coset spaces $\{G/H\}$. A torus-bundle means that tori are being fibered over the base space, whence simply connected C-spaces are just the $\{G/H\}$. For all the coset spaces in this construction, G is a compact simple Lie group and H is a regular semi-simple subgroup.

A little Lie algebra and group theory reminder: As customary, for a semi-simple Lie group G, $\mathfrak{g}$ will denote the algebra spanned by the generators $\mathfrak{t}$. For a complex(ified) $\mathfrak{g}$, the generators may be partitioned into[¶] $\mathfrak{h} \oplus \mathfrak{r}_+ \oplus \mathfrak{r}_-$, where $\mathfrak{h}$ are the Cartan generators and $\mathfrak{r}_\pm$ are the positive (negative) roots, that is, $\mathfrak{h}$ are the charge operators and $\mathfrak{r}_\pm$ are the raising (lowering) operators.

The prototype or *standard Borel subalgebra* is $\mathfrak{b} = \mathfrak{h} \oplus \mathfrak{r}_+$. For example, for $\mathfrak{sl}(n, \mathbb{C})$, $\mathfrak{h}$ consists of all the diagonal and $\mathfrak{r}_+$ of all strictly upper triangular $n \times n$ matrices. All Borel subalgebras are G-conjugate to $\mathfrak{b}$. A *parabolic subalgebra* $\mathfrak{p}$ then satisfies $\mathfrak{b} \subset \mathfrak{p} \subset \mathfrak{g}$. Schematically, $\mathfrak{p}$ consists of some negative roots (from $\mathfrak{r}_-$) in addition to all elements in $\mathfrak{b} = \mathfrak{h} \oplus \mathfrak{r}_+$.

By exponentiation, these subalgebras give rise to subgroups B and P of the Lie group $G_\mathbb{C}$; all Borel (parabolic) subgroups of $G_\mathbb{C}$ are equivalent by $G_\mathbb{C}$-conjugation.

For a given complex semi-simple Lie group $G_\mathbb{C}$ and some parabolic subgroup P thereof, one defines the homogeneous coset space $\mathbb{F}_{[P \subset G]} = G_\mathbb{C}/P$ and we state here without proof that it is simply connected, compact, has a natural complex structure induced from the Hermitian structure on $G_\mathbb{C}$ and is in fact a Kähler manifold of complex dimension $\frac{1}{2}(\dim G_\mathbb{C} - \dim P)$. Often, there exists a maximal regular subgroup $H \subset G$ such that $G/H \approx G_\mathbb{C}/P$; in fact, the simplest C-space is $\mathbb{P}^n = \left\{\frac{U(n+1)}{U(1)\times U(n)}\right\}$.

More generally, we will write $N = \sum_{f=1}^{F} n_f$ and

$$\mathbb{F}_{[n_1,\dots,n_f]} \stackrel{\text{def}}{=} \left\{\frac{U(N)}{U(n_1) \times \cdots \times U(n_F)}\right\} ; \tag{3.3.1}$$

¶ We forego a mathematically rigorous exposition here; chapters 2 and 3 of Ref. [6] provide all the necessary details. Note however that without $\mathfrak{g}$ being complex(ified), $\mathfrak{r}_+$ and $\mathfrak{r}_-$ would not be possible to split asunder; write $G_\mathbb{C}$ to remind of this.

Baston and Eastwood show that specifying the inclusion $P \subset G_{\mathbb{C}}$ characterizes $\mathbb{F}_{[P\subset G]}$. Next, we note that

$$\mathbb{F}_{[n_1,\dots,n_f]} \approx \{V_{n_1} \subset V_{(n_1+n_2)} \subset \cdots \subset V_N = V\}, \tag{3.3.2}$$

where V_n is an n-dimensional linear subspace of V. A particular choice of such nested subspaces is called a *flag* and the space of such flags, identifiable with the quotient $\mathbb{F}_{[\mathbf{n}]}$, is called the *flag space*. The notion of a *generalized flag variety* of Baston and Eastwood stems from extending this to general quotients $\{G_{\mathbb{C}}/P\}$ of compact complex semi-simple Lie groups by parabolic subgroups, thus covering Wang's classification of Kähler C-spaces.

An important simplification occurs if we require in addition that $b_2(\mathbb{F}) = 1$; such C-spaces were called *simple* in Ref. [180]. Then $\mathbb{F}_{[\mathbf{n}]} = G_{\mathbb{C}}/P$, where P is a *maximal* parabolic subgroup of the simple complex Lie group $G_{\mathbb{C}}$. The list of such spaces is of course infinite since the list of Lie groups is, providing us therefore with an enormous arena for constructing Calabi–Yau 3-folds (and other varieties at the Reader's preference). Nevertheless, Ref. [180] reports that a generalization of the proof presented in § 2.3 ensures that there is a finite number of Calabi–Yau 3-fold hypersurfaces in products of simple Kähler C-spaces.

3.3.2 Homogeneous bundles over flag spaces

In general, a holomorphic vector bundle over some complex base space X is a holomorphically varying family of vector spaces parametrized by the points of X. That is, to every point in X we associate a vector space (fiber) and require that as we move from point to point in X, the corresponding vector spaces move into each other in holomorphic manner.

Take now $X = \{G_{\mathbb{C}}/P\}$, think of it as the space of P-orbits in $G_{\mathbb{C}}$ and consider a vector bundle $\mathcal{V}$ over $G_{\mathbb{C}}$, with fibers V_g at every $g \in G_{\mathbb{C}}$. For $\mathcal{V}$ to qualify as a vector bundle over $\{G_{\mathbb{C}}/P\}$, we recall that two different points in $G_{\mathbb{C}}$, g and g', are considered the same point in $\{G_{\mathbb{C}}/P\}$ if they are related by the action of P, $g = p^{-1}gp$. But, if g and g' are considered the same, then V_g and $V_{p^{-1}gp}$ must be identified also. For the latter to make sense, the fibers V_g must admit an action of P, $V_{p^{-1}gp} = p(V_g)$, and are therefore labeled by representations of P. Such vector bundles are homogeneous and, conversely, all homogeneous vector bundles can be obtained this way. There might be other holomorphic vector bundles on $\{G_{\mathbb{C}}/P\}$ which do not arise this way, but we will not have use for them.

So, homogeneous holomorphic vector bundles over $\mathbb{F}_{[P\subset G]} = \{G_{\mathbb{C}}/P\}$ are labeled by representations of P and we recall that a representation can be written as a direct sum of *irreducible* ones; accordingly, we need to focus on irreducible homogeneous holomorphic vector bundles. As each irreducible representation is uniquely labeled by its highest weight, this information suffices also to distinguish the vector bundles of our interest. Amongst the several different notations [5, 53], the Young tableaux are perhaps best known to physicists and we adhere to them. These become a little cumbersome for $SO(n)$, $Sp(n)$ and the exceptional Lie groups, but Ref. [125] provides all required information in detail. Alternatively, one may use the Dynkin labels $[d_1, \ldots, d_r]$, where r is the rank of the group or even the Dynkin diagrams themselves as in Ref. [6].

We will quite exclusively deal with unitary groups $G = U(n)$, which allows the simpler notation used here and we adopt the following conventions. We write $(b_1, \ldots, b_n)$, where $b_r \leq b_{r+1}$ is the number of boxes in the rth row of the tableau, the bottom row being $r = 1$. For $b_r > 0$ (< 0), the boxes are strung to the right (left) of the 'spine'. In tensorial notation, $(-1, 0, \ldots, 0)$ labels a covariant vector, v_μ, while $(0, \ldots, 0, 1) \sim v^\mu$ is contravariant. All other representations are obtained from these by multiplication and then decomposition into a direct sum of irreducible components through symmetrizing, anti-symmetrizing and taking traces with the invariant tensor, the Kronecker symbol, δ_μ^ν. Note that the Levi-Civita alternating symbol $\epsilon^{\mu_1 \cdots \mu_n}$ is *not invariant*; it transforms with a phase

$$\epsilon^{\mu_1 \cdots \mu_n} \longmapsto \det(\Lambda)\, \epsilon^{\mu_1 \cdots \mu_n} , \qquad \Lambda \in U(n) . \tag{3.3.3}$$

Since locally $U(n) \approx U(1) \times SU(n)$, the phase belongs to this $U(1)$ and it follows that the homogeneous holomorphic vector bundle the fibers of which transform like $\epsilon^{\mu_1 \cdots \mu_n}$ is holomorphically—but not homogeneously—trivial. So, when discussing bundles over $\mathbb{F}_{[n_1, \ldots, n_F]}$ which were defined in Eq. (3.3.1), we will be careful not to cancel the common $U(1)$ factor and will not pass to $\{\frac{SU(N)}{S[U(n_1) \times \cdots \times U(n_F)]}\}$. Not forgetting this $U(1)$ phase will prove useful later, especially when comparing with other, more stringy ways of analysis.

As mentioned above, specifying the subgroup $H = \prod_{f=1}^{F} U(n_f)$ of $G = U(N)$ suffices to characterize $\mathbb{F}_{[\mathbf{n}]}$ and we note that, because of Eq. (3.3.2) the order of the factors $U(n_f)$ *is essential*. Since a representation of such a Lie group H is obtained as a tensor product of representations of the factor $U(n_f)$'s, we write

$$(a_1, \ldots, a_{n_1} | b_1, \ldots, b_{n_2} | \cdots | d_1, \ldots, d_{n_F}) \tag{3.3.4}$$

for the corresponding Young tableau or, to save some space, stack the partitions atop each other.

Examples: • For $\mathbb{P}^n = \{U(n+1)U(1) \times U(n)\}$, we have that

$$\mathcal{T}_{\mathbb{P}^n} \sim (-1|0,\ldots,0,1)\ , \qquad \mathcal{T}^*_{\mathbb{P}^n} \sim (1|-1,0,\ldots,0)\ , \tag{3.3.5}$$

are the holomorphic tangent and cotangent bundles, respectively. The bundle corresponding to degree-k polynomials is

$$(\mathcal{S}^*_{\mathbb{P}^n})^{\otimes k} = \mathcal{O}_{\mathbb{P}^n}(k) \sim (-k|0,\ldots,0)\ . \tag{3.3.6}$$

So, for example,

$$(-1|0,\ldots,0,1) \otimes (1|-1,0,\ldots,0) = (0|0,\ldots,0) \oplus (0|-1,0,\ldots,0,1) \tag{3.3.7}$$

corresponds to $\mathcal{T}_{\mathbb{P}^n} \otimes \mathcal{T}^*_{\mathbb{P}^n} = \mathcal{O}_{\mathbb{P}^n} \oplus \operatorname{End} \mathcal{T}_{\mathbb{P}^n}$, where the latter is the bundle of traceless endomorphisms of the holomorphic tangent bundle. Also,

$$\wedge^2 \mathcal{T}_{\mathbb{P}^n} \sim \wedge^2 (-1|0,\ldots,0,1) = (-2|0,\ldots,0,1,1)\ , \tag{3.3.8}$$

and so on.

• On the 6-dimensional Grassmannian $G_{(2,5)} = \mathbb{F}_{[2,3]} = \{\frac{U(5)}{U(2)\times U(3)}\}$,

$$\mathcal{T}_{\mathbb{F}_{[2,3]}} \sim (-1,0|0,0,1)\ , \quad \mathcal{T}^*_{\mathbb{F}_{[2,3]}} \sim (0,1|-1,0,0)\ , \quad \mathcal{S}_{\mathbb{F}_{[2,3]}} \sim (-1,0|0,0,0)\ ,$$

$$\wedge^2 \mathcal{T}_{\mathbb{F}_{[2,3]}} \sim (-1,-1|0,0,2) \oplus (-2,0|0,1,1)\ , \qquad \mathcal{K}^*_{\mathbb{F}_{[2,3]}} \sim (-3,-3|2,2,2)\ . \tag{3.3.9}$$

As regards the first two examples in Table 3.2, we now see that $\mathcal{L} = \det \mathcal{S}^* = (1,1|0,0,0)$. The anti-canonical bundle of $[G_{(2,5)} \| 3\mathcal{L}]$ is (with $k\mathcal{L} \stackrel{\text{def}}{=} \oplus^k \mathcal{L}$)

$$(-3,-3|2,2,2) \otimes \det\left[3\,(1,1|0,0,0)\right] = (0,0|2,2,2)\ , \tag{3.3.10}$$

which is non-trivial and has 50 sections. For $\left[G_{(2,5)} \middle\| 2\mathcal{L} \oplus \mathcal{L}^{\otimes 2}\right]$, $\mathcal{K}^* = (1,1|2,2,2)$, and similarly for $[G_{(2,6)} \| 5\mathcal{L}]$, $\mathcal{K}^* = (1,1|2,2,2,2)$.

3.3.3 The (anti)canonical bundle

The last few results present an immediate application of the preceding information about bundles over flag spaces $\mathbb{F}_{[P \subset G]}$: we can now determine the canonical bundle of any complete intersection submanifold in products of flag spaces.

Generalizing the complete intersections in (products of) complex projective spaces, we now consider an arbitrary product of flag varieties

$\mathcal{X} \stackrel{\text{def}}{=} \prod_r \mathbb{F}^{(r)}$ for the embedding space. Clearly, vector bundles over $\mathcal{X}$ are simply the products of vector bundles over the factors. We again consider complete intersection spaces $\mathcal{M}$, defined just as in Eq. (2.1.3), as the common zero-set of a system of constraints, $\xi^a = 0$, where ξ^a is a *sufficiently generic* section of $\mathcal{V}_a$. Note, however that if rank $\mathcal{V}_a > 1$, the corresponding $\xi^a = 0$ is itself a system of constraints; by 'sufficiently generic,' we will imply that these constraints are independent. This then also implies that

$$\mathcal{N}_{\mathcal{X}/\mathcal{M}} = \bigoplus_{a=1}^{k} \mathcal{V}_a, \quad \dim \mathcal{M} = \dim \mathcal{X} - \operatorname{rank}(\bigoplus_{a=1}^{k} \mathcal{V}_a) . \tag{3.3.11}$$

Then, there is again a short exact sequence such as (1.3.7) and

$$\det(\mathcal{T}_{\mathcal{M}}) = \det(\mathcal{T}_{\mathcal{X}}) \otimes \det(\bigoplus_a \mathcal{V}_a) = \mathcal{K}^*_{\mathcal{X}} \otimes \bigotimes_a \det(\mathcal{V}_a) . \tag{3.3.12}$$

As we are interested in a characteristic class property, that is, in the ampleness of the anti-canonical bundle, we may also apply the splitting principle and think of the system of vector bundles as a system (or systems) of line bundles.

This is clearly the straightforward generalization of the formula (2.1.16). However, this time we need a little more information. Surely, $\otimes_a \det(\mathcal{V}_a)$ is given easily in terms of a Kronecker product of Young tableaux (or any other mnemonic tool for multiplying representations of Lie groups), as these bundles were open to our choice. All we need to know is that these bundles have sections; as we shall see later, they do have sections whenever the symbol $(\ldots|\ldots|\ldots)$ labels a valid representation of $U(N)$, that is whenever *all* integers in $(\ldots|\ldots|\ldots)$ are non-decreasing from left to right.

However, $\mathcal{K}_{\mathcal{X}}$ and so $\mathcal{T}_{\mathcal{X}}$ must be determined independently; the price to pay for the simplicity of this technique is that for flag spaces of the type (3.3.1) with more than two factors in the denominator, the $\mathcal{T}_{\mathcal{X}}$ is no longer irreducible and it of course becomes more and more complicated. $\mathcal{T}_{\mathcal{X}}$ becomes an *extension*$^{(L)}$ of two or more irreducible homogeneous bundles $\mathcal{T}_i$. Luckily, for the purpose of characteristic class computations, we may replace $\mathcal{T}_{\mathcal{X}}$ with the direct sum (although not holomorphically!) of $\mathcal{T}_i$. The determinant formula will of course hold also holomorphically. So we remain 'only' with the task to determine $\mathcal{T}_i$.

3.3.4 Bundles for free

While discussing the simplest C-spaces, the $\mathbb{P}^n$'s, the relation (1.3.20), i.e., that $\mathcal{K}_{\mathbb{P}^n} \approx \mathcal{O}_{\mathbb{P}^n}(-n-1)$ was simple to derive. The analogous is also possible for Grassmannians, although the derivation is a little more involved. The reason for this is that the total Chern class of $\mathbb{P}^n$ was a simple power series in a single variable, while the total Chern class of a Grassmannian will have more algebraically independent generators (see Ref. [11], p. 292–294). Clearly, this gets more and more out of hand as we allow more and more 'stripes,' n_f, in the flags of our flag spaces $\mathbb{F}_{[n_1,\dots,n_F]}$.

Let us reexamine the relation (1.3.20) in a little more detail so as to be able to generalize for all $\mathbb{F}_{[\mathbf{n}]}$'s. We may think of $\mathbb{P}^n$ as the space of all lines $L \approx \mathbb{C}^1$ through the origin of $\mathbb{C}^{n+1}$. Each line L is defined as the zero-set of some linear polynomial, $\ell(x)$ over $\mathbb{C}^{n+1}$. Of all tangent vectors on C^{n+1}, those written as

$$\ell^i(x)\frac{\partial}{\partial x^i} = x^j\,\lambda_j{}^i\frac{\partial}{\partial x^i}\;, \qquad \ell^i(x) \in H^0(\mathbb{P}^n, \mathcal{O}_{\mathbb{P}^n}(1))\;, \tag{3.3.13}$$

will span $\mathcal{T}_{\mathbb{P}^n}$ if taken modulo the homoteties, that is, $x^i\frac{\partial}{\partial x^i} \simeq 0$, i.e., $\mathrm{Tr}[\boldsymbol{\lambda}] = \lambda_i{}^i = 0$. We have just obtained the so-called Euler (short exact) sequence:

$$0 \to \mathcal{O}_{\mathbb{P}^n} \to \mathcal{O}_{\mathbb{P}^n}(1)^{\oplus(n+1)} \to \mathcal{T}_{\mathbb{P}^n} \to 0\;. \tag{3.3.14}$$

From a related point of view, at every point $p \in \mathbb{P}^n$, vectors of $\mathcal{T}_{\mathbb{P}^n}$ act by mapping $p \in \mathbb{P}^n$ into a nearby point. In other words, the line $L \subset \mathbb{C}^{n+1}$ is being rotated around the origin into another line, whence[◊] $\mathcal{T}_{\mathbb{P}^n} = \mathrm{Hom}(L, \mathbb{C}^{n+1}/L)$, i.e., $\mathcal{T}_{\mathbb{P}^n} \otimes L = \{\mathbb{C}^{n+1}/L\}$. This relation is equivalent to the exactness of the short sequence, which we write as:

$$0 \to \mathcal{S} \to \mathbb{C}^N \to \mathcal{Q} \to 0\;, \qquad \mathcal{Q} = \mathcal{T}_{\mathbb{F}_{[\mathbf{n}]}} \otimes \mathcal{S}\;, \tag{3.3.15}$$

where in our present case, $\mathbb{F}_{[\mathbf{n}]} = \mathbb{F}_{[1,n]} = \mathbb{P}^n$, $N = n{+}1$ and $\mathcal{S} = L$. Of course, $\mathbb{C}^N$ is here considered as the trivial rank-N bundle over $\mathbb{F}_{[\mathbf{n}]} = \mathbb{P}^n$, $\mathbb{P}^n \times \mathbb{C}^N$, while $\mathcal{S}$ is the *universal sub-bundle* of $\mathbb{F}_{[\mathbf{n}]} \times \mathbb{C}^N$,

$$\mathcal{S} = \{(\mathcal{S}_p, x) \in \mathbb{F}_{[\mathbf{n}]} \times \mathbb{C}^N \;:\; x \in \mathcal{S}_p\}\;. \tag{3.3.16}$$

The quotient bundle $\mathcal{Q} = \{\mathbb{C}^N/\mathcal{S}\}$ is called the *universal quotient bundle* and the sequence (3.3.15) is called the *tautological sequence.* A little thought reveals that $L = \mathcal{O}_{\mathbb{P}^n}(-1)$, whence the exact sequence (3.3.14) can also be obtained from Seq. (3.3.15) through tensoring the latter with $L^* = \mathcal{O}_{\mathbb{P}^n}(1)$.

Note: since $L^* = \mathcal{O}_{\mathbb{P}^n}(1) \sim (-1|0,\dots,0)$ in the above notation, it follows that $\mathcal{T}_{\mathbb{P}^n} \sim (-1|0,\dots,0,1)$.

◊ See the Hom$(\mathcal{A},\mathcal{B})$ entry in the Lexicon.

Next we need to reinterpret $\mathcal{S}$ for more complicated cases. For a Grassmannian $G_{(k,N)} = \mathbb{F}_{[k,(N-k)]}$, the universal sub-bundle is $\mathcal{S} = \{(H_p, x) \in G_{(k,N)} \times \mathbb{C}^N : x \in H_p\}$, where H_p is the k-plane through the origin of $\mathbb{C}^N$ corresponding to the point $p \in G_{(k,N)}$, since $G_{(k,N)}$ is the space of k-planes through the origin of $\mathbb{C}^N$. Clearly, this interpretation needs to be iterated inclusion after inclusion for $\mathbb{F}_{[n_1,\dots,n_F]}$, recalling the relation (3.3.2). As a consequence, $\mathcal{T}_{\mathbb{F}_{[\mathbf{n}]}}$ will be given in terms of a sequence of sequences and will fail to be given as a single irreducible factor. Nevertheless, the above formalism of homogeneous bundles over simply connected C-spaces, which we have adopted from Eastwood [18], deals with this easily. Suffice it here to give some examples; the general pattern will be clear:

$$\mathcal{T}_{\mathbb{F}_{[n_1,n_2]}} \sim (-1,0,\dots,0|0,\dots,0,1) \; ; \tag{3.3.17}$$

$$\mathcal{T}_{\mathbb{F}_{[n_1,n_2,n_3]}} \sim (-1,0,\dots,0|0,\dots,0|0,\dots,0,1) + \begin{matrix} (-1,0,\dots,0|0,\dots,0,1|0,\dots,0) \\ (0,\dots,0|-1,0,\dots,0|0,\dots,0,1) \end{matrix} \; ; \tag{3.3.18}$$

$$\begin{aligned} \mathcal{T}_{\mathbb{F}_{[n_1,n_2,n_3,n_4]}} \sim\; & (-1,0,\dots,0|0,\dots,0|0,\dots,0|0,\dots,0,1) + \begin{matrix} (-1,0,\dots,0|0,\dots,0|0,\dots,0,1|0,\dots,0) \\ (0,\dots,0|-1,0,\dots,0|0,\dots,0|0,\dots,0,1) \end{matrix} \\ & + \begin{matrix} (-1,0,\dots,0|0,\dots,0,1|0,\dots,0|0,\dots,0) \\ (0,\dots,0|-1,0,\dots,0|0,\dots,0,1|0,\dots,0) \\ (0,\dots,0|0,\dots,0|-1,0,\dots,0|0,\dots,0,1) \end{matrix} \; , \end{aligned} \tag{3.3.19}$$

and so on. Summands in a direct sum of vector bundles were stacked vertically to save some space. The symbol '+' here means '$\oplus$' in the C^∞ category, but not holomorphically; see *exact sequence*$^{(L)}$. The various brackets in the above examples represent the irreducible summands $\mathcal{T}_i$ of $\mathcal{T}_{\mathbb{F}_{[\mathbf{n}]}}$.

3.4 Calabi–Yau Submanifolds in C-Spaces

Because $\mathcal{K}_X = \det \mathcal{T}^*_X$, once we have the irreducible summands of $\mathcal{T}_X$, $\mathcal{K}_X$ is easy to obtain using that

$$\mathcal{K}_{\mathbb{F}_{[n_1,\dots,n_F]}} \approx \det\Big(\bigoplus_{i=1}^{\binom{F}{2}} \mathcal{T}^*_i\Big) = \bigotimes_{i=1}^{\binom{F}{2}} \det\big(\mathcal{T}^*_i\big) \; , \tag{3.4.1}$$

and that

$$\det(\underbrace{0,\dots,0,1}_{p}|\underbrace{-1,0,\dots,0}_{q}) = (q,\dots,q|-p,\dots,-p) \; . \tag{3.4.2}$$

We now have all we need to verify for any particular case the (non)triviality of the canonical bundle of the complete intersection (*if non-empty*) in an arbitrary product of flag spaces.

3.4.1 Some examples

For $\mathcal{M} \in [5\|4\ 2]$, this becomes: $\mathcal{V}_1 \sim (-4|0,0,0,0,0)$ and $\mathcal{V}_2 \sim (-2|0,0,0,0,0)$ are both line bundles, corresponding to a quartic and a quadric constraint.

$$\begin{aligned}\mathcal{K}_{\mathcal{M}} &= \det \mathcal{T}^*_{\mathbb{P}^5} \otimes \mathcal{V}_1 \otimes \mathcal{V}_2 \\ &= \det(1|-1,0,0,0,0) \otimes (-4|0,0,0,0,0) \otimes (-2|0,0,0,0,0) \\ &= (5|-1,-1,-1,-1,-1) \otimes (-6|0,0,0,0,0) \\ &= (-1|-1,-1,-1,-1,-1) \;=\; (1|1,1,1,1,1)^{-1}\ .\end{aligned} \tag{3.4.3}$$

Note that $\mathcal{K}_{\mathcal{M}}$ is holomorphically trivial, although not homogeneously so; its fibers transform by a phase, as the $\epsilon_{i_1\cdots i_6}$ of $\mathbb{P}^5$.

Consider now the 4-dimensional Grassmannian $G_{(2,4)} = \mathbb{F}_{[2,2]}$, with $\mathcal{T}^*_{\mathbb{F}_{[2,2]}} \sim (0,1|-1,0)$, so $\mathcal{K}_{\mathbb{F}_{[2,2]}} \sim (2,2|-2,-2)$. To make the canonical bundle on a 3-dimensional subspace $\mathcal{M}$ holomorphically trivial, we choose to define $\mathcal{M}$ as the zero-set of a section of the line bundle $\mathcal{L} \sim (-2,-2|2,2)$, since then $\mathcal{K}_{\mathcal{M}} = \mathcal{K}_{\mathbb{F}_{[2,2]}} \otimes \mathcal{L} \sim (0,0|0,0)$, which is holomorphically and in fact also homogeneously trivial; $\mathcal{M}$ is therefore a Calabi–Yau hypersurface in the Grassmannian $G_{(2,4)}$.

As another example of this kind, consider embedding a Calabi–Yau 3-fold in the 5-dimensional flag space $\mathbb{F}_{[1,2,1]}$. We have

$$\mathcal{T}_{\mathbb{F}_{[1,2,1]}} \sim (-1|00|1) + \frac{(-1|01|0)}{(0|-10|1)}\ , \qquad \Longrightarrow \qquad \mathcal{K}_{\mathbb{F}_{[1,2,1]}} \sim (3|00|-3)\ . \tag{3.4.4}$$

✎ Prove this. ❧

For a 3-fold $\mathcal{M} \hookrightarrow \mathbb{F}_{[1,2,1]}$, we need two constraints and these may be chosen as sections of $(-3|00|0) \oplus (0|00|3)$ or of $(-2|00|1) \oplus (-1|00|2)$. This produces two (deformation classes of) Calabi–Yau manifolds; it can be shown moreover that these can also be embedded via the configurations

$$\left[\begin{matrix}3\\3\end{matrix}\middle\|\begin{matrix}3&0&1\\0&3&1\end{matrix}\right] \quad \text{and} \quad \left[\begin{matrix}3\\3\end{matrix}\middle\|\begin{matrix}2&1&1\\1&2&1\end{matrix}\right]\ , \tag{3.4.5}$$

respectively, since $\mathbb{F}_{[1,2,1]}$ is isomorphic to a generic member of $\left[\begin{smallmatrix}3\\3\end{smallmatrix}\middle\|\begin{smallmatrix}1\\1\end{smallmatrix}\right]$, defined by $h_{\alpha\beta}\, x^\alpha y^\beta = 0$. Note, however, that special (singular) hypersurfaces in $\left[\begin{smallmatrix}3\\3\end{smallmatrix}\middle\|\begin{smallmatrix}1\\1\end{smallmatrix}\right]$, when $\mathrm{rank}[h_{\alpha\beta}] < 4$, are not isomorphic to $\mathbb{F}_{[1,2,1]}$. Since the cubics can be arranged to avoid the singularity of the hypersurface, there are smooth members of $\left[\begin{smallmatrix}3\\3\end{smallmatrix}\middle\|\begin{smallmatrix}3&0&1\\0&3&1\end{smallmatrix}\right]$ which cannot be embedded in $\mathbb{F}_{[1,2,1]}$ (see Ref. [121, 76]).

As a final example of this kind, consider embedding $\mathcal{M} \hookrightarrow \mathbb{F}_{[2,3]} = G_{(2,5)}$, $\mathcal{K}_{\mathbb{F}_{[2,3]}} \sim (3,3|-2,-2,-,2)$. This Grassmannian is 6-dimensional, so we need three locally independent constraints. We may choose two of them to be the components of a 2-vector section of the rank-2 vector bundle $\mathcal{S}^* \sim (-1,0|000)$. Since $\det(-1,0|000) = (-1,-1|000)$, we may choose for the remaining line bundle $\mathcal{L} \sim (-2,-2|2,2,2)$ so that

$$\mathcal{K}_{\mathcal{M}} \sim (3,3|-2,-2,-,2) \otimes (-1,-1|000) \otimes (-2,-2|2,2,2) = (00|000) . \tag{3.4.6}$$

Note that $\mathcal{L}$ is holomorphically equivalent to $\big(\det \mathcal{S}^*\big)^{\otimes 4}$:

$$\mathcal{L} = \big(\det \mathcal{S}^*\big)^{\otimes 4} \otimes (2,2|2,2,2) . \tag{3.4.7}$$

So, using a 2-vector section of $\mathcal{S}^*$ and a section of $\big(\det \mathcal{S}^*\big)^{\otimes 4}$, we define another Calabi–Yau 3-fold, $\mathcal{M}'$ with $\mathcal{K}_{\mathcal{M}'} \sim (-2,-2|-2,-2,-2)$. For appropriate choices of the defining sections, $\mathcal{M}$ and $\mathcal{M}'$ are isomorphic as complex manifolds, although as submanifolds of the *homogeneous* space $\mathbb{F}_{[2,3]}$, $\mathcal{M} \not\approx \mathcal{M}'$: $\mathcal{K}_{\mathcal{M}}$ is trivial under overall rescalings of $\mathbb{F}_{[2,3]}$ while $\mathcal{K}_{\mathcal{M}'}$ is not*.

We will return later to 3-folds embedded in products of Kähler simply connected compact homogeneous spaces, when we turn to discuss computations of cohomology groups. The interested Reader should by now be well equipped for some preliminary independent exploration in the wilderness of this uncharted wealth of new constructions and—quite expectedly—some hitherto unknown beasts.

3.5 Pfaffian Constraint Systems

Considering the constructions discussed so far, the Reader may wonder if embeddings other than complete intersections can be analyzed. Somewhat as a curiosity at the moment, we now turn to a simple construction of a Calabi–Yau 3-fold which is not a complete intersection [103].

3.5.1 A Pfaffian 3-fold

Start with an skew-symmetric 7×7 matrix of linear polynomials $\mathbf{A}(x) \stackrel{\text{def}}{=} a_{[ij]}(x)$, where $(x^0{:}\ldots{:}x^6) \in \mathbb{P}^6$. Let $\mathbf{A}_k$, $k = 0,\ldots,6$, denote the seven principal 6×6 minors

$$\mathbf{A}_k \stackrel{\text{def}}{=} \{\, a_{[ij]}(x) : \; i,j \neq k \,\} . \tag{3.5.1}$$

* It does not seem to be clear at this time whether such a distinction is in fact relevant in the immediate physics application.

Then $\det(\mathbf{A}_k)$ are seven sextic polynomials over $\mathbb{P}^6$ and since the $\mathbf{A}_k$ are skew-symmetric,

$$\det(\mathbf{A}_k) \; = \; f_k(x)^2 \, , \qquad f_k(x) \; = \; \mathrm{Pf}\,[\,\mathbf{A}_k\,] \, , \quad k = 0, \ldots, 6 \, . \tag{3.5.2}$$

$\mathrm{Pf}\,[\,\mathbf{A}_k\,]$ denotes the *Pfaffian* of $\mathbf{A}_k$. Clearly, this provides a system $\{f_k(x)\}$ of seven rather special cubic polynomials over $\mathbb{P}^6$.

We define now our prospective 3-fold $\mathfrak{X}$ as the vanishing set of all seven cubics $f_k(x)$:

$$\mathfrak{X} \; \stackrel{\text{def}}{=} \; \{\, x \in \mathbb{P}^6 \mid f_k(x) = 0 \, , \; k = 0, \ldots, 6 \,\} \subset \mathbb{P}^6 \, . \tag{3.5.3}$$

This clearly cannot be a complete intersection, for if it were, $\mathfrak{X}$ would have been -1-dimensional. In fact, for a general choice of the initial 7×7 matrix $a_{[ij]}(x)$, we shall see that $\mathfrak{X}$ is in fact a smooth Calabi–Yau 3-fold.

To begin with, we want to show that $\mathfrak{X}$ is a 3-fold. To this end, note that since $\mathbf{A}$ is skew-symmetric, it must be of even rank, that is of rank 0, 2, 4 or 6.

Now, at a generic point $p \in \mathbb{P}^6$, we may choose local coordinates so that

$$\mathbf{A}(p) \; = \; \left[\begin{array}{c|c|c} \mathbb{O} & \mathbb{1}_3 & \mathbf{0} \\ \hline -\mathbb{1}_3 & \mathbb{O} & \mathbf{0} \\ \hline \mathbf{0} & \mathbf{0} & 0 \end{array}\right] , \tag{3.5.4}$$

where the matrix is divided into blocks of 3–3–1 elements on each side; $\mathbb{1}_3$ denotes the 3×3 identity matrix. Then $\mathrm{rank}(\mathbf{A}) = 6$ for all generic $p \in \mathbb{P}^6$ and not all the cubics in the system $\{f_k(x)\}$ can vanish. So, $\mathfrak{X}$ is a proper subspace of $\mathbb{P}^6$.

Next, at a sub-generic point of $\mathbb{P}^6$, local coordinates may be chosen so that

$$\mathbf{A} \; = \; \left[\begin{array}{c|c|ccc} \mathbb{O} & \mathbb{1}_2 & \mathbf{0} & \mathbf{0} & \mathbf{0} \\ \hline -\mathbb{1}_2 & \mathbb{O} & \mathbf{0} & \mathbf{0} & \mathbf{0} \\ \hline \mathbf{0} & \mathbf{0} & 0 & a_{[56]} & a_{[57]} \\ \mathbf{0} & \mathbf{0} & -a_{[56]} & 0 & a_{[67]} \\ \mathbf{0} & \mathbf{0} & -a_{[57]} & -a_{[67]} & 0 \end{array}\right] . \tag{3.5.5}$$

So, to obtain $\mathrm{rank}(\mathbf{A}) = 4$, we need to impose three locally independent conditions, $a_{[56]}(x) = a_{[57]}(x) = a_{[67]}(x) = 0$. This implies that the dimension of $\mathfrak{X}$ is not more than 3. That it is also precisely 3 at each point of $\mathfrak{X}$, we need to check the possibilities of $\mathrm{rank}\,\mathbf{A} = 0, 2$. By a similar argument, to bring the rank of $\mathbf{A}$ down to 2, we would need to impose 10 locally independent conditions, that is, given that the initial 21 linear polynomials $a_{[ij]}(x)$ were chosen generically. On $\mathbb{P}^6$, this is four constraints too many. Finally, to achieve $\operatorname{rank}\mathbf{A} = 0$, we would need to impose 21 locally independent conditions.

Alternatively, locally at $p \in \mathbb{P}^6$, the choice of coordinates as in Eq. (3.5.4) implies that $f_i(x) \equiv 0$ for $i = 1, \ldots, 6$ and $f_7(x) = 1 \neq 0$, whence $p \notin \mathfrak{X}$. Similarly, Eq. (3.5.5) implies that locally $f_i(x) \equiv 0$ for $i = 1, \ldots, 4$ and $f_5 = a_{[67]}$, $f_6 = a_{[57]}$, $f_7 = a_{[56]}$. So, at $\mathfrak{X}$ where $f_k(x) = 0$, for $k = 1, \ldots, 7$, $\operatorname{rank}\mathbf{A} = 4$.

Thus, we have proved that $\operatorname{rank}\mathbf{A}(x) = 4$ at the subspace $\mathfrak{X} \subset \mathbb{P}^6$, implying that $\dim \mathfrak{X} = 3$. Analyzing the system of gradients in a similar fashion, it should be clear that $\mathfrak{X}$ will also be smooth for a generic choice of the initial matrix $\mathbf{A}(x)$.

To amuse the Reader, we pause to show that the similar construction with a 3×3 skew-symmetric matrix of linear polynomials $\mathbf{B}(x)$ in $\mathbb{P}^2$ yields $\varnothing$. Since a skew-symmetric 3×3 matrix has precisely three independent entries, by choosing coordinates on $\mathbb{P}^2$, we have that

$$\mathbf{B}(x) \stackrel{\text{def}}{=} \begin{bmatrix} 0 & z & -y \\ -z & 0 & x \\ y & -x & 0 \end{bmatrix} . \tag{3.5.6}$$

Then $\mathbf{B}_k = \{x\tau_2, -y\tau_2, z\tau_2\}$, where $\tau_2 = \left(\begin{smallmatrix} 0 & 1 \\ -1 & 0 \end{smallmatrix}\right)$. Clearly,

$$\operatorname{Pf}[\mathbf{B}_k] = \{x, y, z\}, \qquad k = 1, 2, 3 . \tag{3.5.7}$$

So, in fact, the system $\operatorname{Pf}[\mathbf{B}_k] = 0$ is immediately seen to have no solutions and this lower dimensional analogue of $\mathfrak{X}$ is empty. However, we find it instructive to show the analogue of the above analysis. In the patch $z \neq 0$, we choose local coordinates $\xi = x/z$, $\eta = y/z$ and set $\zeta = z/z = 1$. At the point $p = (1{:}0{:}0) \in \mathbb{P}^2$, we indeed have $\mathbf{B}(p) = \begin{pmatrix} 0 & 1 & 0 \\ -1 & 0 & 0 \\ 0 & 0 & 0 \end{pmatrix}$, as above, and $\operatorname{rank}\mathbf{B}(x) \leq 2$. In fact, $\operatorname{rank}\mathbf{B}(x) = 2$ precisely since, in analogy to the case (3.5.5), for $\operatorname{rank}\mathbf{B}(x) < 2$ we need $x = y = z = 0$ which cannot happen on $\mathbb{P}^2$.

✎ Show that the analogous construction based on a 5×5 skew-symmetric matrix of linear polynomials in $\mathbb{P}^4$ provides a Pfaffian embedding of the torus $T^2 \hookrightarrow \mathbb{P}^4$. ❧

3.5.2 Triviality of the canonical bundle

To verify that the 3-fold $\mathfrak{X}$ constructed above is Calabi–Yau, we would need to verify that it is Ricci-flat. Equivalently, we would need to show that the first Chern class vanishes, that is, that the canonical bundle is holomorphically trivial.

To see why this might be so, recall that we started from the skew-symmetric matrix $\mathbf{A}$, with 21 different non-zero entries. However, these were chosen to be linear polynomials over $\mathbb{P}^6$, that is, sections of $\mathcal{O}_{\mathbb{P}^6}(1)$. Since there are only seven global homogeneous coordinates, not more than seven entries of the matrix $\mathbf{A}$ can be linearly independent. Furthermore, the defining functions, $f_k(x)$, were found as Pfaffians of the principal minors. Now, although the normal bundle is not spanned by these defining functions globally (there are too many of them), *locally* some three (combinations) of these seven defining functions do span the normal space. The determinant of the normal bundle therefore does correspond to a product of the independent elements of $\mathbf{A}$, with some constant coefficient. If $\mathbf{A}$ is chosen generically, then it has seven independent elements and $\det \mathcal{N}_{\mathbb{P}^6/\mathfrak{X}} = \mathcal{O}_{\mathbb{P}^6}(7)$, so that $\mathcal{K}_{\mathfrak{X}} = \mathcal{K}_{\mathbb{P}^6} \otimes \det \mathcal{N}_{\mathbb{P}^6/\mathfrak{X}} \approx \mathbb{C}$.

It turns out that certain techniques which we will discuss later allow an analysis of such a relation in quite more detail and in a rather more general framework. We therefore return to the proof of $\mathcal{K}_{\mathfrak{X}} \approx \mathbb{C}$ in § 9.5; it will also indicate a generalization of Pfaffian constraint systems as a framework for constructing novel Calabi–Yau n-folds.

Chapter 4

Group Actions, Quotients and Singularities

Most Calabi–Yau manifolds constructed by the methods reviewed so far are inappropriate for realistic compactification, since they are simply connected. Recalling the wish list in § 0.8 and the discussion in § 0.4 and 0.7, we see that—most likely—we need our Calabi–Yau 3-fold to be multiply connected.

4.1 Free Holomorphic Actions

The simplest method of constructing multiply connected varieties is to construct a variety $\mathcal{Y}$ which admits a *free action*$^{(L)}$ of some group G and then pass to the quotient space $\mathcal{M} \stackrel{\text{def}}{=} \{\mathcal{Y}/G\}$. In fact, this is possible for all $\mathcal{M}$ with $c_1 = 0$ [79, 7]:

Theorem 4.1 (Bogomolov). *Let $\mathcal{M}$ be a Kähler n-fold with $h^{1,0} = h^{0,1} = k$ and $\mathcal{K}_{\mathcal{M}} \approx \mathbb{C}$. Then $\mathcal{Y} \times T^k$ is a finite cover of $\mathcal{M}$, where $\mathcal{Y}$ is simply connected, $\mathcal{K}_{\mathcal{Y}} \approx \mathbb{C}$ and T^k is a complex k-dimensional torus.*

Clearly, $\dim \mathcal{Y} = \dim \mathcal{M} - k$ meaning that $\mathcal{Y}$ is a point if $\dim \mathcal{M} = k$. Most of the time, we concern ourselves with the $k = 0$ case (which ensures $SU(n)$ holonomy) and so, in principle, the multiply connected $\mathcal{M}$ can always be obtained by constructing the simply connected covering space $\mathcal{Y}$ first.

There are however several requirements on the action of G, which come from our goal to construct complex (in fact Kähler) and Ricci-flat 3-folds.

4.1.1 Holomorphic actions

The first of these requirements, namely that $\mathcal{M} = \{\mathcal{Y}/G\}$ is a complex space is easily fulfilled if $\mathcal{Y}$ is itself complex *and the action of G is holomorphic.* The complex structure of $\mathcal{Y}$ will then be inherited by $\mathcal{M}$. Using the foregoing ideas to construct $\mathcal{Y}$ as embedded in a (product of) homogeneous space(s), $\mathcal{Y} \hookrightarrow \mathcal{X}$, and to preserve the homogeneity of the defining equations (so that these make sense), the action of G must preserve this homogeneity. So, unless the defining constraints are very special (and then typically singular), we expect the action of G to be linear in terms of coordinates on the embedding space $\mathcal{X}$. For future reference, we write

$$G:\ x^\mu \longmapsto \mathrm{g}^\mu{}_\nu\, x^\nu\ , \qquad \mathrm{g}^\mu{}_\nu \textit{ is constant}\ . \tag{4.1.1}$$

Now, Calabi–Yau n-folds ($n>1$) admit no continuous holomorphic symmetries*, so one seeks a finite group G with a holomorphic and free action on $\mathcal{Y}$.

4.1.2 Multiple connectedness and c_1=0

If the holomorphic n-form Ω on $\mathcal{Y}$ is invariant under the free action of G on $\mathcal{Y}$, then $G^*(\Omega)$ is a nowhere zero holomorphic n-form on $\mathcal{M} = \{\mathcal{Y}/G\}$, so $\mathcal{M}$ is manifestly a Calabi–Yau manifold. This, however, provides no new condition:

Lemma 4.1. *Let $\mathcal{Y}$ be an n-fold with $h^{0,q} = 1$ if $q \equiv 0 \,(\mathrm{mod}\, n)$ and $h^{0,q} = 0$ otherwise. If $G : \mathcal{Y} \to \mathcal{Y}$ has a free action, the holomorphic n-form Ω transforms as $G^* : \Omega \mapsto (-)^{n+1}\Omega$.*

This is in fact an easy consequence of the holomorphic *Lefschetz fixed point formula* $^{(L)}$. So, for Calabi–Yau 3-folds, the invariance of the holomorphic 3-form Ω with respect to a freely acting symmetry is automatic. The converse, however, is not true; it is not sufficient to check the invariance of Ω as the contributions on the right hand side of Eq. (L.40) can cancel out.

As an elementary consequence of the definition, the structure group $\pi_1(\mathcal{M})$ equals G if $\mathcal{Y}$ is simply connected. In other words, for every element $\alpha \in G$, there will exist a non-contractible loop $\Gamma_\alpha \subset \mathcal{M}$ such that if $\alpha^k = \mathbb{1}$ then $\circlearrowleft^k \Gamma_\alpha$ is contractible, that is, looping k times along Γ_α produces a

* If they did, there would exist nowhere zero holomorphic Killing vector fields. Contracting this with the nowhere zero holomorphic n-form, there would also exist harmonic $(n-1)$-forms, which contradicts our general condition for Calabi–Yau n-folds: $h^{n-1,0} = 0$ except for $n = 1$; recall Eq. (1.2.13).

contractible closed path. Moreover, Γ_α is *homotopic*$^{(L)}$ to Γ_β precisely if α and β are in the same conjugacy class of G. Of course, if $\mathcal{Y}$ was multiply connected already, then passing to the quotient by the free action of G increases the multiple connectedness, so to speak.

For a freely acting symmetry group $G : \mathcal{Y} \to \mathcal{Y}$, we have that various (holomorphic) bundles $\mathcal{V}$ over $\mathcal{Y}$ give rise to corresponding bundles over $\mathcal{M} = \{\mathcal{Y}/G\}$ if the fibers V_x and $V_{\mathbf{g}(x)}$ can be made to agree. That is, if $V_{\mathbf{g}(x)} = \mathbf{g}(V_x) \approx V_x$ (holomorphically). By their construction, this will automatically be true of $\mathcal{T}_\mathcal{Y}$ and $\mathcal{T}^*_\mathcal{Y}$, giving rise to $\mathcal{T}_\mathcal{M}$ and $\mathcal{T}^*_\mathcal{M}$, respectively. So, in particular, $H^{p,q}(\mathcal{M})$ is obtained from $H^{p,q}(\mathcal{Y})$ by projecting out all elements which are not invariant under G; the same is true of $H^q(\mathcal{M}, \mathrm{End}\,\mathcal{T}_\mathcal{M})$. Of course, to carry this through in practice, we need to know the action of the symmetry group on the various cohomology groups and we will return to this issue later.

4.1.3 Numerical characteristics

Certain quantities, however, behave much simpler in the process $\mathcal{Y} \xrightarrow{/G} \mathcal{M}$. From the very definition of the Euler characteristic as an integral (over the manifold) of the Euler density (which is a product of curvature forms), it follows that

$$\chi_E(\mathcal{M}) = \chi_E(\mathcal{Y}/G) = \frac{\chi_E(\mathcal{Y})}{|G|}\,, \tag{4.1.2}$$

where $|G|$ denotes the order of G; for example, $|\mathbb{Z}_n| = n$. By the same token, the *indices*$^{(L)}$ of the other three classical *complexes*$^{(L)}$—χ^h, τ_H and $\hat{A}$—have the same divisibility property [see Eq. (2.5.1)–(2.5.4)] and provide some necessary conditions for the action of a symmetry to be free.

Now, $\tau_H(\mathcal{X})$ and $\hat{A}(\mathcal{X})$ vanish identically when $\dim_\mathbb{C}\mathcal{X} \not\equiv 0 \pmod 2$ and χ^h vanishes also if $\mathcal{X}$ is in addition Calabi–Yau†. So, for Calabi–Yau 3-folds, only the divisibility of the Euler characteristics remains a non-trivial condition for the action of a symmetry to be free.

However, one can modify these indices as shown in Eq. (2.5.5)–(2.5.7), 'twisting' them by a bundle $\mathcal{V}$ over the n-fold in question. In general, it is difficult to compute such twisted indices with $\mathcal{V}$ a bundle intrinsic to the Calabi–Yau n-fold $\mathcal{Y}$. On the other hand, when $\mathcal{Y}$ is given as a submanifold of some better understood embedding space $\mathcal{X}$, we may twist the indices by

† Just recall Eq. (2.5.1)–(2.5.4) and the expressions (1.5.13), (1.5.17), (1.5.19).

$\mathcal{T}_{\mathcal{X}}^{\otimes k} \otimes \mathcal{N}_{\mathcal{X}/\mathcal{Y}}^{\otimes \ell}$, for arbitrary k, ℓ. It is important to note that such indices are characteristics of the *embedding* $\mathcal{Y} \hookrightarrow \mathcal{X}$, not the n-fold $\mathcal{Y}$ by itself.

Furthermore, as we study properties of characteristic classes, we may use the *splitting principle* (1.3.19) freely to decompose $\mathcal{N}_{\mathcal{X}/\mathcal{Y}}$ into line bundles $\mathcal{L}_i$, even if not holomorphically, and then twist by the various products $\bigotimes_{\text{some } i} \mathcal{L}_i$. In doing so, of course, one has to restrict to those products which admit a proper action of the candidate symmetry.

Such twisted indices were explicitly computed for the 7868 complete intersection Calabi–Yau 3-folds in products of complex projective spaces [95]. Their divisibility properties were sufficient to isolate a handful of models which may admit a freely acting symmetry, G, such that χ_E is properly divisible by $|G|$, with quotient -6, that is, yielding 3 generations of Standard Model particles, as desired. [2024] ⌈*Since 1992, several deformation families of multiply connected* $\chi_E = -6$ *CICY quotients have been found [349, 334, 348]. However, more involved compactifications that involve gauge fluxes and modify the "standard Ansatz" of § 0.2.3 by gauge instantons,* $(\mathcal{T}_{\mathcal{M}} \rightsquigarrow \mathcal{V}) \subset E_8$, *provide astoundingly many additional realizations of the minimally supersymmetric Standard Model [373]—their numbers estimated in* moles *and a* heptigoogols *(10^{700}) of moles, respectively [365]. In such a way, even* $SO(32)$*-heterotic strings can harbor the Standard Model [449]. In the last decade or so, searches throughout this bewildering bounty of candidate constructions have been mechanized by means of machine-learning magic (AI and ML), which has ushered in a transformative change in the subject; see the Afterword, p. 349 and, e.g., Refs. [266, 370, 472, 400] for starters.*⌋

4.2 Fixed Points, Singularities and Smoothings

Freely acting symmetries are of course rather more difficult to find than symmetries which have fixed point sets. So, in lieu of more smooth quotient models (of potential physical interest), we turn to quotient models which involve non-freely acting symmetries.

4.2.1 Which fixed points produce singularities?

When passing to the quotient by a symmetry $G : \mathcal{Y} \to \mathcal{Y}$ which fixes some points, $\mathring{p}_i \in \mathcal{Y}$, we expect $\mathcal{M} = \{\mathcal{Y}/G\}$ to be singular at each $p_i = G(\mathring{p}_i)$. If the fixed points $\mathring{p}_i$ are isolated, there is a local coordinate neighborhood

$U_i \approx \mathbb{C}^n$ with $\mathring{p}_i$ at the origin of U_i and $\mathring{p}_i \notin U_j$ for all $j \neq i$. Then, U_i folds up under the division by the action of G as the cone $\{\mathbb{C}^n/G\}$, leaving $p_i = G(\mathring{p}_i)$ at the (singular) vertex.

More correctly, however, if the symmetry $G : \mathbb{C}^n \to \mathbb{C}^n$ fixes (only) the origin, $\{\mathbb{C}^n/G\}$ is singular at the origin—if $n > 1$. In each patch of a 1-fold, containing a 'singular' point, one re-defines $z \mapsto (z - z_i)^k$ where z_i are the 'singular' loci and k is a suitable integer. The resulting coordinate patches are smooth and the transition functions are holomorphic: the singularity was fake, i.e., removable.

In general, essentially the same argument shows that if a (finite!) symmetry $G\colon \mathcal{Y} \to \mathcal{Y}$ fixes (among others) a set $\mathcal{S} \subset \mathcal{Y}$ of codimension 1, then the quotient $\{\mathcal{Y}/G\}$ is not singular at $\mathcal{S}$. As we are mostly interested in the case $\dim \mathcal{Y} = 3$, we will have to consider only the cases when the fixed point sets (and so the singular sets in the quotient space) have dimension 0 and/or 1.

— ❦ —

More precisely, one finds the fixed point for each element $\mathbf{g}_\alpha \in G$. The union of these is the fixed point set of the symmetry group G and it may be decomposed, for $\dim \mathcal{Y} = 3$, as

$$\text{f.p.}(G : \mathcal{Y}) \;=\; \big\{\textstyle\bigcup_i p_i\big\} \,\cup\, \big\{\textstyle\bigcup_j \mathcal{C}_j\big\} , \tag{4.2.1}$$

where p_i are the isolated fixed points and $\mathcal{C}_j$ are the fixed curves. This decomposition is done in such a way that there is no overlap. To this end, (a) whenever two fixed curves intersect, $\mathcal{C}'_1 \cap \mathcal{C}'_2 = p$, the above decomposition lists p as a fixed point and $\mathcal{C}_1 \stackrel{\text{def}}{=} (\mathcal{C}'_1 \smallsetminus p)$ and $\mathcal{C}_2 \stackrel{\text{def}}{=} (\mathcal{C}'_2 \smallsetminus p)$ as two separate (punctured) curves. Similarly, (b) if a point q is fixed by $\mathbf{g}_\alpha \in G$ and lies on a curve $\mathcal{C}'_3$ which is fixed by $\mathbf{g}_\beta \in G$ for $\alpha \neq \beta$, the above decomposition lists q as a fixed point and $\mathcal{C}_3 \stackrel{\text{def}}{=} (\mathcal{C}'_3 \smallsetminus q)$ as a (punctured) fixed curve. Once such special points have been separated, the $\mathcal{C}_j$ are said to be of *pure dimension 1.*

In passing to the quotient $\mathcal{M} = \{\mathcal{Y}/G\}$, the disjoint union on the right hand side of Eq. (4.2.1) lists the irreducible components of the singular set

$$\text{Sing}(\mathcal{M}) \;=\; \big\{\textstyle\bigcup_i p_i\big\} \,\cup\, \big\{\textstyle\bigcup_j \mathcal{C}_j\big\} , \qquad G(p_i) = p_i , \quad G(\mathcal{C}_j) = \mathcal{C}_j . \tag{4.2.2}$$

The decomposition of the singularity in Eq. (4.2.2) is useful in determining the possible desingularization(s) of the quotient space $\mathcal{M}$. For isolated singular points, we study the singularity locally, choosing a suitable $\mathbb{C}^3$-like coordinate chart U_i which places the singular point p_i at the origin and contains no other singularity. For the singular curves $\mathcal{C}_j$, we choose a tubular neighborhood "built" from $\mathbb{C}^2$-like discs $D(x_j)$ such that x_j is the origin of $D(x_j)$ and x_j sweep out $\mathcal{C}_j$. Note that this would not be possible if the fixed point set (4.2.1) had not been disjoined as it was‡.

As it is, one may smooth each component of the singularity separately and then make sure that the resulting patches do fit together. Generally, such resolutions will be possible: If the fixed set contains a curve $\mathcal{C}'$ and a point p on it, and if $\mathcal{C}'-p$ is fixed by $\mathbf{g}_\alpha \in G$ while $\mathbf{g}_\beta \in G$ fixes p, then $\mathbf{g}_\beta = (\mathbf{g}_\alpha)^k$ for some integer k. This induces a relation between the possible resolutions of p and $\mathcal{C}'-p$, respectively, which can be used to patch everything back together.

4.2.2 Smoothings in general ...

The general idea, of course, is to replace the singular point $p \in \mathcal{M}$ with something bigger—the *exceptional set*, E. Such sets can typically be regarded as formal sums of non-contractible subspaces of the resolved space $\widehat{\mathcal{M}}$; that is, we smooth $\mathcal{M}$ into $\widehat{\mathcal{M}}$ by "stretching" the singular points into some bigger-than-zero dimensional exceptional sets.

Elementary logic dictates that, in a real 6-dimensional space, an isolated singular point may be smoothed essentially in three different ways: by replacing it with an exceptional set, E, of (real) dimension 2, 3 or 4.

The cases $\dim_{\mathbb{R}} E = 1$ and $\dim_{\mathbb{R}} E = 5$ are related through Poincaré duality. Introducing (real) 1-dimensional exceptional sets, which essentially must be non-contractible circles, would imply that $H_1(\widehat{\mathcal{M}})$—and therefore also $H^1(\widehat{\mathcal{M}})$—would become non-zero, the holonomy of $\widehat{\mathcal{M}}$ would have to be a proper subgroup of $SU(3)$, implying either $N=2$ or no supersymmetry, both of which do not seem to have physical application (see Chapter 0).

Now, consider the case when a $\mathcal{M}$ has a single singular point, p, which is smoothed by replacing it with some (real) 3-dimensional E. Simple local analysis shows that the non-compact manifolds $\mathcal{M} \smallsetminus p$ and $\widehat{\mathcal{M}} \smallsetminus E$ are distinct as complex manifolds, although they are diffeomorphic. Thus, smoothing in $\dim_{\mathbb{R}} E=3$ deforms the complex structure; we refer to this generally as *deformation*.

‡ The gastronomically inclined Reader will find this rather familiar: the preparation of many a culinary delight involves disjoining, separate and not infrequently rather elaborate treatment, and only then putting everything back together.

We remain with cases $\dim_{\mathbb{R}} E=2$ and $\dim_{\mathbb{R}} E=4$; that is $\dim_{\mathbb{C}} E=1$ and $\dim_{\mathbb{C}} E=2$, respectively—assuming that the exceptional sets are in fact complex subsets of the smoothed complex 3-dimensional manifold $\widehat{\mathcal{M}}$. Because of this, essentially, $\mathcal{M}\smallsetminus p$ and $\widehat{\mathcal{M}}\smallsetminus E$ are the same complex (non-compact) manifold. One says: *resolving in codimension 2 and 1*, respectively. Of course, a combination of these two is possible; as usual however, we discuss them separately. Quite generally, the $\dim_{\mathbb{C}} E=2$ case is called *blowing up*, while the $\dim_{\mathbb{C}} E=1$ case is called *small resolution*.

Blowing up and *small resolutions* of complex 3-folds share some similarities but also feature some important differences, as we will see below.

4.2.3 Divisors and line bundles

We have so far discussed (holomorphic) line bundles $\mathcal{L}$ over complex n-folds $\mathcal{X}$ and some elementary facts about $\mathcal{L}$-valued cohomology—especially sections of $\mathcal{L}$, i.e., elements of $H^0(\mathcal{X},\mathcal{L})$. Now we come to the homological analogue of $\mathcal{L}$.

Definition 4.1. *On a complex (not necessarily compact) n-fold $\mathcal{M}$, a locally finite linear combination $D \stackrel{\text{def}}{=} \sum_i a_i\mathcal{Y}_i$ of irreducible analytic hypersurfaces $\mathcal{Y}_i$ of $\mathcal{M}$ is called a* divisor.

An "analytic hypersurface" means that $\dim \mathcal{Y}_i = n-1$ and that every $\mathcal{Y}_i$ is (locally at every point) given as the zero-set of a single holomorphic function. "Locally finite" means that for any $p \in \mathcal{M}$, there exists a neighborhood of p meeting only a finite number of the $\mathcal{Y}_i$'s appearing in D; if $\mathcal{M}$ is compact, this simply means that the defining sum for D is finite. A divisor is *effective* if $a_i \geq 0$ for all i.

Now, every holomorphic function $g(x)$ which vanishes at a point $p \in \mathcal{Y}_i$ is locally a holomorphic multiple of (some power of) the defining function of $\mathcal{Y}_i$; at p, $g(x) = h(x)\cdot f^a(x)$, h holomorphic. The exponent a turns out to be independent of the point, characterizes $\mathcal{Y}_i$ and is called the *order of g along $\mathcal{Y}_i$*; $a = \mathrm{ord}_{\mathcal{Y}_i}(g)$. Clearly,

$$\mathrm{ord}_{\mathcal{Y}_i}(gh) = \mathrm{ord}_{\mathcal{Y}_i}(g) + \mathrm{ord}_{\mathcal{Y}_i}(h)\ . \tag{4.2.3}$$

Also, $\mathrm{ord}_{\mathcal{Y}_i}(g/h) = \mathrm{ord}_{\mathcal{Y}_i}(g) - \mathrm{ord}_{\mathcal{Y}_i}(h)$ for the meromorphic function g/h. With this, to each meromorphic function, $\mu(x)$, we can associate a divisor

$$\mu(x) = \frac{g(x)}{h(x)} \quad\Longleftrightarrow\quad (\mu) \stackrel{\text{def}}{=} \sum_{\mathcal{Y}\subset\mathcal{M}} \mathrm{ord}_{\mathcal{Y}}(\mu)\mathcal{Y}\ . \tag{4.2.4}$$

Similarly, $(\mu)_0 \stackrel{\text{def}}{=} \sum_{\mathcal{Y}\subset\mathcal{M}} \text{ord}_{\mathcal{Y}}(g)$ is the divisor of zeros and $(\mu)_\infty \stackrel{\text{def}}{=} \sum_{\mathcal{Y}\subset\mathcal{M}} \text{ord}_{\mathcal{Y}}(h)$ is the divisor of poles of μ; $(\mu) = (\mu)_0 - (\mu)_\infty$. Loosely then, divisors correspond to meromorphic sections of certain line bundles (more accurately, of rank-1 *sheaves*$^{(L)}$).

As a consequence of the definition, the set of divisors in $\mathcal{M}$ form an additive group, denoted $\text{Div}(\mathcal{M})$. Consider two simple divisors D_1 and D_2, each of which consists of a suitable irreducible hypersurface. Then D_i is the zero-set of a (perhaps meromorphic) section of some line bundle $\mathcal{L}_i$, $i = 1, 2$. The zero-set of the product section is the union of the two hypersurfaces, that is the divisor $D_1 + D_2$. The product section is, of course, a section of the tensor product line bundle $\mathcal{L}_1 \otimes \mathcal{L}_2$. This and the foregoing discussion of (μ) indicates the correspondence

$$\begin{array}{ccc} \text{Div}(\mathcal{M}) \stackrel{\text{def}}{=} & (\{\,\text{divisors in } \mathcal{M}\,\}, +) \\ \updownarrow & \updownarrow \\ \text{Pic}(\mathcal{M}) \stackrel{\text{def}}{=} & (\{\,\text{line bundles over } \mathcal{M}\,\}, \otimes). \end{array} \tag{4.2.5}$$

This correspondence is in fact an isomorphism, as can be seen from the more precise description in Ref. [23], p. 129–134. Suffice it here just to note: $[D]$ stands for the line bundle associated to the divisor D; two divisors are *equivalent*, $D \sim D'$, if $[D] = [D']$. This permits us to freely toggle between line bundle and divisor parlance.

For any complex n-fold $\mathcal{X}$, the canonical line bundle $\mathcal{K}_{\mathcal{X}} = \det \mathcal{T}^*_{\mathcal{X}}$ comes for free and associated to it is the *canonical divisor* $K_{\mathcal{X}}$. In view of the above identification, these are often used interchangeably.

4.3 Blowing Up

This is a very general process, in which a point p in a complex n-fold $\mathcal{X}$ is replaced by a (complex) $n-1$-dimensional exceptional set E. The result of blowing up, the *blowup* $\widetilde{\mathcal{X}}$, is such that $\mathcal{X} \smallsetminus p$ and $\widetilde{\mathcal{X}} \smallsetminus E$ are equal as complex manifolds. One writes $\sigma : \widetilde{\mathcal{X}} \to \mathcal{X}$ for the process indicating that σ is a proper holomorphic map except at $E \subset \widetilde{\mathcal{X}}$, which is being collapsed to the point $p \in X$.

4.3.1 The template

As a simple example which will be useful to keep in mind, we recall the complex projective space. As discussed in § 3.3.4, over $\mathbb{P}^n$ we have the universal line bundle

$$\mathcal{O}_{\mathbb{P}^n}(-1) = \mathcal{L} \stackrel{\text{def}}{=} \{\, (x, v) \in \mathbb{P}^n \times \mathbb{C}^{n+1} \;:\; v^{[i}x^{j]} = 0 \,\} \tag{4.3.1}$$

where $(v^0, \ldots, v^n) \in \mathbb{C}^{n+1}$, $(x^0 : \ldots : x^n) \in \mathbb{P}^n$. The right hand side of the defining equality specifies the total space of $\mathcal{L}$, which is isomorphic to $\mathbb{C}^{n+1}$ *away* from the origin. At the origin, however, the $v^{[i}x^{j]} = 0$ conditions places no restriction on the x's, so that $(0, \ldots, 0) \in \mathbb{C}^{n+1}$ is replaced by all of $\mathbb{P}^n$. The origin of $\mathbb{C}^{n+1}$ is said to have been blown up; the total space of the universal line bundle $\mathcal{L}$ is the blowup of $\mathbb{C}^{n+1}$ at the origin and is denoted—following the general notation—by $\widetilde{\mathbb{C}^{n+1}}$. The copy of the $\mathbb{P}^n$ which replaces the origin is the *exceptional divisor* of the blowup.

Note that the origin of $\mathbb{C}^{n+1}$ was a smooth point. So, by blowing up a smooth point[§] p in an n-fold $\mathcal{X}$, we mean considering a $\mathbb{C}^n$-like neighborhood centered at p and replacing it with $\widetilde{\mathbb{C}^n}$, the total space of $\mathcal{O}_{\mathbb{P}^{n-1}}(-1)$. With this in mind, blowing up is essentially a local operation—indeed, one of the very few surgical techniques available in the holomorphic category.

We note here without proof that

Theorem 4.2. *If $\sigma : \widetilde{\mathcal{X}_p} \to \mathcal{X}$ is the blowup of the n-fold $\mathcal{X}$ at a smooth point $p \in \mathcal{X}$ and $E \subset \widetilde{\mathcal{X}_p}$ the resulting exceptional divisor, then*

$$\mathcal{K}_{\widetilde{\mathcal{X}_p}} \;=\; \sigma^*\mathcal{K}_{\mathcal{X}} \otimes [E]^{\otimes(n-1)}\,. \tag{4.3.2}$$

Here we use the multiplicative notation as appropriate for line bundles; this statement is often seen written in the additive notation, akin to divisors (see Ref. [23], p. 187, to which we also refer for a proof).

Now, consider the total space of the kth tensor power of $\mathcal{L}$. In terms of coordinate patches and for simplicity for $\mathcal{O}_{\mathbb{P}^1}(-1)$ *vs.* $\mathcal{O}_{\mathbb{P}^1}(-k)$,

$\mathbb{P}^1$	$\mathcal{O}_{\mathbb{P}^1}(-1)$		$\mathcal{O}_{\mathbb{P}^1}(-k)$	
patch	**base**	**fiber**	**base**	**fiber**
$(x^0 \neq 0)$:	$\xi_{(0)} = x^1/x^0$	$V_{(0)} = v\,x_0$	$\xi_{(0)} = x^1/x^0$	$W_{(0)} = w\,(x_0)^k$
$(x^1 \neq 0)$:	$\xi_{(1)} = x^0/x^1$	$V_{(1)} = v\,x_1$	$\xi_{(1)} = x^0/x^1$	$W_{(1)} = w\,(x_1)^k$

(4.3.3)

Comparing the two bundles, we see that we can go from $\mathcal{O}_{\mathbb{P}^1}(-1)$ to $\mathcal{O}_{\mathbb{P}^1}(-k)$ simply by replacing $V_{(i)} \mapsto {V_{(i)}}^k = v^k(x_i)^k$. For ${V_{(i)}}^k = W_{(i)}$

[§] § 3.1 shows that—except for the Hirzebruch surfaces—almost all del Pezzo surfaces are blowups of $\mathbb{P}^2$ at a number of smooth points; hence the exceptional divisors are $E_i \approx \mathbb{P}^1$.

and so $v^k = w$ to be 1–1, we have to identify $(V_{(0)}, V_{(1)})$ with $(e^{(2i\pi/k)}V_{(0)}, e^{(2i\pi/k)}V_{(1)})$ away from the origin of $\mathbb{C}^2$, that is, away from $\mathbb{P}^1 \subset \mathcal{O}_{\mathbb{P}^1}(-k)$. In other words, away from the "origin," the total space of $\mathcal{O}_{\mathbb{P}^1}(-k)$ is isomorphic to $\{\mathbb{C}^2/\mathbb{Z}_k\}$ rather than $\mathbb{C}^2$.

More generally, away from the "origin," the total space of $\mathcal{O}_{\mathbb{P}^{n-1}}(-k)$ is isomorphic to $\{\mathbb{C}^n/\mathbb{Z}_k\}$ rather than $\mathbb{C}^n$. Therefore, a $\{\mathbb{C}^n/\mathbb{Z}_k\}$-like singular neighborhood in an n-fold $\mathcal{X}$ can be replaced by the smooth total space of $\mathcal{O}_{\mathbb{P}^{n-1}}(-k)$ and thereby the $\mathbb{Z}_k$ quotient singularity at the origin is replaced by a $\mathbb{P}^{n-1}$.

Of course, the result of the above theorem needs to be corrected now; it is not hard to see that the canonical bundle changes into

$$\mathcal{K}_{\widetilde{(\mathcal{X}/\mathbb{Z}_k)_p}} = \sigma^*\mathcal{K}_{\mathcal{X}} \otimes [E]^{\otimes(n-k)} . \tag{4.3.4}$$

This is an example of holomorphic local surgery. However, this *local* process induces *global* changes in that the canonical (global, holomorphic) line bundle, that is, the first Chern class is being altered. Also, this affects a number of numerical characteristics such as the Euler characteristic, the Hodge numbers...

So if we started from a Calabi–Yau space $\mathcal{Y}$, where $\mathcal{K}_{\mathcal{Y}} \approx \mathbb{C}$, Eq. (4.3.4) indicates that each $\mathbb{Z}_3$ quotient singularity in $\mathcal{M} = \{\mathcal{Y}/\mathbb{Z}_3\}$ can be blown up yielding $\mathcal{K}_{\widetilde{\mathcal{M}}} \approx \mathbb{C}$. So, one might ask what can one do with other $\mathbb{Z}_k$ singularities?

4.3.2 Blowing up quotient singular curves

In the case $\mathbb{C}^2/G$, the situation is quite simple. Let G be a finite subgroup of $SU(2)$ with the action that fixes only the origin. The quotient singularities $\{\mathbb{C}^2/G\}$ can then all be resolved by successive blowups.

A detailed analysis would take us far afield; instead, the interested Reader is advised to consult Ref. [1] as a general reference. Suffice it here to describe briefly the case of abelian $G \approx \mathbb{Z}_r$ when the technique of so-called *toric resolutions* is applicable [42, 20]. In suitable local coordinates, consider the action

$$\mathbf{g} : (x, y) \longmapsto (\omega^a x, \omega^b y) , \qquad \omega \stackrel{\text{def}}{=} e^{(2i\pi/r)} . \tag{4.3.5}$$

It is useful to define the lattice of exponents invariant under this action of $\mathbb{Z}_r$:

$$\mathfrak{C}_{\mathbf{g}} \stackrel{\text{def}}{=} \{(\xi, \eta) \in \mathbb{Z}^2 : a\xi + b\eta \equiv 0 \,(\mathrm{mod}\, r)\} . \tag{4.3.6}$$

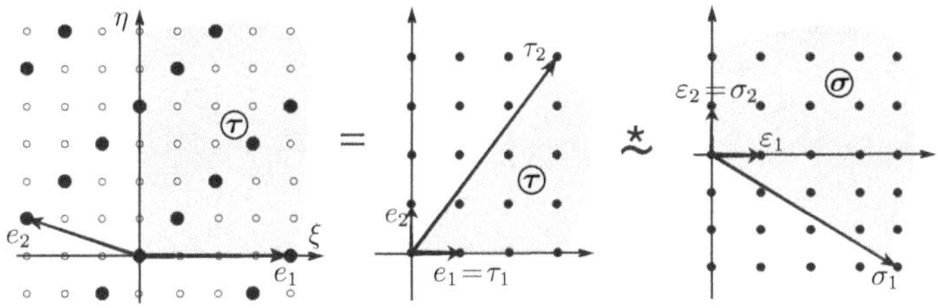

Figure 4.1: The invariant lattice of exponents $\mathfrak{E}_{\mathbf{g}}$, the cone of positive exponents, $\boldsymbol{\tau}$ and its dual, $\boldsymbol{\sigma}$.

Without loss of generality, we may take $(a,b) = (1, r{-}1)$, so that $\xi \equiv \eta \,(\mathrm{mod}\, r)$. A suitable integral basis in $\mathfrak{E}_{\mathbf{g}}$ is given by

$$e_1 = (r, 0) \ , \quad e_2 = (1{-}r, 1) \ , \tag{4.3.7}$$

and sketched in Figure 4.1. In the basis (e_1, e_2), the real cone $\boldsymbol{\tau}$, which "fills" the integral cone of positive exponents, has an opening smaller than $\frac{\pi}{2}$. The real cone $\boldsymbol{\tau}$ may be taken to be spanned by the vectors $\tau_1 = (1,0)$ and $\tau_2 = (r{-}1, r)$, given in terms of the basis elements e_i; write $\boldsymbol{\tau} = \left[\begin{smallmatrix} 1 & 0 \\ r-1 & r \end{smallmatrix}\right]$.

The dual cone is defined as

$$\boldsymbol{\sigma} \stackrel{\text{def}}{=} \{a_1\varepsilon_1 + a_2\varepsilon_2 : \ a_i \in \mathbb{R}, \ a_i \geq 0 \ \forall i\} \ , \tag{4.3.8}$$

has an opening greater than $\frac{\pi}{2}$ and may be taken to be spanned by the vectors $\sigma_1 = \left[\begin{smallmatrix} r \\ 1-r \end{smallmatrix}\right]$ and $\sigma_2 = \left[\begin{smallmatrix} 0 \\ 1 \end{smallmatrix}\right]$; write $\boldsymbol{\sigma} = \left[\begin{smallmatrix} r & 0 \\ 1-r & 1 \end{smallmatrix}\right]$. Clearly, $\boldsymbol{\sigma} = r\,\boldsymbol{\tau}^{-1}$.

Given a collection of dual cones $\boldsymbol{\sigma}_i$, one for each coordinate patch, it is possible to reconstruct the space the singularities of which these cones describe [42, 20]. The fact that $\det(\boldsymbol{\tau}) = \det(\boldsymbol{\sigma}) = r$ indicates that these cones describe a single coordinate patch with a $\mathbb{Z}_r$ quotient singularity at the origin. Therefore, to obtain a smooth model we simply need to divide $\boldsymbol{\sigma}$ into several subcones $\boldsymbol{\sigma}_k$, in such a way that $\det(\boldsymbol{\sigma}_k) = 1$, for each i. To that end, for $k = 1, \dots, r$, we define

$$\begin{aligned} r\boldsymbol{\sigma}_k &\stackrel{\text{def}}{=} k \begin{bmatrix} r & 0 \\ (1{-}r) + (r{-}k) & 0 \end{bmatrix} + (k{-}1) \begin{bmatrix} 0 & r \\ 0 & (1{-}r) + (r{-}k) \end{bmatrix} \\ &= r \begin{bmatrix} k & (k{-}1) \\ (1{-}k) & (2{-}k) \end{bmatrix} \ . \end{aligned} \tag{4.3.9}$$

$\det(\boldsymbol{\sigma}_k) = 1$ follows straightforwardly, so the $r{-}1$ new vectors $\vec{s}_k \stackrel{\text{def}}{=} \left[\begin{smallmatrix} k \\ 1-k \end{smallmatrix}\right]$, for $k = 1, \dots, (r{-}1)$, divide the $\boldsymbol{\sigma}$-cone into r subcones each of which corresponds to a smooth coordinate patch. Accordingly, the set $\{\vec{s}_k\}_{k=1,\dots,(r-1)}$

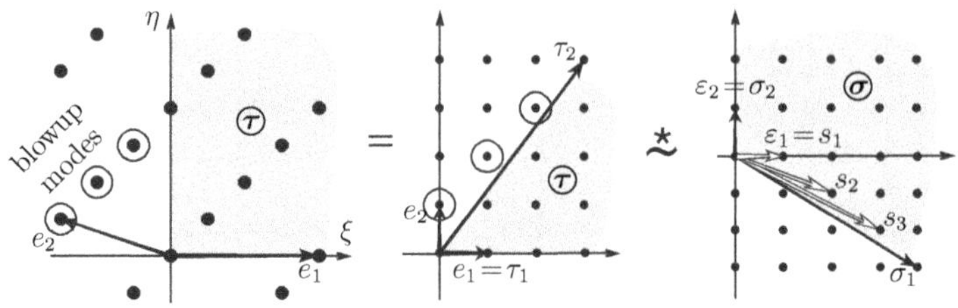

Figure 4.2: Three (toric) renditions of the blowup of $\mathbb{C}^2/\mathbb{Z}_r$.

corresponds to the exceptional set, which has $r-1$ components: $E \stackrel{\text{def}}{=} \{E_k\}_{k=1,\ldots,(r-1)}$, $E_k \approx \mathbb{P}^1$. For a $\mathbb{Z}_r$-fixed curve $\mathcal{C}$ in a 3-fold, we blow up each point in the $\mathbb{C}^2$-like transversal neighborhood. Remember, however, that the whole exceptional set is then $E \times \mathcal{C}$.

Pulling the $\vec{s}_k$ back to the original integral exponent lattice $\mathfrak{E}_\mathbf{g}$, we see that the exceptional divisors E_k correspond to $\mathbb{Z}_r$-invariant rational monomials $\{(x^{k-r}y^k)\}_{k=1,\ldots,(r-1)}$. Note the similarity to the parametrization used in the proof of Proposition 3.2, concerning blowups of $\mathbb{P}^2$. Figure 4.2 shows the exceptional set E from three different (toric) vantage points.

Let us just note that the intersection matrix (E_i, E_j) equals minus the Cartan matrix of the A_{r-1} Lie algebra (the Coxeter element of the Weyl group of A_{r-1} generates a $\mathbb{Z}_r$). In fact, the analogous is true also of other discrete groups: the exceptional lines in the blowup of $\mathbb{C}^2/G$ are related to the D_r, E_6, E_7 and E_8 Lie algebra for G the dihedral, tetrahedral, cubic and icosahedral groups, respectively [1].

4.3.3 Blowing up isolated quotient singular points

Even a semi-comprehensive treatment of isolated 3-fold quotient singularities is quite beyond our scope; the Reader is referred to Ref. [46]. We will just sketch the blowing up of $\mathbb{C}^3/G$ for abelian G, where again toric methods prove effective [42, 20, 206, 104, 189, 57].

The analysis is very similar to that of the $\mathbb{C}^2/G$ case. Given the action

$$\mathbf{g} \;:\; (x,y,z) \;\longmapsto\; (\omega^a x, \omega^b y, \omega^c z)\;, \qquad \omega \stackrel{\text{def}}{=} e^{(2i\pi/r)}\;, \tag{4.3.10}$$

we start from the integral exponent lattice

$$\mathfrak{E}_\mathbf{g} \stackrel{\text{def}}{=} \{(\xi,\eta,\zeta) \in \mathbb{Z}^3 : a\xi + b\eta + c\zeta \equiv 0\,(\text{mod}\, r)\}\;. \tag{4.3.11}$$

Again, without loss of generality, we take $(a, b, c) = (1, b, c)$ and end up with

$$\boldsymbol{\sigma} \stackrel{\text{def}}{=} \begin{bmatrix} r & 0 & 0 \\ -b & 1 & 0 \\ -c & 0 & 1 \end{bmatrix} . \tag{4.3.12}$$

$\boldsymbol{\sigma}$ is now a 3-dimensional triangular cone in $\mathbb{R}^3$, which we may conveniently represent by its base. As in the 2-dimensional case, $\det(\boldsymbol{\sigma}) = r$ indicates that $\boldsymbol{\sigma}$ describes a single coordinate patch with a $\mathbb{Z}_r$ quotient singularity at the origin. The division of $\boldsymbol{s}$ into subcones $\boldsymbol{s}_k$ with $\det(\boldsymbol{\sigma}_k) = 1$ is then conveniently described by a sub-triangulation of the triangular base of $\boldsymbol{\sigma}$. Generalizing Eq. (4.3.9), the new vertices of this sub-triangulation are found in the form

$$\vec{s}_k(\alpha_k, \beta_k, \gamma_k) \stackrel{\text{def}}{=} \frac{1}{\alpha_k+\beta_k+\gamma_k} \left(\alpha_k \begin{bmatrix} r \\ -b \\ -c \end{bmatrix} + \beta_k \begin{bmatrix} 0 \\ 1 \\ 0 \end{bmatrix} + \gamma_k \begin{bmatrix} 0 \\ 0 \\ 1 \end{bmatrix} \right) , \tag{4.3.13}$$

such that $\alpha_k, \beta_k, \gamma_k$ and also the components of $\vec{s}_k$ are integral (the aid of a computer is quite welcome here). The loci of $\vec{s}_k$ in the base of the cone specify the vertices of the desired sub-triangulation. As in the 2-dimensional case, these vectors (vertices) correspond to exceptional divisors E_k (codimension-1).

To complete the sub-triangulation, we introduce edges which join the vertices (representing plaquettes). Now, the plaquette spanned between $\vec{s}_k$ and $\vec{s}_\ell$ represents the intersection $E_k \cap E_\ell$. These exceptional curves (codimension-2) are a novel feature of the 3-dimensional case.

Generally, if $C \subset \mathcal{M}$ is a curve in a 3-fold, then $\det(\mathcal{N}_{\mathcal{M}/C}) = \mathcal{K}_{\mathcal{M}}|_C \otimes \mathcal{K}_C^*$ and so if $\mathcal{M}$ is Calabi–Yau, $\det(\mathcal{N}_{\mathcal{M}/C}) = \mathcal{K}_C^* = \mathcal{T}_C^*$. The latter equality holds since $\dim C = 1$. Now

Theorem 4.3 (Grothendieck). *All holomorphic vector bundles over $\mathbb{P}^1$ are isomorphic to the direct sum of line bundles $\bigoplus_i \mathcal{O}_{\mathbb{P}^1}(\ell_i)$.*

This in fact is the only (non-trivial) case when the splitting principle (1.3.19) is valid also holomorphically. The pattern $(\ell_1, \ldots, \ell_k)$ is called the *splitting type* of the rank-k vector bundle over $\mathbb{P}^1$.

For the exceptional lines $C \approx \mathbb{P}^1$ occurring in toric resolutions, the splitting type of the normal bundle $\mathcal{N}_{\widetilde{\mathcal{M}}/C}$ can be determined from the "sub-fan" (the adjoining of central 3-dimensional cones over $\langle \vec{s}_1, \vec{s}_2, \vec{s}_3 \rangle$ and

$\langle \vec{s}_2, \vec{s}_1, \vec{s}_3 \rangle)$

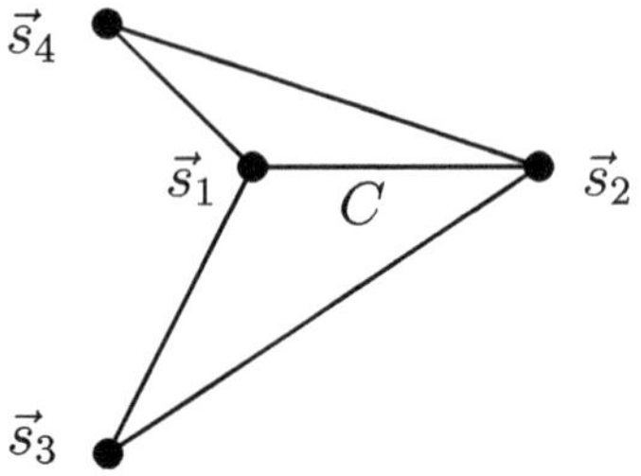

(4.3.14)

We have that the ℓ_i in $\mathcal{N}_{\widetilde{\mathcal{M}}/C} = \mathcal{O}_C(\ell_1) \oplus \mathcal{O}_C(\ell_2)$ satisfy the (overdetermined) vectorial equation

$$\ell_1 \vec{s}_1 + \ell_2 \vec{s}_2 + \vec{s}_3 + \vec{s}_4 = 0 \ . \tag{4.3.15}$$

Now, if our $C \subset \widetilde{\mathcal{M}}$ is a line and $\widetilde{\mathcal{M}}$ is Calabi–Yau, then also $\ell_2 = \ell_1 - 2$, since for $C \approx \mathbb{P}^1$, $\mathcal{T}^*_{\mathbb{P}^1} = \mathcal{O}_{\mathbb{P}^1}(2)$.

— ❦ —

Blowing up, in general, introduces and exceptional set which contains both surfaces (codimension-1) and curves (codimension-2). Thus, we see that it affects both the canonical divisor—the effect of which is given in Eq. (4.3.4)—and the Kähler class. The latter happens because for a harmonic $(1,1)$-form J to be a Kähler form, we need its positivity, i.e., that

$$\int_{\mathcal{C}} J > 0 \ , \qquad \int_{\mathcal{S}} J^2 > 0 \ , \qquad \int_{\mathcal{M}} J^3 > 0 \ , \tag{4.3.16}$$

for all curves $\mathcal{C}$ and all surfaces $\mathcal{S}$ in the manifold $\mathcal{M}$. From another point of view, recall that

$$H_4(\mathcal{M},\mathbb{Z}) \overset{*}{\approx} H_2(\mathcal{M},\mathbb{Z}) \overset{*}{\approx} H^2(\mathcal{M},\mathbb{Z}) \quad \Rightarrow \quad H_4(\mathcal{M},\mathbb{Z}) \approx H^2(\mathcal{M},\mathbb{Z}) \tag{4.3.17}$$

(see *duality*$^{(L)}$) and that $H^2(\mathcal{M},\mathbb{Z})\otimes\mathbb{C} \approx H^{1,1}(\mathcal{M})$ since there are no (2,0)- or (0,2)-forms. Since blowing up introduces new elements in H_4, these reflect in new harmonic (1,1)-forms and thereby lead to new variations of the Kähler class. So, one naturally inquires if the toric resolutions of singularities can distort the Kähler form 'beyond repair,' so that the resulting space—although smooth and Calabi–Yau—may no longer admit any Kähler form.

In fact, given the triangular base and the 'new' vertices $\vec{s}_k$, there may well be more than one sub-triangulation. All of these are valid resolutions of the singularity, but not all of them may turn out to be Kähler. (Note that this is *not* a local issue; each coordinate patch may support a Kähler form, but it may be that these cannot be extended throughout the compact manifold.) For the case of a simple blowup, $\sigma : \widetilde{\mathcal{X}}_p \to \mathcal{X}$, where the exceptional set is a single $\mathbb{P}^2$, it is straightforward to show that the blown up space is again Kähler (see Section II.7 of Ref. [27]). Moreover, the Kähler form on $\widetilde{\mathcal{X}}_p$ can be obtained from that on $\mathcal{X} \smallsetminus p$ and the Eguchi–Hanson metric [19] on the exceptional $\mathbb{P}^2$. It then follows that a toric resolution leads to a Kähler manifold if it is a sequence of blowups. In toric terms, the resolution is Kähler if the sub-triangulation of the triangular base of $\boldsymbol{\sigma}$ can be obtained by successive sub-triangulations for each of the $\vec{s}_k$, one at a time.

4.4 Small Resolution

Unlike blowing up and toric resolutions, small resolutions always occur purely in codimension-2 and are thus a novelty with respect to the (algebraic or otherwise) geometry of curves and surfaces.

Consider a pair of 3-folds, $\widehat{\mathcal{X}}$ and $\mathcal{X}$, such that $\mathcal{X}$ has some isolated singular points p_i and $\widehat{\mathcal{X}}$ is smooth and is a small resolution of $\mathcal{X}$. Let E_i denote the (complex 1-dimensional!) exceptional set introduced to smooth out the p_i. As usual then, the non-compact complex manifolds $(\mathcal{X} \smallsetminus \{p_i\})$ and $(\widehat{\mathcal{X} \smallsetminus \{E_i\}})$ are equal. This means that the canonical divisor $K_{\mathcal{X}}$ (complex 2-dimensional!) of $\mathcal{X}$ equals $K_{\widehat{\mathcal{X}}}$. Thus, small resolutions do not change c_1 and if we started from a singular Calabi–Yau space, its small resolution is also Calabi–Yau.

The language of divisors allows us to conveniently generalize the notion of Ricci-flatness to singular spaces. While the condition $c_1 = 0$ is independent of the specific choice of the metric, and the triviality of the canonical bundle refers to no metric at all, these concepts make sense only if the underlying manifold is smooth. The triviality of the canonical divisor, however, allows the underlying space to be singular, although not very pathologically so. It is at present quite unclear just how singular spaces one should be prepared to consider—neither from the mathematical, nor from the physical point of view. Certainly, however, hyperquotient singularities which are essentially quotient and hypersurface singularities perhaps atop each other [46] should be allowed from the physical point of view [170]; see also § F.7. This is the rationale for the present selection of topics regarding singularities.

One of the simplest occasions for small resolutions is not a quotient singularity, but a *hypersurface singularity*. That is, we describe the singular point by specifying its local neighborhood as the solution of a single constraint in $\mathbb{C}^4$. Of course, for the resulting space to be singular, we require the four gradients to have a common zero—and precisely one—if we discuss isolated hypersurface singular points. By a simple change of coordinates, the singular point is typically arranged to be at the origin.

4.4.1 A node: the complex story

The simplest isolated hypersurface singular point in a 3-fold is called a *node*, or a *double point*. It occurs at the vertex of the cone in $\mathbb{C}^4$, defined by the simple equation $xy = zt$.

Let $\psi \stackrel{\text{def}}{=} (xy - zt)$ be the defining polynomial of the cone containing the node. Clearly, $\mathrm{d}\psi = 0$ precisely at the origin, were the node is. To obtain the (complex) base of this cone, we simply projectivize, interpreting $(x{:}y{:}z{:}t) \in \mathbb{P}^3$. $\psi = 0$ is then a non-degenerate quadric in $\mathbb{P}^3$, since the only degenerate point, $(0,0,0,0) \in \mathbb{C}^4$ is not in $\mathbb{P}^3$. As discussed above, a smooth quadric in $\mathbb{P}^3$ equals $\mathbb{P}^1 \times \mathbb{P}^1$ and the cone has $\mathbb{C}^1$-like generators sweeping over a $\mathbb{P}^1 \times \mathbb{P}^1$ base.

It should be obvious that the node can be resolved most simply by replacing it with a copy of a $\mathbb{P}^1$ [2]. This happens entirely in codimension-2, as promised. Since the base is a product of two $\mathbb{P}^1$'s, we have a topological ambiguity in this process; typically, we will deal with 3-folds with many nodes and at each of them this ambiguity shows up. Thus, in principle, a 3-fold with N nodes has 2^N distinct small resolutions.

It is sometimes useful to think of the small resolution as a halfway blowup. Indeed, in the present case, if we blew the vertex of the quadratic cone up, the exceptional divisor would have been the $\mathbf{F}_0 = \mathbb{P}^1 \times \mathbb{P}^1$ surface. One or the other small resolution is then obtained by collapsing one of the two $\mathbb{P}^1$'s in $\mathbf{F}_0$. This situation sometimes occurs in the course of toric resolutions and is then described in the diagram in Figure 4.3. The process by which we pass from one small resolution of a node to the blowup and then to the other small resolution is called a *flop* and is indicated in the diagram.

4.4.2 A node: the real story

Alternatively, we may change coordinates into $\psi = \sum_{A=1}^{4} (\zeta^A)^2$, think of ζ^A as a four-vector and write this equation as $\vec{\zeta}^2 = 0$. To find the base of this real cone, separate $\vec{\zeta}$ into its real and imaginary parts $\vec{\zeta} = \vec{\xi} + i\vec{\eta}$,

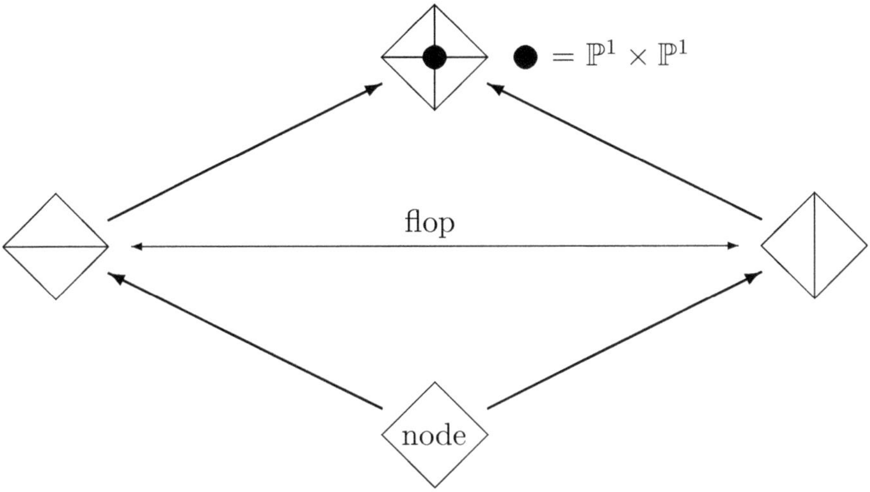

Figure 4.3: The blowup and two distinct small resolutions of a node.

whence $\psi = 0$ becomes the pair

$$\vec{\xi}^2 - \vec{\eta}^2 = 0 , \qquad \vec{\xi} \cdot \vec{\eta} = 0 . \tag{4.4.1}$$

The base is the intersection of the cone with a 7-sphere centered at the origin of $\mathbb{C}^4 \approx \mathbb{R}^8$. We take the sphere to have radius $r\sqrt{2}$ so that $\vec{\xi}^2 + \vec{\eta}^2 = 2r^2$. The base is thus described by the equations

$$\vec{\xi}^2 = \vec{\eta}^2 = r^2 , \qquad \vec{\xi} \cdot \vec{\eta} = 0 . \tag{4.4.2}$$

The space of ξ's is an S^3. For each point of this S^3, say $\vec{\xi} = (r, 0, 0, 0)$, $\vec{\xi} \cdot \vec{\eta} = 0$ restricts $\vec{\eta}$ to be the 3-vector $\vec{\eta} = (0, \eta^2, \eta^3, \eta^4)$. $\vec{\eta}^2 = r^2$ then defines an S^2. Thus the base of the cone a is fiber bundle with base S^3 and fiber S^2. Now, all bundles over S^3 are trivial¶ so the base must in fact be the product $S^2 \times S^3$; the generators are copies of the real semi-axis, $\mathbb{R}_{\geq 0}$.

> The fact that the base of the real cone with a node at the vertex has a base which is a product $S^2 \times S^3$ can be related to the fact that S^3 is parallelizable. Consequently, the same is true only in two more cases: a nodal 1-fold and a nodal 7-fold. In the former case, the base is $S^0 \times S^1$, that is, two circles—as is well known. The complex 7-dimensional nodal cone has $S^6 \times S^7$ for its real base.

¶ To see this, choose two B^3-like neighborhoods to cover S^3. Their overlap is a "thick S^2," that is $S^2 \times \mathrm{I}$ and this is where the transition functions of any bundle have to live. Now if G is the structure group of the vector bundle, the transition functions map $(S^2 \times \mathrm{I}) \to G$ and are therefore elements of $\pi_2(G)$. Finally, it is a classic result that $\pi_2(G) = 0$ for any Lie group G [12].

Thus, besides replacing the vertex of the cone with an S^2, we can also replace it with an S^3. As discussed earlier, this amounts to deformation as the exceptional set is (real) 3-dimensional. Since there are two $\mathbb{P}^1$'s in the base of the complex cone, one may inquire if there are in fact two different choices of S^3's also, fibering the phase of the generator of the complex cone over one or the other $\mathbb{P}^1$. It is straightforward to show that these two S^3's can in fact be rotated nicely one into another, so that there is no topological ambiguity in the deformation of a node [92]. We will return to this below.

4.4.3 Other small-resolvable singularities

Without much ado, Table 4.1 just lists the generalizations of the nodal defining polynomial and their small resolutions. These again correspond to a list of (some) Lie algebras in the sense that their small resolution involves a chain of $\mathbb{P}^1$'s which touch one another as described by the Dynkin diagram of the corresponding Lie algebra. Of course, a node corresponds to A_1. All these sport the same (in)affection to the canonical divisor: K_X is completely untouched in the process of small resolution.

The Kähler class is however disturbed by small resolutions. In fact, if X has a single node, then the small resolution cannot be Kähler. To see this, recall that the base of the real cone with the node at the apex is $S^3 \times S^2$. Now the $\mathbb{R}_{\geq 0}$-like generators of the cone and the S^2 from the base form a (real) B^3-like cap, which is completed elsewhere in the manifold into a 3-cycle—in fact, precisely the 3-cycle, $c^{(3)}$, which is dual to the S^3 in the base of the cone (✎ prove this ❧). When the apex of the cone is replaced by a $E \approx \mathbb{P}^1 = S^2$, the 3-cycle $c^{(3)}$ has been punctured and has become

Table 4.1: Simple hypersurface singularities and their small resolutions.

Defining polynomial	**Lie algebra**	**Small resolution**
$x^{k+1} + y^2 + z^2 + t^2$	A_k	
$x^{k-1} + xy^2 + z^2 + t^2$	D_k	
$x^4 + y^3 + z^2 + t^2$	E_6	
$x^3y + y^3 + z^2 + t^2$	E_7	
$x^5 + y^3 + z^2 + t^2$	E_8	

a 3-chain, with the exceptional $\mathbb{P}^1$ as the boundary. Now any candidate Kähler form, J, would have to satisfy the positivity conditions (4.3.16). However, $\int_E J$ would have to vanish by Stokes' theorem, because E is a boundary.

If there are several nodes and they are replaced by S^2's, these S^2's are components of the boundary of at least one such 3-chain. Now Stokes' theorem requires that the integral of J vanishes over the linear combination of S^2's which constitute the boundary of each such 3-chain. But in this boundary, the S^2 may occur with just such relative orientations that the total integral vanishes although the separate contributions do not. In such a case, the candidate J is not obstructed to be a Kähler form. It is therefore certain that, amongst 2^N small resolutions of a 3-fold with N nodes, only very few *may* admit a Kähler form and it is in general difficult to determine which are—if any.

— ❦ —

We refer to Ref. [1] and Ref. [46] for the general analysis of these singularities and details of their resolutions; see also [2, 382, 465]. Suffice it here just to note that they are of the more general type called *compound Du Val singularities* [(L)], written *cDV*. The general cDV defining equation is

$$F(x,y,z,t) \quad = \quad f(x,y,z) \; + t\, g(x,y,z,t)\ , \tag{4.4.3}$$

where $f(x,y,z) = 0$ defines the Du Val singularities in 2-folds and $g(x,y,z,t)$ is an arbitrary (holomorphic) polynomial. The list of Du Val polynomials $f(x,y,z)$ may be read off the Table 4.1, upon setting $t = 0$. For Du Val singularities in 2-folds, the diagrams in Table 4.1 denote the respective blowups. Thus, cDV singularities in 3-folds may be viewed as a one-parameter (t) deformation of the Du Val singularities. Alternatively, a cDV singularity $p^\sharp$ appears as a Du Val singularity in the hyperplane section through p. Because of the relation to the Lie algebras of the A, D and E type, the cDV singularities are also called ADE-singularities.

4.5 Numerical Characteristics

It is now not hard to determine the Euler characteristic of a resolution of a quotient $\mathcal{M} = \{\mathcal{Y}/G\}$, where the symmetry $G : \mathcal{Y} \to \mathcal{Y}$ has a fixed set, $F \subset \mathcal{Y}$:

$$\chi_{E}(\widetilde{\mathcal{M}}) \quad = \quad \frac{\chi_{E}(\mathcal{Y}) - \chi_{E}(F')}{|G|} + \chi_{E}(E)\ , \tag{4.5.1}$$

where $\widetilde{\mathcal{M}}$ is the blowup or small resolution of $\mathcal{M}$ (as appropriate), E is the exceptional set into which F has been resolved and F' is the fixed point set, taken however with some multiplicities. More precisely, let $\mathbf{g}_i$ denote the element of G which fixes the component $F_i \subset F$ and let $|\mathbf{g}_i|$ denote the order of $\mathbf{g}_i$ (which must divide $|G|$ or be equal to it). Then

$$\chi_E(F') = \sum_i \frac{|G|}{|\mathbf{g}_i|} \chi_E(F_i) . \tag{4.5.2}$$

Another rendition of the above formula is then

$$\chi_E(\widetilde{\mathcal{M}}) = \frac{\chi_E(\mathcal{Y})}{|G|} - \sum_i \frac{\chi_E(F_i)}{|\mathbf{g}_i|} + \sum_k \chi_E(E_k) , \tag{4.5.3}$$

where the sum on i runs over all irreducible components of the fixed point set (see Eq. (4.2.1)) and k labels the irreducible components of the exceptional set.

As another important issue, it should be clear that all resolutions considered in this chapter leave the structure group as it was. This is not hard to show; one has to analyze the multiple connectedness of the neighborhood were the singular point has been cut out and the also the structure group of the exceptional divisor used for smoothing. In all the cases considered above, no change is being introduced to π_1. For explicit calculations to this end in some models, we refer to Ref. [189].

With the detailed information presented in the analysis of toric resolutions or small resolution, we have the complete information on new 2-cycles and new 4-cycles which have been introduced in the smoothing process. This, in principle, provides the dimensions of $H_2(\mathcal{M})$ and $H_4(\mathcal{M})$, and therefore also $h^{1,1}$ and $h^{2,1}$. One only has to be cautious to take into account various homology relations between the exceptional sets; this may in fact become a global problem, as we will see later.

Chapter 5

Embeddings in Weighted Projective Spaces

In Chapter 3, we have studied some generalizations of complete intersections in products of *complex projective spaces*$^{(L)}$, namely complete intersections in products including now also *almost del Pezzo surfaces*$^{(L)}$, *almost Fano 3-folds*$^{(L)}$ and (generalized) *flag spaces*$^{(L)}$.

5.1 Weighted Projective Spaces

Already the simple *complex projective spaces*$^{(L)}$, $\mathbb{P}^n$, may be generalized in another important fashion as follows. Recall that in $\mathbb{P}^n$, there exist global homogeneous coordinates z_i, $i = 0, \ldots, n$, which are sections of the *hyperplane bundle*$^{(L)}$ $\mathcal{O}_{\mathbb{P}^n}(1)$ and obey the projectivity relation*

$$(z_0 : \cdots : z_n) \cong (\lambda z_0 : \cdots : \lambda z_n) = \lambda(z_0 : \cdots : z_n) . \tag{5.1.1}$$

The fact that this scaling is isotropic suggests an immediate generalization; modifying the projectivity relation into

$$(z_0 : \cdots : z_n) \cong (\lambda^{w_0} z_0 : \cdots : \lambda^{w_n} z_n) \tag{5.1.2}$$

defines a *weighted complex projective space* with *weights* $(w_0 : \cdots : w_n)$. We will write $\mathbb{P}^n_{(w_0:\cdots:w_n)}$ or $\mathbb{P}^n_{\mathbf{w}}$ for the space spanned by the $n+1$ *quasi-homogeneous coordinates* z_i, taken modulo the equivalence (5.1.2).

Notation: As with (quasi)homogeneous coordinates of a (weighted) projective space, only the relative ratios of the weights are really of consequence. To remind the Reader of this, we write $(z_0 : \cdots : z_n)$ and $(w_0 : \cdots : w_n)$ for the projective space, $\mathbb{P}^n_{(w_0:\cdots:w_n)}$. In contrast, we write $(z_0, \ldots, z_n)$ and $(w_0, \ldots, w_n)$

* Throughout this chapter, for ease of notation, we subscript coordinates as if they were co-variant rather than contra-variant.

for the affine space $\mathbb{C}^{n+1}_{(w_0:\cdots:w_n)}$; of course, $\mathbb{P}^n_{(w_0:\cdots:w_n)} = \{\mathbb{C}^{n+1}_{(w_0,\dots,w_n)} \smallsetminus (0,\dots,0)\}/\sim$, where "$\sim$" denotes the equivalence relation (5.1.2). We also write $\mathbf{w}$ for both $(w_0:\dots:w_n)$ and $(w_0,\dots,w_n)$: the context should clarify which of the two is meant.

2024⌈*Analogously, there also exist weighted versions of Grassmannians, more general flag spaces and C-spaces, mentioned in §§ 3.2.3 and 3.3, which may also be used to embed Calabi–Yau algebraic varieties [452].*⌋

5.1.1 Some basics

Now, it is clear from the projectivity relation (5.1.2) that if all weights are divisible by an integer, $w_i = kw'_i$, by redefining λ, $\mathbb{P}^n_{\mathbf{w}} \approx \mathbb{P}^n_{\mathbf{w}'}$. This is already encompassed in our notation $\mathbf{w} = (w_0:\cdots:w_n)$. However, there are further identities among weighted complex projective spaces, and to this effect we quote from Ref. [16], without proof,

Lemma 5.1 (Delorme). *Let* $\mathbf{w} = (w_0:\cdots:w_n)$ *and*

$$\begin{aligned} d_i &\stackrel{\text{def}}{=} \gcd(w_0,\dots,w_{i-1},w_{i+1},\dots,w_n)\ , \\ m_i &\stackrel{\text{def}}{=} \operatorname{lcm}(d_0,\dots,d_{i-1},d_{i+1},\dots,d_n)\ . \end{aligned} \tag{5.1.3}$$

Then,

$$\mathbb{P}^n_{\mathbf{w}} \approx \mathbb{P}^n_{\mathbf{w}'}\ , \qquad w'_i \stackrel{\text{def}}{=} (w_i/m_i)\ , \quad i = 0,\dots,n. \tag{5.1.4}$$

Note that for each i*,* m_i *is divisible by* w_i*,* d_i *and* m_i *are coprime and* $d_i m_i = \operatorname{lcm}(d_0,\dots,d_n)$ *(no summation on* i*).*

Some easy examples are $\mathbb{P}^2_{(1:2:2)} \approx \mathbb{P}^2$ (the isotropic projective space), $\mathbb{P}^3_{(1:2:4:4)} \approx \mathbb{P}^3_{(1:1:2:2)}$, $\mathbb{P}^4_{(2:3:6:12:18)} \approx \mathbb{P}^4_{(1:1:1:2:3)}$ and so on. In particular, $\mathbb{P}^1_{\mathbf{w}} \approx \mathbb{P}^1$ for any $\mathbf{w}$. This clearly cuts down on the number of $\mathbb{P}^n_{\mathbf{w}}$'s that need to be considered.

Definition 5.1. *A weighted projective space,* $\mathbb{P}^n_{\mathbf{w}}$*, is called* well-formed *if each* n *out of* $n+1$ *weights are coprime, that is, when*

$$\gcd(w_0,\dots,w_{i-1},w_{i+1},\dots,w_n) = 1\ , \qquad i = 0,\dots,n. \tag{5.1.5}$$

In view of the above lemma, we need to consider only well-formed $\mathbb{P}^n_{\mathbf{w}}$'s.

There is an important aspect in which $\mathbb{P}^n_{\mathbf{w}}$ (with $\mathbf{w} \neq \vec{1}$) differs from $\mathbb{P}^n$: most weighted projective spaces are singular. Consider the simple weighted projective space, $\mathbb{P}^4_{(1:1:1:2)}$. Then,

$$(z_0 : z_1 : z_2 : z_3 : z_4) \cong (\lambda z_0 : \lambda z_1 : \lambda z_2 : \lambda z_3 : \lambda^2 z_n) \ . \tag{5.1.6}$$

In the neighborhood of $(0\!:\!0\!:\!0\!:\!0\!:\!1)$ and for $\lambda = -1$, we have

$$(z_0 : z_1 : z_2 : z_3 : 1) \cong (-z_0 : -z_1 : -z_2 : -z_3 : 1) \ . \tag{5.1.7}$$

Therefore, this coordinate patch is isomorphic to $\mathbb{C}^4/\mathbb{Z}_2$ and has a $\mathbb{Z}_2$-quotient singular point. Similarly, $\mathbb{P}^4_{(1:1:1:2:2)}$ has a $\mathbb{Z}_2$-quotient singular line, parametrized by $(0\!:\!0\!:\!0\!:\!z_4\!:\!z_5)$. From their definition, we have [16, 110]:

Corollary 5.1. *The singular set of a well-formed $\mathbb{P}^n_{\mathbf{w}}$ has codimension at least 2 (dimension at most $n-2$) and consists of cyclic quotient singularities.*

Suffice it here just to note that weighted projective spaces are also projective, that is, they can be embedded in some big enough $\mathbb{P}^N$. This also implies that they admit a Kähler class, represented by a Kähler form which degenerates only at the singular points of $\mathbb{P}^n_{\mathbf{w}}$. As a matter of fact, one can even use the well-known Fubini–Study type Kähler potential

$$\Phi_{\text{F.S.}}(z_i, \overline{z_i}) = \log \|z\|^2 \stackrel{\text{def}}{=} \log \Big[\sum_{i=0}^{n} |z_i|^{2q_i} \Big] \ , \quad q_i \stackrel{\text{def}}{=} \frac{\text{lcm}\{w_j\}_{j=0,\dots,n}}{w_i} \tag{5.1.8}$$

much the same as one would do with the Fubini–Study Kähler potential for the (isotropic, $w_i = 1$) projective space.

5.1.2 Canonical stuff

Being so similar to (isotropic) projective spaces, one surmises that the "naturally bundled hardware" (see § 3.3.4) of homogeneous spaces should come for free with weighted (quasi-homogeneous) spaces also. Indeed, recall that the homogeneous coordinates on $\mathbb{P}^n$ are all sections of the same bundle, $\mathcal{O}_{\mathbb{P}^n}(1)$, and so have the same weight—$w_i = 1$. Clearly then, for $\mathbb{P}^n_{\mathbf{w}}$, the coordinate z_i of weight w_i is a section of the bundle $\mathcal{O}_{\mathbb{P}^n_{\mathbf{w}}}(w_i)$. With this, we have the weighted analogue of the *Euler sequence*$^{(L)}$ (Adjunction formula 1)

$$0 \longrightarrow \mathcal{O}_{\mathbb{P}^n_{\mathbf{w}}} \longrightarrow \bigoplus_{i=0}^{n} \mathcal{O}_{\mathbb{P}^n_{\mathbf{w}}}(w_i) \longrightarrow \mathcal{T}_{\mathbb{P}^n_{\mathbf{w}}} \longrightarrow 0 \ , \tag{5.1.9}$$

which is equivalent to the relation $\mathcal{T}_{\mathbb{P}^n_{\mathbf{w}}} = \{\bigoplus_{i=0}^n \mathcal{O}_{\mathbb{P}^n_{\mathbf{w}}}(w_i)/\mathcal{O}_{\mathbb{P}^n_{\mathbf{w}}}\}$. Using the determinant formula (1.1.1), we easily obtain the Adjunction formula 2,

$$c(\mathbb{P}^n_{(w_0:\cdots:w_n)}) = \prod_{i=0}^{n}(1 + w_i J) . \tag{5.1.10}$$

Furthermore, a polynomial $f_a(z)$ in the quasi-homogeneous coordinates of $\mathbb{P}^n_{\mathbf{w}}$ of total weight q_a is a section of the line bundle $\mathcal{O}_{\mathbb{P}^n_{\mathbf{w}}}(q_a)$, and we therefore have a straightforward generalization of Eq. (2.1.15):

$$c\,[\mathcal{X} \,\|\, \mathcal{E}] = \frac{\prod_{r=1}^m \prod_{i=0}^{n_r}(1 + w_i^{(r)} J_r)}{\prod_{a=1}^K (1 + \sum_{s=1}^m q_a^s J_s)} , \tag{5.1.11}$$

where $\mathcal{X}$ is a product of m weighted projective spaces, labeled by r, the weights of which are $w_i^{(r)}$. $\mathcal{E}$ is the direct sum of line bundles

$$\mathcal{E} = \bigoplus_{a=1}^{K} \mathcal{E}_a, \qquad \mathcal{E}_a = \bigotimes_{r=1}^{m} \mathcal{O}_{\mathbb{P}^{n_r}_{\mathbf{w}^{(r)}}}(q_a^r) . \tag{5.1.12}$$

As we discuss below, even the generic members of such configurations are singular. Thus, the analogue of the vanishing first Chern class (2.1.17), which now reads

$$\sum_{i=0}^{n_r} w_i^{(r)} = \sum_{a=1}^{K} q_a^r, \qquad r = 1, \ldots, m, \tag{5.1.13}$$

says that the (possibly singular) space $\mathcal{M} \in [\mathcal{X}\|\mathcal{E}]$ has trivial canonical divisor class, $K_{\mathcal{M}} \approx 0$. The next task is therefore to determine the singularity types expected to be found in $\mathcal{M}$ and to see if they can be resolved in such a way as to (1) preserve the triviality of the canonical divisor and (2) preserve the projectivity (the possibility to embed in some big enough $\mathbb{P}^N$). The triviality of the canonical divisor implies that the canonical bundle over the nonsingular (and non-compact) part of $\mathcal{M}$ is trivial. The Calabi–Yau desingularization can therefore be seen as one that enables the trivial canonical bundle to extend over the exceptional set used to smooth out the singularity.

— ❦ —

Clearly, the Euler characteristic computed from the above expression for $c[\mathcal{X}\|\mathcal{E}]$ and a straightforward generalization of Eq. (2.1.31) refers to the generic member of this configuration and we accordingly write $\chi_E[\mathcal{X}\|\mathcal{E}]$. Since the generic member will most certainly inherit some of the singularity of the embedding weighted space $\mathcal{X}$, we consider its resolution $\widetilde{[\mathcal{X}|\mathcal{E}]}$

instead, and χ_{E} will surely be affected in this process. Assuming that all singularities are of cyclic quotient type, using Eq. (4.5.3), the Euler characteristic becomes

$$\chi_{E}\big(\widetilde{[\mathcal{X}|\mathcal{E}]}\big) = \chi_{E}\big([\mathcal{X}|\mathcal{E}] - \sum_{i} \frac{\chi_{E}(F_i)}{|\mathbf{g}_i|} + \sum_{k} \chi_{E}(E_k)\ , \tag{5.1.14}$$

where F_i are the irreducible components of the singular set fixed locally in $\mathcal{X}$ by $\mathbf{g}_i$ and E_k the irreducible components of the exceptional set introduced to smooth the singularity.

— ❦ —

Consider now a space, $\mathcal{M}$, embedded in a (product of) weighted projective space(s), $\mathcal{X}$, by means of a (system of) defining constraint(s), just as we did in the isotropic case, in Chapter 2. $\mathcal{M}$ may turn out to be singular in two different ways: by having a non-transversal defining (system of) constraint(s) and/or by meeting the singular set of the embedding space.

5.2 Acquired Singularities

Besides the hereditary singularities described below, the defining equations of our subspace $\mathcal{M} \hookrightarrow \mathcal{X}$ may lack transversality. Of course, this can always happen because of a bad choice of polynomials on the human part. More seriously, however, a sufficiently generic choice of defining polynomials may just not be possible: because of the different weights, there are generally fewer monomials available at any given degree.

Now, a glance at Eq. (5.1.13) tells that, as compared to embeddings in (isotropic) projective spaces, embeddings in weighted projective spaces require defining polynomials of higher degree. One might hope that this will allow a sufficient number of different polynomials. As a simple example, to obtain a Calabi–Yau hypersurface in $\mathbb{P}^4$, we need a (smooth) quintic; a Calabi–Yau hypersurface in $\mathbb{P}^4_{(2:3:5:7:11)}$ is of (total) degree 28. Nevertheless, while there are 126 quintic monomials on $\mathbb{P}^4$, there are only 53 monomials of weight 28 on $\mathbb{P}^4_{(2:3:5:7:11)}$. However, the question still remains whether the generic linear combination of these 53 monomials is actually transversal.

The answer turns out to be negative, as can be shown remembering *Bertini's Theorem*$^{(L)}$ and the definition of the base locus. By Bertini, a generic element of a linear system is smooth away from the base locus; so we first need to determine the latter. The base locus is the common zero-set of all monomials from which the linear system is constructed. Note that this system will contain the simple monomials ${z_0}^{14}$ and ${z_3}^4$; these two vanish when $z_0 = z_3 = 0$, so that the base locus must be in the $\mathbb{P}^2_{\mathbf{w}}$ parametrized by $(0\!:\!z_1\!:\!z_2\!:\!0\!:\!z_4)$. Straightforwardly, there are only four monomials of total weight 28 in only these three variables:

$$z_1\, {z_2}^5\,, \quad {z_1}^2\, {z_4}^2\,, \quad {z_1}^4\, z_2\, z_4\,, \quad {z_1}^6\, {z_2}^2\,. \tag{5.2.1}$$

All of these will vanish if either $z_1 = 0$ or $z_2 = 0 = z_4$, and so the base locus consists of the $\mathbb{P}^1$ parametrized by $(0\!:\!0\!:\!z_2\!:\!0\!:\!z_4)$ and the point $(0\!:\!1\!:\!0\!:\!0\!:\!0)$. This is where all 53 degree-28 monomials vanish. Direct computation then verifies that their general linear combination, $p(z)$, is in fact singular ($\mathrm{d}p(z) = 0$) only at $(0\!:\!0\!:\!0\!:\!0\!:\!1)$.

Although these may turn out not to be simple singular points, it is quite possible that a non-transversal (complete intersection of) hypersurface(s) in a (product of) weighted projective space(s) contains at worst cyclic quotient and/or cDV hypersurface (a.k.a. hyperquotient) singularities.

— ❦ —

For the rest of this section, we turn to the general problem of characterizing transversal (quasi-)homogeneous polynomials in n (weighted) variables. Recall that transversality means that the gradients of the polynomial vanish only at the origin.

For a few variables, such as $n \leq 3$, this problem has been analyzed in great detail and solutions can be found in the literature; for example, Chapter 13 of the first volume by Arnold, Gusein-Zade and Varchenko [1] presents a complete discussion. Following this, we state here without proof:

Theorem 5.1 (Arnold). *Quasi-homogeneous polynomials $\phi(x, y, z)$ which contain either of the following five groups of three monomials are transverse for arbitrary positive a, b, c:*

$$\begin{gathered} \{\, x^a\,, \, y^b\,, \, z^c \,\}\,, \qquad \{\, x^a\,, \, y^b\,, \, z^c y \,\}\,, \qquad \{\, x^a y\,, \, y^b z\,, \, z^c x \,\}\,, \\ \{\, x^a\,, \, y^b z\,, \, z^c y \,\}\,, \qquad \{\, x^a\,, \, y^b z\,, \, z^c x \,\}\,. \end{gathered} \tag{5.2.2}$$

Also:

Theorem 5.2 (Arnold). *Quasi-homogeneous polynomials $\phi(x, y, z)$ which contain either of the following two groups of four monomials are transverse for positive a, b, c and almost all ε:*

$$\begin{aligned} &\{\, x^a\,, \, y^b x\,, \, z^c x\,, \, \varepsilon y^p z^q \,\}\,, \qquad && (a{-}1) \,\big|\, \mathrm{lcm}(b, c), \\ &\{\, x^a y\,, \, y^b x\,, \, z^c x\,, \, \varepsilon y^p z^q \,\}\,, \qquad && [(a{-}1)\gcd(b, c)] \,\big|\, (b{-}1)c. \end{aligned} \tag{5.2.3}$$

The conditions on the right merely assure quasi-homogeneity[†], *for which, of course,* p, q *have to be chosen suitably.*

Now, following this pattern, it is not hard to compile a list of groups of monomials in more than three variable, such that a generic linear combination of the monomials from each one group at a time provides a transverse polynomial. Such a straightforward extension of Arnold's list for the case of five variables can be found in Refs. [96, 184, 376, 448][‡]. It is quite unclear how such a procedure could (algorithm) be proven to be complete in general. We enclose however some remarks that might prove helpful in explorations of this kind with not too many variables.

Consider the first of the two cases in Theorem 5.2 (without loss of generality, we may consider the non-zero coefficients to be unity):

$$\phi(x, y, z) = x^a + y^b x + z^c x + \varepsilon y^p z^q. \tag{5.2.4}$$

For $\varepsilon = 0$, this becomes $\phi_{\varepsilon=0} = x(x^{a-1} + y^b + z^c)$, which is singular when $x = 0$ and $(y^b + z^c) = 0$, that is, $y = (-1)^{1/b} z^{c/b}$. Alternatively, we say that $\phi_{\varepsilon=0}$ vanishes there to second order since $\phi_{\varepsilon=0}$ factorizes, and having in mind its Taylor expansion around the singular set. To change this polynomial into one which would be singular only at the origin, we need to add a monomial which does not vanish at the singular set of $\phi_{\varepsilon=0}$; thereby, we will have moved the zero away from the singular set. Indeed, by adding the monomial $y^p z^q$ with a non-zero (generic) coefficient, ε, we obtain $\phi_{\varepsilon\neq0}$, which is transversal.

This "trick" can clearly be applied to polynomials in more variables also, the additional variables can just 'tag along,' added to the polynomial in the form of pure powers:

$$\phi(x, y, z, t) = x^a + y^b x + z^c x + \varepsilon y^p z^q \; + t^d \tag{5.2.5}$$

is certainly transverse. Moreover, we can also set $\varepsilon = t$, leading to

$$\phi(x, y, z, t) = x^a + y^b x + z^c x + t y^p z^q \; + t^d, \tag{5.2.6}$$

which has been omitted in Ref. [96].

The monomials $t y^p z^q$ are rather different from those used in the above two theorems of Arnold's and the straightforward extensions in Ref. [96].

[†] Recall, $x|y$ means that y is divisible by x, i.e., that $(y/x) \in \mathbb{Z}$.

[‡] The authors of Ref. [96] also restrict themselves to monomials in no more than two of the five variables and so omit a number of transversal polynomials; some of these are given here.

Those not only had at most two variables, but were of degree either $(a, 0, \ldots, 0)$ or $(a, 1, 0, \ldots, 0)$. Note the latter of these two, say 'elementary,' types can be obtained from a suitable power of a single variable by a *linear* change of variables. The monomials $y^p z^q$ and $ty^p z^q$, however, cannot be obtained in such a way, except $x^p y^q$ with $p = 1$ or $q = 1$. Simultaneously, note that the number of monomials needed to make the polynomials in Theorem 5.2 and also our example 5.2.6 is one more than the number of variables. It would appear that such 'less elementary' monomials are required to occur in addition to the 'elementary ones' of which there are always as many as there are variables.

2024 *Clearly, with five variables, the list diversifies even more, and soon after the original publication of the Bestiary, the endeavor of constructing Calabi–Yau 3-folds, and preferably in mirror-pairs, took on a substantially broader scope, owing to the ground-breaking works of Refs. [68, 241, 309]: embeddings in toric varieties, which subsume projective and weighted-projective spaces. By the turn of the millennium, a database of toric (almost) Fano 4-folds was completed [295, 430, 431], with almost half a billion of certain 4-dimensional polytopes, most of which had multiple triangulations, each specifying a toric 4-fold and a deformation family of Calabi–Yau hypersurfaces. Of course, complete intersections of hypersurfaces in higher-dimensional toric varieties are just as well-defined, but their combinatorial complexity quickly runs beyond current means of systematic study. We will return to toric embeddings at the end of Chapter F.*

5.3 Hereditary Singularities

Given that most weighted projective spaces have finite quotient (orbifold-like) singularities, hypersurfaces and intersections therein are likely to meet these singular sets and inherit singular points, curves and so on. It is therefore important to know just how badly singular spaces should one expect in our $\mathcal{M} \hookrightarrow \mathcal{X}$.

It is gratifying to know that there is a just absolute bound as to how singular $\mathcal{M}$ may become. For simplicity, consider a complete intersection of K polynomials $f_a(z)$ in a single $\mathbb{P}^{K+3}_{\mathbf{w}}$. Following Ref. [110], we define

$$m(p) \stackrel{\text{def}}{=} \#\{\, i : p|w_i \,\} , \qquad k(p) \stackrel{\text{def}}{=} \#\{\, a : p|w(f_a(z)) \,\} , \tag{5.3.1}$$

where $w(f_a)$ is the total weight of $f_a(z)$ and p is any prime.

Theorem 5.3. *Let* $\mathcal{M} = \bigcap_{a=1}^{K} f_a^{-1}(0) \subset \mathbb{P}^{K+n}_{\mathbf{w}}$, *for a generic choice of the defining polynomials* $f_a(z)$. *Then (with* $\dim \mathcal{M} = n$*),*

$$\mathrm{Sing}(\mathcal{M}) \;=\; \mathcal{M} \cap \mathrm{Sing}(\mathbb{P}^{K+n}_{\mathbf{w}}) , \qquad \textit{and} \qquad \dim\big(\mathrm{Sing}(\mathcal{M})\big) \leq n-2 \tag{5.3.2}$$

precisely if $k(p)+n-m(p)-1\geq 0$ *for any prime* p. *Also,*

$$\dim\big(\mathrm{Sing}(\mathcal{M})\big)\leq m(p)+n-k(p)-4\ . \tag{5.3.3}$$

Clearly, the above criterion needs to be checked only for $p\leq\max(w_i,q_a)$.

Thus, when defining a (Calabi–Yau or not) 3-fold as a complete intersection in a weighted projective space, generic such models have—at worst—isolated cyclic quotient singular points and/or curves. Precisely, the resolutions of these singularities have been discussed in the previous chapter, and the formula (5.1.14) is correct. We are now ready to construct new smooth 3-folds in a much larger set of available embedding spaces; for comparison, Chapter 1 listed precisely five Calabi–Yau 3-folds constructed as complete intersections of hypersurfaces in a single (isotropic) $\mathbb{P}^n$. By contrast, even as simple transversal hypersurfaces in single $\mathbb{P}^4_{\mathbf{w}}$'s (for various choices of $\mathbf{w}$), Ref. [96] constructs some 6000 Calabi–Yau 3-folds!

Now, the bounding value from Theorem 5.3, $[m(p)-k(p)-1]$, can in fact become as large as 2; take for example $[\,\mathbb{P}^4_{(1:1:3:3:3)}\,\|\,11\,]$: $m(3)=3$, $k(3)=0$, while for all other primes p, $[m(p)-k(p)-1]=-1$. The bound indicates that the degree-11 Calabi–Yau 3-folds in $[\,\mathbb{P}^4_{(1:1:3:3:3)}\,\|\,11\,]$ may inherit all of the singular surface $\mathbb{P}^3_{(3:3:3)}\subset\mathbb{P}^4_{(1:1:3:3:3)}$. Note however that since 11 is not divisible by 3, any cubic $C(z_2,z_3,z_4)$ must in fact be multiplied by a quadric $Q(z_0,z_1)$. Therefore, the "best bet" for our defining equation is something like

$$\phi(z)\ \stackrel{\text{def}}{=}\ {z_0}^{11}+{z_1}^{11}+\sum_i Q_i(z_0,z_1)C_i(z_2,z_3,z_4)\ =\ 0\ , \tag{5.3.4}$$

and obviously $\mathrm{d}f(z)=0$ when $z_0=z_1=0$. In fact, general genericity arguments (of the type we used when discussing Bertini's theorem) lead us to expect that

$$\dim\big(\,\mathcal{M}\cap\mathrm{Sing}\,\big)\ <\ 2 \tag{5.3.5}$$

if $\mathcal{M}$ is a *transversal* intersection. For single hypersurfaces in $\mathcal{X}=\mathbb{P}^4_{\mathbf{w}}$, this has been proven in Ref. [96]. For $\mathcal{X}$ a product of weighted projective spaces and $\mathcal{M}$ a complete intersection of several hypersurfaces, we still expect the same result, as a little thought and perhaps a few experiments will assure the Reader.

On the other hand, if $\mathcal{M}$ does contain a singular surface—it is a consequence of non-transversality, as in the $[\,\mathbb{P}^4_{(1:1:3:3:3)}\,\|\,11\,]$ case, rather than the intersection with the singular set of $\mathbb{P}^4_{(1:1:3:3:3)}$. We will come back to this in § 5.4.1.

5.4 Weighted Calabi–Yau Complete Intersections

The foregoing discussion clearly indicates that complete intersections in embedding spaces which include weighted projective factors are on the one hand a straightforward generalization of the constructions considered so far. On the other hand, they typically describe singular spaces. We then inquire if it is always possible to resolve these singularities in such a way that the resolution is Ricci-flat.

In answer to this, we quote from Ref. [157]:

Theorem 5.4. *For a quasismooth weighted complete intersection 2- or 3-fold X in a product of weighted projective spaces, the minimal desingularization $\widetilde{X}$ of X is a smooth Calabi–Yau manifold as long as Eq. (5.1.13) is satisfied.*

Quasismooth here means that the complete intersection is transversal, so that the only singularities are those inherited from intersecting the singular set of the (weighted) embedding spaces. An extension of this theorem to include mildly non-transversal cases would be desirable, but remains for now an open question.

— ❦ —

Let us mention also that similarly to weighted projective spaces, one can consider weighted Grassmannians and, furthermore, weighted flag spaces, generalizing even more the embeddings considered thus far. We leave the exploration of these uncharted territories to the adventurous Reader and proceed with some simple examples which relate to our earlier study.

5.4.1 Some simple (smooth) samples

We recall the two simple constructions mentioned in § 3.2.3:

$$\left[\mathbb{P}^4_{(1:1:1:1:2)}\middle\|4\right] \cong \overline{\mathbb{P}^3}^{(2)}_{[3\|4]}~, \qquad \left[\mathbb{P}^4_{(1:1:1:1:3)}\middle\|6\right] \cong \overline{\mathbb{P}^3}^{(2)}_{[3\|6]}~. \tag{3.2.21$'$}$$

As we have computed there, the anti-canonical bundles are $\mathcal{O}_{\mathbb{P}^3}(2)$ and $\mathcal{O}_{\mathbb{P}^3}(1)$, respectively. Consulting again Theorem 3.4, we see that we can obtain in rather similar fashion two Calabi–Yau 3-folds:

$$\left[\mathbb{P}^4_{(1:1:1:1:2)}\middle\|6\right] \cong \overline{\mathbb{P}^3}^{(3)}_{[3\|6]}~, \qquad \left[\mathbb{P}^4_{(1:1:1:1:4)}\middle\|8\right] \cong \overline{\mathbb{P}^3}^{(2)}_{[3\|8]}~. \tag{5.4.1}$$

On one hand, these are a triple cover of $\mathbb{P}^3$, branched over a smooth sextic surface and a double cover of $\mathbb{P}^3$ branched over a smooth octic surface, respectively. At the same time, they can be viewed as simple hypersurfaces in the weighted projective space $\mathbb{P}^4_{(1:1:1:1:2)}$ and $\mathbb{P}^4_{(1:1:1:1:4)}$, respectively.

— ❦ —

Now, we are ready for yet another description of such spaces (not necessarily Calabi–Yau). Compare the isotropic projectivity relation of a usual projective space and the anisotropic projectivity relation of a weighted projective space, both of which are represented by

$$(z_0 : \cdots : z_n) \cong (\lambda^{w_0} z_0 : \cdots : \lambda^{w_n} z_n) \tag{5.4.2}$$

in the sense that $\mathbf{w} = \vec{1}$ characterizes $\mathbb{P}^n$, while $\mathbf{w} \neq \vec{1}$ specifies $\mathbb{P}^n_{\mathbf{w}}$. Quite obviously, we can map

$$\mathbb{P}^n(x_0 : \cdots : x_n) \xrightarrow{x_i \longmapsto z_i = x_i{}^{w_i}} \mathbb{P}^n_{\mathbf{w}}(z_0 : \cdots : z_n)\ . \tag{5.4.3}$$

This however makes sense only if we identify $x_i \cong \alpha_{(i)}^{w_i} x_i$, where $\alpha_{(i)}$ is a $w_i{}^{th}$ primitive root of unity. Otherwise, the map would not be one-to-one, or in other words, the inverse would fail to be single-valued. This means that

$$\mathbb{P}^n_{\mathbf{w}} = \frac{\mathbb{P}^n}{\prod_{i=0}^n \mathbb{Z}_{w_i}}\ . \tag{5.4.4}$$

Let us write $\varpi_{\mathbf{w}} : \mathbb{P}^n \longrightarrow \mathbb{P}^n_{\mathbf{w}}$ for the quotient map.

— ❦ —

With this in mind, we now reconsider the first (family of) 3-fold(s) described by Eq. (5.4.1), writing $\phi(z) = 0$ for the defining equation. We now have the 'commutative diagram'[§]

$$\begin{array}{ccc} \mathbb{P}^4(x_0 : \cdots : x_4) & \xrightarrow{\phi(x)=0} & \left[\mathbb{P}^4 \,\|\, 6\right] \\ \Big\downarrow z_i = x_i{}^{w_i} & & \Big\downarrow \varpi_{(1:1:1:1:2)} \\ \mathbb{P}^4_{(1:1:1:1:2)}(z_0 : \cdots : z_4) & \xrightarrow{\phi(z)=0} & \left[\mathbb{P}^4_{\mathbf{w}} \,\|\, 6\right] \cong \overline{\mathbb{P}^3}^{(3)}_{[3\|6]}\ . \end{array} \tag{5.4.5}$$

So, a Calabi–Yau space embedded as a sextic hypersurface in the weighted $\mathbb{P}^4_{(1:1:1:1:2)}$, alternatively a double cover of $\mathbb{P}^3$, branched over a smooth sextic surface in $\mathbb{P}^3$, can also be described as a finite quotient (by $\varpi_{\mathbf{w}}$) of a sextic in an ordinary (isotropic) $\mathbb{P}^4$! Note that the anti-canonical bundle of $[4\|6]$

§ A 'commutative diagram' is, quite simply, exactly what it says: there's more than one way to get from 'object' A to 'object' B, but 'all roads lead to Rome'; you end up with the same result no matter which way you go about it.

is $\mathcal{O}_{\mathbb{P}^4}(-1)$, that is, it is negative. In other words, the canonical bundle is $\mathcal{K}_{[4\|6]} = \mathcal{O}_{\mathbb{P}^4}(1)$ and has five sections, rather than the unique holomorphic 3-form for Calabi–Yau manifolds. Indeed, according to the tentative taxonomy suggested in Table 9.1, the sextic in $\mathbb{P}^4$ is sort of opposite to an ample 3-fold and is there called *scant*, just like in § 2.1.2, after Table 2.1.

— ❦ —

Now, we have seen that there are just four hypersurfaces in $\mathbb{P}^4$ with negative canonical bundle, one (the quintic) with a trivial canonical bundle and there are infinitely many hypersurfaces with positive canonical bundle. The right most vertical arrow in the diagram (5.4.5) therefore indicates a process which could—in principle—produce Calabi–Yau 3-folds from any (almost) scant 3-fold. Let us therefore examine this a little closer.

To that end, note that the action of $\varpi_{\mathbf{w}}$ is non-trivial only on the fifth coordinate, x_4, which is essentially being squared. So the intersection of the $x_4 = 0$ hyperplane in $\mathbb{P}^4$ with the sextic defining equation $\phi(x) = 0$ specifies a surface which then belongs to a configuration $[4\|1,6]$. At that surface, $\varpi_{\mathbf{w}}$ maps one-to-one, while away from it, it is a two-to-one map. That is, we have a $\mathbb{Z}_2$ action with a codimension-1 fixed set; as discussed in § 4.2.1, such a fixed set does not lead to a singularity; a simple local change of variables suffices to eliminate this (apparent) coordinate singularity. However, since neighboring points to this surface are covered twice, unlike the surface, we do have a branching set, $\mathcal{B}_{\varpi_{\mathbf{w}}}$, at this surface and the "ramification formula" (3.2.19) is applicable; we only need to specify $\mathcal{L}(\mathcal{B}_{\varpi_{\mathbf{w}}})$, the bundle which defines the branching locus.

As discussed in the sketch of proof of Theorem 3.4, the branching divisor for a k-fold covering is taken with a $k-1$-fold multiplicity. Here, we have $k = 2$, so the branching divisor $\mathcal{B}_{\varpi_{\mathbf{w}}}$ is taken at face value, with no multiplicity. Furthermore, since it belongs to the configuration $[4\|1,6]$ and is a surface on a 3-fold which belongs to $[4\|6]$, i.e., the sextic defining equation is in common, it should be obvious that $\mathcal{L}(\mathcal{B}) = \mathcal{O}_{\mathbb{P}^4}(1)$, and so

$$\mathcal{O}_{\mathbb{P}^4}(-1) \;=\; \mathcal{K}^*_{[\mathbb{P}^4\|6]} \;\sim\; \mathcal{K}^*_{[\mathbb{P}^4_{\mathbf{w}}\|6]} \otimes \mathcal{L}^*(\mathcal{B}_{\varpi_{\mathbf{w}}}) \;=\; \mathcal{K}^*_{[\mathbb{P}^4_{\mathbf{w}}\|6]} \otimes \mathcal{O}_{\mathbb{P}^4}(-1)\,, \quad (5.4.6)$$

from which it follows that $\mathcal{K}^*_{[\mathbb{P}^4_{\mathbf{w}}\|6]} \approx \mathbb{C}$. In a way, this just manifests what has perhaps been obvious to the erudite Reader from the beginning—that the process of passing to a finite quotient is in a way inverse to constructing finite covering spaces, where the fixed points of the former correspond to the branching sets of the latter.

✎ Calculate $\chi_E[\mathbb{P}^4 \| 6]$ and relate it to $\chi_E[\mathbb{P}^4_{\mathbf{w}} \| 6]$ and Eq. (3.2.16) and (4.5.3). ❧

✎ Repeat the above analysis for $[\mathbb{P}^4_{(1:1:1:1:4)} \| 8]$ (caution: $k = 4$ here). ❧

— ❦ —

In view of the recent discussion, we look back at the example defined in Eq. (5.3.4), a degree-11 hypersurface in $\mathbb{P}^4_{(1:1:3:3:3)}$. By pulling it back to the covering hypersurface in (the isotropic) $\mathbb{P}^4$, following the diagram (5.4.5), the defining equation becomes

$$\phi(x) \stackrel{\text{def}}{=} {x_0}^{11} + {x_1}^{11} + \sum_i Q_i(x_0, x_1) C_i(x_2^3, x_3^3, x_4^3) \quad = \quad 0 \;, \tag{5.4.7}$$

which is clearly singular at the projective surface $(0{:}0{:}x_2{:}x_3{:}x_4) \subset \mathbb{P}^4$. The fact that $\phi(x)$ is a singular choice in the scant configuration $[\mathbb{P}^4 \| 11]$ is the "true cause" for the singularity of the example (5.3.4), although the singular locus of course coincides with $\mathrm{Sing}(\mathbb{P}^4_{(1:1:3:3:3)})$.

We conclude that if the multiple cover of the type described by the diagram (5.4.5) is smooth, the weighted complete intersection is singular only by

intersecting the singularity set of the (weighted) embedding space. The converse is, however, not true, as can be seen from the following very simple example:

$$\{\, {z_0}^2 + z_0 {z_1}^2 + {z_2}^4 + {z_3}^4 + {z_4}^4 \,\} \subset \left[\mathbb{P}^4_{(2:1:1:1:1)} \middle\| 4\right] \tag{5.4.8}$$

is smooth while its cover,

$$\{\, {x_0}^4 + {x_0}^2 {x_1}^2 + {x_2}^4 + {x_3}^4 + {x_4}^4 \,\} \subset \left[\mathbb{P}^4 \middle\| 4\right], \tag{5.4.9}$$

is singular at $(0{:}1{:}0{:}0{:}0)$. Although weighted complete intersections can be analyzed in their own terms, it is often useful to employ the relations indicated in the diagram (5.4.5).

This situation is not at all unlike that of Riemann surfaces: there is one with negative (g−1), one with vanishing (g−1) and then an infinite number of those with positive (g−1). The Riemann surfaces with higher genus can be described as branched covers of Riemann surfaces with lower genus simply by passing to a finite quotient. One of the simplest examples is provided by the $\mathbb{Z}_2$-symmetric torus, $T^2_{\mathbb{Z}_2}$, the $\mathbb{Z}_2$ quotient of which is a sphere; see Figure 5.1.

— ❦ —

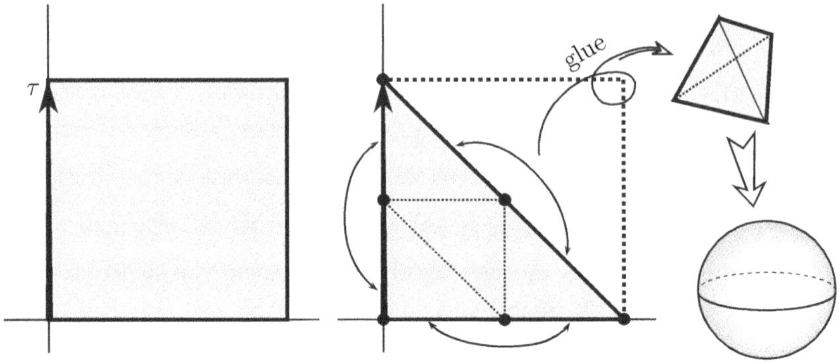

Figure 5.1: The fundamental domain of a $\mathbb{Z}_2$-symmetric torus, its $\mathbb{Z}_2$ quotient and another view thereof. The second transition only requires local coordinate changes at the vertices.

It is quite straightforward in principle, although formidable in notation, to extend these relations to arbitrary complete intersections in products including weighted homogeneous spaces. In a way though, because of this relation between complete intersections in weighted projective spaces and finite quotients of intersections in (isotropic) homogeneous spaces, it may appear that the whole excursion regarding weighted spaces is unnecessary. While it is true that these various constructions are related by 'commutative diagrams' and so it ultimately does not matter which way one goes about it, one or the other route might prove useful in some particular computations. For instance, hypersurfaces in weighted $\mathbb{P}^4_{\mathbf{w}}$ turn out to be in a very simple correspondence with certain supersymmetric field theories and allow in many cases exact quantum calculations (we will return to this briefly later; see also Refs. [152, 185, 191, 227], for example).

5.5 Calabi–Yau from Almost Fano 3-Folds (II)

Recall again the relation $[\mathbb{P}^4_{(1:1:1:1:4)}\|8] \cong \overline{\mathbb{P}^3}^{(2)}_{[3\|6]}$ and the simple defining equation

$$z^2 + B(x) = 0 , \qquad B(x) \stackrel{\text{def}}{=} \sum_{i=0}^{3}(x_i)^8, \tag{5.5.1}$$

like in Eq. (3.2.15), except that now we chose the weights and the degree of the hypersurface to obtain a Calabi–Yau 3-fold. As already discussed in

§ 5.4.1, manifolds of this type are double covers of $\mathbb{P}^3$ branched over $\mathcal{B}$, a smooth octic surface. Note that $\mathcal{B}$ can also be described as the zero-set of a section of $\mathcal{K}^{*}_{\mathbb{P}^3}{}^{\otimes 2}$.

5.5.1 An easy corollary

Now, more generally, given any almost Fano 3-fold, $\mathfrak{F}$, we can take its double cover branched over the zero-set, $\mathcal{B}$, of a generic section of $\mathcal{K}^{*}_{\mathfrak{F}}{}^{\otimes 2}$—we obtain a Calabi–Yau 3-fold, $\mathcal{M} \stackrel{\text{def}}{=} \overline{\mathfrak{F}}^{(2)}_{\mathcal{B}}$. The Euler characteristic of such a Calabi–Yau 3-fold is straightforwardly given by Eq. (3.2.16) in terms of $\chi_{E}(\mathfrak{F})$ and $\chi_{E}(\mathcal{B})$. Since the branching surface was there already in the almost Fano 3-fold, no 4-cycle or 2-cycle has been added in this construction, and since $H_4(\mathcal{M}) \stackrel{*}{\approx} H_2(\mathcal{M}) \stackrel{*}{\approx} H^2(\mathcal{M})$, $h^{1,1}(\mathcal{M}) = h^{1,1}(\mathfrak{F})$ (neither $\mathcal{M}$ nor $\mathfrak{F}$ have any (2,0)- or (0,2)-forms, so that $(H^2 \otimes \mathbb{C}) \approx H^{1,1}$).

In fact, even more generally,

Corollary 5.2. *Let $\mathfrak{F}$ be any almost Fano 3-fold with index r and q equal to 1 or some positive integer that divides r. Then, the $(q+1)$-fold covering of $\mathfrak{F}$, branched over a generic surface $\mathcal{B} \in [\,\mathfrak{F} \,\|\, \mathcal{K}^{*}_{\mathfrak{F}}{}^{\otimes(\frac{q+1}{q})}\,]$ is a Calabi–Yau 3-fold $\mathcal{M}$ with Euler characteristic*

$$\chi_{E}(\mathcal{M}) \;=\; (q+1)\chi_{E}(\mathfrak{F}) - q\chi_{E}(\mathcal{B}) \;. \tag{5.5.2}$$

Proof. To begin with, recall that by definition of the index (see page 83), there exists a line bundle $\mathcal{L}$ such that $\mathcal{L}^{\otimes r} \approx \mathcal{K}^{*}_{\mathfrak{F}}$, where r is the index. This means that $\mathcal{K}^{*}_{\mathfrak{F}}{}^{\otimes(\frac{q+1}{q})} \approx \mathcal{L}^{\otimes(q+1)(r/q)}$, which is a well-defined holomorphic bundle since $(r/q) \in \mathbb{Z}$, and so $\mathcal{B} \in [\,\mathfrak{F} \,\|\, \mathcal{L}^{\otimes(q+1)(r/q)}\,]$ is a well-defined surface. The statement now follows as a straightforward application of Theorem 3.4, which is why this result is labeled as a corollary. ☑

It is amusing to note that, since the index of $\mathbb{P}^3$ is 4, a (suitably branched) double, triple or five-fold cover will be a Calabi–Yau 3-fold. The first two of these have been identified in Eq. (5.4.1). A little thought reveals that the third one is the best known Calabi–Yau space, the quintic in $\mathbb{P}^5$.

5.5.2 Coverings, complete intersections and resolutions

The main reason for postponing this result until here is an important relation between the Calabi–Yau manifolds constructed in Corollary 5.2 and those constructed in § 3.2.5, which however involves small resolutions.

For simplicity, compare

$$\mathcal{M} \in \begin{bmatrix} 3 \\ 1 \end{bmatrix}\!\!\begin{Vmatrix} 4 \\ 2 \end{Vmatrix} \qquad \text{and} \qquad \mathcal{M}' \in \overline{\mathbb{P}^3}{}^{(2)}_{[3\|8]} \ . \tag{5.5.3}$$

For the record, $\chi_E(\mathcal{M}) = -168$, while $\chi_E(\mathcal{M}') = -296$; also $h^{1,1}(\mathcal{M}) = 2$ while $h^{1,1}(\mathcal{M}') = 1$. The relation becomes apparent when we consider the defining equation for the configuration on the left. Quite like in the 2-fold case considered on page 79, we write

$$\mathcal{M}: \qquad Q_{00}(x){z_0}^2 + 2Q_{01}(x)z_0z_1 + Q_{11}(x){z_1}^2 \;=\; 0 \ , \tag{5.5.4}$$

where the $Q_{ij}(x)$ are generic quartics over $\mathbb{P}^3$. For a generic point $x \in \mathbb{P}^3$, the defining equation is a non-degenerate quadric and has therefore two distinct solutions. $\mathcal{M}$ therefore covers the generic part of $\mathbb{P}^3$ twice. However, over some sub-generic points $x \in \mathbb{P}^3$, the discriminant

$$\varDelta(x) \;\stackrel{\text{def}}{=}\; 2\big(\,Q_{00}(x)Q_{11}(x) - Q_{01}(x)^2\,\big) \tag{5.5.5}$$

of the quadric equation vanishes; being a single octic equation in $\mathbb{P}^3$, this defines an octic surface, $\mathcal{B}$. So, $\mathcal{M}$ is a double cover of $\mathbb{P}^3$, branched over the octic surface¶ $\mathcal{B} \in \mathbb{P}^3$. This appears identical to $\mathcal{M}'$, so how can they have so different numerical characteristics?

The answer lies in the form of $\varDelta(x)$: the surface $\mathcal{B}$, defined by $\varDelta(x) = 0$, is in fact singular. Indeed, $\mathrm{d}\varDelta = 0$ whenever all three $Q_{ij}(x) = 0$. These being three quartic equations in $\mathbb{P}^3$, this is *bound* to happen at $4{\cdot}4{\cdot}4 = 64$ points. But, from Bertini's theorem, we know that the generic $\mathcal{M}$ is smooth (and we have assumed a generic choice of the $Q_{ij}(x)$'s)! Again, a second look at the quadratic equation—for the case when all three $Q_{ij} = 0$—tells that the variables of the $\mathbb{P}^1$ are completely unconstrained.

Finally, then, we have a complete description of $\mathcal{M}$: it is a double cover of $\mathbb{P}^3$, branched over a *singular* octic surface, the 64 singular points of which have been small-resolved into a $\mathbb{P}^1$ each. In turn, $\mathcal{M}'$ is a double cover of $\mathbb{P}^3$, branched however over a *smooth* octic surface—which needs no smoothing by small resolution or otherwise. The differing Euler numbers are indeed related by

$$\chi_E(\mathcal{M}) - \chi_E(\mathcal{M}') \;=\; 64{\cdot}\chi_E(\mathbb{P}^1) \ . \tag{5.5.6}$$

Furthermore, the singular octic in $\mathbb{P}^3$ used as the branching locus for $\mathcal{M}$ is clearly a singularizing deformation of the smooth octic used for $\mathcal{M}'$.

Clearly, any almost Fano 3-fold could have been used in place of $\mathbb{P}^3$ in the relation (5.5.3).

¶ Note that, when thought of as a special subset of the zero-set of the quadratic equation for $\mathcal{M}$, the points of the branching set come with multiplicity 2!

Chapter 6

Fibered Products

After having reviewed a large number of constructions of Calabi–Yau 3-folds, we here briefly mention one more class of constructions—not quite unrelated to those discussed before. The geometric description however will be somewhat novel, which is our primary reason to examine them.

6.1 Fibering Elliptic Curves

To begin with, we describe a simple, 1-dimensional model rather than a Calabi–Yau 3-fold. Consider the family of cubics in $\mathbb{P}^2$

$$\phi_\alpha(x, y, z) \stackrel{\text{def}}{=} x^3 + y^3 + z^3 \ - 3\alpha\, x\, y\, z \ = 0\ . \tag{6.1.1}$$

Unless α is chosen to satisfy $\alpha^3 = 1$ (or in the limit $\alpha \to \infty$), this defines a smooth complex 1-dimensional torus, also called a smooth elliptic curve, as a hypersurface in $\mathbb{P}^2$. As the parameter α is varied, the resulting torus varies and we have a family of elliptic curves, parametrized by α.

In a way, the whole picture is rather similar to the general image one has about vector bundles, in that α of course spans a complex plane and we have associated to every point of that plane an elliptic curve, that is, a torus. Moreover, as is clear from the defining equation, these tori vary holomorphically with α and we have a holomorphic fiber space of which the α-plane is the base and the particular torus parametrized by a given value of α is the *fiber* at α. The present situation is however markedly different from the standard notion of a bundle: at special values of α, the fiber *is* singular and we are unable to construct appropriate transition functions.

— ❦ —

On first impulse, one would expect that therefore the total (complex 2-dimensional) space of this fibration (base and fibers together), while

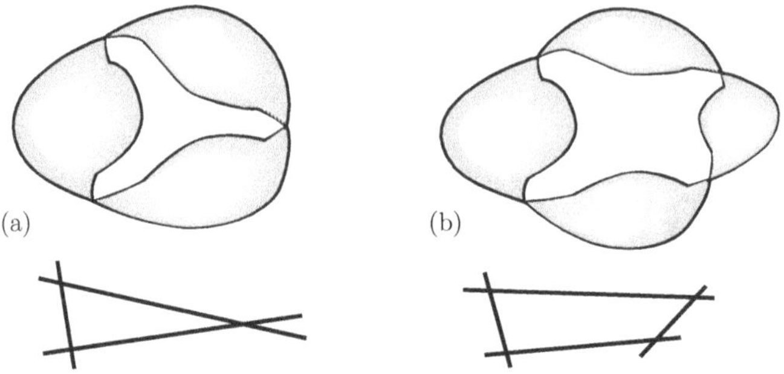

Figure 6.1: Degenerations of elliptic curves: (a) cubic, (b) bi-quadric.

obviously a complex space, is singular where the singular tori are. But, is that so?

First of all, the tori *are* singular when $\alpha^3 = 1$, as can be verified since, e.g.,

$$\tfrac{1}{3}\mathrm{d}\phi_{\alpha=1} \;=\; (x^2 - yz)\mathrm{d}x + (y^2 - xz)\mathrm{d}y + (z^2 - xy)\mathrm{d}z, \tag{6.1.2}$$

which vanishes at the $(1{:}1{:}1)$, $(1{:}\omega{:}\omega^2)$ and $(1{:}\omega^2{:}\omega)$ points of $\mathbb{P}^2$; $\omega = e^{2i\pi/3}$. These cubic elliptic curves degenerate into a connected union of three lines ($\mathbb{P}^1$'s) each two of which touch at a point—the three singular points; see Figure 6.1 (a).

However, to determine the singularity of the total space, we have to stop treating α as a constant parameter and regard it as a coordinate instead. This changes the total differential of $\phi_\alpha(x, y, z)$ into

$$\tfrac{1}{3}\mathrm{d}\phi \;=\; (x^2 - \alpha yz)\mathrm{d}x + (y^2 - \alpha xz)\mathrm{d}y + (z^2 - \alpha xy)\mathrm{d}z - (xyz)\mathrm{d}\alpha, \tag{6.1.3}$$

which is, quite clearly, non-vanishing at any of the singular points on the singular tori. Instead, at those points, $\mathrm{d}\phi$ is purely in the α direction.

The above equation may appear a little awkward as it is homogeneous in the x, y, z variables but not in α; this we remedy easily by declaring $\alpha = (v/u)$ and multiplying through with u (assumed non-zero here), to obtain

$$\phi(x, y, z; u, v) \;\stackrel{\text{def}}{=}\; u(x^3 + y^3 + z^3) \;-\; 3v(x\,y\,z) \;=\; 0\,. \tag{6.1.4}$$

By adding the point at $\alpha = \infty$, that is $u = 0$, we obtain the standard compactification of the α-plane ($\mathbb{C}$) into the $(u{:}v)$ projective space, $\mathbb{P}^1$. Of course, the torus at $u = 0$ is also singular, again at three points in

$\mathbb{P}^2$: (1:0:0), (0:1:0) and (0:0:1). It can be readily checked that the total differential, with respect to x, y, z, but also u, v, is non-zero everywhere.

— ❦ —

On second thought, Eq. (6.1.4) clearly describes a hypersurface in $\mathbb{P}^2 \times \mathbb{P}^1$, of bi-degree (3,1) and so corresponds to the configuration $\left[\begin{smallmatrix}2\\1\end{smallmatrix}\middle\|\begin{smallmatrix}3\\1\end{smallmatrix}\right]$. A quick glance back to section § 3.1 and the list (3.1.20) reminds us that the generic compact surfaces which belong to this configuration are almost del Pezzo surfaces of $\chi_E = 12$. That is, the total space of elliptic curves fibered linearly over $\mathbb{P}^1$ is in fact this almost ample surface (which we recall from § 3.1 is the only one which is *not* ample). Note that the elliptic curves which provide the fibers are themselves embedded in a del Pezzo surface (here quite simply $\mathbb{P}^2$) by means of a section of its anti-canonical bundle; $\mathcal{K}^*_{\mathbb{P}^3} \approx \mathcal{O}_{\mathbb{P}^3}(3)$—hence cubics.

The thought of replacing $\mathbb{P}^2$ with another del Pezzo surface comes readily and, indeed, any surface belonging to any of the configurations in the list (3.1.20) could be used*. Of course, all the so obtained almost del Pezzo surfaces will have $\chi_E = 12$, as that is the only one which is not ample†. Yet, there are some interesting differences.

Consider, as compared to Eq. (6.1.4) the fiber space again over $\mathbb{P}^1$, but this time with the fiber being given not as a cubic in $\mathbb{P}^2$, but rather as a bi-quadric in $\mathbb{P}^1 \times \mathbb{P}^1$. The defining equation is now

$$A(x^0 : x^1; y^0 : y^1)\, u + B(x^0 : x^1; y^0 : y^1)\, v \;=\; 0\;, \qquad (6.1.5)$$

where A and B are quadratic in the first pair of arguments and also in the second pair. A quick analysis of this says that there are three values of (u/v) where the bi-quadric elliptic curve is singular and, at each of those values of (u/v), one finds that the bi-quadric curve degenerates into a connected union of four lines ($\mathbb{P}^1$'s), as in Figure 6.1 (b).

In fact, general theory (see Section V.7 of Ref. [4]) shows that elliptic curves can be fibered over $\mathbb{P}^1$ much more generally than the above two cases. In particular, there are fibrations in which there occur fibers of type I_n, that is, where the elliptic curve had degenerated into n copies of $\mathbb{P}^1$ in a sort of a circular chain, with n singular points where two $\mathbb{P}^1$'s meet

---

* Remember that all del Pezzo surfaces can be represented as complete intersections in a product of complex projective spaces.

† This fibration representation of the $\chi_E = 12$ almost del Pezzo surfaces lets us see explicitly why they cannot be ample: the canonical bundle is trivial along all (except perhaps finitely many) fibers.

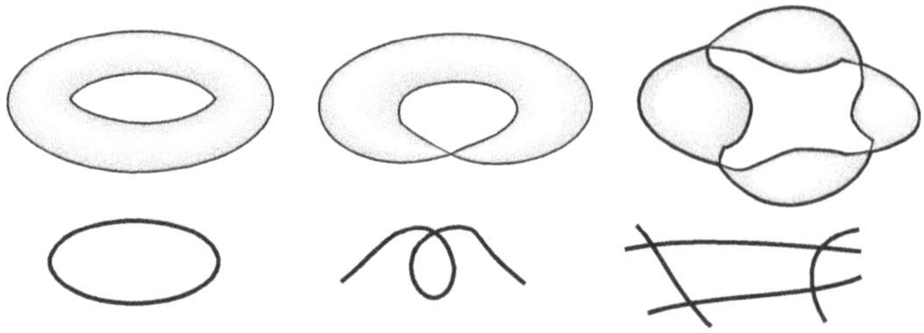

Figure 6.2: The simplest types of elliptic fibers: I_0–non-degenerate, I_1–irreducible and I_n–n-reducible.

(see Figure 6.2). Our above examples thus have fibers of types I_3 and I_4, respectively.

— ❦ —

By the way, just to dispel some possible misconception about the cubics in $\mathbb{P}^2$, consider

$$\phi'(x,y,z;u,v) \stackrel{\text{def}}{=} u(x\,y\,z) - v\big(\tfrac{1}{3}(x^3+y^3+z^3) - x^2y + y^2x\big) = 0\ . \quad (6.1.6)$$

Unlike the symmetric case above, this fibration has seven singular tori:

1. $u=1$, $v=0$: a single fiber with three singular points, at $(1{:}0{:}0)$, $(0{:}1{:}0)$ and $(0{:}0{:}1)$;
2. $u=2\omega$, $v=1$: at $\omega \in \{1, e^{2i\pi/3}, e^{4i\pi/3}\}$, there are three fibers, with two singularities each, at $(1{:}\mp i{:}\omega^2(\pm i{-}1))$;
3. $u=4^{1/3}\omega$, $v=1$: at $\omega \in \{1, e^{2i\pi/3}, e^{4i\pi/3}\}$, there are three fibers, with one singularity each, at $(1{:}1{:}{-2\omega^2 2^{1/3}})$.

Thus, by breaking the symmetry of the most popular cubic (6.1.4), it *is* possible to obtain several different elliptic curves, including here I_1 and I_2. Still, rather few of the possible I_n's seem to be realized in such an explicit way.

— ❦ —

Since we have obtained an almost del Pezzo surface, one naturally comes to think of fibering elliptic curves in a way that would produce a Calabi–Yau surface, that is, a model of K3. Looking at the configuration which describes the total space of the above fibrations as a complete intersection, we see that we can simply consider, for example,

$$\begin{bmatrix}2\\1\end{bmatrix}\!\!\begin{Vmatrix}3\\2\end{Vmatrix} : \quad A(x,y,z)\,u^2 \,+\, 2B(x,y,z)\,u\,v \,+\, C(x,y,z)\,v^2 \,=\, 0\ . \quad (6.1.7)$$

Nothing really changes in the geometrical interpretation, except that the elliptic curve races *quadratically* over the base $\mathbb{P}^1$ instead of linearly as in the above examples. The Reader may wonder why we have written the defining equation in terms of three, rather than just two summands as in the linear case. It turns out that without having all three terms—as a generic quadratic equation in u, v has them—the defined surface is singular.

✎ Prove this. ❧

Unlike in the above less-than-maximally symmetric example (6.1.6), we are now *forced* to introduce a term which will break at least some of the many symmetries of (6.1.4). Consequently, some of the singular fibers again 'decompose' into I_n, $n < 3$.

— ❦ —

Suppose, for the moment, that instead of these non-trivial fibrations, we were considering a simple product $T \times \mathbb{P}^1$. The Euler characteristic is multiplicative, so

$$\chi_{_E}(T \times \mathbb{P}^1) = \chi_{_E}(T) \cdot \chi_{_E}(\mathbb{P}^1) = 0 \cdot 2 = 0 \; . \tag{6.1.8}$$

This would pertain to be true even if the fibration were non-trivial but all fibers were smooth. Instead, in the above examples, we know that a number of fibers has become singular—now we see that comparing the Euler characteristic of the total space of the fibration with the trivial product in fact tells the number of singular fibers, counting multiplicities. In the fibered cubic case, $\chi_{_E} \left[\begin{smallmatrix} 2 \\ 1 \end{smallmatrix} \middle\| \begin{smallmatrix} 3 \\ 1 \end{smallmatrix}\right] = 12$ and there were four fibers with three singular points each. In the other example, there were three fibers with four singular points each—again twelve, as predicted.

This calculation of course seems trivial here, but the general principle holds in higher dimensions: The difference between the Euler characteristic of the total space of a fibration and the Euler characteristic of the trivial product counts the singular fibers, with multiplicities counted. More precisely, of course,

$$\chi_{_E}(\text{non-trivial}) - \chi_{_E}(\text{trivial}) \;=\; \sum_i \chi_{_E}(\text{singularity}_i) \; , \tag{6.1.9}$$

where "trivial" and "non-trivial" refer to the fibrations and "singularity" is the fiber-wise singularity, i.e., while keeping the locus in the base fixed (treat $\alpha = (v/u)$ again as a parameter).

6.2 Fibering Elliptic Surfaces

Now, just as the foregoing showed that the almost del Pezzo surface (which is not ample) may be regarded as a fibration in which the fibers have vanishing first Chern class and the base is $\mathbb{P}^1$, could it be that (at least some) almost Fano 3-folds can be regarded as a similar fibration?

Indeed, the answer is yes. In fact, understanding the example (6.1.4) in this way, it is not hard to see that Eq. (3.2.23),

$$\widetilde{\mathfrak{F}} \in \begin{bmatrix} \mathfrak{F} \\ \mathbb{P}^1 \end{bmatrix}\!\!\begin{Vmatrix} \mathfrak{f}_1 \\ J \end{Vmatrix} , \tag{6.2.1}$$

represents just that. The defining equation being

$$A(x)\,u \;+\; B(x)\,v \;\; = \; 0 \;, \qquad x \in \mathfrak{F} \;, \quad (u:v) \in \mathbb{P}^1 \;, \tag{6.2.2}$$

at any given point of $\mathbb{P}^1$, we have a particular equation in variables over $\mathfrak{F}$. As dictated by the configuration, $A(x)$ and $B(x)$ are sections of a line bundle of which the first Chern class equals $\mathfrak{f}_1 = c_1(\mathfrak{F})$, i.e., they are sections of the anti-canonical bundle. Consequently, at every point of $\mathbb{P}^1$, we have a surface with $c_1 = 0$, which can only be a model of K3; it is embedded as a hypersurface in $\mathfrak{F}^\ddagger$.

Just as with the elliptic curves, these elliptic surfaces (K3's) vary holomorphically over the base $\mathbb{P}^1$ and form a fibered space. Again some special points of $\mathbb{P}^1$ carry singular K3's, as can be revealed from the Euler characteristic computation; $\chi_{_E}(\widetilde{\mathfrak{F}})$ is given, as a consequence of Eq. (3.2.23), in terms of numerical characteristics of $\mathfrak{F}$ and so

$$2 \cdot 24 - \chi_{_E}(\mathfrak{F}) + C_1{}^3(\mathfrak{F}) \tag{6.2.3}$$

is the number of fiber-wise singularities (the sign changes with the dimension).

— ❦ —

Instead of fibering a surface over a $\mathbb{P}^1$, can we form a fibered space using for the base a surface, and if so, what type of surface do we need so that the total space would turn out to be a Calabi–Yau manifold?

Note that in all the previous cases, we used a copy of $\mathbb{P}^1$ as the base, which is the only almost ample curve (in fact it is ample). We thus expect

‡ The Reader unsettled with such generality may replace $\mathbb{P}^3$ for the almost Fano threefold $\mathfrak{F}$ and consequently have $\mathcal{O}_{\mathbb{P}^3}(4)$ for the anti-canonical bundle; the K3 surface is then simply one of the well known quartics in $\mathbb{P}^3$.

that (almost) ampleness is required for the base space. There are of course more than one almost ample surfaces, so there will be more possible constructions. Indeed, W.-W. Sung [221] proves that our intuition has not failed us; we quote the main theorem without proof as a proper rendition would take us too deep into technical details.

Theorem 6.1. *For $\mathcal{M}$ a smooth Calabi–Yau (projective) 3-fold and $\varpi : \mathcal{M} \to \mathcal{S}$ a proper, surjective holomorphic map with connected fibers, onto a smooth compact complex surface $\mathcal{S}$. Then, $\mathcal{S}$ is an almost del Pezzo surface.*

So, indeed, for a Calabi–Yau 3-fold to be represented as a fibration over a surface $\mathcal{S}$, it must be that $\mathcal{S}$ is one of those defined in § 3.1.

6.3 Fibering Double Elliptic Curves

Finally, we come to the closing of the part of this volume where the primary concern was how to construct (families of) Calabi–Yau 3-folds and carry through some of the most rudimentary analysis. To this end, we consider double elliptic curve fibrations studied by C. Schoen [212] and which also relates to the complete intersections discussed in Chapter 2.

Following Schoen, we consider a class of Calabi–Yau 3-folds which may be constructed from a singular bi-cubic hypersurface in $\mathbb{P}^2_1 \times \mathbb{P}^2_2$, defined by

$$\mathcal{M}_S : \quad A_{10}(x_{(1)})\, A_{21}(x_{(2)}) - A_{11}(x_{(1)})\, A_{20}(x_{(2)}) = 0 \ , \tag{6.3.1}$$

where $A_{ij}(x_{(i)})$ are cubic polynomials in the homogeneous coordinates $x_{(i)} \in \mathbb{P}^2_i$, $i = 1, 2$. At any particular point in $\mathbb{P}^2_1 \times \mathbb{P}^2_2$, we may use the four variables A_{ij} as local holomorphic coordinates and note that the defining equation becomes $xy - zt = 0$ or, after a linear change of variables, $\sum_{i=1}^4 \zeta_i{}^2 = 0$. This is precisely the equation for the simplest of all hypersurface singularities, the node (see Table 4.1); the location of the singularity is at the origin, $x = y = z = t = 0$, that is, at $A_{ij} = 0$. These are four cubic equations, two in each $\mathbb{P}^2$ and, for a generic choice of the A_{ij}, they have $(3{\cdot}3)(3{\cdot}3) = 91$ nodes for the space of solutions.

Thus, $\mathcal{M}_S$ is a (purely) nodal Calabi–Yau 3-fold. From the discussion in § 4.4, we know that each node can be resolved into a $\mathbb{P}^1$ in two distinct ways, so that there are 2^{81} (!) small resolutions of this singular 3-fold. What is not obvious is whether any of these is Kähler. It turns out—yes, and it is moreover embeddable as a smooth complete intersection mentioned

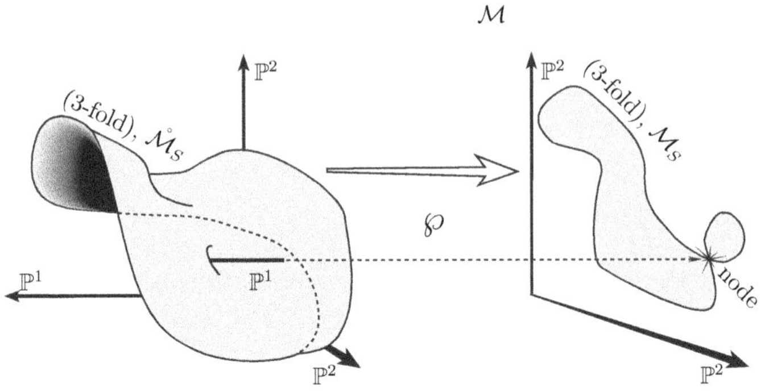

Figure 6.3: A rough sketch of the projection $\mathring{\mathcal{M}}_S \xrightarrow{\wp} \mathcal{M}_S$.

in (3.1.28),

$$\mathring{\mathcal{M}}_S \in \begin{bmatrix} 2 \\ 2 \\ 1 \end{bmatrix}\!\!\begin{Vmatrix} 3 & 0 \\ 0 & 3 \\ 1 & 1 \end{Vmatrix} . \tag{3.1.28$'$}$$

To prove this, consider the generic system of defining equations in this configuration:

$$\begin{aligned} A_{10}(x_{(1)})\, y^0 \;+\; A_{11}(x_{(1)})\, y^1 \;&=\; 0\,, \\ A_{20}(x_{(2)})\, y^0 \;+\; A_{21}(x_{(2)})\, y^1 \;&=\; 0\,. \end{aligned} \tag{6.3.2}$$

Regard this as a system of two linear homogeneous equations in two variables, y^0 and y^1, which must not vanish simultaneously. The solvability condition of course is the vanishing of the determinant of the system, which is exactly Eq. (6.3.1). This relates the Calabi–Yau manifold $\mathring{\mathcal{M}}_S$ and the Calabi–Yau nodal 3-fold $\mathcal{M}_S$.

In other words, $\mathring{\mathcal{M}}_S$ is embedded in $\mathbb{P}^2 \times \mathbb{P}^2 \times \mathbb{P}^1$. If we project along the $\mathbb{P}^1$ factor (eliminate the y^i variables), we obtain the projection on $\mathbb{P}^2 \times \mathbb{P}^2$, which is precisely $\mathcal{M}_S$ as defined in Eq. (6.3.1)

$$\mathcal{M}_S = \wp(\,\mathring{\mathcal{M}}_S\,)\ . \tag{6.3.3}$$

On a more local level, the situation is as follows. $\mathcal{M}_S$ is singular only when all four A_{ij} vanish. On the other hand, $\mathring{\mathcal{M}}_S$ is not singular when all four A_{ij} vanish; rather, it contains a full copy of the embedding $\mathbb{P}^1$ as the variables y^i are not constrained at all. One says that the inverse image of each node in $\mathcal{M}_S$, under $\wp$, is a $\mathbb{P}^1$ in $\mathring{\mathcal{M}}_S$ (see Figure 6.3). Thus, $\mathring{\mathcal{M}}_S$ is a small resolution of $\mathcal{M}_S$!

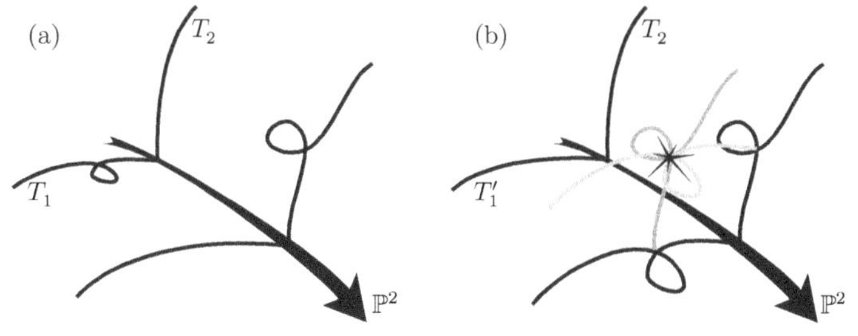

Figure 6.4: Double elliptic fibrations: (a) smooth vs. (b) singular.

From the nodal Calabi–Yau 3-fold (6.3.2), Schoen then constructs a collection of Calabi–Yau manifolds by varying the coefficients A_{ij} so that the generic member is $\mathcal{M}_S$ while less generic members involve special choices of $A_{ij}(x_{(i)})$.

Note that, for a given point in $\mathbb{P}^1$, Eq. (3.1.28) define two separate cubics in a product of two $\mathbb{P}^2$'s, that is, a product of two elliptic curves parametrized by the variables y^i. Similarly to the fibrations above, we have that $\mathring{\mathcal{M}}_S$ is a double elliptic fibration over $\mathbb{P}^1$. To describe the rest of this collection, note that as these elliptic curves vary along the $\mathbb{P}^1$ (see Figure 6.4), at certain points of the $\mathbb{P}^1$, these tori degenerate having a number of their "small" circles pinched to points. Of course, for a generic choice of $A_{ij}(x_{(i)})$, the two tori do not degenerate simultaneously and the resulting fiber space is a smooth manifold, $\mathcal{M}_S$.

Choosing special cubics for which the two elliptic curves do degenerate (but no worse than nodal) simultaneously at some special points of $\mathbb{P}^1$ gives rise to nodal limits of $\mathcal{M}_S$, classifying them in terms of the classification of singular elliptic surfaces. The nodes of nearly all of these conifolds (see § D.2) can be repaired by small resolution leading to other Kähler manifolds [212]; we state here without proof

Lemma 6.1. *Let $\mathcal{M}$ denote the double elliptic fibration over $\mathbb{P}^1$ with fibers $I_{m,n} \stackrel{\text{def}}{=} I_m \times I_n$, where I_n is an elliptic curve with n nodes. Let S and S' denote the set of points in $\mathbb{P}^1$ where, respectively, one and the other elliptic curve degenerate.*
(0) $\mathcal{M}$ is singular only over $S \cap S'$;
(i) if $\mathcal{M}$ contains only $I_{0,0}$, it is smooth and Kähler;
(ii) if $\mathcal{M}$ contains only $I_{m,m}$, it has a smooth and Kähler resolution;
(iii) if $\mathcal{M}$ contains a fiber $I_{m,1}$ and also $I_{p,0}$ or $I_{0,q}$, with $m, p, q \neq 0$, it has no smooth and Kähler resolution.

Note: the reason for the negative result in *(iii)* is essentially the failure of the positivity conditions (4.3.16) as described in § 4.4.3.

This provides a collection of Calabi–Yau manifolds, with many values $0 < \chi_E \leq +168$, as a list of small resolutions of various nodal deformations of $\wp(\mathcal{M}_S)$.

Part II
Cohomology

Chapter 7

(Co)homology Basics

The Reader is most probably at least acquainted with the basic notion of cohomology, differential forms and we forego introducing these in detail. Should need arise, some of the basic texts on this subject should be consulted, starting perhaps with the relevant sections from the physicist's favorite Ref. [19], Ref. [25, 52] and then carrying on with Ref. [11] ...

We will however, review the barest essentials of some (co)homology theories which are ultimately those of most physical relevance, in my own personal opinion.

7.1 Cohomology $1 \wedge 1$

7.1.1 Some generalities

Generally, we think of a cohomology group as an abelian additive group of certain objects for which we can specify two integers called degrees* so that the total cohomology group takes the form

$$H^{\star,\star} = \bigoplus_{p,q} H^{p,q} \ , \tag{7.1.1}$$

where p and q take some integer values. One can think of the (p,q)-pair as a *bi-grading* of the total cohomology group $H^{\star,\star}$. Usually, $H^{p,q}$ vanishes identically outside of some finite range of p and q.

The elements of these cohomology groups can generally be obtained from a collection of objects called 'forms,' α. These also have a well-defined degree, so we speak of (p,q)-forms (which vanish identically outside of some

* This is also often called 'dimension' which may lead to silly conflicts in nomenclature and possible confusion, so we try to stick to 'degree'. Also, we immediately deal with (p,q)-cohomology, having in mind the application throughout this volume.

finite range of p and q) and generally denote their space by $A^{p,q}$. Bundled with these $A^{p,q}$ comes a nilpotent operator ∂, called a differential of type $(1,0)$, which acts on these forms and so maps

$$\partial \,:\, A^{p,q} \longmapsto A^{p+1,q}\,, \qquad \partial\circ\partial = 0\,. \tag{7.1.2}$$

The conjugate operator, $\bar{\partial}$, sends $A^{p,q}$ to $A^{p,q+1}$ and is similarly nilpotent. Now, we may define the $\bar{\partial}$-cohomology† as

$$H^{p,q}_{\bar{\partial}} \stackrel{\text{def}}{=} \{\,\{\alpha\in A^{p,q} : \bar{\partial}\alpha = 0\,\}\,/\,\{\,\bar{\partial}\eta\,,\ \eta\in A^{p,q-1}\,\}\,\}\,. \tag{7.1.3}$$

That is, the elements of $H^{p,q}$ are identified with classes of elements in $A^{p,q}$, where two α's are considered equivalent if they differ by something that is annihilated by $\bar{\partial}$. Because of the nilpotency of $\bar{\partial}$, $\alpha \cong \alpha + \bar{\partial}\eta$ (see also *Hodge decomposition*$^{(L)}$ of forms).

Finally, if we are given a Hermitian scalar product $\langle\alpha,\beta\rangle$, we can define the adjoint operator $\partial^\dagger$ *with respect to this Hermitian product* by

$$\langle\bar{\partial}^\dagger\alpha,\beta\rangle \stackrel{\text{def}}{=} \langle\alpha,\bar{\partial}\beta\rangle\,, \qquad \forall\alpha,\beta. \tag{7.1.4}$$

The analogous goes for $\partial^\dagger$. Then

$$\triangle_{\bar{\partial}} \stackrel{\text{def}}{=} \bar{\partial}\bar{\partial}^\dagger + \bar{\partial}^\dagger\bar{\partial} \tag{7.1.5}$$

is called the $\bar{\partial}$-Laplacian and the ∂-Laplacian is defined analogously; clearly, the two are related by complex conjugation and if that is a symmetry, $\triangle_{\bar{\partial}} = \triangle_{\partial}$. The forms annihilated by the Laplacian are called "harmonic"; there generally exist natural isomorphisms between the cohomology groups defined as above and the space of harmonic forms.

With these general facts, the definition of some particular cohomology theory boils down to selecting a suitable collection of 'forms,' $A^{p,q}$, and finding an appropriate (conjugate pair of) nilpotent operator(s) for the above definition.

— ❦ —

In our standard situation, the '(p,q)-forms' are defined over some complex n-fold $\mathcal{X}$ and may be represented in some local coordinate chart by

$$\omega_{\mu_1\cdots\mu_p\bar{\nu}_1\cdots\bar{\nu}_q}(z,\bar{z})\,\mathrm{d}z^{\mu_1}\wedge\ldots\wedge\mathrm{d}z^{\mu_p}\wedge\mathrm{d}z^{\bar{\nu}_1}\wedge\ldots\wedge\mathrm{d}z^{\bar{\nu}_q}\,. \tag{7.1.6}$$

† Clearly, nothing stops us from defining the ∂-cohomology either. The above notation is just aimed to remind of the Dolbeault cohomology, which refers to the antiholomorphic derivative, $\bar{\partial}$. Complex conjugation relates these two cohomology groups and provides an isomorphism if it is a symmetry.

Here $\omega_{\mu_1\cdots\mu_p\bar{\nu}_1\cdots\bar{\nu}_q}(z,\bar{z})$ is the *tensor coefficient* of the (p,q)-form the components of which are local functions of the coordinates z and $\bar{z}$, and '$\wedge$' is the usual antisymmetric product of differentials.

> The skew symmetry of '$\wedge$' is not at all a frivolous novelty of differential geometry; it is as old as the notion of surface integration. For, if we perform the change of coordinates $(x,y) \mapsto (y,x)$, the integration measure changes as $\mathrm{d}x\mathrm{d}y \mapsto -\mathrm{d}y\mathrm{d}x$ owing to the Jacobian of the transformation.

The differential ∂ is, of course, then just the wedge product from left with $\mathrm{d}z^\mu\partial_\mu$, where ∂_μ denotes partial derivation with respect to z^μ; the analogous holds for $\bar{\partial}=\mathrm{d}z^\mu\partial_\mu$.

In $N=2$ supersymmetric models in 2-dimensional spacetime, $A^{p,q}$ are field-function(al)s p-linear in the spinor field ψ, q-linear in the Hermitian conjugate, $\overline{\psi}$ and more or less of arbitrary dependence on the bosonic fields. The spinorial derivatives, $D_\pm$ and $\bar{D}_\pm$ are all nilpotent and satisfy

$$D_\pm\bar{D}_\pm + \bar{D}_\pm D_\pm \;=\; \partial_{\pm\pm}\ , \qquad D_\pm\bar{D}_\mp + \bar{D}_\mp D_\pm \;=\; 0\ . \tag{7.1.7}$$

With the usual scalar product in the field space, $\bar{D}_\pm = {D_\pm}^\dagger$ and the *first order* light-cone partial derivatives ∂_{++} and ∂_{--} play the rôle of the Laplacians.

Another, structurally almost identical, example of cohomology which is found in field theory is the BRST cohomology. There however, all usual BRST operators (differentials) anti-commute, so that the "Laplacians" are zero. There are also many rather differently defined *cohomology groups*$^{(L)}$ in 'standard mathematics,' but we leave that to the said references.

7.1.2 Cohomology ring

By definition, we have addition between forms, with respect to which they form an abelian group. Forms can be multiplied by scalars (elements of the *base field*, $\Bbbk = \mathbb{Z}, \mathbb{Q}, \mathbb{R}, \mathbb{C}\ldots$). They can also be "$\mathcal{V}$ valued" if they map into (the coefficients in (7.1.6) take values in) $\mathcal{V}$, where $\mathcal{V}$ may be a vector *bundle*$^{(L)}$ or a *sheaf*$^{(L)}$ (or some other suitable locally valued structure); by default, if no $\mathcal{V}$ is specified, the base field $\Bbbk$ is understood.

In addition to this structure, as a rather important characteristic of all cohomology groups, they give rise to a *ring*$^{(L)}$ structure. Namely, we can also multiply two forms and obtain a third one and this multiplication satisfies the axioms of the ring structure.

The simplest and in many ways the natural *choice* is, of course, the $\wedge$ product which takes[‡]

$$\begin{array}{rcccccl} \wedge: & A^{p,q} & \otimes & A^{r,s} & \to & A^{p+r,q+s} \, , \\ & H^{p,q} & \otimes & H^{r,s} & \to & H^{p+r,q+s} \, . \end{array} \tag{7.1.8}$$

In particular, note that the subset $\bigoplus_p H^{p,p}$ (and more generally, $\bigoplus_p A^{p,p}$) is closed under $\wedge$. In general, we can write

$$\omega_i \wedge \omega_j = C_{ij}{}^k \, \omega_k \, , \tag{7.1.9}$$

where the labels i, j, k range over all (harmonic) forms; the $C_{ij}{}^k$ are the *structure constants* and characterize the ring structure. One should however be aware that there could (and often do) exist other possible choices, which may involve more structure but appear more 'natural' from some particular point of view.

— ❦ —

For complex 3-folds, the wedge product of two harmonic (1,1)-forms produces a harmonic (2,2)-form. On 3-folds, $H^{1,1}$ is dual to $H^{2,2}$ That is, the wedge product of harmonic (1,1)-forms and harmonic (2,2)-forms produces harmonic (3,3)-forms, the integral of which over the 3-fold $\mathcal{M}$ depends only on the cohomology class of the forms involved. So, with any basis $\{e_A\}$ for $H^{1,1}$ and a basis $\{e^A\}$ for $H^{2,2}$ such that $\int_{\mathcal{M}} e_A \wedge e^B = \delta_A{}^B$, we can write

$$e_A \wedge e_B = \mathring{\kappa}_{ABC} \, e^C \, , \quad \text{i.e.} \quad \mathring{\kappa}_{ABC} \stackrel{\text{def}}{=} \int_{\mathcal{M}} e_A \wedge e_B \wedge e_C \, , \tag{7.1.10}$$

which is precisely the expression of the tree-level Yukawa couplings in Eq. (0.5.5). Therefore, the world-sheet instanton corrections to the Yukawa couplings (0.6.3) provide a correction to the ring structure on $H^{1,1}$; this new structure could therefore be called the *quantum ring structure*, implying the name *classical ring structure* for the one defined by Eq. (0.5.5).

This quantum ring structure depends on the tangle of holomorphic images, L, of $\mathbb{P}^1$ in the 3-fold and is weighted exponentially by $-S[J; L]$, the 2-dimensional action evaluated on L with J the chosen Kähler class.

‡ Note that a product of two harmonic forms need not be harmonic, i.e., a product of two non-trivial elements of a cohomology group need not be also non-trivial. For example, on Calabi–Yau 3-folds, a product of a harmonic (2,1)-form and a harmonic (1,1)-form is clearly a (3,2)-form, of which there are no non-trivial harmonic ones.

Therefore, the "smallest" such curves dominate the correction while "large" ones may be neglected[§]. In a situation where all such curves are large, the quantum ring structure attains the classical limit, as should be the case. Conversely, the quantum ring structure may be regarded as a (quantum) deformation of the classical one. Thus, we may define the "quantum cohomology product" by requiring that the structure constant in

$$e_A \diamond_{J,L} e_B \;=\; \kappa_{ABC}\, e^C\;, \tag{7.1.11}$$

be the quantum Yukawa coupling

$$\kappa_{ABC} \;\stackrel{\text{def}}{=}\; \mathring{\kappa}_{ABC} \;+\; \sum_L e^{-S_{\mathcal{M}}[J;L]} \Big(\int_L \omega_A\Big)\Big(\int_L \omega_B\Big)\Big(\int_L \omega_C\Big)\;. \tag{0.6.3$'$}$$

— ❦ —

For a Calabi–Yau n-fold $\mathcal{M}$, the degree-n cohomology can also be given a ring structure which depends on the (projectively) unique element $\Omega \in H^{n,0}(\mathcal{M})$. By means of Eq. (0.3.13), $H^{n-1,1}(\mathcal{M})$ may be identified with the holomorphic tangent bundle valued cohomology $H^1(\mathcal{M},\mathcal{T}_{\mathcal{M}})$. With a basis $\{\varphi^{\mu}_{\alpha}\}$ for $H^1(\mathcal{M},\mathcal{T}_{\mathcal{M}})$ and a basis $\{\varphi^{\alpha}_{\mu}\}$ for $H^2(\mathcal{M},\mathcal{T}^*_{\mathcal{M}}) = H^{1,2}(\mathcal{M})$ such that $\int_{\mathcal{M}} \varphi^{\mu}_{\alpha} \wedge \varphi^{\beta}_{\mu} = \delta_{\alpha}{}^{\beta}$, we can write

$$(\varphi^{\mu}_{\alpha} \wedge \varphi^{\nu}_{\beta})\Omega_{\mu\nu\rho} \;=\; \kappa_{\alpha\beta\gamma}\varphi^{\gamma}_{\rho}\;, \quad \text{i.e.} \quad \kappa_{\alpha\beta\gamma} \;=\; \int_{\mathcal{M}} \Omega \wedge (\varphi^{\mu}_{\alpha} \wedge \varphi^{\nu}_{\beta} \wedge \varphi^{\rho}_{\gamma})\Omega_{\mu\nu\rho}\;, \tag{7.1.12}$$

which is precisely the expression of the Yukawa couplings in Eq. (0.5.7). These Yukawa couplings receive no quantum corrections (see § 0.6) and so the classical ring structure is the same as the quantum one. Through this roundabout procedure, we have defined a product "$\diamond_{\Omega}$":

$$\begin{array}{ccccc}
H^1(\mathcal{M},\mathcal{T}_{\mathcal{M}}) & \otimes & H^1(\mathcal{M},\mathcal{T}_{\mathcal{M}}) & \xrightarrow{\;\wedge\,,\,\cdot\Omega\;} & H^2(\mathcal{M},\mathcal{T}^*_{\mathcal{M}}) \\
\big\downarrow{\scriptstyle \cdot\Omega} & & \big\downarrow{\scriptstyle \cdot\Omega} & & \big\downarrow{\scriptstyle \text{Dolbeault}} \\
H^{2,1}(\mathcal{M}) & \otimes & H^{2,1}(\mathcal{M}) & \xrightarrow{\;\diamond_{\Omega}\;} & H^{1,2}(\mathcal{M})
\end{array} \tag{7.1.13}$$

which certainly is *not* the wedge product. In fact, the usual wedge product is mostly zero on the degree-3 cohomology.

[§] The "size" of such a curve, L, is measured by integrating over L the pull-back of the chosen Kähler class on the 3-fold; this is, essentially, $S_{\mathcal{M}}[J;L] = \int_{L\hookrightarrow\mathcal{M}} \|\mathrm{d}z\|^2_J = i\int_L J$.

7.2 Homology $1 \cap 1$

Here, we establish some notation and review some less well known facts about homology computations in Kähler manifolds.

Let $\mathcal{X}$ be a Kähler manifold of complex dimension n and let $\mathcal{Y}$ be a complex submanifold of complex dimension k then $\mathcal{Y}$ represents a homology class, $[\mathcal{Y}] \in H_{2k}(\mathcal{X})$. Note that elements of $H_0(\mathcal{Y}, \mathbb{k})$ may be identified with corresponding elements of the field $\mathbb{k}$. In particular, for example, integers n are identified with elements of $H_0(\mathcal{X}, \mathbb{Z})$ and in the case $n \geq 0$ correspond to n points of $\mathcal{X}$.

If $\mathcal{Y}$ and $\mathcal{Y}'$ are complex submanifolds of $\mathcal{X}$ with complex dimensions respectively k and k', we say $\mathcal{Y}$ and $\mathcal{Y}'$ intersect *transversely* if $T_p(\mathcal{Y}) + T_p(\mathcal{Y}') = T_p(\mathcal{X})$ at every point p of their intersection. In other words, transversality implies that, at every point of intersection, the tangent space of $\mathcal{X}$ is spanned by vectors tangent to $\mathcal{Y}$ and those tangent to $\mathcal{Y}'$. Now, if $k + k' < n$, this requires $\mathcal{Y}$ and $\mathcal{Y}'$ to be disjoint. This may be understood intuitively in the sense that for $k + k' < n$, $\mathcal{Y}$ and $\mathcal{Y}'$ may be deformed and 'moved away'¶ within $\mathcal{X}$ until they have no point in common. If $\mathcal{Y}$ and $\mathcal{Y}'$ intersect transversely then $[\mathcal{Y} \cap \mathcal{Y}'] \in H_{2k+2k'-2n}(\mathcal{X})$. Note that the *codimension*(L) (with respect to $\mathcal{X}$) is additive:

$$\begin{aligned} \operatorname{codim}(\mathcal{Y} \cap \mathcal{Y}') &= (n - (k + k' - n)) \\ &= (n - k) + (n - k') = \operatorname{codim}(\mathcal{Y}) + \operatorname{codim}(\mathcal{Y}') \ . \end{aligned} \tag{7.2.1}$$

This intersection product can be generalized to a bilinear product of homology classes whether or not they can be represented by submanifolds so that we have

$$[\mathcal{Y}] \cap [\mathcal{Y}'] = [\mathcal{Y} \cap \mathcal{Y}'] \ . \tag{7.2.2}$$

Note also in this context that, besides being a submanifold of $\mathcal{X}$ as are $\mathcal{Y}$ and $\mathcal{Y}'$, $\mathcal{Y} \cap \mathcal{Y}'$ is in fact a submanifold of both $\mathcal{Y}$ and $\mathcal{Y}'$ and so determines elements of $H_{2k+2k'-2n}(\mathcal{Y})$ and $H_{2k+2k'-2n}(\mathcal{Y}')$. The homomorphism

$$H_m(\mathcal{X}) \xrightarrow{\cap[\mathcal{Y}]} H_{m+2k-2n}(\mathcal{Y}) \tag{7.2.3}$$

¶ There is no obstruction to doing this smoothly but it is not necessarily possible if one requires that the move also be complex analytic. Fortunately, the computation of $\mathring{\kappa}_{ABC}$ does not require such holomorphicity; as elements of $H_*(\mathcal{X})$, $[\mathcal{Y}]$ and $[\mathcal{Y}']$ may be equivalent even thought they are distinct as complex submanifolds of $\mathcal{X}$.

also generalizes to the full homology of $\mathcal{X}$ regardless of representability as submanifolds. In this context, we have the identity

$$(A \cap_{\mathcal{X}} [\mathcal{Y}]) \cap_{\mathcal{Y}} (B \cap_{\mathcal{X}} [\mathcal{Y}]) = (A \cap_{\mathcal{X}} B) \cap_{\mathcal{X}} [\mathcal{Y}]. \tag{7.2.4}$$

Similarly, if $\mathcal{X}$ and $\mathcal{X}'$ are two Kähler manifolds of dimension n and n' respectively, and $\mathcal{Y}$ and $\mathcal{Y}'$ are submanifolds respectively of $\mathcal{X}$ and $\mathcal{X}'$ of respective dimensions k and k', then $[\mathcal{Y} \times \mathcal{Y}'] \in H_{2k+2k'}(\mathcal{X} \times \mathcal{X}')$. Again this external product generalizes to all homology classes regardless of representability so that we have

$$[\mathcal{Y}] \times [\mathcal{Y}'] = [\mathcal{Y} \times \mathcal{Y}']. \tag{7.2.5}$$

The two '$\times$'-products are related by the identity

$$(A \times B) \cap (C \times D) = (-1)^{\dim(B)\cdot\dim(C)} (A \cap C) \times (B \cap D). \tag{7.2.6}$$

Now let $\mathcal{X} \stackrel{\text{def}}{=} \mathcal{X}_1 \times \cdots \times \mathcal{X}_m$, where $n_r = \dim \mathcal{X}_r$ and $N = \dim \mathcal{X}$. Denote $x_r \in H^2(\mathcal{X}_r, \mathbb{Z})$ and let $\chi^r \in H_{(2n_r-2)}(\mathcal{X}_r, \mathbb{Z})$ correspond to x_r so that

$$x_r[\chi^s] \stackrel{\text{def}}{=} \int_{\chi^s} x_r = \delta_r{}^s. \tag{7.2.7}$$

Write

$$\hat{\chi}^r \stackrel{\text{def}}{=} [\mathcal{X}_1] \times \cdots \times [\mathcal{X}_{r-1}] \times \chi^r \times [\mathcal{X}_{r+1}] \times \cdots \times [\mathcal{X}_m] \in H_{(2N-2)}(\mathcal{X}, \mathbb{Z}) \tag{7.2.8}$$

for the homology element in $\mathcal{X}$ which projects to $\chi^r \in H_{(2n_r-2)}(\mathcal{X}_r, \mathbb{Z})$ under $\pi_r \colon \mathcal{X} \to \mathcal{X}_r$. If $\mathcal{Y}$ is a hypersurface in $\mathcal{X}$ defined as the zero set of a section of a line bundle with Chern class $\sum_{r=1}^m \pi_r^*(\chi^r)$, we then have that

$$H_{(2N-2)}(\mathcal{X}, \mathbb{Z}) \ni [\mathcal{Y}] = \sum_{r=1}^{m} \hat{\chi}^r. \tag{7.2.9}$$

If $\mathcal{Y}$ is an intersection of K hypersurfaces, then $[\mathcal{Y}] \in H_{(2N-2K)}(\mathcal{X}, \mathbb{Z})$ is the intersection product of terms of the form of the right hand side of Eq. (7.2.9).

7.3 The Kähler Package

In Chapter 0, we have identified the massless fields in superstring compactification with certain harmonic forms (cohomology classes) on the compactifying space $\mathcal{M}$. While $\mathcal{M}$ is smooth, we may use any of the standard cohomology theories in explicit computations, as standard theorems ensure that all these rather differently defined cohomologies yield isomorphic results. On singular spaces, however, all Hell seems to break loose; the different cohomology theories easily yield rather different results, so—which one does the physics of compactification choose?

Based on general properties of field theories and in particular, (2,2)-supersymmetric 2-dimensional ones, we expect several important characteristics of the cohomology which is to correspond to the massless fields. Consulting Chapter 0 for the dictionary between field theory and cohomology, we see that the prospective cohomology must feature:

1. *Hodge decomposition*$^{(L)}$: $H^r(\mathcal{M}, \mathbb{C}) \approx \bigoplus_{p+q=r} H^{p,q}(\mathcal{M})$, with the additional requirement of
2. *Complex conjugation*$^{(L)}$: $H^{p,q}(\mathcal{M}) = \overline{H^{q,p}(\mathcal{M})}$; this follows directly from CPT-conjugation in the field theory.
3. *Poincaré duality*$^{(L)}$, that is, *Hodge star duality*$^{(L)}$, so that, in particular $H^{1,1} \stackrel{*}{\sim} H^{2,2}$.
4. *Lefschetz Hard Theorem*$^{(L)}$, which amounts to a decomposition of forms into irreducible representations of the holonomy group.
5. *Künneth formula*$^{(L)}$, which says that in a product of two spaces, harmonic forms are products of a harmonic form from one space and a harmonic form from the other space, with their degrees added.

These features are all parts of what is called the "Kähler package" [15]. We therefore inquire if there is a cohomology which maintains the above features even when the underlying space singularizes (and we keep in mind that at least hypersurface and finite quotient singularities should be included).

Note that the standard (co)homology theories experience difficulties even with the simplest of the singularities. For example, consider the complex torus with one of its "small" cycles pinched to a point. The 1-cycle represented by the small circles therefore becomes trivial as it manifestly can be contracted to a point. The 1-cycle represented by the "big" circle cannot be contracted to a point and so remains non-trivial. The standard homology of the pinched torus is therefore

$$H_q(T_{\bowtie}, \Bbbk) = \Bbbk, \quad q=0,1,2\ , \quad \Bbbk = \mathbb{Z}, \mathbb{Q}, \ldots \tag{7.3.1}$$

If the $H_* \stackrel{*}{\sim} H^\star$ duality holds, H^1 is odd-dimensional and there can be no Hodge decomposition! If we simply excise the singular point, the "big" circle becomes an open interval and thus contractible, but now the "small" circle is non-contractible; in fact, the space is a cylinder and we are back to an odd-dimensional H^1 and so no Hodge decomposition.

7.3.1 L^2-cohomology

The solution to the above requirements turns out to exist and moreover be defined by the very axioms of field theory. Namely, the 2-dimensional field theory which is used to describe the propagation of (super)strings through spacetime—including the compactification space—requires *all physical states to be square-integrable.*

The cohomology theory which incorporates square-integrability is called simply L^2-cohomology and denoted $H^{p,q}_{(2)}$; it has been developed mainly by Cheeger [15]. The standard definition is changed in that candidate (p,q)-forms are now required to be square-integrable with respect to a chosen norm. This in general means that some of the usual cohomology classes of (p,q)-forms do not exist in $H^{p,q}_{(2)}$. On the other hand, in the equivalence relation $\omega \cong \omega + \bar{\partial}\eta$, *both* ω *and* η must be square-integrable. This restricts the equivalence relations so that, if $\omega - \omega' = \bar{\partial}\eta$ where η is not square-integrable, ω and ω' are not equivalent in $H^{p,q}_{(2)}$. This makes it possible for $H^{\star,\star}_{(2)}$ to be bigger than $H^{\star,\star}$. Clearly, both effects are typically present and $H^{p,q}_{(2)}$ can be rather different from $H^{p,q}$.

Most importantly however, $H^{p,q}_{(2)}(\mathcal{M}) = H^{p,q}(\mathcal{M})$ on any smooth manifold $\mathcal{M}$, and on a rather large class of singular spaces, $H^{p,q}_{(2)}$ remains "well-behaved" and still possesses the above listed properties.

— ❦ —

By definition, $H^{p,q}_{(2)}$ does depend on the norm chosen on the space. This dependence is however rather weak. Given two norms $N[\omega]$ and $N'[\omega]$ on the space of forms, the L^2-cohomology computed using one will equal the cohomology computed with the other norm precisely if there exists a constant $K \in \mathbb{R}$ such that

$$\tfrac{1}{K}N[\omega] \;<\; N'[\omega] \;< K\,N[\omega]\;. \tag{7.3.2}$$

This defines the universality class of norms for which $H^{p,q}_{(2)}$ is constant and therefore provides a computational trick: typically, the desired norm (metric) is too complicated or even unknown. It may be possible, however, to find another norm in the same universality class such that the required computation becomes easier.

7.3.2 Intersection (co)homology

Just as in all classical cohomology theories, the explicit computation of $H^{p,q}_{(2)}$ may be a very difficult task. In the classical case, rescue often comes in the form of some homology theory, in that the desired cohomology group can be related to some homology group (by a duality relation or even an isomorphism), which in turn is easy to compute (see Chapter 8).

There exists a homology theory, called *middle perversity*☿ *intersection homology theory* [26, 34], denoted $IH_*(\mathcal{X})$, which has been proven to satisfy the *De Rham duality*$^{(L)}$ pairing with $H^{p,q}_{(2)}(\mathcal{X})$ for a class of singular spaces $X \hookrightarrow \mathbb{P}^N$ [15], where $H^{p,q}_{(2)}(\mathcal{X})$ refers to the restriction to X of the Fubini–Study metric on the $\mathbb{P}^N$. In particular, this class includes spaces with isolated conical singularities—precisely those which have occurred in Chapters 4–6.

Let us just mention though that IH_* may be defined for a very broad class of so-called *(stratified) pseudomanifolds.*✱

Definition 7.1. *A (stratified*$^{(L)}$*) pseudomanifold* $\mathcal{X}$ *of dimension* n *is a space which admits a* stratification

$$\mathcal{X} \quad = \quad \mathcal{X}_n \supset \mathcal{X}_{n-2} \supset \cdots \supset \mathcal{X}_1 \supset \mathcal{X}_0 \ , \tag{7.3.3}$$

such that $(\mathcal{X}_n \smallsetminus \mathcal{X}_{n-2})$ *is an oriented dense* n*-dimensional manifold and* $(\mathcal{X}_k \smallsetminus \mathcal{X}_{k-1})$ *for* $k \leq n-2$ *is a* k*-dimensional manifold along which the normal structure of* $\mathcal{X}$ *is trivial.*

Akin to n-folds, we will also say "pseudo-n-fold". The last requirement roughly means that for each point $p \in (\mathcal{X}_k \smallsetminus \mathcal{X}_{k-1})$, there exists a distinguished neighborhood $N_{p \in X}$ which looks like an open cone, the base of which is again (iteratively) nicely stratified. In other words, an n-dimensional pseudomanifold (recall that all dimensions are complex) looks like a collection of manifolds of all dimensions not bigger than n and not $(n-1)$, which are glued together in an orderly fashion. Because of the 'emptiness' requirement in $n-1$ dimensions, pseudomanifolds are really different from ordinary manifolds once $n > 2$.

There is no *a priori* reason why such objects would be banned from string theory, and it would seem worthwhile trying to extend the current

☿ So, ... not only physicists indulge in silly nomenclature; the 'middle perversity' adjective is usually omitted, though.

✱ ⌈*There also exists a complementary,* combined (co)homology *[250], which in fact better fits mirror symmetry and the conifold (and other singularizing) transitions; see § D.2 and Ref. [410]. Corresponding physical models include 'branes and other types of* defects, *as befits pseudomanifold geometry, presaged already in 1994 [304, Fig. 19]; see also Refs. [413]. Square-integrability and related (world-sheet) physical requirements then need to be adapted.*⌋

understanding of what a "Calabi–Yau 3-fold" means, so as to also include pseudomanifolds. We leave this to the adventurous Reader and turn now to some results which can be of help in explicit computations.

7.3.3 Some helpful facts

CW complexes: In many cases of interest, the desired homology groups of a space can be determined by iterating a simple procedure, called *attaching cells* [11]. By an n-cell, one simply means a real n-dimensional ball together with its boundary. This then is glued to another space (not necessarily of the same dimension!) in a way prescribed by a "function" f. With $\mathsf{C}^{(n)}$ denoting the n-cell, we write $Y = X \cup_f \mathsf{C}^{(n)}$, meaning that the space Y is obtained by attaching $\mathsf{C}^{(n)}$ to X.

The pet example for this process is the relation

$$S^n = \{\,\text{point}\,\} \cup \mathsf{C}^{(n)} \,. \tag{7.3.4}$$

That is, we obtain the n-sphere by attaching a bounded n-ball to a point (see Figure 7.1). Usually, one thinks of making an n-sphere from $\mathbb{R}^n$ by attaching the point at infinity and 'attaching cells' seems sort-of the same but the other way around: we first add the (S^{n-1}-like) boundary to $\mathbb{R}^n$, producing $\mathsf{C}^{(n)}$ and then glue all of the boundary together, to the point "at infinity". (Since all of the boundary is being mapped to a single point, the specification f is not really a function.)

The following is an important property of this procedure.

Theorem 7.1. *Attaching an n-cell to a space $\mathcal{X}$ does not alter the homotopy in dimension strictly less than $n{-}1$, but may kill elements in $\pi_{n-1}(\mathcal{X})$; more precisely, the inclusion $\mathcal{X} \hookrightarrow \mathcal{X} \cup \mathsf{C}^{(n)}$ induces isomorphisms*

$$\pi_q(\mathcal{X}) \xrightarrow{\sim} \pi_q(\mathcal{X} \cup \mathsf{C}^{(n)}) \,, \qquad q < n{-}1 \,, \tag{7.3.5}$$

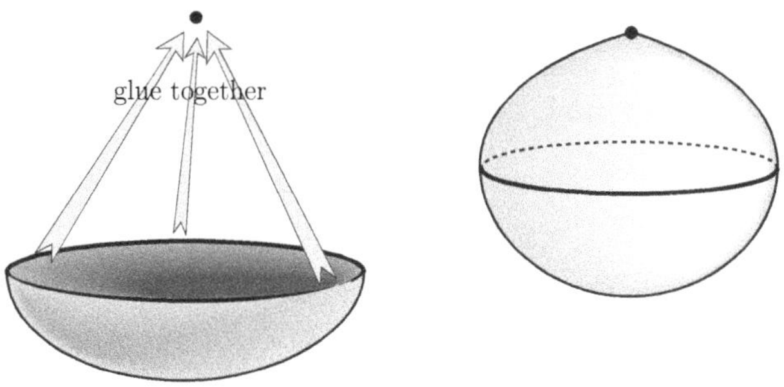

Figure 7.1: Building a sphere by attaching a cell.

and a surjection

$$\pi_{n-1}(\mathcal{X}) \longrightarrow \pi_{n-1}(\mathcal{X} \cup \mathsf{C}^{(n)}) \ . \tag{7.3.6}$$

Compare this with the *Lefschetz Hyperplane Theorem*$^{(L)}$; see (1.6.1).

Definition 7.2. *A* CW complex *is a space Y built up from a collection of points by the successive attaching of cells, where the cells are attached in the order of increasing dimensions.*

*The cells of dimension at most n in a CW complex Y together comprise the n-*skeleton *of Y.*

Clearly, every triangularizable space and also every manifold is a CW complex. The well known example is the "telescoping definition,"

$$\mathbb{P}^n = \{\,\text{point}\,\} \cup \mathsf{C}^{(2)} \cup \mathsf{C}^{(4)} \cup \ldots \cup \mathsf{C}^{(2n)} \ . \tag{7.3.7}$$

Or, the other way around: $\mathbb{P}^n$ may be thought of as defined recursively

Definition 7.3 (Recursive). *$\mathbb{P}^n$ is a copy of $\mathbb{C}^n$ with a $\mathbb{P}^{n-1}$ glued at the "infinity," and where $\mathbb{P}^0$ is defined to be a point.*

— ❦ —

Relating the homotopy and homology groups, we have the

Theorem 7.2 (Hurewicz). *If $\mathcal{X}$ is a simply connected path-connected CW complex, the first non-zero homotopy and homology groups are equal. That is, if $\pi_k = 0$ for $k < n$, then $\pi_n \approx \oplus_{p+q=n} H^{p,q}$.*

Also worth knowing is that π_k are all abelian groups for $k > 1$ ($H^{p,q}$ are all abelian). For further relations, we refer to Ref. [11].

Computing L^2 and intersection (co)homology: For all the cases considered in this volume, the following few results will suffice to obtain the intersection homology and the L^2-cohomology from the usual (classical) (co)homology.

Theorem 7.3. *Let $\mathcal{X}$ be a n-fold with a single isolated singularity, χ. Then*

$$IH_k(\mathcal{X}) \quad = \quad \begin{cases} H_k(\mathcal{X}) & k > n, \\ \mathrm{im}[H_n(\mathcal{X} \smallsetminus \chi) \to H_n(\mathcal{X})] & k = n, \\ H_k(\mathcal{X} \smallsetminus \chi) & k < n. \end{cases} \tag{7.3.8}$$

The map in the middle dimension simply means the following: (**1**) *discard* all would-be n-cycles from $H_n(\mathcal{X})$ which contract to χ and (**2**) *discard* all n-cycles which break open upon excising χ. Note that $IH_k(\mathcal{X})$ need not be calculated for $k<n$ at all—Poincaré duality relates it to $IH_{2n-k}(\mathcal{X})$.☆

As a simple example, note that on the torus with one of the "small" circles pinched to a point, the "small" circles are no longer 1-cycles since they contract to the (singular) point. As part 2, excise the singularity, thereby turning the "big" circles into open intervals, which are contractible and so have to be discarded. We remain with

$$IH_q(T^2_{\bowtie},\Bbbk) = \begin{cases} \Bbbk, & \text{for } q=0,2, \\ 0, & \text{for } q=1; \end{cases} \quad \text{with } \Bbbk = \mathbb{Z},\mathbb{Q},\dots \tag{7.3.9}$$

In other words, $IH_*(T_{\bowtie})$ equals $H_*(S^2)$. And more generally, for a Riemann surface of genus g and with k handles pinched at a "small" circle, the intersection homology equals that of a Riemann surface of genus $g-k$. Now, in (super)string theory, when the world-sheet of the (super)string pinches in this fashion, the dynamics effectively replaces it with a lower genus sheet (we refer to Ref. [22] for details of this process).

Of course, in the preceding example, the intersection homology and the L^2-cohomology do satisfy the formal duality requirements, and we safely conclude that (super)strings speak the L^2-cohomology, at least as far as the dynamics of the world-sheet goes. 2024⌈*The key exception to this reliance on L^2-cohomology is in the middle dimension: Superstrings compactified on Calabi–Yau manifolds that singularize in certain ways (see § D.2) have massive modes that become massless in the singular limit [467]. This* **adds** *to the otherwise expected count, opposing that IH_n in (7.3.8)* **discards** *elements that appear only on one side of the indicated mapping. This discovery prompted a "working definition" of a stringy (co)homology, wherein $SH_n(\mathcal{X})$* **retains** *elements that appear on either side of the mapping in (7.3.8) [410]. Such a "complementary" combined (co)homology indeed exists and is developed rigorously (with complete superstring application!) in Ref. [250]. Already in the simple case of a "pinched torus"—and unlike (7.3.9)—$SH_*(T^2_{\bowtie}) \approx H_*(T)$ identifies also this singular model as a Calabi–Yau space, suggesting that SH_* is better suited than IH_* for the target space in (super)string models.*⌋

— ❦ —

☆ 2024⌈*The combined homology, SH_*, of [250, pp. 166–168] effectively replaces both instances of "discard" with "include," and so is complementary to intersection homology.*⌋

The next result is somewhat trivial for curves and surfaces and becomes really non-trivial when the dimension is at least 3.

Definition 7.4. *Given a singular complex n-fold* $\mathcal{X}$ *and a non-singular complex n-fold* $\mathcal{Y}$, *such that* $(\mathcal{Y}\smallsetminus E)\xrightarrow{f}(\mathcal{X}\smallsetminus\chi)$ *is an isomorphism and* $E\xrightarrow{f}\chi$ *is surjective* $^{(L)}$, f *is called a* small map *if*

$$\operatorname{codim}_{\mathbb{C}}\big(x\in\mathcal{X}\big)\quad>\quad\dim_{\mathbb{C}}\big(f^{-1}(x)=E\big)\ . \tag{7.3.10}$$

This means that in curves and surfaces, a small map is finite, that is, each point $x\in\mathcal{X}$ must be replaced by finitely many copies of x in $\mathcal{Y}$. In 3- and 4-folds, a small map is finite, except perhaps for a number of isolated points which may be replaced at most by curves. In 5- and 6-folds, a small map is finite, except perhaps for a number of isolated curves C_i, each of which may be replaced at most by a surface E_i; each E_i is obtained by replacing each point on C_i by a curve so that E_i is a fibered space over C_i.

Actually, both $\mathcal{X}$ and $\mathcal{Y}$ may be allowed to be singular, in that $\mathcal{Y}$ is a partial resolution of $\mathcal{X}$. This also provides a general definition: *small resolution* is a small map $f:\mathcal{Y}\to\mathcal{X}$ which provides at least a partial resolution of singularities in $\mathcal{X}$.

Since divisors and, in particular, the canonical divisor are all in (complex) codimension-1, the foregoing yields

Theorem 7.4 (Small Map). *If* $f:\mathcal{Y}\to\mathcal{X}$ *provides a small resolution of the singular 3-fold* $\mathcal{X}$, *then* $IH_*(\mathcal{X})=H_*(\mathcal{Y})$ *and* $K_{\mathcal{X}}=K_{\mathcal{Y}}$.

7.4 The Poincaré Polynomial

As a collective notation for the cohomology of a manifold $\mathcal{M}$, one has❖

Definition 7.5. *The* Poincaré series *of a manifold* $\mathcal{M}$ *is defined to be*

$$P_t(\mathcal{M})\ \stackrel{\text{def}}{=}\ \sum_r t^r\dim H^r(\mathcal{M})\ . \tag{7.4.1}$$

Of course, all finite dimensional manifolds have $H^r(\mathcal{M})=0$ for $r<0$ and for $r>\dim\mathcal{M}$, so that the Poincaré series truncates to a finite polynomial.

❖ For a fuller account on Poincaré series, we refer the Reader to Ref. [11] and Ref. [3, 49]; here we only list some possibly useful results.

As a few important examples, we list

$$P_t(\mathbb{P}^n) = \frac{(1-t^{2n})}{(1-t^2)}\,, \tag{7.4.2}$$

$$P_t(G_{(k,n)}) = \frac{\prod_{i=n-k+1}^{n}(1-t^{2i})}{\prod_{i=1}^{k}(1-t^{2k})}\,, \tag{7.4.3}$$

$$P_t(\mathbb{F}_{[n_1,\cdots,n_f]}) = \frac{\prod_{i=1}^{n}(1-t^{2i})}{\prod_{r=1}^{f}\prod_{j=1}^{n_r}(1-t^{2j})}\,, \tag{7.4.4}$$

which are the Poincaré series of the complex projective space, the Grassmannian and the generalized flag space, respectively. Recall that

$$\mathbb{F}_{[n_1,\cdots,n_f]} \stackrel{\text{def}}{=} \frac{U(N)}{U(n_1)\times\cdots\times U(n_f)}\,, \qquad N = \sum_{r=1}^{f} n_r \tag{7.4.5}$$

is the generalized (unitary) flag space, so that $\mathbb{P}^n = \mathbb{F}_{[1,n]}$ and $G_{(k,n)} = \mathbb{F}_{[k,n-k]}$.

Of course, one can similarly define the $\mathcal{V}$-valued Poincaré series, where the cohomology $H^r(\mathcal{M})$ is replaced in the above definition with the $\mathcal{V}$-valued cohomology. 2024⌈*Very general line-bundle valued cohomology generalization of (7.4.1) turns out to admit analogous generating functions [362]. The same should be true at least for vector bundles defined in terms of systems of line bundles, such as (3.3.14). Since the Koszul resolution (9.3.7) is also defined in terms of systems of line bundles (3.3.14), cohomology groups* $H^1(\mathcal{M},(\mathcal{T}_r)^{\ell_r}\otimes\mathcal{O}_r(k_r^a))$ *should also be computable.*⌋

— ❦ —

In a more general context, one can define the Poincaré series of a graded algebra $\mathfrak{A} = \bigoplus_i \mathfrak{A}_i$ as

$$P_t(\mathfrak{A}) \stackrel{\text{def}}{=} \sum_i t^i \dim(\mathfrak{A}_i)\,. \tag{7.4.6}$$

Useful examples of this kind are the Poincaré series of polynomial rings. If x is a single variable of degree d, then

$$P_t(\mathbb{C}[x]) = 1 + t^d + t^{2d} + \cdots = \frac{1}{1-t^d}\,, \tag{7.4.7}$$

where $\mathbb{C}[x]$ denotes the polynomial ring generated by x (of course, simple monomials), with complex coefficients. Owing to the well known properties of polynomials, we have that for several independent variables x_i, of degree d_i,

$$\prod_i \mathbb{C}[x_i] = \mathbb{C}[\{x_i\}] , \quad \Longrightarrow \quad P_t(\mathbb{C}[\{x_i\}]) = \prod_i \frac{1}{1-t^{d_i}} . \tag{7.4.8}$$

If a single relation, I, of degree q is imposed on the polynomial ring $\mathbb{C}[x_1, \ldots, x_n]$, we have

$$P_t(\mathbb{C}[x_1, \ldots, x_n]/I) = P_t(\mathbb{C}[x_1, \ldots, x_n]) \cdot (1-t^q) = \frac{(1-t^q)}{\prod_{i=1}^n (1-t^{d_i})} . \tag{7.4.9}$$

It would be desirable to extend this result to the case of several relations. There, however, we need the following:

Definition 7.6. *Let $\mathcal{R}$ be a ring. A non-zero element $z \in \mathcal{R}$ is called a* zero-divisor *if there exists another non-zero element $a \in \mathcal{R}$ such that $az = 0$ in $\mathcal{R}$.*
A sequence of elements $r_1, \ldots, r_k$ is regular *if*

1. *r_1 is not a zero-divisor in $\mathcal{R}$ and*
2. *for each $i > 1$, the image of r_i in $\{\mathcal{R}/(r_1, \ldots, r_{i-1})\}$ is not a zero-divisor.*

Roughly, the elements of a regular sequence are non-zero and are supposed not to annihilate their successors.

Then we have the following results [11]:

Lemma 7.1. *If $r_1, \ldots, r_k, a$ and $r_1, \ldots, r_k, b$ are both regular sequences, then so is $r_1, \ldots, r_k, ab$.*

And finally,

Theorem 7.5. *Let $\mathcal{R}$ be a graded ring over a field $\mathbb{k}$ and $r_1, \ldots, r_k$ a regular sequence of elements $r_i \in \mathcal{R}$, with degrees $q_1, \ldots, q_k$. Then*
1. *any permutation of $r_1, \ldots, r_k$ is also regular;*
2. $P_t(\{\mathcal{R}/(r_1, \ldots, r_k)\}) = P_t(\mathcal{R}) \cdot \prod_{i=1}^k (1-t^{q_i})$.

These results will have an obvious application in Chapter C, where we will see that, for all complete intersections $\mathcal{M}$ in products of flag spaces, the cohomology on $\mathcal{M}$ can be given the structure of a quotient ring $\mathcal{R} = \{\mathbb{C}[x_1, \ldots, x_k] / \Im\}$, where $\mathbb{C}[x_1, \ldots, x_k]$ is the polynomial ring on the embedding space, generated by homogeneous coordinates, and $\Im$ is an ideal, generated by gradients of the defining constraints, modulo perhaps some relations among those.

Chapter 8

Topological Triple Couplings

In the previous chapters, we have constructed Calabi–Yau 3-folds in several different ways, including complete intersections of hypersurfaces in products of complex projective spaces and hypersurfaces in a products of two compact complex surfaces. Three such constructions,

$$
\begin{aligned}
\mathcal{M}_{-18} &\in \left[\begin{array}{c\|ccc} 3 & 3 & 0 & 1 \\ 3 & 0 & 3 & 1 \end{array}\right]^{14,23}_{-18} &&\xrightarrow{/\mathbb{Z}_3} \mathcal{M}_{-6}\ [223],\\
\mathcal{M}_{-54} &\in \left[\begin{array}{c\|cc} 3 & 3 & 1 \\ 2 & 0 & 3 \end{array}\right]^{8,35}_{-54} &&\xrightarrow{/(\mathbb{Z}_3)^2} \mathcal{M}_{-6}\ [210],\\
\mathcal{M}_{-162} &\in \left[\begin{array}{c\|c} 2 & 3 \\ 2 & 3 \end{array}\right]^{2,83}_{-162} &&\xrightarrow{/(\mathbb{Z}_3)^3} \mathcal{M}_{-6}\ [223],
\end{aligned}
\tag{8.0.1}
$$

provide the earliest attempts at providing phenomenologically acceptable models (see § 0.8 for the requirements) and have been analyzed extensively in the past third of a century. [2024] *While relatively few CICY quotients by free group actions lead to 3-generation models [349, 334, 348] via the standard Ansatz of Chapter 0, the number of promising models relying on non-standard (extended bundle) Ansätze is estimated in* moles *(10^{23}); see the discussion on p. 108. Nevertheless, the relevant techniques and computational methods turn out to be closely related, and we continue in the framework of the standard Ansatz as an effective showcase.*

The Euler characteristic, χ_E, of such manifolds is easily computed from Eq. (2.1.30) and is given, respectively, in the subscripts above. The Hodge numbers $h^{p,q}$ are then obtained using the Lefschetz hyperplane theorem to yield $h^{1,1}$ as indicated by the first superscript above; the second is $h^{2,1}$.

The three 3-folds (8.0.1) are hypersurfaces in products of two surfaces: $\Sigma \times \Sigma$, $\Sigma \times \mathbb{P}^2$ and $\mathbb{P}^2 \times \mathbb{P}^2$, respectively, where $\Sigma \in [3\|3]$ is a generic cubic surface in $\mathbb{P}^3$. This structure turns out to allow a rather detailed description of the (1,1)-cohomology and we now turn to it.

8.1 The Structure of $H^{1,1}(\mathcal{M})$

We will now be interested in evaluating the triple product integral (0.5.5):

$$\mathring{\kappa}_{ABC} \stackrel{\text{def}}{=} \int_{\mathcal{M}} e_A \wedge e_B \wedge e_C \;, \qquad e_A \in H^{1,1}(\mathcal{M}) \;.$$

For all Calabi–Yau n-folds ($n{\neq}2$) $h^{2,0} = h^{0,2} = 0$ so that $H^{1,1}(\mathcal{M}) \approx H^2(\mathcal{M}, \mathbb{C})$. By the same argument, we can choose an integral basis and work with $H^2(\mathcal{M}, \mathbb{Z})$; that is, any integral harmonic 2-form must also be a $(1,1)$-form.

Remark: On K3, the Calabi–Yau 2-fold, $H^{2,0} \approx H^{0,2} \approx \mathbb{C}$, so $H^{1,1}$ is a codimension-2 subset of $H^2(\mathbb{C})$ and so has lots of room to miss $H^2(\mathbb{Z}) \subset H^2(\mathbb{C})$ completely (just think of a real line in $\mathbb{R}^2$ with an irrational slope, missing thus all non-zero integral points). In fact, if we do not require projectivity (the possibility of embedding in some $\mathbb{P}^n$) and define Calabi–Yau 2-folds as simply connected compact complex 2-space with trivial canonical bundle, a generic K3 is not projective and $H^{1,1} \cap H^2(\mathbb{Z}) = 0$. ⌈*In turn, projective K3 surfaces together with a* **primitive** *(not a power of another) ample line bundle are called* **polarized** *and have been classified [340].*⌋

Now, $H^2(\mathcal{M}, \mathbb{Z})$ is naturally isomorphic to $H_4(\mathcal{M}, \mathbb{Z})$ since both are dual to $H^4(\mathcal{M}, \mathbb{Z})$ (see *duality*$^{(L)}$). Clearly, this natural identification of $H^2(\mathcal{M})$ with $H_4(\mathcal{M})$ is equally valid over $\mathbb{Q}, \mathbb{R}$ or $\mathbb{C}$ as well. Under this identification, the triple products (0.5.5) correspond to triple intersections,

$$\mathring{\kappa}_{ABC} = \mathring{\kappa}(A, B, C) = [A] \cap [B] \cap [C] \;, \quad \text{where } e_A \cdot [B] \stackrel{\text{def}}{=} \int_B e_A = \delta_A{}^B \;, \tag{8.1.1}$$

and which are often easily computed and yield intersection numbers. In particular, when $\mathcal{M}$ is an n-fold (Calabi–Yau or not), given as a complete intersection in a product of manifolds whose homology and intersection structure we know, we can use the ideas presented in § 7.2 to compute all n-fold intersection products on $\mathcal{M}$ of classes of the form $\hat{x}^r \cap [\mathcal{M}]$.

8.1.1 Wall's classification theorem

There is another numerical invariant of $H^{1,1}(\mathcal{M})$, which has so far not surfaced in the physical application but plays an important rôle in classification. It is

$$\mathrm{p}_1[\omega] \stackrel{\text{def}}{=} \int_{\mathcal{M}} \mathrm{p}_1 \wedge \omega = -2\int_{\mathcal{M}} c_2 \wedge \omega, \qquad \omega \in H^{1,1}(\mathcal{M}). \tag{8.1.2}$$

We have used that $c_1(\mathcal{M}) = 0$, so that the first Pontrjagin class $\mathrm{p}_1 = c_1^2 - 2c_2 \in H^4(\mathcal{M})$ simplifies. Computed for elements of $H^2(\mathcal{M}, \mathbb{Z})$, the cubic form $\mathring{\kappa}(\ ,\ ,\ )$ and linear form $\mathrm{p}_1[\ ]$ are together with the Hodge numbers $h^{1,1}$ and $h^{2,1}$ sufficient for classifying Calabi–Yau manifolds up to homotopy type [50].

More precisely, we have

Theorem 8.1 (Wall). *The homotopy types of complex compact 3-folds are classified by the following numerical characteristics: $h^{p,q}$, $\mathring{\kappa}_{ABC}$ and $\mathrm{p}_1[\omega_{(A)}]$.*

Clearly, two cubic forms $\mathring{\kappa}(\ ,\ ,\ )$ and $\mathring{\kappa}'(\ ,\ ,\ )$ are equivalent if there is a change of bases under which $\mathring{\kappa}(\ ,\ ,\ ) \to \mathring{\kappa}'(\ ,\ ,\ )$; simultaneously, we must also have $\mathrm{p}_1[\ ] \to \mathrm{p}_1'[\ ]$.

Remark: Generally, the linear form $\mathrm{p}_1[\ ]$ on H^2 of some (not necessarily Calabi–Yau) 3-fold $\mathcal{M}$ is given in terms of the cubic form $\mathring{\kappa}(\ ,\ ,\ )$ and $c_1(\mathcal{M})$. To see this, restrict to a surface $\mathcal{S} \subset \mathcal{M}$; write $c(\mathcal{M}|_{\mathcal{S}}) = 1 + c_1 + c_2$, $c(\mathcal{S}) = 1+s_1+s_2$, and use $\mathcal{S}$ also to denote the $H_4(\mathcal{M}) \approx H^2(\mathcal{M})$ element it represents. Then $c[\mathcal{N}_{\mathcal{M}/\mathcal{S}}] = 1 + [c_1 - s_1]$ and so

$$c(\mathcal{S}) = \frac{c(\mathcal{M}|_{\mathcal{S}})}{c[\mathcal{N}_{\mathcal{M}/\mathcal{S}}]} \qquad \Longleftrightarrow \qquad c_2 = s_2 - {s_1}^2 + c_1 \wedge s_1 \ . \tag{8.1.3}$$

Integrating over $\mathcal{S}$ and using Eq. (3.1.3), it is not hard to verify that

$$\mathrm{p}_1[\mathcal{S}] = 4\mathring{\kappa}(\mathcal{S},\mathcal{S},\mathcal{S}) - 24\chi^h(\mathcal{S}) + \mathring{\kappa}(K_{\mathcal{M}}, K_{\mathcal{M}}, \mathcal{S}) - 2\mathring{\kappa}(K_{\mathcal{M}}, \mathcal{S}, \mathcal{S}) \ , \tag{8.1.4}$$

where $K_{\mathcal{M}}$ is the *canonical divisor*$^{(L)}$ in $\mathcal{M}$. For a Calabi–Yau 3-fold $\mathcal{M}$, this simplifies since $c_1(\mathcal{M}) = 0$, and the last two terms drop out.

In order for two Calabi–Yau manifolds $\mathcal{M}$ and $\mathcal{M}'$ to have the same homotopy type it is necessary that they have the same Hodge numbers and that there is an isomorphism $f\colon H_4(\mathcal{M}, \mathbb{Z}) \to H_4(\mathcal{M}', \mathbb{Z})$ such that:

$$\mathring{\kappa}(x,y,z) \ = \ \mathring{\kappa}(f(x), f(y), f(z)) \ , \qquad \mathrm{p}_1[x] \ = \ \mathrm{p}_1[f(x)] \ . \tag{8.1.5}$$

This criterion can be difficult to use in practice both because it is often hard to find an explicit integral basis for the homology, and also because it is generally far from obvious whether or not two cubic forms are equivalent up to a change of basis. For this reason it is useful to define some readily computable numerical invariants of the pair, $(\mathring{\kappa}(\ ,\ ,\ ), \mathrm{p}_1[\ ])$, which will serve to distinguish homotopy types.

— ❦ —

The simplest divisibility invariants are

$$\begin{aligned} d_1 &= \gcd\{\mathring{\kappa}(x,y,z)\} ,\\ d_2 &= \gcd\{\mathring{\kappa}(x,x,y)\} ,\\ d_3 &= \gcd\{\mathring{\kappa}(x,x,x)\} , \end{aligned} \tag{8.1.6}$$

where x, y and z range all over $H_4(\mathcal{M},\mathbb{Z})$. It suffices to compute respectively

$$\begin{aligned} d_1 &= \gcd\{\mathring{\kappa}(a,b,c)\} ,\\ d_2 &= \gcd\{\mathring{\kappa}(a,a,c)\, ,\, 2\mathring{\kappa}(a,b,c)\} ,\\ d_3 &= \gcd\{\mathring{\kappa}(a,a,a)\, ,\, 3(\mathring{\kappa}(a,a,c) \pm \mathring{\kappa}(a,c,c))\, ,\, 6\mathring{\kappa}(a,b,c)\} , \end{aligned} \tag{8.1.7}$$

where a, b and c range over any integral basis of $H_4(\mathcal{M})$. We note the evident fact that d_1 divides d_2, which in turn divides d_3.

Similarly, we have $d_{\mathrm{P}} = \gcd\{\mathrm{p}_1[x]\} = \gcd\{\mathrm{p}_1[a]\}$.

Next, we consider the symmetric quadrilinear form obtained by symmetrizing

$$\langle x,y,z,w\rangle \overset{\text{def}}{=} \Big(\mathring{\kappa}(x,y,z)\,\mathrm{p}_1[w] + \text{ cyclic permutations}\Big).$$

Clearly there is an analogous hierarchy of divisibility invariants for $\langle\ ,\ ,\ ,\ \rangle$:

$$\begin{aligned} d_4 &= \gcd\{\langle a,b,c,d\rangle\} ,\\ d_5 &= \gcd\{\langle a,a,c,d\rangle\, ,\, 2\langle a,b,c,d\rangle\} ,\\ d_6 &= \gcd\{\langle a,a,a,d\rangle\, ,\, 3(\langle a,a,c,d\rangle \pm \langle a,c,c,d\rangle)\, ,\, 6\langle a,b,c,d\rangle\} ,\\ d_7 &= \gcd\{\langle a,a,a,a\rangle\, ,\, 2(2\langle a,a,a,d\rangle \pm 3\langle a,a,d,d\rangle \pm 2\langle a,d,d,d\rangle)\, ,\\ &\qquad 12(\langle a,a,c,d\rangle \pm \langle a,c,c,d\rangle \pm \langle a,c,d,d\rangle)\, ,\, 24\langle a,b,c,d\rangle\} . \end{aligned} \tag{8.1.8}$$

Additionally, we can interpret $\langle\ ,\ ,\ ,\ \rangle$ as a quadratic form $\mathcal{Q}(\ ,\ )$ on the symmetric square of $H_4(\mathcal{M},\mathbb{Z})$. The rank and the signature of $\mathcal{Q}(\ ,\ )$ are invariants over the real homology of $\mathcal{M}$. Further, let $d_{\mathcal{Q}}$ denote the determinant of $\mathcal{Q}$ and note that, under a change of basis $x_i \to \Lambda_i^j\, x_j$, for $x_i \in H_4(\mathcal{M},\Bbbk)$ the symmetric 'square' transforms as:

$$x_{(i}x_{j)} \to \Lambda_i^k \Lambda_j^\ell x_{(k}x_{\ell)} \ , \quad \mathbf{\Lambda} \in \mathrm{GL}(h^{1,1};\Bbbk) \ , \tag{8.1.9}$$

whence

$$d_{\mathcal{Q}} \to (\det \mathbf{\Lambda})^{h^{1,1}+1} d_{\mathcal{Q}} \ . \tag{8.1.10}$$

If we have an integral basis for $H_4(\mathcal{M})$, $\mathbf{\Lambda}$ must be an integral linear transformation and so of unit determinant. Then, $d_{\mathcal{Q}}$ is strictly invariant. However, even if we compute $\mathcal{Q}$ over $H_4(\mathcal{M}, \mathbb{Q})$, it retains certain invariance. $d_{\mathcal{Q}}$ is invariant up to multiplicatve $(h^{1,1}+1)^{\text{th}}$ powers of rational numbers and we may reduce $d_{\mathcal{Q}}$ by such powers; write $d_{\mathcal{Q}} \cong_{\mathbb{Q}} d_{\mathcal{Q}}^{(\text{reduced})}$. Just like the formula for the Euler characteristic (2.1.30), the divisibility properties are well suited for machine computation, although of course d_1, d_2 and d_3 are often easily calculated also by hand.

8.2 Favorable Configurations, Again

Consider again the set of complete intersection varieties in products of complex projective spaces as defined in Chapter 2. In the notation (2.1.5):

$$\mathcal{M} \in [\mathcal{X} \| \mathcal{E}] \equiv \left[\begin{array}{c||ccc} n_1 & q_1^1 & \cdots & q_K^1 \\ \vdots & \vdots & \ddots & \vdots \\ n_m & q_1^m & \cdots & q_K^m \end{array} \right], \tag{8.2.1}$$

where $\mathcal{E}$ denotes the direct sum of K line bundles $\mathcal{E}_a$ over $\mathcal{X}$, the sections of which, ξ_a, define $\mathcal{M}$: $\xi_a = 0$.

We recall that a generic $\mathcal{M} \in [\mathcal{X} \| \mathcal{E}]$, corresponding to a generic choice of polynomials, is a Kähler manifold of dimension $\dim \mathcal{M} = \dim \mathcal{X} - K$. The Euler characteristic, $\chi_{_E}$, of such a smooth n-fold is computed as the coefficient in the formal expansion in J_r, the generators of $H^{1,1}(\mathbb{P}_r^{n_r}, \mathbb{Z})$:

$$\left[\frac{\prod_{r=1}^m (1 + J_r)^{n_r+1}}{\prod_{a=1}^K (1 + \sum_{r=1}^m q_a^r J_r)} \right] \cdot \prod_{a=1}^K \left(\sum_{r=1}^m q_a^r J_r \right) = \ldots + \chi_{_E}(\mathcal{M}) \cdot \prod_{r=1}^m (J_r)^{n_r} \ . \tag{8.2.2}$$

The quantity in the square brackets is $c(\mathcal{M})$ whence the series is truncated to an order-$n(= \dim \mathcal{M})$ polynomial in J_r; also, we have that $(J_r)^{n_r+1} \equiv 0$. In the extreme case $\dim \mathcal{M} = 0$, i.e., when $\mathcal{M}$ is discrete, $c(\mathcal{M}) = 1$ and $\chi_{_E}(\mathcal{M})$ reduces to the coefficient of $\prod_{r=1}^m (J_r)^{n_r}$ in $\prod_{a=1}^K (\sum_{r=1}^m q_a^r J_r)$.

The first Pontrjagin class $\mathrm{p}_1 = {c_1}^2 - 2c_2$ is computed straightforwardly and for Calabi–Yau 3-folds ($c_1 = 0$):

$$\mathrm{p}_1 = -2c_2 = \sum_{r,s=1}^m \left[(n_r + 1)\, \delta^{r,s} - \sum_{a=1}^K q_a^r\, q_a^s \right] J_r J_s \ . \tag{8.2.3}$$

8.2.1 Computing integrals by counting

In the sector of $H^{1,1}(\mathcal{M})$ generated by the Kähler forms of the $\mathbb{P}_r^{n_r}$ factors in $\mathcal{X}$, $\mathring{\kappa}(\ ,\ ,\ )$ and $\mathrm{p}_1[\ ]$ may be computed as follows:

1. The Kähler form $J_r \in H^{1,1}(\mathbb{P}_r^{n_r})$ may be represented by a generic hyperplane in $X_r \subset \mathbb{P}_r^{n_r}$ (corresponds to a linear constraint on $\mathbb{P}_r^{n_r}$ and independent of the other factors in $\mathcal{X}$); we denote such a constraint by X_r as well.
2. The triple product (Yukawa coupling) $\mathring{\kappa}(X_r, X_s, X_t)$ is evaluated as

$$\mathring{\kappa}(X_r, X_s, X_t) = \chi_E\,[\mathcal{X}\|\mathcal{E}\; X_r\, X_s\, X_t]\;, \tag{8.2.4}$$

   enforcing the three constraints X_r, X_s and X_t in addition to the defining constraints of $\mathcal{M}$. This makes $[\mathcal{X}\|\mathcal{E}\; X_r\, X_s\, X_t]$ into a collection of points and $\mathring{\kappa}(X_r, X_s, X_t)$ is their number. (For an n-fold, one considers the n-fold intersection.)
3. The $\mathrm{p}_1[X_r]$ evaluation is computed as in Eq. (8.2.2), replacing the term in the square brackets with $\mathrm{p}_1\, X_r$.

The foregoing is sufficient to compute all $(1,1)^3$ couplings and p_1-evaluations whenever $H^2(\mathcal{M})$ is given completely by a restriction from $H^2(\mathcal{X})$. As we show in Chapter 9, the kernel of this restriction is

$$\ker\Big[\,H^{1,1}(\mathcal{X}) \to H^{1,1}(\mathcal{M})\Big] \;=\; H^1(\mathcal{M}, \mathcal{E}^*), \qquad \mathcal{E}^* = \bigoplus_{a=1}^{K} \mathcal{E}_a{}^*\;, \tag{8.2.5}$$

and if non-zero, indicates that not all generators J_r are independent when restricted to $H^{1,1}(\mathcal{M})$. The cohomology group on the left hand side can be computed with the techniques discussed below. However, if $H^1(\mathcal{M}, \mathcal{E}^*)$ is non-zero, the embedding of $\mathcal{M}$ may be modified using Lemma 2.2, in such a way that $H^1(\mathcal{M}, \mathcal{E}^*)$ becomes zero. Thus we incur no loss of generality by considering embeddings for which the restriction $H^{1,1}(\mathcal{X}) \to H^{1,1}(\mathcal{M})$ is 1–1, that is, all the generators J_r restrict to independent elements of $H^{1,1}(\mathcal{M})$.

On the other side, there might be (1,1)-forms on $\mathcal{M}$ which do not stem from J_r; the cycles corresponding to such forms are usually called "vanishing cycles". However, if $b_2(\mathcal{X}) = b_2(\mathcal{M})$, the restriction $H^{1,1}(\mathcal{X}) \to H^{1,1}(\mathcal{M})$ is an isomorphism over the complex numbers and hence over the reals and rationals as well, so, we can choose $\{J_r\}_{r=1,\dots,m}$ as a basis for $H^{1,1}(\mathcal{M}, \mathbb{Q})$.

In a number of cases, we can verify that $H^{1,1}(\mathcal{X}) \to H^{1,1}(\mathcal{M})$ is actually an isomorphism also over the integers by relying on Bott's proof [10] of the Lefschetz hyperplane theorem (see § 1.6). As described in § 2.4, one applies this theorem iteratively [145] provided positivity of the line bundles $\mathcal{E}_a$ is satisfied in every iteration.

8.3 Some Examples

As a quick application of our computations to homotopy classification and to clarify the technique, we include a few simple examples.

8.3.1 Simple hypersurfaces

Consider $\mathcal{M} \in [5\|3\,3]^1_{-144}$ *vs.* $\mathcal{M}' \in [6\|3\,2\,2]^1_{-144}$. In both cases $h^{1,1} = 1$ and $h^{1,2} = 73$ so that their Hodge diamonds are identical. The Lefschetz hyperplane theorem applies in both cases and implies that the only element of the (1,1)-cohomology on the Calabi–Yau manifold is the pullback of J, the Kähler form of the ambient space $\mathcal{X}$. Moreover, J generates the *integral* cohomology on the Calabi–Yau manifold and a generic element of this cohomology is $X = kJ$, an integral multiple of J. Thus we have:

$$\begin{aligned} [5\|3\ 3]^{1,73}_{-144} &: \quad \mathring{\kappa}(X,X,X) = 9k^3, \quad \mathrm{p}_1[X] = -12k, \\ [6\|3\,2\,2]^{1,73}_{-144} &: \quad \mathring{\kappa}(X,X,X) = 12k^3, \quad \mathrm{p}_1[X] = -10k. \end{aligned} \tag{8.3.1}$$

It is quite obvious that there is no linear transformation of $H^2(\mathcal{M})$ which would bring $\mathring{\kappa}(X,X,X)$ and $\mathrm{p}_1[X]$ of $\mathcal{M}$ in the form of those of $\mathcal{M}'$. Therefore, they are distinct Calabi–Yau manifolds. In point of fact, $\mathcal{M}$ and $\mathcal{M}'$ differ in *all* of the invariants discussed in § 8.1.1 except, of course, the rank and signature of $\mathcal{Q}(\ ,\ )$.

— ❦ —

8.3.2 Three complete intersections

To illustrate further the application of the divisibility invariants of $\mathring{\kappa}(\ ,\ ,\ )$ and $\mathrm{p}_1[\ ]$, we present the detailed computations for three related examples.

Consider the three configurations:

$$\mathcal{A} \in \begin{bmatrix} 2 & \| & 2 & 1 & 0 \\ 2 & \| & 1 & 1 & 1 \\ 2 & \| & 1 & 1 & 1 \end{bmatrix}^{3,51}_{-96}, \quad \mathcal{B} \in \begin{bmatrix} 2 & \| & 2 & 1 \\ 2 & \| & 2 & 1 \\ 1 & \| & 0 & 2 \end{bmatrix}^{3,51}_{-96}, \quad \mathcal{C} \in \begin{bmatrix} 2 & \| & 1 & 2 \\ 2 & \| & 2 & 1 \\ 1 & \| & 0 & 2 \end{bmatrix}^{3,51}_{-96}. \tag{8.3.2}$$

In all three cases, the Lefschetz hyperplane theorem can be applied to guarantee that $H^2(\mathcal{A},\mathbb{Z})$, $H^2(\mathcal{B},\mathbb{Z})$ and $H^2(\mathcal{C},\mathbb{Z})$ respectively are generated by the restrictions of the J_r.

For $\mathcal{A}$, we compute $\mathring{\kappa}(\ ,\ ,\ )$ as follows:

$$A, B, C \in H_4(\mathcal{A},\mathbb{Z}), \qquad A := \left|\begin{matrix}1\\0\\0\end{matrix}\right], \quad B := \left|\begin{matrix}0\\1\\0\end{matrix}\right], \quad C := \left|\begin{matrix}0\\0\\1\end{matrix}\right], \tag{8.3.3}$$

so that we have:

$$\mathring{\kappa}(A,A,A) = \chi_E \left[\begin{array}{c||ccc\,ccc} 2 & 2&1&0 & 1&1&1 \\ 2 & 1&1&1 & 0&0&0 \\ 2 & 1&1&1 & 0&0&0 \end{array}\right] = 0,$$

$$\mathring{\kappa}(A,A,B) = \chi_E \left[\begin{array}{c||ccc\,ccc} 2 & 2&1&0 & 1&1&0 \\ 2 & 1&1&1 & 0&0&1 \\ 2 & 1&1&1 & 0&0&0 \end{array}\right] = 3,$$

$$\mathring{\kappa}(A,A,C) = \chi_E \left[\begin{array}{c||ccc\,ccc} 2 & 2&1&0 & 1&1&0 \\ 2 & 1&1&1 & 0&0&0 \\ 2 & 1&1&1 & 0&0&1 \end{array}\right] = 3,$$

$$\mathring{\kappa}(A,B,B) = \chi_E \left[\begin{array}{c||ccc\,ccc} 2 & 2&1&0 & 1&0&0 \\ 2 & 1&1&1 & 0&1&1 \\ 2 & 1&1&1 & 0&0&0 \end{array}\right] = 3,$$

$$\mathring{\kappa}(A,B,C) = \chi_E \left[\begin{array}{c||ccc\,ccc} 2 & 2&1&0 & 1&0&0 \\ 2 & 1&1&1 & 0&1&0 \\ 2 & 1&1&1 & 0&0&1 \end{array}\right] = 6,$$

$$\mathring{\kappa}(A,C,C) = \chi_E \left[\begin{array}{c||ccc\,ccc} 2 & 2&1&0 & 1&0&0 \\ 2 & 1&1&1 & 0&0&0 \\ 2 & 1&1&1 & 0&1&1 \end{array}\right] = 3,$$

$$\mathring{\kappa}(B,B,B) = \chi_E \left[\begin{array}{c||ccc\,ccc} 2 & 2&1&0 & 0&0&0 \\ 2 & 1&1&1 & 1&1&1 \\ 2 & 1&1&1 & 0&0&0 \end{array}\right] = 0,$$

$$\mathring{\kappa}(B,B,C) = \chi_E \left[\begin{array}{c||ccc\,ccc} 2 & 2&1&0 & 0&0&0 \\ 2 & 1&1&1 & 1&1&0 \\ 2 & 1&1&1 & 0&0&1 \end{array}\right] = 2,$$

$$\mathring{\kappa}(B,C,C) = \chi_E \left[\begin{array}{c||ccc\,ccc} 2 & 2&1&0 & 0&0&0 \\ 2 & 1&1&1 & 1&0&0 \\ 2 & 1&1&1 & 0&1&1 \end{array}\right] = 2,$$

$$\mathring{\kappa}(C,C,C) = \chi_E \left[\begin{array}{c||ccc\,ccc} 2 & 2&1&0 & 0&0&0 \\ 2 & 1&1&1 & 0&0&0 \\ 2 & 1&1&1 & 1&1&1 \end{array}\right] = 0,$$

and of course the permutations of these.

Using Eq. (8.2.3) we have, denoting $X_r = A, B, C$, for $r = 1, 2, 3$:

$$\mathrm{p}_1 = \sum_{r,s=1}^{3} \left(3\delta^{r,s} - \sum_{a=1}^{3} q_a^r q_a^s\right) X_r X_s = -2[A^2 + 3AB + 3AC + 3BC]\ . \tag{8.3.4}$$

Some of the homotopy invariants introduced above are $d_1=1$, $d_2=1$, $d_3=6$, $d_{\rm P}=72$ and $d_{\mathcal{Q}}=-2^{30}\,3^{14}\cong_{\mathbb{Q}}-2^2\,3^2$.

By analogous choice of basis and computation in the case of $\mathcal{B}$, we obtain the non-vanishing $\mathring{\kappa}(\ ,\ ,\ )$ couplings:

$$\begin{array}{lll}\mathring{\kappa}(A,A,B)=4, & \mathring{\kappa}(A,A,C)=2, & \mathring{\kappa}(A,B,B)=4,\\ \mathring{\kappa}(A,B,C)=4, & \mathring{\kappa}(B,B,C)=2, & \end{array} \tag{8.3.5}$$

and the first Pontrjagin class,

$$\mathrm{p}_1=-2[A^2+B^2+5AB+2AC+2BC]\ . \tag{8.3.6}$$

Now we have $d_1=2$, $d_2=2$, $d_3=6$, $d_{\rm P}=24$ and $d_{\mathcal{Q}}=-2^{32}\,3^{10}\cong_{\mathbb{Q}}-3^2$.

Finally, in the case of $\mathcal{C}$, we obtain the non-vanishing $\mathring{\kappa}(\ ,\ ,\ )$ couplings,

$$\begin{array}{lll}\mathring{\kappa}(A,A,B)=4, & \mathring{\kappa}(A,A,C)=2, & \mathring{\kappa}(A,B,B)=2,\\ \mathring{\kappa}(A,B,C)=5, & \mathring{\kappa}(B,B,C)=2, & \end{array} \tag{8.3.7}$$

and the first Pontrjagin class,

$$\mathrm{p}_1=-2[A^2+B^2+4AB+4AC+2BC]\ . \tag{8.3.8}$$

This time $d_1=1$, $d_2=2$, $d_3=6$, $d_{\rm P}=24$ and $d_{\mathcal{Q}}=-2^{18}\,3^{12}\,13^2\cong_{\mathbb{Q}}-2^2\,13^2$.

Comparing the divisibility properties of $\mathring{\kappa}(X,X,X)$ and $\mathrm{p}_1[X]$ shows that these three Calabi–Yau manifolds indeed belong to three distinct homotopy types. While $d_{\mathcal{Q}}$ alone suffices to distinguish all three, it is also harder to compute; d_1 and d_2 are of course easier to compute and provide the same resolution in the present case.

✎ Compute d_4, d_5, d_6 and d_7. ❧

8.4 Hypersurfaces in Products of Surfaces, Again

Let $\mathcal{M}$ be a Calabi–Yau manifold which is the zero locus of a section of the anti-canonical bundle $\mathcal{K}^*_{\mathcal{S}}\otimes\mathcal{K}^*_{\mathcal{S}'}$ of the product of two almost del Pezzo surfaces, $\mathcal{S}$ and $\mathcal{S}'$. By Eq. (7.2.9), we have

$$[\mathcal{M}]=K_{\mathcal{S}}\times[\mathcal{S}']\ +\ [\mathcal{S}]\times K_{\mathcal{S}'}, \tag{8.4.1}$$

where $K_{\mathcal{S}}\in H_2(\mathcal{S})$ is dual to $c_1(\mathcal{S})\in H^{1,1}(\mathcal{S})$. Since $c_1(\mathcal{S})=c_1[\mathcal{K}_{\mathcal{S}}]$, the homology class $K_{\mathcal{S}}$ is called the *canonical class*$^{(L)}$. $\mathcal{M}$ being a hypersurface in $\mathcal{S}\times\mathcal{S}'$, $H_4(\mathcal{M})$ is generated by terms of the form $(x\times\mathcal{S}')\cap[\mathcal{M}]$ and $(\mathcal{S}\times x')\cap[\mathcal{M}]$, where x and x' denote elements of $H_2(\mathcal{S})$ and $H_2(\mathcal{S}')$ respectively. By Eq. (7.2.4), the computation of triple intersection products of these terms is equivalent to the computation of quadruple products in $\mathcal{S} \times \mathcal{S}'$ with $[\mathcal{M}]$ as the fourth factor:

$$\bigcap\nolimits_{i=1}^{3} (x_i \times \mathcal{S}') \cap [\mathcal{M}] \;=\; \Big((x_1 \cap x_2 \cap x_3) \times [\mathcal{S}']\Big) \cap [\mathcal{M}] = 0 \;, \tag{8.4.2}$$

$$\bigcap\nolimits_{i=1}^{2} (x_1 \times \mathcal{S}') \cap (\mathcal{S} \times x') \cap [\mathcal{M}] \;=\; (x_1 \cap x_2) \times (x' \cap K_{\mathcal{S}'}) \;, \tag{8.4.3}$$

with corresponding modifications for two primed terms and one unprimed term.

For Wall's classification theorem 8.1, it is necessary to provide $\mathrm{p}_1[x^i]$, the elements $x^i \in H^2(\mathcal{M}, \mathbb{Z})$ integrated against the first Pontrjagin class of $\mathcal{M}$. We thus compute the Chern class of $\mathcal{M} \hookrightarrow \mathcal{S} \times \mathcal{S}'$, related to the Chern classes of $\mathcal{S}$ and $\mathcal{S}'$:

$$\begin{aligned} c\begin{bmatrix} \mathcal{S} \\ \mathcal{S}' \end{bmatrix}\!\!\left\|\begin{matrix} c_1 \\ c_1' \end{matrix}\right] &= \frac{c(\mathcal{S}) \cdot c(\mathcal{S}')}{(1 + c_1 + c_1')} = \frac{(1 + c_1 + c_2) \cdot (1 + c_1' + c_2')}{(1 + c_1 + c_1')} \;, \\ &= 1 + [c_2 + c_1 c_1' + c_2'] - [c_1 c_2 + c_1 c_1'(c_1 + c_1') + c_1' c_2'] \;. \end{aligned} \tag{8.4.4}$$

Using that the homology class corresponding to $c_2(\mathcal{S})$ may be identified with $\chi_E(\mathcal{S})$, the homology class corresponding to $c_2(\mathcal{M})$ is

$$\left(\chi_E(\mathcal{S}) \times [\mathcal{S}'] \;+\; K_{\mathcal{S}} \times K_{\mathcal{S}'} \;+\; [\mathcal{S}] \times \chi_E(\mathcal{S}')\right) \cap [\mathcal{M}] \;. \tag{8.4.5}$$

Using Eq. (8.4.1) and Eq. (3.1.3) yields

$$c_2(\mathcal{M}) \;\overset{*}{\sim}\; 12(K_{\mathcal{S}} \times 1 \;+\; 1 \times K_{\mathcal{S}'}) \tag{8.4.6}$$

as a homology class on $\mathcal{S} \times \mathcal{S}'$. If we now write x^i indifferently for an element of $H^2(\mathcal{S})$ and also for the corresponding element of $H^2(\mathcal{M})$, and x_i for the corresponding element of $H_2(\mathcal{S})$, we have

$$\int_{\mathcal{M}} c_2 \wedge x^i \;=\; 12(K_{\mathcal{S}} \cap x_i) \;. \tag{8.4.7}$$

Hence

$$\mathrm{p}_1[x^i] = \int_{\mathcal{M}} \mathrm{p}_1 \wedge x^i \;=\; -\,24(K_{\mathcal{S}} \cap x_i) \;, \tag{8.4.8}$$

and similarly for elements of $H^2(\mathcal{S}')$.

8.5 The Homotopy Type of Surfaces

In § 3.1, we have considered Calabi–Yau 3-folds embedded as a hypersurface in the product of some two almost del Pezzo surfaces. Among these, we distinguish the class of *generic* almost del Pezzo surfaces which includes $\mathbf{F}_0 \stackrel{\text{def}}{=} \mathbb{P}^1 \times \mathbb{P}^1$ and blowups of $\mathbb{P}^2$ at 0, …, 7 or 9, distinct points in general position (see Proposition 6 of Ref. [147] for exact conditions).

Since the Calabi–Yau manifolds obtained as hypersurfaces in a product of two generic *vs.* non-generic almost del Pezzo surfaces are diffeomorphic, we shall analyze only the generic case in this section. Then we discuss the application of our analysis in the non-generic case and conclude with some 'stringy' issues.

— ❦ —

To begin with, we will compute the intersection matrix (x, y) and $c_1[x]$ for all generic almost del Pezzo surfaces $\mathcal{S}$, for $x, y \in H_2(\mathcal{S}, \mathbb{Z})$. We write (x, y) for the homology intersection, and $c_1[x]$ denotes the integral of $c_1(\mathcal{S})$ over a cycle representing x or, equivalently, $c_1[x] = (K_{\mathcal{S}}, x)$. Note that $(x, y) \in H_0(\mathcal{S}, \mathbb{Z})$ is an integer and can be computed as

$$(x, y) = \chi_E\{x \cap y\} \tag{8.5.1}$$

provided the cycles representing the homology classes x and y can be chosen to be transversely intersecting smooth curves.

8.5.1 The homology of $\mathbf{F}_0$

The most natural basis for $H_2(\mathbf{F}_0, \mathbb{Z})$, $\mathbf{F}_0 = \mathbb{P}^1 \times \mathbb{P}^1$ (see § 8.6), consists of copies of the $\mathbb{P}^1$ factors themselves. Since a hyperplane in $\mathbb{P}^1$ is simply a point, we can obtain a copy of one factor $\mathbb{P}^1$ by imposing a linear constraint on the other one and have:

$$a \stackrel{\text{def}}{=} \mathbb{P}^1 \times \{\cdot\} = \left[\begin{matrix}1\\1\end{matrix}\middle\|\begin{matrix}0\\1\end{matrix}\right], \qquad \text{and} \qquad b \stackrel{\text{def}}{=} \{\cdot\} \times \mathbb{P}^1 = \left[\begin{matrix}1\\1\end{matrix}\middle\|\begin{matrix}1\\0\end{matrix}\right]. \tag{8.5.2}$$

The intersection number of two generators is computed as:

$$\begin{aligned}(a, a) &:= \chi_E\left[\begin{smallmatrix}1\\1\end{smallmatrix}\middle\|\begin{smallmatrix}0&0\\1&1\end{smallmatrix}\right] = 0,\\ (a, b) &:= \chi_E\left[\begin{smallmatrix}1\\1\end{smallmatrix}\middle\|\begin{smallmatrix}0&1\\1&0\end{smallmatrix}\right] = 1,\\ (b, b) &:= \chi_E\left[\begin{smallmatrix}1\\1\end{smallmatrix}\middle\|\begin{smallmatrix}1&1\\0&0\end{smallmatrix}\right] = 0.\end{aligned} \qquad \begin{array}{c|cc}(\ ,\ ) & a & b\\ \hline a & 0 & 1\\ b & 1 & 0\end{array} \tag{8.5.3}$$

Note that the intersection matrix for $\mathbf{F}_0$ is of Type II, i.e., for a general element $x \in H_2(\mathbf{F}_0)$, $x = \alpha a + \beta b$,

$$(x, x) = 2\alpha\beta \quad \in 2\mathbb{Z}, \quad \alpha, \beta \in \mathbb{Z}. \tag{8.5.4}$$

In other words, $\mathbf{F}_0$ contains only curves of even self-intersection.

The anti-canonical class of $\mathbb{P}^1$ is the square of its hyperplane bundle and corresponds to generic quadratic polynomials in the homogeneous coordinates of $\mathbb{P}^1$. The homology class dual to the Chern class of the anti-canonical bundle of $\mathbb{P}^1$ is $2 \in H_0(\mathbb{P}^1)$, i.e., two points of the 0-locus of a quadric equation. It follows from Eq. (7.2.9) that the canonical class of $\mathbf{F}_0$ is given by

$$K_{\mathbf{F}_0} = 2a + 2b, \qquad \Rightarrow \qquad c_1[a] = 2,\ c_1[b] = 2. \tag{8.5.5}$$

8.5.2 Successive blowups of $\mathbb{P}^2$

For the generic almost del Pezzo surfaces which are ρ-fold ($\rho = h^{1,1}-1$) iterated blowups of $\mathbb{P}^2$, the homology classes $\{E_i\}_{1 \leq i \leq r}$, corresponding to the exceptional divisors (lines) into which the ρ points were blown up, offer a natural choice of basis elements of $H_2(\mathcal{S}, \mathbb{Z})$. These are disjoint and each has self-intersection -1.

For the remaining basis element of $H_2(\mathcal{S}, \mathbb{Z})$ we choose the class E_0 carried by a generic line on $\mathbb{P}^2$, which misses all the points which were blown up. Thus we have:

$(\, , \,)$	E_0	E_j
E_0	1	0
E_i	0	$-\delta_{i,j}$

(8.5.6)

Note that, by the first line of (8.5.6), all these intersection matrices are of Type I, that is they do contain curves with an odd self-intersection.

Remark: The fact that their intersection matrices are of different types serves to distinguish between the homotopy types of $\mathbf{F}_0 = \mathbb{P}^1 \times \mathbb{P}^1$ and $\mathbf{F}_1 = \widetilde{\mathbb{P}}^2_{p_1}$, i.e., $\mathbb{P}^2$ blown up at one point although they have the same Hodge numbers.

On this basis, the divisor $K_{\mathcal{S}}$ corresponding to the anti-canonical class is [23]:

$$K_{\mathcal{S}} = 3E_0 - \sum_{i=1}^{\rho} E_i, \quad \Rightarrow \quad c_1[E_0] = 3, \quad c_1[E_i] = 1,\ i = 1, \ldots, \rho. \tag{8.5.7}$$

For all almost del Pezzo surfaces the intersection matrices have rank $h^{1,1}$ and signature $(1, h^{1,1}-1)$ (i.e., one plus sign and $(h^{1,1}-1)$ minus signs).

8.5.3 Topological Yukawa couplings computed

For the Calabi–Yau manifolds embedded as hypersurfaces in the product of two del Pezzo surfaces, the following cases arise:

1. $\mathcal{M}$ is a hypersurface in $\mathbf{F}_0 \times \mathbf{F}_0$. In this case we have $\{a', b', a'', b''\}$ as a basis for $H^{1,1}(\mathcal{M}) \approx H_4(\mathcal{M})$, and the non-vanishing triple couplings are:

$$\begin{aligned} \mathring{\kappa}(a', b', a'') &= 2 \;, & \mathring{\kappa}(a', b', b'') &= 2 \;, \\ \mathring{\kappa}(a', a'', b'') &= 2 \;, & \mathring{\kappa}(b', a'', b'') &= 2 \;. \end{aligned} \tag{8.5.8}$$

   We readily compute $d_1 = 2$, $d_2 = 4$, $d_3 = 12$ and $d_Q = 48$.

2. $\mathcal{M}$ is a hypersurface in $\mathbf{F}_0 \times \mathcal{S}$ where $\mathcal{S}$ is a ρ-fold iterated blowup of $\mathbb{P}^2$ where $0 \leq \rho \leq 7$, i.e., $3 \leq \chi_E(\mathcal{S}) \leq 10$. In this case we have $\{a, b, E_i\}$, $i = 0, \ldots, \rho$, as a basis for $H_4(\mathcal{M}) \approx H^{1,1}(\mathcal{M})$, and the non-vanishing triple couplings are:

$$\begin{aligned} \mathring{\kappa}(a, b, E_0) &= 3 \;, & \mathring{\kappa}(a, b, E_i) &= 1 \;, \\ \mathring{\kappa}(a, E_0, E_0) &= 2 \;, & \mathring{\kappa}(a, E_i, E_j) &= -2\delta_{i,j} \;, \\ \mathring{\kappa}(b, E_0, E_0) &= 2 \;, & \mathring{\kappa}(b, E_i, E_j) &= -2\delta_{i,j} \;. \end{aligned} \tag{8.5.9}$$

   This time $d_1 = 1$, $d_2 = 2$, $d_3 = 6$ and $d_Q = 24$.

3. $\mathcal{M}$ is a hypersurface in $\mathcal{S} \times \mathcal{S}'$ where $\mathcal{S}$ and $\mathcal{S}'$ is a ρ-fold and a ρ'-fold iterated blowup of $\mathbb{P}^2$ where $0 \leq \rho, \rho' \leq 7$, i.e., $3 \leq \chi_E(\mathcal{S}), \chi_E(\mathcal{S}') \leq 10$. In this case we have $\{E_a, E'_\alpha\}$, $a = 0, \ldots, \rho$ and $\alpha = 0, \ldots, \rho'$, as a basis for $H_4(\mathcal{M}) \approx H^{1,1}(\mathcal{M})$, and the non-vanishing triple couplings are:

$$\begin{aligned} \mathring{\kappa}(E_0, E_0, E'_0) &= 3 \,, & \mathring{\kappa}(E_0, E'_0, E'_0) &= 3 \,, \\ \mathring{\kappa}(E_0, E_0, E'_\alpha) &= 1 \,, & \mathring{\kappa}(E_a, E'_0, E'_0) &= 1 \,, \\ \mathring{\kappa}(E_a, E_b, E'_0) &= -3\delta_{a,b} \,, & \mathring{\kappa}(E_0, E'_\alpha, E'_\beta) &= -3\delta_{\alpha,\beta} \,, \\ \mathring{\kappa}(E_a, E_b, E'_\gamma) &= -\delta_{a,b} \,, & \mathring{\kappa}(E_c, E'_\alpha, E'_\beta) &= -\delta_{\alpha,\beta} \,. \end{aligned} \tag{8.5.10}$$

   Now $d_1 = 1$, $d_2 = 1$, $d_3 = 6$ and $d_Q = 24$.

Remark: The computation of d_2 shows that for $\mathcal{S}$ any almost del Pezzo surface with $\chi_E \leq 10$, Calabi–Yau hypersurfaces in $\mathbf{F}_0 \times \mathcal{S}$ are of a different homotopy type from those in $\mathbf{F}_1 \times \mathcal{S}$, although they have the same Hodge numbers. This completes the homotopy classification of Calabi–Yau hypersurfaces in products of almost del Pezzo surfaces.

Finally, let $\mathcal{Z}$ denote an iterated 9-fold blowup of $\mathbb{P}^2$, $\mathcal{Z} = \widetilde{\mathbb{P}}^2_{p_1,\dots,p_9}$ and $\chi_E(\mathcal{Z}) = 12$. It has been proven (Chapter 3, also Ref. [147]) that for any almost del Pezzo surface $\mathcal{S}$, the Calabi–Yau manifold $\mathcal{M}$ embedded in $\mathcal{Z}\times\mathcal{S}$ has $\chi_E(\mathcal{M}) = 0$ and is biholomorphic to one embedded in $\mathcal{Z} \times \mathcal{Z}'$, where $\mathcal{Z}'$ is an almost del Pezzo surface with $\chi_E(\mathcal{Z}') = 12$, and we focus now on the latter embedding.

Firstly, $h^{1,1}(\mathcal{M}) = 19$ (see Ref. [212]) and the set of 20 elements

$$\{E_0, \dots, E_9, E'_0, \dots, E'_9\} , \tag{8.5.11}$$

which would follow from the construction above, is redundant. Indeed, using the above results, it can be shown that $\mathring{\kappa}(x,y,z) \equiv 0\ \forall y,z \in H^2(\mathcal{M})$ precisely if x is a multiple of

$$I = 3(E_0 - E'_0) - \left(\sum_{a=1}^{9} E_a - \sum_{\alpha=1}^{9} E'_\alpha\right) . \tag{8.5.12}$$

Therefore $I \cong 0 \in H_4(\mathcal{M})$, leaving only 19 non-trivial independent generators.

Secondly, all the formulae (8.5.10) of case 3. are valid modulo, of course, the relation (8.5.12). In order to show that the triple products and Pontrjagin class satisfy the corresponding divisibility condition, it would however be necessary to know that $H^2(\mathcal{Z}\times\mathcal{Z}',\mathbb{Z})$ restricts *onto* $H^2(\mathcal{M},\mathbb{Z})$. Since $\mathcal{K}^*_{\mathcal{Z}}$ is not positive, the Lefschetz hyperplane theorem cannot be used. The triple product computation shows that the restriction is onto over the rationals, so that $d_{\mathbb{Q}}$, *modulo multiplicative* 20th *powers of integers*, is still a valid and rather precise invariant.

✎ Calculate $d_{\mathbb{Q}}$. ❧

8.6 Non-Generic Surfaces

The results of § 8.5 were derived for generic almost del Pezzo surfaces and we discuss now how to apply our analysis including non-generic and other auxiliary models.

8.6.1 Hirzebruch surfaces

It was demonstrated in § 3.1.2 that $\mathbf{F}_0 \approx \mathbb{P}^1 \times \mathbb{P}^1$ is a deformation of $\mathbf{F}_2$ and that they both may be constructed as members of the configuration $\left[\begin{smallmatrix}3\\1\end{smallmatrix}\middle\|\begin{smallmatrix}1&1\\1&1\end{smallmatrix}\right]$ (see Eq. (3.1.15) and the related discussion). For generators of H_2, we choose $A := \left|{}^{1}_{0}\right]$ and $B := \left|{}^{0}_{1}\right]$, so that:

$$
\begin{aligned}
(A,A) &= \chi_E\left[\begin{smallmatrix}3\\1\end{smallmatrix}\middle\|\begin{smallmatrix}1&1&1&1\\1&1&0&0\end{smallmatrix}\right] = 2\,,\\
(A,B) &= \chi_E\left[\begin{smallmatrix}3\\1\end{smallmatrix}\middle\|\begin{smallmatrix}1&1&1&0\\1&1&0&1\end{smallmatrix}\right] = 1\,,\\
(B,B) &= \chi_E\left[\begin{smallmatrix}3\\1\end{smallmatrix}\middle\|\begin{smallmatrix}1&1&0&0\\1&1&1&1\end{smallmatrix}\right] = 0\,.
\end{aligned}
\qquad
\begin{array}{c|cc}
(\;,\;) & A & B\\ \hline
A & 2 & 1\\
B & 1 & 0
\end{array}
\tag{8.6.1}
$$

The anti-canonical class is now represented by $\left|{}^{2}_{0}\right]$ (check: $c_1(\left[\begin{smallmatrix}3\\1\end{smallmatrix}\middle\|\begin{smallmatrix}1&1&2\\1&1&0\end{smallmatrix}\right]) = 0$) so that:

$$
c_1[A] = \chi_E\left[\begin{smallmatrix}3\\1\end{smallmatrix}\middle\|\begin{smallmatrix}1&1&2&1\\1&1&0&0\end{smallmatrix}\right] = 4, \qquad c_1[B] = \chi_E\left[\begin{smallmatrix}3\\1\end{smallmatrix}\middle\|\begin{smallmatrix}1&1&2&0\\1&1&0&1\end{smallmatrix}\right] = 2. \tag{8.6.2}
$$

Note that this computation makes no homotopy distinction between $\mathbf{F}_0$ and $\mathbf{F}_2$, as should be the case since the former is a deformation of the latter.

By the Lefschetz hyperplane theorem (see § 1.6), $\{A, B\}$ generate $H_2(\mathcal{Z}, \mathbb{Z})$ and note that:

$$
\begin{bmatrix}A\\B\end{bmatrix} = \begin{bmatrix}1&1\\0&1\end{bmatrix}\begin{bmatrix}a\\b\end{bmatrix}, \quad \begin{bmatrix}a\\b\end{bmatrix} = \begin{bmatrix}1&-1\\0&1\end{bmatrix}\begin{bmatrix}A\\B\end{bmatrix}, \quad \det\begin{bmatrix}1&1\\0&1\end{bmatrix} = 1\,, \tag{8.6.3}
$$

so that one can freely choose between the $\{a, b\}$ and the $\{A, B\}$ basis. In (8.5.3), we have used the former for convenience; the integral basis-change (8.6.3) relates (8.5.3) and (8.6.1).

2021 *As noted regarding (3.1.17), the deformation family of hypersurfaces $\left[\begin{smallmatrix}2\\1\end{smallmatrix}\middle\|\begin{smallmatrix}1\\2\end{smallmatrix}\right]$ harbors $\mathbf{F}_0$ as the generic elements, while $\mathbf{F}_2$ is its special, $\{p_0(x,y) \overset{\text{def}}{=} x_0 y_0{}^2 + x_1 y_1{}^2 = 0\}$, case. Here also, $A = \left|{}^{1}_{0}\right]$, $B = b = \left|{}^{0}_{1}\right]$ and $a = \left|{}^{1}_{-1}\right]$ recover the intersection tables (8.6.1) and (8.5.3). Furthermore, denoting $e = \left|{}^{1}_{-2}\right]$, we easily compute*

$$
\begin{aligned}
(a,a) &= \chi_E\left[\begin{smallmatrix}2\\1\end{smallmatrix}\middle\|\begin{smallmatrix}1&1&1\\2&0&0\end{smallmatrix}\right] = 2\,,\\
(a,e) &= \chi_E\left[\begin{smallmatrix}2\\1\end{smallmatrix}\middle\|\begin{smallmatrix}1&1&1\\2&0&-2\end{smallmatrix}\right] = 0\,,\\
(e,e) &= \chi_E\left[\begin{smallmatrix}2\\1\end{smallmatrix}\middle\|\begin{smallmatrix}1&1&1\\2&-2&-2\end{smallmatrix}\right] = -2.
\end{aligned}
\qquad
\begin{array}{c|cc}
(\;,\;) & a & e\\ \hline
a & 2 & 0\\
e & 0 & -2
\end{array}
\tag{8.6.4}
$$

This identifies the equivalence class $[e] = [a - 2b]$ as the exceptional divisor class of self-intersection -2. Whereas this formal equivalence class may be defined both for $\mathbf{F}_0$ and $\mathbf{F}_2$, only in the latter is it represented by an irreducible, holomorphic submanifold, the directrix *[23]. Variously motivated in Refs. [296, 319, 386, 321], we define the* equivalence class of sections*:*

$$
\mathfrak{s}(x,y;\lambda) \overset{\text{def}}{=} \left[\frac{x_0}{y_1{}^2} - \frac{x_1}{y_0{}^2} + \lambda\frac{p_0(x,y)}{(y_0y_1)^2}\right] = \begin{cases} 2\frac{x_0}{y_1{}^2}, & \text{for } \lambda = +1, \text{ when } y_1 \neq 0,\\ -2\frac{x_1}{y_0{}^2}, & \text{for } \lambda = -1, \text{ when } y_0 \neq 0.\end{cases} \tag{8.6.5}
$$

This $\mathfrak{s}(x,y)$ *is holomorphic and well-defined on the hypersurface* $\mathbf{F}_2 = \{p_0(x,y)=0\}$, *since:* (**1**) $\mathfrak{s}(x,y;\lambda)$ *has a holomorphic representative on each coordinate chart of* $\mathbb{P}^1$, *and* (**2**) $\mathfrak{s}(x,y;+1) - \mathfrak{s}(x,y;-1) \propto p_0(x,y)$ *vanishes on* $\mathbf{F}_2 := p_0^{-1}(0)$. *Note that* $\deg \mathfrak{s}(x,y;\lambda) = (1,-2)$, *so the zero-locus* $\mathfrak{s}^{-1}(0) \subset \mathbf{F}_2$ *is a (holomorphic, irreducible) representative of* e *in (8.6.4).*
✎ *Find the defining section of the directrix in* $\mathbf{F}_2 \in \left[\begin{smallmatrix}3\\1\end{smallmatrix}\middle\|\begin{smallmatrix}1&1\\1&1\end{smallmatrix}\right]$. ❧

8.6.2 Non-generic blowups of $\mathbb{P}^2$

In § 8.5.2 we have assumed that $\mathrm{dP}_n \stackrel{\text{def}}{=} \mathrm{Bl}^n[\mathbb{P}^2] = \widetilde{\mathbb{P}^2}_{p_1,\dots,p_n}$ was constructed by blowing up n points in general position on $\mathbb{P}^2$. More generally, if any p_i is on the exceptional line obtained by blowing up another point p_j, $j \neq i$, we have

$$(E_j, E_j) = -2, \quad (E_i, E_i) = -1, \quad (E_i, E_j) = +1\ . \tag{8.6.6}$$

To recover the basis of Subsection 8.5.2, we replace E_j by $\hat{E}_j := E_j - E_i$ and have

$$(\hat{E}_j, \hat{E}_j) = -1\ , \quad (E_i, E_i) = -1\ , \quad (E_i, \hat{E}_j) = 0\ , \tag{8.6.7}$$

which clearly generalizes for any other non-generic almost del Pezzo surface of this type, and (dropping the hats) we recover the basis with the intersection matrix (8.5.6).

We also recall that

$$\widetilde{\mathbb{P}^2}_{p_1,p_2,p_3} \in \left[\begin{matrix}1\\1\\1\end{matrix}\middle\|\begin{matrix}1\\1\\1\end{matrix}\right] \tag{8.6.8}$$

precisely if the three points are not collinear. To include the non-generic case when the three points *are* collinear, one embeds

$$\widetilde{\mathbb{P}^2}_{p_1,p_2,p_3} \in \left[\begin{matrix}2\\1\\1\\1\end{matrix}\middle\|\begin{matrix}1&1&1\\1&0&0\\0&1&0\\0&0&1\end{matrix}\right]\ . \tag{3.1.19$'$}$$

Interestingly enough, $b_2(\widetilde{\mathbb{P}^2}_{p_1,p_2,p_3}) = 4$, and explicit computation verifies that $H_2(\widetilde{\mathbb{P}^2}_{p_1,p_2,p_3})$ is completely spanned by (the pull-backs of) the four Kähler forms. This could not have been predicted using the Lefschetz hyperplane theorem—because it does not apply.

2024 ⌈*As noted in Ref. [337], multiple blowups of* $\mathbb{P}^2$ *may indeed be found in configurations generalizing (3.1.19). For example,*

$$\underset{\stackrel{\text{def}}{=}\widetilde{\mathbb{P}^2}_{p_1,\cdots,p_4}}{\mathrm{dP}_4} \in \left[\begin{matrix}2\\1\\1\\1\\1\end{matrix}\middle\|\begin{matrix}1&1&1&1\\1&0&0&0\\0&1&0&0\\0&0&1&0\\0&0&0&1\end{matrix}\right], \quad e_0 = \left[\begin{matrix}2\\1\\1\\1\\1\end{matrix}\middle\|\begin{matrix}1\\0\\0\\0\\0\end{matrix}\right], \quad e_1 = \left[\begin{matrix}2\\1\\1\\1\\1\end{matrix}\middle\|\begin{matrix}1\\-1\\0\\0\\0\end{matrix}\right], \quad e_2 = \left[\begin{matrix}2\\1\\1\\1\\1\end{matrix}\middle\|\begin{matrix}1\\0\\-1\\0\\0\end{matrix}\right], \quad \textit{etc.} \tag{8.6.9}$$

As above, we easily compute: $(e_0, e_0) = 1$, $(e_0, e_i) = 0$ *and* $(e_i, e_j) = -\delta_{i,j}$ *recovering (8.5.6). The anticanonical bundle evidently has* deg $= (-1, 1, 1, 1, 1)$, *is limping (see Table 9.1, below), and corresponds to the divisor class* $[K] = [3e_0 - \sum_{i=1}^{4} e_i]$. *This yields* $(K, e_0) = 3$, $(K, e_i) = 1$ *for* $i = 1, \ldots, 4$ *and* $(K, K) = 5$. *Since* $\chi_E(\mathrm{dP}_4) = 7$, *the sum* $\chi_E + (K, K) = 12$ *recovers Noether's formula (3.1.3) and (3.1.4). It remains "only" to find defining equations for* dP_n *such that:* ***(1)*** *their complete intersection is transverse, and* ***(2)*** *the* e_i *can be represented by directrix-like, (8.6.5), holomorphic defining sections.* ✎ *We leave that to the industrious Reader.* ❧

8.7 Physics Application

In order to employ the above formulae and derive certain physical observables several further steps of analysis are needed and we now turn briefly to discuss these.

8.7.1 Normalization matrix of the kinetic term

The lowest order term in the superpotential of the effective low-energy theory that involves only E_6 $\mathbf{27}^*$-plet superfields is

$$d_{\bar{I}\bar{J}\bar{K}}\ \mathring{\kappa}_{ABC}\ \overline{\Psi}{}^{\bar{I}}_A(x)\overline{\Psi}{}^{\bar{J}}_B(x)\overline{\Psi}{}^{\bar{K}}_C(x), \tag{8.7.1}$$

where $d_{\bar{I}\bar{J}\bar{K}}$ is the E_6 $\left[\mathbf{27}^* \otimes \mathbf{27}^* \otimes \mathbf{27}^*\right]_{\mathbf{1}}$ Clebsch–Gordan coefficient and $\mathring{\kappa}_{ijk}$ is (the point-field limit) Yukawa coupling (0.5.5).

The kinetic term for these superfields is however

$$\sum_{A,\bar{B},\bar{I}} N_{A\bar{B}}\ \left(\overline{\Psi}{}^{\bar{I}}_B(x)\right)^\dagger \overline{\Psi}{}^{\bar{I}}_A(x)\ , \tag{8.7.2}$$

where the normalization matrix $N_{A\bar{B}}$ is (in the point-field limit) given [218] by the inner product,

$$N_{A\bar{B}} \equiv N(e_A, e_B) = \tfrac{1}{2}\int_{\mathcal{M}} e_A \wedge (*e)_{\bar{B}}\ , \qquad e_A \in H^2(\mathcal{M})\ , \tag{8.7.3}$$

and $(*e)_{\bar{B}} \in H^4(\mathcal{M})$ is the dual of e_B. For the triple couplings $\mathring{\kappa}(\ ,\ ,\ )$ to have usual interpretation as Yukawa couplings, the superfields $\overline{\Psi}{}^{\bar{I}}_A(x)$ have to be (re)normalized to $(N_{A\bar{B}})$-eigenvectors, so that

$$\sum_{A,B,\bar{I}} N_{A\bar{B}}\left(\overline{\Psi}{}^{\bar{I}}_B\right)^\dagger \overline{\Psi}{}^{\bar{I}}_A \quad\to\quad \sum_{A,\bar{I}}\left(\widetilde{\overline{\Psi}}{}^{\bar{I}}_A\right)^\dagger \widetilde{\overline{\Psi}}{}^{\bar{I}}_A. \tag{8.7.4}$$

Denoting the Kähler class of $\mathcal{M}$ by J, $N_{A\bar{B}}$ may be expressed as [218]

$$N_{A\bar{B}} = \frac{3}{8}\frac{\mathring{\kappa}(e_A, J, J)\,\mathring{\kappa}(e_B, J, J)}{\mathring{\kappa}(J, J, J)} - \tfrac{1}{4}\mathring{\kappa}(e_A, e_B, J). \tag{8.7.5}$$

Using the above results for $\mathring{\kappa}(\ ,\ ,\ )$, this normalization matrix can be obtained straightforwardly. As should be manifest from Eq. (8.7.5), $N_{A\bar{B}}$ depends on the choice of the cohomology class of the Kähler form

$$J = \sum_{i=1}^{h^{1,1}-1} v^A e_A\ , \qquad e_A \in H^{1,1}(\mathcal{M})\ , \quad v^A \in \mathbb{R}_+, \tag{8.7.6}$$

where the coefficients v^A are restricted so that J is a positive (1,1)-form and $\mathring{\kappa}(J,J,J) \in \mathbb{R}_+$ since this is, up to a positive numerical coefficient, the volume of the Calabi–Yau manifold.

As an illustration, consider $\mathcal{M}$ defined as a hypersurface in $\mathcal{X} = \mathbf{F}_0 \times \mathbf{F}_0$. We choose the Kähler class of $\mathcal{M}$ to be represented by the sum of the first Chern classes of the two surfaces F_0 in $\mathcal{X}$. Using Eq. (8.5.8) and Eq. (8.7.5) produces

$2^3{\cdot}N(\,,)$	a	b	a'	b'
a	9	1	1	1
b	1	9	1	1
a'	1	1	9	1
b'	1	1	1	9

$$\xrightarrow[\text{similarity tr.}]{\left[\begin{smallmatrix} 1 & 1 & 1 & 1 \\ -1 & 0 & 0 & 1 \\ -1 & 0 & 1 & 0 \\ -1 & 1 & 0 & 0 \end{smallmatrix}\right]}$$

$2{\cdot}N(\,,)$	n_0	n_1	n_2	n_3
n_0	3	0	0	0
n_1	0	1	0	0
n_2	0	0	1	0
n_3	0	0	0	1

(8.7.7)

⌈*The four E_6 **27***-plet superfields representing the "mirror" particles in a low-energy model compactified on $\mathcal{M}$ are represented by four linearly independent combinations of $\{a, b, a', b'\}$, i.e., the normal modes, $\{n_0, \dots, n_3\}$. This choice of basis elements to represent physical fields clearly depends on the normalization matrix/metric (8.7.5), but also on the symmetries of the defining equations and other details; we refer the interested Reader to Refs. [114, 73] for further details in a particular, well-studied model. While a good start for the relevant metric in the underlying world-sheet theory, the expression (8.7.5) will become renormalized, so (8.7.7) may well be far from the physically relevant metric. When the mirror of $\mathcal{M}$ is known, the metric $N_{A\bar{B}}$ for the 27*'s is the mirror (pre)image of the Weil–Petersson metric for the **27**'s in the mirror, which is much better understood and can even be computed as a function of moduli; see § 0.8 (the 6th veil), [120], and also [165, 345] for substantial computational improvements in this direction.*⌋

8.7.2 Corrections

The value of $\mathring{\kappa}(\ ,\ ,\ )$ and therefore also $N(\,,)$ will receive corrections from quantum effects. While corrections to $\mathring{\kappa}(\ ,\ ,\ )$ are governed by certain non-renormalization theorems [111], it is not clear if $N(\, , \,)$ receives quantum corrections other than those induced, through Eq. (8.7.5), by corrections to $\mathring{\kappa}(\, , \, , \,)$.

In particular, $\mathring{\kappa}(\, , \, , \,)$ will receive corrections from world-sheet instantons. Such an instanton is represented by a non-trivial holomorphic map of the world-sheet into the "internal" Calabi–Yau manifold $\mathcal{M}$ and is therefore represented by embeddings of non-trivial holomorphic curves in $L \subset \mathcal{M}$. By definition then, $[L]$ is an element of $H_2(\mathcal{M})$.

The contribution to the full Yukawa coupling $\kappa(e_A, e_B, e_C)$, Eq. (0.6.3), of a world-sheet instanton represented by the holomorphic curve $L \subset \mathcal{M}$ is

$$\delta_L \, \mathring{\kappa}(e_A, e_B, e_C) \\ = \Delta e^{-M_{Pl}/M_C} \cdot \Big(\int_L e_A\Big)\Big(\int_L e_B\Big)\Big(\int_L e_C\Big) \,, \quad e_A \in H^2(\mathcal{M}) \,, \qquad (8.7.8)$$

where M_{Pl} and M_C are the Planck mass and the compactification mass-scale and Δ is a factor independent of e_A. Denoting by e^A, the four dimensional homology class corresponding to e_A, we have

$$\int_L e_A = [L] \cap e^A \,. \qquad (8.7.9)$$

Therefore, $\delta_L \, \mathring{\kappa}(e_A, e_B, e_C)$ can be computed by a straightforward application of our results presented in § 8.2–8.6, once $[L]$ has been identified. Needless to say, one has to add up/integrate all the contributions of all world-sheet instanton configurations L. In a particular model, for a particular Calabi–Yau manifold, the specific choice of the defining equations may simplify this task considerably and one may be able to derive phenomenologically relevant facts without actually having to complete these computations.

2024 *The discovery of mirror symmetry [156] (see also § F.8 and Refs. [291, 270]) made the world-sheet instanton-corrected* $(\mathbf{27^*})^3$*-couplings (0.5.5) and normalization matrix (8.7.3) exactly calculable, identifying them with the* $\mathbf{27}^3$*-couplings (0.5.7) and the Weil–Petersson metric in the complex structure moduli space of the mirror Calabi–Yau 3-fold; see [94, 120]. This feature enabled the first known* **exact solution** *of a 1-parameter deformation family of superstring compactification models already in 1991 [89], producing en route—but most spectacularly—a* **generating function** *for producing the count of rational curves in Calabi–Yau 3-folds of any degree! Soon replicated for several similar models [194, 435, 88], this analysis grew into a more general framework of analysis of the special geometry of Calabi–Yau moduli spaces [219], involving the periods of the hallmark holomorphic volume form (1.1.3) [82,*

*92, 71] and the classical Picard–Fuchs equations**; Refs. [346, 359, 165, 182, 327] should provide a good start on this topic; see also §§ E.2–E.3.*

In the same few years after the first edition of the Bestiary [272], another of the subsequent enormous strides provided a method for computing the cumulative effect of the world-sheet instanton corrections (8.7.8) for all GLSMs [443]—with no appeal to mirror symmetry! This type of analysis describes the "enlarged" (complexified and complete) Kähler structure moduli space [304], which involves "Horn uniformization" [408, 469], the A-hypergeometric system of Gel'fand, Kapranov and Zelevinski [264], and has been extended to non-abelian GLSMs [393]. This description of the Kähler structure moduli space turns out to fully reflect the geometry of the complex structure moduli space for the mirror models—even in the recently explored embedding of Calabi–Yau models in non-Fano (and pre-complex† *) toric varieties and their conjectured "transposition" mirrors [320, 321, 322]; see also § F.8.*

Finally, physically normalized $\mathbf{27}^3$ *and* $(\mathbf{27}^*)^3$ *Yukawa couplings for explicitly parametrized deformation families of heterotic 3-generation Calabi–Yau compactifications have only recently (in 2024) been computed in the "standard" embedding [345] and a "non-standard" embedding [363]. These proof-of-concept computations also indicate ways to compute the likewise physically normalized* $\mathbf{1}^3$ *and* $\mathbf{27}{\cdot}\mathbf{27}^*{\cdot}\mathbf{1}$ *Yukawa couplings. In addition, recently developed ML methods [315] can identify Calabi–Yau spaces with specific high-curvature regions, which may well be key to provide a physically realistic Minkowski-to-de Sitter uplift [314]. These developments are jointly framing a way to organize and deploy systematic (ML/AI) searches for physically realistic Standard Model candidates throughout the dauntingly bountiful web of Calabi–Yau compactifications; see commentary on p. 108.*

The successful confluence of these various developments is unquestionably exciting and provides additional motivation for the updates in this edition. ⌟

---

* Picard-Fuchs equations also turn up in computations of Feynman amplitudes [446], at least some of which do turn out to be related to Calabi–Yau varieties [339].

† On a "pre-complex" space, the complex structure may degenerate at positive codimension loci.

Chapter 9

(Co)homological Algebra

We now turn to discuss some techniques of cohomological algebra which can be used to compute a variety of cohomological data for complete intersections, but also in some more general situations.

9.1 A Short Exact Sequence

The basic idea was introduced already in (1.3.7) for the simple case of hypersurfaces in $\mathbb{P}^n$ and then generalized in Chapter 2.

If $\mathcal{X}$ is a complex manifold of dimension n and $\mathcal{M}$ an m-dimensional submanifold, the *normal bundle* $^{(L)}$, $\mathcal{N}_{\mathcal{X}/\mathcal{M}}$, of $\mathcal{M}$ in $\mathcal{X}$ is *defined* to be the quotient bundle $\{\mathcal{T}_{\mathcal{X}}|_{\mathcal{M}}/\mathcal{T}_{\mathcal{M}}\}$, where $\mathcal{T}_{\mathcal{M}}$ and $\mathcal{T}_{\mathcal{X}}$ denote the (holomorphic) tangent bundles of $\mathcal{M}$ and $\mathcal{X}$ respectively. Let $\mathcal{E}$ be a holomorphic bundle over $\mathcal{X}$ and ξ a holomorphic section of $\mathcal{E}$ such that $\mathcal{M}$ is the zero locus of ξ, i.e., $\xi^{-1}(0) = \mathcal{M} \hookrightarrow \mathcal{X}$. Then we have the map $\nabla\xi : \mathcal{T}_{\mathcal{X}}|_{\mathcal{M}} \to \mathcal{E}|_{\mathcal{M}}$, which is independent of any choice of a connection (as discussed in § 1.3.1) and is a holomorphic linear mapping of bundles over $\mathcal{M}$. As in § 1.3.1, $\ker\nabla\xi = \mathcal{T}_{\mathcal{M}}$. If, in addition, $\nabla\xi$ is onto, i.e., covers all of $\mathcal{E}|_{\mathcal{M}}$, we have the short *exact sequence* $^{(L)}$:

$$0 \to \mathcal{T}_{\mathcal{M}} \xrightarrow{i} \mathcal{T}_{\mathcal{X}}|_{\mathcal{M}} \xrightarrow{\nabla\xi} \mathcal{E}|_{\mathcal{M}} \to 0 \ , \qquad (9.1.1)$$

which allows us to identify $\mathcal{E}|_{\mathcal{M}}$ with $\mathcal{N}_{\mathcal{X}/\mathcal{M}}$.

This is precisely the situation in the case of non-singular complete intersections, as exemplified in Chapters 1, 2 and 3.

9.1.1 The accompanying cohomology sequence

An important consequence of the exactness of the short sequence (9.1.1) is that it induces another exact sequence, one which links $H^\star(\mathcal{M}, \mathcal{T}_{\mathcal{M}})$ with

$H^\star(\mathcal{M},\mathcal{T}_\mathcal{X})$ and $H^\star(\mathcal{M},\mathcal{E})$:

$$\begin{array}{rlrlrl} 0 \ \to & H^0(\mathcal{M},\mathcal{T}_\mathcal{M}) & \xrightarrow{i_0} & H^0(\mathcal{M},\mathcal{T}_\mathcal{X}) & \xrightarrow{j_0} & H^0(\mathcal{M},\mathcal{E}) \ \xrightarrow{\delta_0} \\ \xrightarrow{\delta_0} & H^1(\mathcal{M},\mathcal{T}_\mathcal{M}) & \xrightarrow{i_1} & H^1(\mathcal{M},\mathcal{T}_\mathcal{X}) & \xrightarrow{j_1} & H^1(\mathcal{M},\mathcal{E}) \ \xrightarrow{\delta_1} \\ \xrightarrow{\delta_1} & H^2(\mathcal{M},\mathcal{T}_\mathcal{M}) & \xrightarrow{i_2} & & \cdots\cdots & \\ & & \cdots\cdots & & \xrightarrow{j_m} & H^m(\mathcal{M},\mathcal{E}) \ \to \ 0\,. \end{array} \tag{9.1.2}$$

Oft quoted as an 'elementary fact' and presented in this still somewhat unusual notation for many physicists, this may just hide the simplicity of the idea and we pause to explain it in some detail. The maps i_q are the cohomology embeddings which may be written in a local coordinate system as

$$i_q \ : \ \mathrm{d}z^{\mu_1}\ldots\mathrm{d}z^{\mu_q}\omega_{\mu_1\ldots\mu_q}{}^{\alpha} \ \longmapsto \ \mathrm{d}z^{\mu_1}\ldots\mathrm{d}z^{\mu_q}\omega_{\mu_1\ldots\mu_q}{}^{a}\Big|_{a=\alpha}\,, \tag{9.1.3}$$

where a and α are holomorphic tangent indices on $\mathcal{X}$ and $\mathcal{M}$, respectively, and the μ's are cotangent indices on $\mathcal{M}$. Similarly, the maps j_q project the tangent index so to speak of a $\mathcal{T}_\mathcal{X}$-valued q-form on $\mathcal{M}$ to the normal space directions.

9.1.2 The dimension changing map

The action of the dimension changing maps δ_q is a little trickier. For a rigorous account, we refer the Reader to Ref. [164], p. 28; suffice it here to give a somewhat heuristic description. Consider, for example, $\mathcal{E}$-valued 0-forms. To construct a 1-form, we take two such 0-forms at nearby points, $\xi(z)$ and $\xi(z+\delta z)$. They are valued, respectively, in the normal vectors $\hat{n}_z$ and $\hat{n}_{z+\delta z}$. Their difference points in the direction $(\hat{n}_z - \hat{n}'_z)$, where $\hat{n}'_z$ is $\hat{n}_{z+\delta z}$, parallel-transported back to z. To first order in δz, $(\hat{n}_z - \hat{n}'_z) \propto \hat{t}_z$, where $\hat{t}_z \in T_z(\mathcal{M})$. Therefore,

$$\mathrm{d}z^\mu\omega_\mu \ \stackrel{\text{def}}{=} \ \mathrm{d}z^\mu \lim_{\delta z\to 0}\frac{\xi(x)-\xi(z+\delta z)}{\delta z^\mu} \tag{9.1.4}$$

is $\mathcal{T}_\mathcal{M}$-valued and belongs to $H^1(\mathcal{M},\mathcal{T}_\mathcal{M})$. This is straightforwardly generalized to all δ_q. A little more work remains to show that the *long exact cohomology sequence* (9.1.2) is exact, i.e., that for the maps δ_q just constructed, $\ker(\delta_q) = \mathrm{im}(j_q)$ and $\ker(i_{q+1}) = \mathrm{im}(\delta_q)$. The Reader should have no trouble filling in these details, bearing in mind that the forms being mapped are harmonic and are hence equivalent to zero if exact.

The main utility of the long exact cohomology sequence (9.1.2) is that it links $H^q(\mathcal{M},\mathcal{T}_\mathcal{M})$—typically the unknown cohomology group—to

$H^q(\mathcal{M}, \mathcal{T}_X)$ and $H^q(\mathcal{M}, \mathcal{E})$, which are valued in bundles over the embedding space $\mathcal{X}$, which is usually better known. However, $H^q(\mathcal{M}, \mathcal{T}_X)$ and $H^q(\mathcal{M}, \mathcal{E})$ are cohomology groups on $\mathcal{M}$, involve the restriction of $\mathcal{T}_X$ and $\mathcal{E}$ to $\mathcal{M}$ and are in general not simply equal to $H^q(\mathcal{X}, \mathcal{T}_X)$ and $H^q(\mathcal{X}, \mathcal{E})$, respectively.

Indeed, the general strategy for computing the cohomology on a submanifold $\mathcal{M}$, valued in a bundle $\mathcal{V}$ intrinsic to the submanifold proceeds as follows: (1) Relate $\mathcal{V}$ to (restrictions of) bundles $\mathcal{E}_i$ over the embedding space $\mathcal{X}$, (2) determine the $\mathcal{E}_i$-valued cohomology on $\mathcal{M}$ as a restriction of certain cohomology groups on $\mathcal{X}$. Then (3) compute all required cohomology on $\mathcal{X}$ and (4) work backwards, from this data, to determine the $\mathcal{V}$ valued cohomology on $\mathcal{M}$. This general scenario for such computations involves some very powerful machinery, for which will need a little digression on *sheaves*$^{(L)}$.

9.2 Sheaves

A *presheaf* $\mathfrak{S}$ over a manifold $\mathcal{X}$ assigns to each open subset U of $\mathcal{X}$, a set (actually a vector space in all the examples we shall consider) $\mathfrak{S}_U$. We require that $\mathfrak{S}_\varnothing = 0$. Also, to each inclusion of open sets $U \subseteq V$, a presheaf assigns a map (for us a homomorphism) $\mathbf{r}_U{}^V : \mathfrak{S}_V \to \mathfrak{S}_U$. These maps can generally be interpreted as restrictions and must satisfy

$$\mathbf{r}_U{}^V \circ \mathbf{r}_V{}^W = \mathbf{r}_U{}^W , \quad U \subseteq V \subseteq W \tag{9.2.1}$$

and, clearly, $\mathbf{r}_U{}^U$ must be the identity. For $\mathfrak{S}$ to be moreover a *sheaf*$^{(L)}$, we need in addition that

1. if U is an open set and $\{V_i\}$ an open cover of U—then $\mathfrak{s} = 0$ is the only element $\mathfrak{s} \in \mathfrak{S}_U$ for which
$$\mathbf{r}_{V_i}{}^U(\mathfrak{s}) = 0 , \qquad \forall i ; \tag{9.2.2}$$
2. if U is an open set, $\{V_i\}$ an open cover of U and there are elements over each neighborhood, $\mathfrak{s}_i \in \mathfrak{S}_{V_i}$, which equal over all pairwise overlaps, $(V_i \cap V_j)$, then there is a unique $\mathfrak{s} \in \mathfrak{S}_U$ such that $\mathfrak{s}_i = \mathbf{r}_{V_i}{}^U(\mathfrak{s})$.

The examples we will need are, in order of increasing generality:

(A) Let $\mathcal{X}$ be a complex manifold and $U \subseteq V$ two open neighborhoods as above. $\mathcal{O}_X(U)$ is defined to be the ring of holomorphic functions on U. Also, with $\{U_i\}$ an open covering of $\mathcal{X}$, $\mathcal{O}_{\mathcal{X}} = \bigcup_i \mathcal{O}_{\mathcal{X}}(U_i)$ is the *structure sheaf*$^{(L)}$.

(B) Let $\mathcal{E}$ be a holomorphic vector bundle over the complex manifold $\mathcal{W}$. We define $\mathcal{E}(U)$ to be the space of holomorphic sections of $\mathcal{E}|_U$ and the sheaf* $\mathcal{E}$ to be the union over some open covering.

(C) Let $\mathcal{M}$ be a submanifold of the complex manifold $\mathcal{X}$ and let $\mathcal{E}$ be a holomorphic vector bundle over $\mathcal{M}$. Since $\mathcal{M}$ is a submanifold in $\mathcal{X}$, $\mathcal{E}$ clearly has measure-zero support in $\mathcal{X}$ and cannot be used to define a vector bundle over $\mathcal{X}$. However, we define $\mathcal{E}_{\mathcal{M}}$ to be the sheaf over $\mathcal{X}$ which assigns to each open $U \subseteq \mathcal{X}$ the space of holomorphic sections of $\mathcal{E}|_{(U\cap\mathcal{M})}$. In this respect, sheaves are 'looser' than vector bundles and this is one of the reasons for introducing them.

In all cases, ${\mathbf{r}_U}^V$ is the appropriate restriction map.

If $\mathfrak{S}$ and $\mathfrak{S}'$ are sheaves over $\mathcal{X}$, a map of sheaves, $f : \mathfrak{S} \to \mathfrak{S}'$ assigns to each open U a map $f_U : \mathfrak{S}_U \to \mathfrak{S}'_U$ which satisfy the obvious compatibility condition

$$f_U \circ \mathbf{r}_U^{(\mathfrak{S})V} = \mathbf{r}_U^{(\mathfrak{S}')V} \circ f_V . \tag{9.2.3}$$

The definition of exactness for a sequence of linear maps of linear sheaves over $\mathcal{X}$ is not precisely what one would assume; a sequence of maps of sheaves is called exact if it is exact over sufficiently small open subsets; to emphasize this, we may write *sheaf-exact.* In particular, an exact sequence of holomorphic vector bundles induces an exact sequence of the corresponding sheaves, but the reverse is not true in general.

There exists a cohomology theory of sheaves which assigns to a manifold $\mathcal{X}$ and a linear sheaf $\mathfrak{S}$ over $\mathcal{X}$ cohomology groups $H^q(\mathcal{X}, \mathfrak{S})$, $q \geq 0$. What we need to know about sheaf cohomology is summarized in the following:

Theorem 9.1. *i) If $\mathcal{E}$ is a holomorphic vector bundle over $\mathcal{X}$, the sheaf cohomology groups $H^q(\mathcal{X}, \mathcal{E})$ may be identified with the $\bar\partial$-cohomology of $\mathcal{X}$ with coefficients*† *in $\mathcal{E}$.*
ii) With $\mathcal{X}$, $\mathcal{M}$ and $\mathcal{E}$ as in example (C) *above, $H^q(\mathcal{X}, \mathcal{E}_{\mathcal{M}}) = H^q(\mathcal{M}, \mathcal{E})$.*
iii) If $0 \to \mathfrak{S} \xrightarrow{f} \mathfrak{S}' \xrightarrow{g} \mathfrak{S}'' \to 0$ is a short exact sequence of sheaves over

* Most of the time, we will make no notational distinction between a vector bundle and the associated sheaf; the choice should be clear from the context.

† These cohomology groups may also be identified with the space of harmonic $\mathcal{E}$-valued $(0, q)$-forms which has a natural (additive) Abelian group structure.

$\mathcal{X}$, then the long cohomology sequence:

$$0 \to \ldots \xrightarrow{\delta_{q-1}} H^q(\mathcal{X}, \mathfrak{S}) \xrightarrow{f_q} H^q(\mathcal{X}, \mathfrak{S}') \xrightarrow{g_q} H^q(\mathcal{X}, \mathfrak{S}'') \xrightarrow{\delta_q} \ldots \to 0 \tag{9.2.4}$$

is exact. The maps f_q and g_q are induced from f and g and δ_q is constructed as usual (see § 9.1).

Proofs of these statements can be found in Refs. [23, 24, 27, 29] ...

9.3 The Koszul Complex

One of the steps in our Programme is to obtain (harmonic) forms on the submanifold $\mathcal{M}$ from (harmonic) forms on the embedding space $\mathcal{X}$. To this end, we will have to ask how to obtain functions on $\mathcal{M}$ from those on $\mathcal{X}$, that is, to study the restriction $\mathcal{O}_{\mathcal{X}} \xrightarrow{\varrho} \mathcal{O}_{\mathcal{M}}$.

9.3.1 The Koszul resolution

Consider, for example, a complete intersection $\mathcal{M} \subset \mathcal{X}$ defined as the zero-set of a system of 3 sections, $f(x) = g(x) = h(x) = 0$, where $f(x)$ is a section of the line bundle $\mathcal{E}_f$, $g(x)$ a section of $\mathcal{E}_g$ and so on. Let us write $\xi = (f, g, h)$, i.e., $\xi^a = f, g, h$ for $a = 1, 2, 3$ and $\mathcal{E} = \mathcal{E}_f \oplus \mathcal{E}_g \oplus \mathcal{E}_h$, accordingly.

Clearly, for any function $\vartheta(x) \in \mathcal{O}_{\mathcal{X}}(U)$ we can simply restrict the range to $x \in \mathcal{M} \cap U$, but this is not sufficient. Because $\xi^a = 0$ on $\mathcal{M}$, $\vartheta(x)$ and $\vartheta(x) + \sum_{a=1}^{3} \beta_a \xi^a$ become indistinguishable when restricted to $\mathcal{M}$. So, it is the *equivalence classes* of functions on $\mathcal{X}$—modulo multiples of the defining sections—which are identified as functions on $\mathcal{M}$. Now, ξ^a are sections of non-trivial line bundles while ϑ are just plain (local) holomorphic functions (when extended globally, they must be constant since $\mathcal{X}$ is compact). Consequently, the products $\beta_1 \xi^1$, $\beta_2 \xi^2$ and so on must be plain functions also, whence β_a must correspond to bundles *dual* to $\mathcal{E}_a$.

Typically, the ξ^a are polynomials[‡] of degree $q(a)$, $\xi^a(x) = \xi^a_{i_1 \cdots i_{q(a)}} x^{i_1} \cdots x^{i_{q(a)}}$, and are represented by the tensor coefficients $\xi^a_{i_1 \cdots i_{q(a)}}$. Then, β^a are represented by contravariant tensor coefficients, $\beta_a^{i_1 \cdots i_{q(a)}}$, so

[‡] These may be polynomials in the homogeneous coordinates of the projective embedding space or simply some suitable variables, representing local coordinates with perhaps some redundancy.

that

$$\langle \beta_a | \xi^a \rangle \;=\; \sum_{i_1,\cdots,i_{q(a)}} \beta_a^{i_1\cdots i_{q(a)}} \xi^a_{i_1\cdots i_{q(a)}} \in \mathcal{O}_{\mathcal{X}} \,, \qquad \text{no sum on } a \,. \quad (9.3.1)$$

The general case requires only more complicated notation, but is otherwise precisely the same. Let us write $\phi^\star$, $\gamma^\star$ and $\eta^\star$ for some generic formal representatives of $\mathcal{E}_f{}^*$, $\mathcal{E}_g{}^*$ and $\mathcal{E}_h{}^*$, respectively, and $\langle\phi|f\rangle$ for the natural contraction (inner product) of $\phi^\star$ with f.

It follows that

$$\mathcal{E}^* \xrightarrow{\xi\cdot} \mathcal{O}_{\mathcal{X}} \xrightarrow{\varrho} \mathcal{O}_{\mathcal{M}} \to 0 \qquad (9.3.2)$$

is sheaf-exact, that is, the sheaf of holomorphic functions on $\mathcal{M}$, $\mathcal{O}_{\mathcal{M}}$, may be regarded as a quotient of $\mathcal{O}_{\mathcal{X}}$ modulo suitable multiples of ξ. Moreover, *away* from $\mathcal{M} \subset \mathcal{X}$ where $\xi \neq 0$, the map provided by contraction with $\xi = (f, g, h)$ covers all of $\mathcal{O}_{\mathcal{X}}$. Indeed: varying $\phi^\star$, $\gamma^\star$ and $\eta^\star$ over all variables which yield a scalar when contracted with f, g and h, respectively, $\sum_{a=1}^{3}\langle\beta_a|\xi^a\rangle$ easily covers all of $\mathcal{O}_{\mathcal{X}/\mathcal{M}}$. Therefore, on $(\mathcal{X}-\mathcal{M})$, the map ξ has no cokernel, ϱ vanishes and so does $\mathcal{O}_{\mathcal{M}}$. On the other hand, at $\mathcal{M} \subset \mathcal{X}$ the triple of sections, ξ, vanishes and the map ξ is null. Hence ϱ identifies $\mathcal{O}_{\mathcal{M}}$ with $\mathcal{O}_{\mathcal{X}}|_{\mathcal{M}}$, the restriction of $\mathcal{O}_{\mathcal{X}}$ to $\mathcal{M}$—precisely as desired.

To complete the above sequence to the left, note that the map ξ has a kernel. In fact, all complex linear combinations of

$$\begin{bmatrix} -\langle\gamma|g\rangle\phi^\star \\ \langle\phi|f\rangle\gamma^\star \\ 0 \end{bmatrix}, \quad \begin{bmatrix} \langle\eta|h\rangle\phi^\star \\ 0 \\ -\langle\phi|f\rangle\eta^\star \end{bmatrix}, \quad \begin{bmatrix} 0 \\ -\langle\eta|h\rangle\gamma^\star \\ \langle\gamma|g\rangle\eta^\star \end{bmatrix} \qquad (9.3.3)$$

in $\mathcal{E}^*$, are (trivially) annihilated in the contraction with ξ; such triples span $\ker(\mathcal{E}^* \xrightarrow{\xi\cdot} \mathcal{O}_{\mathcal{X}})$. The very same elements of $\mathcal{E}^*$ are also obtained from

$$\wedge^2\mathcal{E}^* \;=\; \mathcal{E}_f{}^* \wedge \mathcal{E}_g{}^* \;\oplus\; \mathcal{E}_f{}^* \wedge \mathcal{E}_h{}^* \;\oplus\; \mathcal{E}_g{}^* \wedge \mathcal{E}_h{}^* \,, \qquad (9.3.4)$$

as the image of the mapping through ξ. To see this, note that $\wedge^2\mathcal{E}^*$ is represented by linear combinations of

$$\begin{bmatrix} 0 & \phi^\star\gamma^\star & 0 \\ -\phi^\star\gamma^\star & 0 & 0 \\ 0 & 0 & 0 \end{bmatrix}, \quad \begin{bmatrix} 0 & 0 & -\phi^\star\eta^\star \\ 0 & 0 & 0 \\ \phi^\star\eta^\star & 0 & 0 \end{bmatrix}, \quad \begin{bmatrix} 0 & 0 & 0 \\ 0 & 0 & \gamma^\star\eta^\star \\ 0 & -\gamma^\star\eta^\star & 0 \end{bmatrix} \qquad (9.3.5)$$

(the antisymmetry of the wedge product, $\wedge$, accounts for the alternating signs). Then, for example,

$$[f, g, h] \cdot \begin{bmatrix} 0 & \phi^\star\gamma^\star & 0 \\ -\phi^\star\gamma^\star & 0 & 0 \\ 0 & 0 & 0 \end{bmatrix} = [-\langle\gamma|g\rangle\phi^\star, \langle\phi|f\rangle\gamma^\star, 0] . \tag{9.3.6}$$

Iterating this procedure, we obtain the *Koszul complex*$^{(L)}$, $K_\star$, represented for a general complete intersection of K hypersurfaces by the sheaf-exact sequence, also called the *resolution* of $\mathcal{O}_\mathcal{M}$,

$$0 \to \wedge^K \mathcal{E}^* \xrightarrow{\xi\cdot} \cdots \xrightarrow{\xi\cdot} \wedge^2\mathcal{E}^* \xrightarrow{\xi\cdot} \mathcal{E}^* \xrightarrow{\xi\cdot} \mathcal{O}_\mathcal{X} \xrightarrow{\varrho} \mathcal{O}_\mathcal{M} \to 0 . \tag{9.3.7}$$

This sequence relates holomorphic functions (scalars) on $\mathcal{M}$ ($\mathcal{O}_\mathcal{M}$) to quantities defined entirely on $\mathcal{X}$; the analogous will also be true if we tensor the whole sequence with a vector bundle over $\mathcal{X}$, such as $\mathcal{T}_\mathcal{X}$, $\mathcal{E}$, ...

Clearly, had we chosen an r-tuple of sections which do not vanish simultaneously over $\mathcal{X}$, the complete intersection is empty, $\mathcal{O}_\mathcal{M} = 0$ and the resulting sequence is exact even as a sequence of vector bundles.

Given the exact sequence 9.3.7, we are tempted to write a long cohomology sequence formally analogous to Seq. (9.1.2) as derived from the short exact sequence (9.1.1). However, looking back at the derivation of δ_q in § 9.1.2, we see that it crucially depended on the fact that the exact sequence (9.1.1) had precisely three non-trivial terms. When $K \neq 1$, Seq. (9.3.7) is longer than three terms and it will soon be clear that no suitable dimension changing map could possibly be found in general.

9.3.2 One hypersurface at a time

We aim to describe the cohomology groups $H^q(\mathcal{M}, \mathcal{O}_\mathcal{M})$ in terms of $H^q(\mathcal{X}, \wedge^k\mathcal{E}^*)$—a task to be generalized for $H^q(\mathcal{M}, \mathfrak{S})$, where $\mathfrak{S}$ is a sheaf over $\mathcal{X}$—in terms of certain cohomology data entirely over $\mathcal{X}$. A simple trick to circumvent the non-shortness of Seq. (9.3.7) is to consider one hypersurface at a time instead of restricting to the complete intersection at once. To simplify the notation, let's take $\mathcal{M}$ to be an intersection of just two hypersurfaces, $\mathcal{Y} \cap \mathcal{Z} \subset \mathcal{X}$, defined respectively as $\mathcal{Y} \stackrel{\text{def}}{=} \{f(x) = 0\}$ and $\mathcal{Z} \stackrel{\text{def}}{=} \{g(x) = 0\}$.

We then first restrict to $\mathcal{Y}$, for example, using the short sheaf-exact sequence (a length-1 Koszul complex)

$$0 \to \mathcal{E}_f^* \xrightarrow{f} \mathcal{O}_\mathcal{X} \to \mathcal{O}_\mathcal{Y} \to 0 . \tag{9.3.8}$$

From this we have the long exact cohomology sequence

$$\begin{array}{rlllll} 0 & \to & H^0(\mathcal{X},\mathcal{E}_f{}^*) & \xrightarrow{f} & H^0(\mathcal{X}) & \xrightarrow{\varrho_y} & H^0(\mathcal{Y}) & \xrightarrow{\delta_0} \\ & \xrightarrow{\delta_0} & H^1(\mathcal{X},\mathcal{E}_f{}^*) & \xrightarrow{f} & \cdots\cdots\cdots \\ & & & & \cdots\cdots\cdots & \xrightarrow{\varrho_y} & H^4(\mathcal{Y}) & \xrightarrow{\delta_4} \\ & \xrightarrow{\delta_4} & H^5(\mathcal{X},\mathcal{E}_f{}^*) & \xrightarrow{f} & H^5(\mathcal{X}) & \xrightarrow{\varrho_y} & 0\,, \end{array} \tag{9.3.9}$$

which determines $H^q(\mathcal{Y})$ in terms of $H^0(\mathcal{X})$ and $H^0(\mathcal{X},\mathcal{E}_f{}^*)$.

Now embed $(\mathcal{M} = \mathcal{Y}\cap\mathcal{Z}) \hookrightarrow \mathcal{Y}$, that is, we impose the $g(x) = 0$ constraint in addition to $f(x) = 0$ which was used to define $\mathcal{Y}\hookrightarrow\mathcal{X}$. This being a simple hypersurface, we again have a short sheaf-exact sequence

$$0 \to \mathcal{E}_g{}^*|_{\mathcal{Y}} \xrightarrow{f} \mathcal{O}_{\mathcal{Y}} \xrightarrow{\varrho_z} \mathcal{O}_{\mathcal{M}} \to 0\,. \tag{9.3.10}$$

From this we have the long exact cohomology sequence

$$\begin{array}{rlllll} 0 & \to & H^0(\mathcal{Y},\mathcal{E}_g{}^*) & \xrightarrow{g} & H^0(\mathcal{Y}) & \xrightarrow{\varrho_z} & H^0(\mathcal{M}) & \xrightarrow{\delta_0} \\ & \xrightarrow{\delta_0} & H^1(\mathcal{Y},\mathcal{E}_g{}^*) & \xrightarrow{g} & & \cdots\cdots \\ & & & \cdots\cdots & & \xrightarrow{\varrho_z} & H^3(\mathcal{M}) & \xrightarrow{\delta_3} \\ & \xrightarrow{\delta_3} & H^4(\mathcal{Y},\mathcal{E}_g{}^*) & \xrightarrow{g} & H^4(\mathcal{Y}) & \xrightarrow{\varrho_z} & 0\,, \end{array} \tag{9.3.11}$$

which determines $H^q(\mathcal{M})$ in terms of $H^0(\mathcal{Y})$ and $H^0(\mathcal{Y},\mathcal{E}_g{}^*)$. Of these last two cohomology groups, the former has been determined just above, while the latter we determine using Seq. (9.3.8), tensored however with $\mathcal{E}_g{}^*$

$$0 \to (\mathcal{E}_f\otimes\mathcal{E}_g)^* \xrightarrow{f} \mathcal{E}_g{}^* \to \mathcal{E}_g{}^*|_{\mathcal{Y}} \to 0 \tag{9.3.12}$$

and the associated long exact cohomology sequence

$$\begin{array}{rlllll} 0 & \to & H^0(\mathcal{X},(\mathcal{E}_f\otimes\mathcal{E}_g)^*) & \xrightarrow{f} & H^0(\mathcal{X},\mathcal{E}_g{}^*) & \xrightarrow{\varrho_y} & H^0(\mathcal{Y},\mathcal{E}_g{}^*) & \xrightarrow{\delta_0} \\ & \xrightarrow{\delta_0} & H^1(\mathcal{X},(\mathcal{E}_f\otimes\mathcal{E}_g)^*) & \xrightarrow{f} & & \cdots\cdots \\ & & & \cdots\cdots & & \xrightarrow{\varrho_y} & H^4(\mathcal{Y},\mathcal{E}_g{}^*) & \xrightarrow{\delta_4} \\ & \xrightarrow{\delta_4} & H^5(\mathcal{X},(\mathcal{E}_f\otimes\mathcal{E}_g)^*) & \xrightarrow{f} & H^5(\mathcal{X},\mathcal{E}_g{}^*) & \xrightarrow{\varrho_y} & 0\,. \end{array} \tag{9.3.13}$$

The end result of this calculation is that $H^q(\mathcal{M})$ is related, through three long exact cohomology sequences (9.3.9), (9.3.11) and (9.3.13) and a number of intermediary results, to $H^q(\mathcal{X})$, $H^q(\mathcal{X},\mathcal{E}_f{}^*)$, $H^q(\mathcal{X},\mathcal{E}_g{}^*)$ and $H^q(\mathcal{X},\mathcal{E}_f{}^*\otimes\mathcal{E}_g{}^*)$. This can straightforwardly be done for every complete intersection, regardless of the number of constraints although, of course, the procedure becomes more and more involved with each new constraint.

As the foregoing cavalcade might have lost the Reader unfamiliar with exact sequences, let us sum up this procedure in terms of the following diagram:

$$
\begin{array}{ccccccccc}
 & & 0 & & 0 & & 0 & & \\
 & & \downarrow & & \downarrow & & \downarrow & & \\
0 & \to & (\mathcal{E}_f \otimes \mathcal{E}_g)^* & \xrightarrow{f} & \mathcal{E}_g^* & \xrightarrow{\varrho_y} & \mathcal{E}_g^*|_{\mathcal{Y}} & \to & 0 \\
 & & \downarrow g & & \downarrow g & & \downarrow g & & \\
0 & \to & \mathcal{E}_f^* & \xrightarrow{f} & \mathcal{O}_{\mathcal{X}} & \xrightarrow{\varrho_y} & \mathcal{O}_{\mathcal{Y}} & \to & 0 \\
 & & \downarrow \varrho_z & & \downarrow \varrho_z & & \downarrow \varrho_z & & \\
0 & \to & \mathcal{E}_f^*|_{\mathcal{Z}} & \xrightarrow{f} & \mathcal{O}_{\mathcal{Z}} & \xrightarrow{\varrho_y} & \mathcal{O}_{\mathcal{M}} & \to & 0 \\
 & & \downarrow & & \downarrow & & \downarrow & & \\
 & & 0 & & 0 & & 0 & &
\end{array}
\tag{9.3.14}
$$

where all row-sequences and all column-sequences are exact. The $\mathcal{O}_{\mathcal{M}}$-valued cohomology on $\mathcal{M}$ (lower right corner) is related to the cohomology on $\mathcal{X}$, valued in the bundles inside the dotted rectangle. Quite clearly, Seq. (9.3.7), for $K = 2$, may be thought of as the diagonal of the diagram (9.3.14) since $\wedge^2\mathcal{E}^* = (\mathcal{E}_f \otimes \mathcal{E}_g)^*$ and $\mathcal{E} = \mathcal{E}_f \oplus \mathcal{E}_g$. Indeed,

Lemma 9.1. *A* 3×3 *square diagram of short exact sequences as in* (9.3.14) *implies a 4-term exact sequence as in* (9.3.7) *for* $K = 2$.

is a standard result and is not hard to generalize beyond $K = 2$; for $K = 3$, in place of (9.3.14), we have a $3 \times 3 \times 3$ cube of short exact sequences which implies the exactness of a 5-term exact sequence obtained as the 'body-diagonal' of the cube of short exact sequences, and so on.

Markedly, whatever the length (K) of the Koszul complex, the exactness of the sequence (9.3.7) is equivalent to saying that the structure sheaf of the submanifold $\mathcal{M}$ is given as a quotient, $\mathcal{O}_{\mathcal{M}} = \{\mathcal{O}_{\mathcal{X}}/\Im_K\}$, where $\Im_{\mathrm{K}}$ might be called the Koszul *ideal*$^{(L)}$. Furthermore, in the simplest case $K=1$ and $\Im_{\mathrm{K}}=\mathcal{E}^*$; for $K=2$, $\Im_{\mathrm{K}}=\{\mathcal{E}^*/\wedge^2\mathcal{E}^*\}$; when $K=3$,

$$
\Im_{\mathrm{K}} = \left\{ \frac{\mathcal{E}^*}{\{\wedge^2\mathcal{E}^*/\wedge^3\mathcal{E}^*\}} \right\} , \tag{9.3.15}
$$

and so on.

The Seq. (9.3.7) may be seen as a diagonally collapsed version of (hyper)cubic diagrams like (9.3.14) where some of the 'intermediate' bundles (sheaves) have been dropped. It is plausible to expect that a similar collapsed version of the tangle of intertwined long exact cohomology sequences (9.3.9), (9.3.11) and (9.3.13) also exists, where no 'intermediate' (and often otherwise unused) cohomology group is ever computed. There indeed does exist such a general tool, called a spectral sequence, and we now turn to describe it.

9.3.3 The Koszul spectral sequence

Given the exact sequence 9.3.7, we are tempted to write a long cohomology sequence formally analogous to Seq. (9.1.2) as derived from the short exact sequence (9.1.1). The corresponding cohomology groups can again be arranged in analogy to (9.1.2)

$$\begin{array}{ccccc|c}
\wedge^K\mathcal{E}^* & \cdots & \wedge^2\mathcal{E}^* & \mathcal{E}^* & \mathcal{O}_\mathcal{W} & \mathcal{O}_\mathcal{M} \\
\hline
H^0(\mathcal{X},\wedge^K\mathcal{E}^*) & \cdots & H^0(\mathcal{X},\wedge^2\mathcal{E}^*) & H^0(\mathcal{X},\mathcal{E}^*) & H^0(\mathcal{X}) & H^0(\mathcal{M}) \\
H^1(\mathcal{X},\wedge^K\mathcal{E}^*) & \cdots & H^1(\mathcal{X},\wedge^2\mathcal{E}^*) & H^1(\mathcal{X},\mathcal{E}^*) & H^1(\mathcal{X}) & H^1(\mathcal{M}) \\
\vdots & \ddots & \vdots & \vdots & \vdots & \vdots
\end{array} \tag{9.3.16}$$

but there is no analogue of δ_q. Instead, there will now be maps ('differentials' d_r) that act $r+1$ steps to the right and r steps up the chart[§]. The horizontal, single-step maps, acting as in (9.1.2), are the $r=0$ case thereof. From the action of these maps, the cohomology in the right-most column is related to the cohomology groups which appear in the lower left quadrant of (9.3.16), somewhat as a summary of the circuitous procedure of § 9.3.2. The precise explanation of how this machinery works in full generality requires some technical details which are beyond the scope of this volume; these can be found in Ref. [23], p. 438–445, for example.

It may be clarifying to actually write down the intertwined six long exact cohomology sequences accompanying each of the nine short exact sequences in the diagram (9.3.14). The three horizontal ones of course extend horizontally, for

§ That all maps must act in such a 'knight tour' like fashion on a spectral sequence chart can be proven in complete generality [23]. In our case, we just note that the differentials of our spectral sequence are induced from the vertical and horizontal maps in the diagram (9.3.14) and can be traced that way.

$q = 0, \ldots, 5$, while the vertical ones extend downwards, wrap around and keep crossing the horizontal ones for every q. It is not hard to draw a model of this tangle of sequences on a cylindrically wrapped sheet of paper and recover the action of the maps in the Koszul spectral sequence for $K=2$. ❧

In practice, one starts from (9.3.16), called the 0th level of the spectral sequence; the entries in the lower left quadrant are usually labeled as $E_0^{q,k} = H^q(\mathcal{X}, \wedge^k \mathcal{E}^*)$. We have the maps induced straightforwardly from the contraction with ξ in the Koszul complex: $E_0^{q,k} \xrightarrow{\mathrm{d}_0} E_0^{q,k-1}$. We then pass to the d_0-cohomology,

$$E_1^{q,k} = \left\{ (\ker(\mathrm{d}_0) \cap E_0^{q,k}) \Big/ \mathrm{d}_0(E_0^{q,k+1}) \right\} . \tag{9.3.17}$$

To deal with the higher 'differentials,' d_r, we determine their action

$$\mathrm{d}_r \ : \ E_r^{q,k} \ \longrightarrow \ E_r^{q-r,k-r-1} \tag{9.3.18}$$

and pass to the cohomology quotients

$$E_{r+1}^{q,k} \stackrel{\text{def}}{=} \left\{ (\ker(\mathrm{d}_r) \cap E_r^{q,k} \Big/ \mathrm{d}_r(E_r^{q+r,k+r+1}) \right\} , \tag{9.3.19}$$

order by order in r. It should be obvious that this sequence of approximations converges for some $r \leq K$; when $r > K$, even for the entries in the left-most column, the maps would point out of the lower left quadrant of (9.3.16).

Now, contributions to $H^q(\mathcal{M})$ consist only of d_r-closed (modulo d_r-exact) forms, for all r. In practice, we advance order by order in r canceling out all pairs (α, β) which satisfy $\mathrm{d}_r\alpha = \beta$ for some r. What remains spans $E_\infty^{q,k}$ and abuts to $H^q(\mathcal{M})$ according to

$$\sum_{k=0}^{K} E_\infty^{q,k} \implies H^q(\mathcal{M}) . \tag{9.3.20}$$

The collection of such contributions does not form a (holomorphic) direct sum in $H^q(\mathcal{M})$ and, as in § 3.3.4, we shall use the '+' symbol instead of '$\oplus$'.

The astute Reader will have noticed that all computations which employ the long exact cohomology sequences (9.1.2), and even more so the spectral sequences (9.3.16), rest upon the ability to discern the action of several maps. This otherwise hopeless task¶ is however straightforwardly

¶ Of course, in fortunate cases, the $E_0^{q,k}$ chart is sufficiently sparse, so that either all differentials vanish and $E_0^{q,k} = E_\infty^{q,k}$ or their action is otherwise elementary to deduce.

accomplished using the coset representation of flag spaces and the tensor algebra associated with representations of Lie groups.

9.4 The Bott–Borel–Weil Theorem

9.4.1 Homogeneous cohomology over flag spaces

One of the main advantages of considering such generalized flag spaces is that all cohomology of all homogeneous vector bundles over them can be given by means of an amusingly simple [18] (see also Ref. [6] for generalizations) algorithmic variant of the Bott–Borel–Weil theorem:

Theorem 9.2 (Bott–Borel–Weil). *Let $\mathbb{F}$ be a flag space as in Eq. (3.3.1)–(3.3.2) and $\mathcal{V} \sim (a_1, \ldots, a_{n_1}| \cdots |b_1, \ldots, b_{n_F})$ a holomorphic homogeneous vector bundle over $\mathbb{F}$. Then:*
(1) Homogeneous vector bundles $\mathcal{V}$ over $\mathbb{F}$ are in 1–1 correspondence with the $U(n_1) \times \cdots \times U(n_F)$ representations.
(2) The $H^q(X, \mathcal{V})$ is non-zero for at most one value of q, in which case it furnishes an irreducible representation of $U(N)$, $H^q(X, \mathcal{V}) \approx (c_1, \ldots, c_N)\mathbb{C}^N$.
(3) $(a_1, \ldots, a_{n_1}| \cdots |b_1, \ldots, b_{n_F})$ determines $(c_1, \ldots, c_N)$ according to the following algorithm:
1. Add the sequence $1, \ldots, N$ to the entries in $(a_1, \ldots, a_{n_1}| \cdots |b_1 \ldots b_{n_F})$.
2. If any two entries in the result of Step 1 are equal, all cohomology vanishes; otherwise proceed.
3. Swap the minimum number $(= q)$ of neighboring entries required to produce a strictly increasing sequence.
4. Subtract the sequence $1, \ldots, N$ from the result of 3, to obtain $(c_1, c_2, \ldots, c_N)$.

In more general flag spaces, where G is an arbitrary Lie group and P and arbitrary parabolic subgroup thereof, the same algorithm applies, with the only distinction that instead of counting "swaps" in Step 3, we consider the action of the Weyl subgroup of G; in fact, "swaps" are the elementary transformations of the permutation group S_n which, in turn, is the Weyl subgroup of $U(n)$. We refer the Reader to Ref. [6] for a more 'serious' wording of this theorem and a rather simple proof. Note that the theorem provides not only the dimensions of the cohomology groups, but in fact the transformation property with respect to G; that will provide one of the most general and powerful computational tools in this neck of the woods.

9.4.2 A general viewpoint

The cohomological computation described here is a combination of (1) the spectral sequence accompanying the Koszul resolution, (2) the Bott–Borel–Weil theorem for embedding spaces which are products of flag spaces and (3) sequences derived like (1.3.7) which relate bundles over the submanifold to bundles over the embedding space. It exhibits one of the goals in algebraic geometry: solving differential equations through algebraic means. Indeed, we see from Table 0.2 that we need $H^q(\mathcal{M}, \mathcal{T}^*_{\mathcal{M}})$ and $H^1(\mathcal{M}, \operatorname{End}\mathcal{T}_{\mathcal{M}})$, that is, we need forms like $\omega = \mathrm{d}z^\mu \mathrm{d}z^{\overline{\nu}} \omega_{\mu\overline{\nu}}$ which are 0-modes of the Laplacian on $\mathcal{M}$.

The purely algebraic technique which we describe in this couple of Chapters and to which we refer simply as 'Koszul spectral sequence' describes the solutions of these Laplace's differential equations on $\mathcal{M}$ without any differential geometry. In a way, this works just like the Laplace transform which converts a system of differential equations into a system of algebraic equations. Essentially, the calculations become manipulations of systems of algebraic equations where the variables and the parameters are certain (irreducible) tensors.

From the point of view of the physics application, the manifold itself is not really required to be known. Rather, it is the rings generated by the cohomology groups $H^1(\mathcal{M}, \mathcal{T}_{\mathcal{M}})$, $H^1(\mathcal{M}, \mathcal{T}^*_{\mathcal{M}})$ and $H^1(\mathcal{M}, \operatorname{End}\mathcal{T}_{\mathcal{M}})$ which are physically relevant, although this information largely determines the underlying compactification manifold $\mathcal{M}$. As usual in algebraic geometry, one will be interested in the specifics of the polynomial ring over $\mathcal{M}$ and its image in the interesting cohomology groups. Many of the answers can be formulated in terms of certain subrings of the residue class ring $\{\mathbb{C}[x]/\Im\}$, where $\mathbb{C}[x]$ is the ring of (complex) polynomials in some suitable variables such as coordinates and $\Im$ is some *ideal*$^{(L)}$.

9.5 Calabi–Yau or Not?

Knowing the resolution sequence of the structure sheaf $\mathcal{O}_{\mathcal{M}}$ of a submanifold $\mathcal{M} \subset \mathcal{X}$ allows one to immediately tell the type of the canonical bundle $\mathcal{K}_{\mathcal{M}}$, that is, whether $\mathcal{M}$ is Calabi–Yau, almost ample, ...

Consider, for simplicity, a smooth complete intersection of a quartic and a quadric hypersurface in $\mathbb{P}^5$, $\mathcal{M} \in [5\|4\,2]$. The resolution, written in the notation of § 9.4, is

$$0 \to (6|00000) \to \begin{matrix}(4|00000)\\(2|00000)\end{matrix} \to (0|00000) \to \mathcal{O}_{\mathcal{M}} \to 0\ ; \tag{9.5.1}$$

again, we stack summands in a direct sum atop each other. Using Theorem 9.2, it is straightforward to fill in the lower left quadrant of the 0th level of the spectral sequence as in (9.3.16), associated to this resolution. The Reader should have no trouble verifying that only the top right hand and bottom left hand entries, $E_0^{0,0}$ and $E_0^{5,2}$ respectively, in the lower left quadrant are non-zero. Moreover, these cohomology groups are 1-dimensional, $\approx \mathbb{C}^1$. It is then straightforward that all differentials, d_r, vanish and $E_0^{q,k} = E_\infty^{q,k}$, for all k, q.

Through (9.3.20), we have that $\dim H^0(\mathcal{M}) = 1 = \dim H^3(\mathcal{M})$. The former of these results simply says that $\mathcal{M}$ is a connected manifold. The latter, however, says that the anti-holomorphic (harmonic) 3-form is projectively unique since

$$H^3(\mathcal{M}) \;=\; H^0(\mathcal{M}, \wedge^3\mathcal{T}^*_{\mathcal{M}}) \;=\; H^0(\mathcal{M}, \mathcal{K}_{\mathcal{M}})\ , \tag{9.5.2}$$

implying that $\mathcal{K}_{\mathcal{M}} \approx \mathbb{C}_{\mathcal{M}}$ is the trivial line bundle over $\mathcal{M}$. That Seq. (9.3.7) always results in such a simple spectral sequence for Calabi–Yau complete intersections in products of complex projective spaces was proven in Ref. [146].

Using the Kodaira–Nakano Theorem 3.3, it is not hard to prove this more generally. The proof rests on a few simple observations. Firstly, $\mathcal{O}_{\mathcal{X}}$ clearly has $H^q(\mathcal{X}, \mathcal{O}_{\mathcal{X}}) \neq 0$ only for $q = 0$. Second, the map ξ in Seq. (9.3.7) is realized as a contraction with a chosen section of the bundle $\mathcal{E}$ over $\mathcal{X}$, whence $\mathcal{E}$ must be non-trivial and positive for otherwise it would have no non-trivial global holomorphic sections. Then, however, $\wedge^k\mathcal{E}^*$ is non-trivial and more and more negative as k is increased from 1 to K. Thus, $H^q(\mathcal{X}, \wedge^k\mathcal{E}^*) = 0$ except perhaps for $q = \dim\mathcal{X}$.

So then, what we need for $\mathcal{M}$ to be Calabi–Yau is that $\wedge^K\mathcal{E}^* \approx \mathcal{K}_{\mathcal{X}}$. Then $\wedge^k\mathcal{E}^* \otimes \mathcal{K}^*_{\mathcal{X}}$ is (almost) positive for $k < K$ and all cohomology in the lower left quadrant of the chart like (9.3.16) vanishes except for the top right and bottom left corners, which are moreover 1-dimensional.

It is straightforward to see that the preceding argument in fact did not require $\mathcal{M}$ to be a complete intersection. Indeed, we recall the simple construction in § 3.5.1. The Pfaffian 3-fold $\mathfrak{X}$ was defined as the zero-set of seven Pfaffian cubic polynomials $f_k(x)$ over $\mathbb{P}^6$. This tells us that the resolution sequence ends as

$$\mathscr{O}_{\mathbb{P}^6}(-3)^{\oplus 7} \xrightarrow{\widehat{f}} \mathscr{O}_{\mathbb{P}^6} \xrightarrow{\varrho} \mathscr{O}_{\mathfrak{X}} \to 0\ , \tag{9.5.3}$$

where the map $\widehat{f}$ obtained by multiplication with the 7-vector of Pfaffians $f_k(x)$, $k = 1, \ldots, 7$. Given that the seven cubics $f_k(x)$ were obtained as the Pfaffians of the principal minors of a single skew-symmetric matrix, $\mathbf{A}$, a little thought reveals that the kernel of the map $\widehat{f}$ is also obtained as the image of multiplication by the matrix $\mathbf{A}$. That map again has a kernel, easily verified to be again obtained as the image of multiplication by (the transpose of) the 7-vector $f_k(x)$.

The result is the resolution

$$0 \to \mathscr{O}_{\mathbb{P}^6}(-7) \xrightarrow{\widehat{f}^T} \mathscr{O}_{\mathbb{P}^6}(-4)^{\oplus 7} \xrightarrow{\mathbf{A}} \mathscr{O}_{\mathbb{P}^6}(-3)^{\oplus 7} \xrightarrow{\widehat{f}} \mathscr{O}_{\mathbb{P}^6} \xrightarrow{\varrho} \mathscr{O}_{\mathfrak{X}} \to 0\ . \tag{9.5.4}$$

Indeed, the left-most bundle in the sequence is the canonical bundle of $\mathbb{P}^6$ and it is a simple exercise to verify that

$$\dim H^q(\mathfrak{X}, \mathscr{O}_{\mathfrak{X}}) = \begin{cases} 1, & \text{for } q = 0, 3, \\ 0, & \text{otherwise}\,; \end{cases} \tag{9.5.5}$$

precisely as required for $\mathfrak{X}$ to be a Calabi–Yau 3-fold. In full detail, we obtain

q:	$\mathscr{O}_{\mathbb{P}^6}(-7)$	$\mathscr{O}_{\mathbb{P}^6}(-4)^{\oplus 7}$	$\mathscr{O}_{\mathbb{P}^6}(-3)^{\oplus 7}$	$\mathscr{O}_{\mathbb{P}^6}$	$\mathscr{O}_{\mathfrak{X}}$
0:	0	0	0	$\mathbb{C}$	$\to H^0(\mathfrak{X}) \approx \mathbb{C}$
1:	0	0	0	0	$H^1(\mathfrak{X}) = 0$
2:	0	0	0	0	$H^2(\mathfrak{X}) = 0$
3:	0	0	0	0	$\to H^3(\mathfrak{X}) \approx \mathbb{C}$
4:	0	0	0	0	$\equiv 0$
5:	0	0	0	0	$\equiv 0$
6:	$\mathbb{C}$	0	0	0	$\equiv 0$

(9.5.6)

Since all $E_0^{q,k}$ vanish except $E_0^{0,0}$ and $E_0^{6,3}$, which are 1-dimensional. The dotted arrows represent the relation (9.3.20) and provide us with the required cohomology of a Calabi–Yau 3-fold.

Table 9.1: A partial taxonomy of submanifolds.

Submanifold $\mathcal{M}$	**Condition on** $\mathcal{E}_K$ ($\mathcal{E}_0{}^* \stackrel{\text{def}}{=} \mathcal{O}_\mathcal{X}$)
Ample	$\mathcal{E}_K \otimes \mathcal{K}_\mathcal{X} < 0$
Almost Ample	$\mathcal{E}_K \otimes \mathcal{K}_\mathcal{X} \leq 0$ ("$<$" along at least some $\mathcal{Y} \subset \mathcal{X}$)
Calabi–Yau	$\mathcal{E}_K \otimes \mathcal{K}_\mathcal{X} = 0$ $(\approx \mathbb{C})$
Almost Scant	$\mathcal{E}_K \otimes \mathcal{K}_\mathcal{X} \geq 0$ ("$>$" along at least some $\mathcal{Y} \subset \mathcal{X}$)
Scant	$\mathcal{E}_K \otimes \mathcal{K}_\mathcal{X} > 0$
Limping	$\mathcal{E}_K \otimes \mathcal{K}_\mathcal{X} \neq 0$ ("$>$" along at least some $\mathcal{Y} \subset \mathcal{X}$, "$<$" along at least some $\mathcal{Z} \subset \mathcal{X}$)

$\mathcal{E}_K{}^*$ is the left-most bundle in the resolution of $\mathcal{O}_\mathcal{M}$.

Clearly, if the left-most bundle turned out to be $\mathcal{O}_{\mathbb{P}^6}(-k)$ with $k < 7$, $\mathfrak{X}$ would have been ample; with $k > 7$, $\mathfrak{X}$ would have been scant; see Table 9.1 for our taxonomy, and Ref. [286] for a more general and foundational discussion. More generally for a submanifold $\mathcal{M} \subset \mathcal{X}$, let $\mathcal{E}_K{}^*$ be the left-most bundle in the resolution of $\mathcal{O}_\mathcal{M}$, such as in (9.5.7) below. The same will of course be said of a *configuration*$^{(L)}$ $[\mathcal{X}\|\mathcal{E}]$, where $\mathcal{X}$ is the embedding space and $\mathcal{E}$ is the positive bundle over $\mathcal{X}$ the sections of which are used as defining constraints for $\mathcal{M} \in [\mathcal{X}\|\mathcal{E}]$. (For a complete intersection $\mathcal{M}$, the left-most bundle in the resolution of $\mathcal{O}_\mathcal{M}$ is $\det \mathcal{E}^*$.)

Recall that for a complete intersection $\mathcal{M}$, the left-most bundle in the resolution (9.3.7) is always the determinant of the dual of the bundle the sections of which we have used to define $\mathcal{M}$. In the present example, this is not so because $\mathfrak{X}$ is not a complete intersection, but may instead be thought of as a generalization thereof. The resolution sequences of both the complete intersections and the Pfaffian 3-fold $\mathfrak{X}$ (and also other similar constructions) do have a common property: in their resolution sequence $(\mathcal{E}_0{}^* \stackrel{\text{def}}{=} \mathcal{O}_\mathcal{X})$

$$0 \to \mathcal{E}_K{}^* \to \mathcal{E}_{K-1}{}^* \to \cdots \to \mathcal{E}_1{}^* \to \mathcal{E}_0{}^* \to \mathcal{O}_\mathcal{M} \to 0 \,, \qquad (9.5.7)$$

we have that

$$\mathcal{E}_k{}^* \otimes \mathcal{E}_{K-k}{}^* \approx \mathcal{K}_\mathcal{X} \,, \qquad k = 0, \ldots, K \,. \qquad (9.5.8)$$

The above criterion for the triviality of the canonical bundle, $\mathcal{K}_{\mathcal{M}}$, (equivalently: the vanishing first Chern class, Ricci-flatness, existence of nowhere vanishing holomorphic 3-form, and so on) remains valid generally:

Theorem 9.3. *A 3-dimensional compact submanifold $\mathcal{M} \subset \mathcal{X}$ of a compact manifold $\mathcal{X}$ is a Calabi–Yau 3-fold if the structure sheaf $\mathcal{O}_{\mathcal{M}}$ fits into a sheaf-exact resolution sequence*

$$0 \to \mathfrak{E}_K{}^* \to \mathfrak{E}_{K-1}{}^* \to \cdots \to \mathfrak{E}_1{}^* \to \mathfrak{E}_0{}^* \to \mathcal{O}_{\mathcal{M}} \to 0\ , \tag{9.5.9}$$

where $\mathfrak{E}_0{}^* = \mathcal{O}_{\mathcal{X}}$ and $\mathfrak{E}_K{}^* = \mathcal{K}_{\mathcal{X}}$.

Remark: By requirements of a resolution sequence, all $\mathfrak{E}_k$ are bundles over $\mathcal{X}$. $\mathcal{O}_{\mathcal{M}}$ defines a sheaf over $\mathcal{X}$ extending by zero; that is, at $\mathcal{M} \subset \mathcal{X}$, $\mathcal{O}_{\mathcal{M}}$ is the sheaf of germs of holomorphic functions over $\mathcal{M}$, while over $\mathcal{X} - \mathcal{M}$, $\mathcal{O}_{\mathcal{M}}$ is defined to be zero.

2024 *The computational (9.3.16)-based framework provides a direct verification of (3.2.1) for the almost (but not) Fano hypersurface $\{p(x,y)=0\} = \mathfrak{F} \in \left[\begin{smallmatrix}3\\1\end{smallmatrix}\middle\|\begin{smallmatrix}4\\1\end{smallmatrix}\right]$ from § 3.2.1. It also provides a simple application of the Bott–Borel–Weil theorem (§ 9.4.1). Recalling the § 3.3.2 notation, the bidegree-(4, 1) defining polynomial, $p(x,y)$, is a section of the bundle $\left(\begin{smallmatrix}-4\\-1\end{smallmatrix}\middle|\begin{smallmatrix}0\,0\,0\\0\,0\end{smallmatrix}\right) \overset{\text{def}}{=} \mathcal{O}_{\mathbb{P}^3}(4)\otimes\mathcal{O}_{\mathbb{P}^1}(1)$. Its dual appears in the left end of the (9.3.7)-like ($K=1$) Koszul resolution, and we stack its induced homology sequence underneath:*

$$\begin{array}{c|c}
\begin{array}{ccc}
q & \left(\begin{smallmatrix}4\\1\end{smallmatrix}\middle|\begin{smallmatrix}0\,0\,0\\0\,0\end{smallmatrix}\right) \xrightarrow{p} & \left(\begin{smallmatrix}0\\0\end{smallmatrix}\middle|\begin{smallmatrix}0\,0\,0\\0\,0\end{smallmatrix}\right) \\ \hline
0: & 0 & \left(\begin{smallmatrix}0\,0\,0\,0\\0\,0\end{smallmatrix}\right)^1_1 \\
1: & 0 & 0 \\
2: & 0 & 0 \\
3: & 0 & 0 \\
4: & 0 & 0
\end{array}
&
\begin{array}{ll}
\mathcal{O}_{\mathfrak{F}} & \\ \hline
\dashrightarrow H^0(\mathfrak{F},\mathcal{O}): & h^{0,0} = 1; \\
\dashrightarrow H^1(\mathfrak{F},\mathcal{O}): & h^{0,1} = 0; \\
\dashrightarrow H^2(\mathfrak{F},\mathcal{O}): & h^{0,2} = 0; \\
\dashrightarrow H^3(\mathfrak{F},\mathcal{O}): & h^{0,3} = 0. \\
\text{—} &
\end{array}
\end{array} \tag{9.5.10}$$

The $\mathcal{K}^*_{\mathfrak{F}}$-valued cohomology is computed similarly (the $m=1$ case of (9.6.2), below) to be generated by the (properly $\mathfrak{F}$-restricted) copies of the homogeneous $\mathbb{P}^1$-coordinates, which cannot both vanish by definition, consistent with the definition of almost Fano varieties in § 3.2.1.

The anti-canonical hypersurface in $\mathfrak{F}$, used in the provided proof of Lemma 3.3, is indeed a smooth K3 surface. Being a complete intersection in $\mathbb{P}^3 \times \mathbb{P}^1$ of the hypersurface $\mathfrak{F} = \{p(x,y)=0\}$ and the zero-locus of a $\mathcal{K}^*_{\mathfrak{F}}$-section, $q(y) = q_0 y_0 + q_1 y_1$, it belongs to the configuration $\left[\begin{smallmatrix}3\\1\end{smallmatrix}\middle\|\begin{smallmatrix}4&0\\1&1\end{smallmatrix}\right]$, which simplifies:

$$\begin{bmatrix}3\\1\end{bmatrix}\!\!\begin{Vmatrix}4&0\\1&1\end{Vmatrix} = \begin{bmatrix}3\\1\end{bmatrix}\!\!\begin{Vmatrix}0&4\\1&1\end{Vmatrix} = \begin{bmatrix}3\\0\end{bmatrix}\!\!\begin{Vmatrix}4\\1\end{Vmatrix} = [3\,\|\,4] \ni K3. \tag{9.5.11}$$

This formal sequence is substantiated by reparametrizing $\mathbb{P}^1$, so that $q(y)=0$ implies, e.g., $y_0=0$ and so $y_1=1$, whereupon the bidegree-$(4,1)$ defining polynomial, $p(x,y)$, of $\mathfrak{F}$ becomes simply a quartic in $\mathbb{P}^3$. Chosen to be transverse, this defines a smooth K3 surface, as expected. Alternatively, ✎ prove that $p(x,y)$, $q(y)$ and $\mathrm{d}p\wedge\mathrm{d}q$ cannot all vanish simultaneously for generic $p(x,y)$ and $q(y)$. ❧

These rather explicit computations provide more than is guaranteed by the Kodaira–Nakano and Kawamata–Viehweg theorems mentioned in § 3.2.4, and so provide direct, practical motivation for this computational framework.

9.6 Plurigenera

With the foregoing technology, we are now in position to complete the calculation of ${\mathcal{K}_X^*}^{\otimes m}$-valued cohomology for arbitrary complete intersections in products of flag-varieties, without having to rely on the Kodaira–Nakano theorem 3.3 and the positivity of the anti-canonical bundle $\mathcal{K}_X^*$. To this end, we multiply each term in the resolution (9.3.7) by ${\mathcal{K}_X^*}^{\otimes m}$ and use the associated spectral sequence to determine the ${\mathcal{K}_X^*}^{\otimes m}$-valued cohomology.

Consider then an almost-Fano 3-fold $\mathfrak{F} \in \left[\begin{smallmatrix}3\\1\end{smallmatrix}\middle\|\begin{smallmatrix}4\\1\end{smallmatrix}\right]$. As discussed at the end of § 3.2.2, the anti-canonical bundle is $\mathcal{K}_{\mathfrak{F}}^* \approx \mathcal{O}_{\mathbb{P}^1}(1)$, and so is not positive; it is null along $\mathbb{P}^3$. That is why the Kodaira–Nakano theorem does not apply and the cohomology $H^q(\mathfrak{F}, \mathcal{K}_{\mathfrak{F}}^*)$ is not restricted to $q=0$ only. The Riemann–Roch–Hirzebruch theorem 3.2 provides the alternating sum of the dimensions of the cohomology groups, but not each of those individually. The above described application of exact and spectral sequences however is straightforward to apply and does completely determine the $\mathcal{K}_{\mathfrak{F}}^{*\otimes m}$-valued cohomology for any m. The almost-Fano 3-fold $\mathfrak{F}$ actually being a simple hypersurface the resolution is a short exact sequence, but we will utilize the spectral sequence format (9.3.16), as done in (9.5.10), so as to indicate how the more involved complete intersections would be dealt with.

As $\mathfrak{F}$ is embedded in $\mathbb{P}^3 \times \mathbb{P}^1$, all relevant bundles will be products of bundles over $\mathbb{P}^3$ and those over $\mathbb{P}^1$. We stack the corresponding Young tableaux (see § 3.3.2) one above the other, as in (9.5.10), and have $\mathcal{K}_{\mathfrak{F}}^{*\otimes m} \approx \mathcal{O}_{\mathbb{P}^1}(m) = \left(\begin{smallmatrix}0\\-m\end{smallmatrix}\middle|\begin{smallmatrix}0\,0\,0\\0\end{smallmatrix}\right)$. Multiplying the resolution sequence (9.5.10) produces

$$0 \to \left(\begin{smallmatrix}4\\1-m\end{smallmatrix}\middle|\begin{smallmatrix}0\,0\,0\\0\end{smallmatrix}\right) \to \left(\begin{smallmatrix}0\\-m\end{smallmatrix}\middle|\begin{smallmatrix}0\,0\,0\\0\end{smallmatrix}\right) \to \mathcal{K}_{\mathfrak{F}}^{*\otimes m} \to 0\,. \tag{9.6.1}$$

Stacking the cohomology under the corresponding bundles in this sequence, we obtain

$$
\begin{array}{c|c|c||c}
q & \left(\begin{smallmatrix}4 & 0\,0\,0\\ 1-m & 0\end{smallmatrix}\right) & \left(\begin{smallmatrix}0 & 0\,0\,0\\ -m & 0\end{smallmatrix}\right) & \mathcal{K}^{*\otimes m}_{\mathfrak{F}} \\
\hline
0: & 0 & \left(\begin{smallmatrix}0 & 0\,0\,0\\ -m & 0\end{smallmatrix}\right)^1_{m+1} \cdots\cdots\to & \left(\begin{smallmatrix}0 & 0\,0\,0\\ -m & 0\end{smallmatrix}\right)^1_{m+1} \\
1: & 0 & 0 \cdots\cdots\to & 0 \\
2: & 0 & 0 \cdots\cdots\to & \left(\begin{smallmatrix}1 & 1\,1\,1\\ 1-m & 0\end{smallmatrix}\right)^1_{m} \\
3: & \left(\begin{smallmatrix}1 & 1\,1\,1\\ 1-m & 0\end{smallmatrix}\right)^1_{m} \nearrow & 0 \cdots\cdots\to & 0 \\
4: & 0 & 0 & \text{—}
\end{array}
\tag{9.6.2}
$$

Adding up the dimensions, we obtain

$$
\begin{aligned}
\mathfrak{P}^m(\mathfrak{F}) &\stackrel{\text{def}}{=} \sum_{q=0}^{3}(-)^q \dim H^q(\mathfrak{F}, \mathcal{K}^{*\otimes m}_{\mathfrak{F}}) \\
&= (m{+}1) + 0 + m + 0 \;=\; 2m{+}1 \;,
\end{aligned}
\tag{9.6.3}
$$

in complete agreement with Eq. (3.2.14), for ${C_1}^3 = 0$. The $m{=}0$ case reproduces and in agreement with the computation (9.5.10) the holomorphic Euler characteristic, i.e., the Todd genus (2.5.2). This makes the plurigenera their generalization.

✎ Calculate the anti-canonical bundle-valued cohomology for analogous almost-ample n-folds, where $\mathbb{P}^3$ is replaced by $\mathbb{P}^n$, and the defining equation has bi-degree $(n{+}1, 1)$ rather than $(4, 1)$. ❧

2024 ⌈*The common property of Koszul resolution exact sequences (9.5.1) and (9.5.4), noted as the (9.5.8)-condition of (9.5.7), is shared by so-called* Gorenstein varieties*. Such Calabi–Yau 3-folds may be studied hierarchically by the length of the resolution (9.5.7) and the codimension of the so-defined Calabi–Yau 3-fold. Codimension-1 (simple hypersurfaces) have been completely classified within toric Fano 4-folds [430, 341], while codimension-2 cases all turn out to be complete intersections of hypersurfaces [460]. Codimension-3 cases are described as Pfaffian constraint systems, as discussed in § 3.5 [460]. In turn, codimension-4 cases are currently actively explored [460], as even the underlying "structure theorems" are still under development [453].*⌋

Chapter A

Tangent Bundle Valued Cohomology

As we have seen in Table 0.2, in a compactification on a Calabi–Yau 3-fold $\mathcal{M}$, elements of $H^1(\mathcal{M}, \mathcal{T}_\mathcal{M})$ represent generations of particles in the Standard model (together with some additional particles completing E_6 **27**-plets and the 27 superpartners), while elements of $H^1(\mathcal{M}, \mathcal{T}^*_\mathcal{M})$ represent generations of their mirror particles. We will shortly discuss the general method for computing these and other related cohomology groups.

Recall that $H^1(\mathcal{M}, \mathcal{T}^*_\mathcal{M}) = H^{1,1}(\mathcal{M})$ is dual to $H^{2,2}(\mathcal{M})$ and that

$$H^{2,2}(\mathcal{M}) = H^2(\mathcal{M}, \wedge^2 \mathcal{T}^*_\mathcal{M}) \approx H^2(\mathcal{M}, \mathcal{K}_\mathcal{M} \otimes \mathcal{T}_\mathcal{M}) \ . \tag{A.0.1}$$

In view of this, we may proceed either by computing $H^1(\mathcal{M}, \mathcal{T}_\mathcal{M})$ and $H^1(\mathcal{M}, \mathcal{T}^*_\mathcal{M})$, or—if $\mathcal{M}$ is Calabi–Yau, so $\mathcal{K}_\mathcal{M} \approx \mathbb{C}$—equivalently relate $H^1(\mathcal{M}, \mathcal{T}^*_\mathcal{M})$ to (the complex conjugate of) $H^2(\mathcal{M}, \mathcal{T}_\mathcal{M})$ and compute only the $\mathcal{T}_\mathcal{M}$-valued cohomology or only the $\mathcal{T}^*_\mathcal{M}$-valued one. Since neither of the computations requires the manifold $\mathcal{M}$ to be Calabi–Yau or a 3-fold, we will describe briefly the computation both of $H^q(\mathcal{M}, \mathcal{T}_\mathcal{M})$ and of $H^q(\mathcal{M}, \mathcal{T}^*_\mathcal{M})$, and also of some related cohomology groups. For Calabi–Yau n-folds, we will have the relation

$$H^q(\mathcal{M}, \mathcal{T}_\mathcal{M}) \approx H^q(\mathcal{M}, \wedge^{(n-1)} \mathcal{T}^*_\mathcal{M}) = H^{n-1,q}(\mathcal{M}) \ , \qquad \dim \mathcal{M} = n \ , \tag{A.0.2}$$

and also that $H^0(\mathcal{M},\mathcal{T}_\mathcal{M})$ and $H^n(\mathcal{M},\mathcal{T}_\mathcal{M})$ have to vanish since $h^{n-1,0} = 0 = h^{n-1,n}$ on any Calabi–Yau n-fold (see § 1.2).

A.1 Computing Tangent Bundle Valued Forms

As discussed in § 9.1, the tangent bundle of the submanifold, $\mathcal{T}_\mathcal{M}$, is related to the tangent bundle of the embedding space $\mathcal{T}_\mathcal{X}$ and the bundle $\mathcal{E}$ a (system of) section(s) of which has been used to define $\mathcal{M} \hookrightarrow \mathcal{X}$. These bundles and their cohomology over $\mathcal{X}$ are supposed to be known—and surely is easy to obtain whenever we can use the Bott–Borel–Weil theorem 9.2.

A.1.1 Representing the normal bundle

The sequence (9.1.1) is exact and specifies $\mathcal{T}_\mathcal{M}$ completely *if the map* $\nabla\xi : \mathcal{T}_\mathcal{X}|_\mathcal{M} \to \mathcal{E}|_\mathcal{M}$ *is onto*, that is, if $\mathcal{E}|_\mathcal{M}$ may be identified with $\mathcal{N}_{\mathcal{X}/\mathcal{M}}$ (compare with the *adjunction formula 1* $^{(L)}$).

Such is the case precisely for non-singular complete intersections, as exemplified in Chapters 1, 2 and 3, but is not so for the Pfaffian Calabi–Yau 3-fold discussed in § 3.5. In general, we have in mind a subspace $\mathcal{M} \subset \mathcal{X}$ defined as the zero-set of a (system of) section(s) ξ of a bundle $\mathcal{E}$ over $\mathcal{X}$. Then locally some combination of the defining section(s) spans the normal space and we expect the normal bundle to be a sub-bundle of $\mathcal{E}$. Then we write

$$0 \to \mathcal{T}_\mathcal{M} \xrightarrow{i} \mathcal{T}_\mathcal{X}|_\mathcal{M} \xrightarrow{\nabla\xi} \mathcal{E}|_\mathcal{M} \to \mathcal{F} \to 0 \,, \tag{A.1.1}$$

where $\mathcal{F} \stackrel{\text{def}}{=} \text{cok}[\mathcal{T}_\mathcal{X}|_\mathcal{M} \xrightarrow{\nabla\xi} \mathcal{E}|_\mathcal{M}]$ so that the sequence is exact. However, unless we can determine $\mathcal{F}$ independently, the relation given as Seq. (A.1.1) is of little practical use. Of course, Seq. (A.1.1) may extend even further to the right, with a sequence $\mathcal{F}_1 \to \mathcal{F}_2 \to \cdots$ replacing the single $\mathcal{F}$.

For simplicity and as it still applies to an large number of constructions, we restrict our further discussion to complete intersections. There Seq. (A.1.1) simplifies in that $\mathcal{F} = 0$ and leads to the long exact cohomology sequence (9.1.2). If $\mathcal{F} \neq 0$, we have to construct a corresponding spectral sequence instead of Seq. (9.1.2) and the computation becomes quite more involved.

A.1.2 An example

For a compact complex n-fold $\mathcal{M}$ embedded in a compact complex space $\mathcal{X}$ as a complete intersection, the computation proceeds as follows: (1) The $H^q(\mathcal{M},\mathcal{E})$ and $H^q(\mathcal{M},\mathcal{T}_\mathcal{X})$ cohomology groups are obtained from the spectral sequence constructed from the resolution (9.3.7) tensored by $\mathcal{E}$ and by $\mathcal{T}_\mathcal{X}$, respectively. (2) This data is then filled into Seq. (9.1.2), determining $H^q(\mathcal{M},\mathcal{T}_\mathcal{M})$, as required.

Consider, for example, the complete intersection (8.0.1)

$$\mathcal{M}_{-54} \in \begin{bmatrix} 3 \\ 2 \end{bmatrix}\!\!\begin{Vmatrix} 3 & 1 \\ 0 & 3 \end{Vmatrix}^{8,35}_{-54} . \tag{A.1.2}$$

We choose this simple case both because of its possible physical importance [210] and because several of the previously studied techniques apply (see § 1.6, § 3.1).

Let us denote the defining polynomials by

$$f(x) \stackrel{\text{def}}{=} f_{abc}\, x^a x^b x^c\,, \qquad g(x,y) \stackrel{\text{def}}{=} g_{a\,\alpha\beta\gamma}\, x^a y^\alpha y^\beta y^\gamma\,, \tag{A.1.3}$$

where $x \in \mathbb{P}^3$ and $y \in \mathbb{P}^2$. Here $f(x)$ and $g(x,y)$ are sections of $\mathcal{E}_f = \left(\begin{smallmatrix} -3 \\ 0 \end{smallmatrix}\middle|\begin{smallmatrix} 0\,0\,0 \\ 0\,0 \end{smallmatrix}\right)$ and $\mathcal{E}_g = \left(\begin{smallmatrix} -1 \\ -3 \end{smallmatrix}\middle|\begin{smallmatrix} 0\,0\,0 \\ 0\,0 \end{smallmatrix}\right)$, respectively, so $\mathcal{E}$ is the rank-two direct sum of these two line bundles. The two lines in this notation for bundles over $\mathcal{X} = \mathbb{P}^3 \times \mathbb{P}^2$ refer to representations of $U(1) \times U(3)$ and $U(1) \times U(2)$, corresponding to the restriction of a bundle to $\mathbb{P}^3$ and $\mathbb{P}^2$, respectively. The tangent bundle on $\mathcal{X}$ is given as

$$\mathcal{T}_\mathcal{X} = \left[\mathcal{T}_x = \left(\begin{smallmatrix} -1 \\ 0 \end{smallmatrix}\middle|\begin{smallmatrix} 0\,0\,1 \\ 0\,0 \end{smallmatrix}\right)\right] \oplus \left[\mathcal{T}_y = \left(\begin{smallmatrix} 0 \\ -1 \end{smallmatrix}\middle|\begin{smallmatrix} 0\,0\,0 \\ 0\,1 \end{smallmatrix}\right)\right] . \tag{A.1.4}$$

The cohomology groups on $\mathcal{M}$ valued in each of these irreducible bundles are computed by using a corresponding spectral sequence. To this end, note that the resolution (9.3.7) now becomes

$$0 \to \left(\begin{smallmatrix} 4 \\ 3 \end{smallmatrix}\middle|\begin{smallmatrix} 0\,0\,0 \\ 0\,0 \end{smallmatrix}\right) \begin{matrix} \overset{f}{\nearrow} \left(\begin{smallmatrix} 1 \\ 3 \end{smallmatrix}\middle|\begin{smallmatrix} 0\,0\,0 \\ 0\,0 \end{smallmatrix}\right) \overset{g}{\searrow} \\ \underset{-g}{\searrow} \left(\begin{smallmatrix} 3 \\ 0 \end{smallmatrix}\middle|\begin{smallmatrix} 0\,0\,0 \\ 0\,0 \end{smallmatrix}\right) \underset{f}{\nearrow} \end{matrix} \left(\begin{smallmatrix} 0 \\ 0 \end{smallmatrix}\middle|\begin{smallmatrix} 0\,0\,0 \\ 0\,0 \end{smallmatrix}\right) \to \left(\begin{smallmatrix} 0 \\ 0 \end{smallmatrix}\middle|\begin{smallmatrix} 0\,0\,0 \\ 0\,0 \end{smallmatrix}\right)_\mathcal{M} \to 0\,. \tag{A.1.5}$$

The arrows labeled by f and g denote maps induced by contraction with the tensors f_{abc} and $g_{a\,\alpha\beta\gamma}$, respectively.

Tensoring by $\mathcal{E}_f = \left(\begin{smallmatrix} -3 & 0\,0\,0 \\ 0 & 0\,0 \end{smallmatrix}\right)$, the resolution sequence (A.1.5) yields

q	$\left(\begin{array}{c\|c} 1 & 0\,0\,0 \\ 3 & 0\,0 \end{array}\right)$	$\left(\begin{array}{c\|c} -2 & 0\,0\,0 \\ 3 & 0\,0 \end{array}\right)$ $\left(\begin{array}{c\|c} 0 & 0\,0\,0 \\ 0 & 0\,0 \end{array}\right) \xrightarrow{f}$	$\left(\begin{array}{c\|c} -3 & 0\,0\,0 \\ 0 & 0\,0 \end{array}\right)$	$\mathcal{E}_f\vert_{\mathcal{M}}$
0:	0	$\left(\begin{smallmatrix} 0\,0\,0\,0 \\ 0\,0\,0 \end{smallmatrix}\right)^1_1 \xrightarrow{f}$	$\left(\begin{smallmatrix} -3\,0\,0\,0 \\ 0\,0\,0 \end{smallmatrix}\right)^{20}_1 \dashrightarrow$	$\Big[\left(\begin{smallmatrix} -3\,0\,0\,0 \\ 0\,0\,0 \end{smallmatrix}\right)^{20}_1 \Big/ f\left(\begin{smallmatrix} 0\,0\,0\,0 \\ 0\,0\,0 \end{smallmatrix}\right)^1_1\Big]$
1:	0	0	$0 \dashrightarrow$	$\left(\begin{smallmatrix} -2\,0\,0\,0 \\ 1\,1\,1 \end{smallmatrix}\right)^{10}_1$
2:	0	$\left(\begin{smallmatrix} -2\,0\,0\,0 \\ 1\,1\,1 \end{smallmatrix}\right)^{10}_1 \dashrightarrow$	0	$\Rightarrow 0$
3:	0	0	0	$\Rightarrow 0$
4:	0	0	0	—
5:	0	0	0	—

(A.1.6)

The map f—induced from the one in the Koszul complex in the upper left quadrant of the above chart—is of maximum rank and yields the only non-vanishing differential in this spectral sequence.

This is completely general; consider the resolution sequence (9.5.7), tensored by any bundle $\mathcal{V}$ over the embedding space $\mathcal{X}$. All differential maps d_r acting in the SW (lower left) quadrant of

$$E_r^{q,k}(\mathcal{V}) \xrightarrow{\ \mathrm{d}_r\ } E_r^{q-r,k-r-1}(\mathcal{V}) \tag{A.1.7}$$

are induced from the corresponding maps $\mathcal{E}_k{}^* \otimes \mathcal{V} \to \mathcal{E}_{k-r-1}{}^* \otimes \mathcal{V}$ between the irreducible summands in the resolution sequence.

In this simple case, the map

$$f \ : \ \left(\begin{smallmatrix} 0\,0\,0\,0 \\ 0\,0\,0 \end{smallmatrix}\right)^1_1 \ \to \ \left(\begin{smallmatrix} -3\,0\,0\,0 \\ 0\,0\,0 \end{smallmatrix}\right)^{20}_1 \tag{A.1.8}$$

is represented by mapping an arbitrary complex number $\vartheta \in \left(\begin{smallmatrix} 0\,0\,0\,0 \\ 0\,0\,0 \end{smallmatrix}\right)$ to the (additive, abelian) group of rank-3 covariant symmetric tensors, simply by multiplying θ with f_{abc}: $(\vartheta f_{abc}) \in \left(\begin{smallmatrix} -3\,0\,0\,0 \\ 0\,0\,0 \end{smallmatrix}\right)$. Clearly, ϑ is not annihilated in doing this, whence $\ker(f) = 0$ here. On the other hand,

$$\mathrm{cok}\Big[\left(\begin{smallmatrix} 0\,0\,0\,0 \\ 0\,0\,0 \end{smallmatrix}\right)^1_1 \xrightarrow{f} \left(\begin{smallmatrix} -3\,0\,0\,0 \\ 0\,0\,0 \end{smallmatrix}\right)^{20}_1\Big] = \Big\{\left(\begin{smallmatrix} -3\,0\,0\,0 \\ 0\,0\,0 \end{smallmatrix}\right)^{20}_1 \Big/ f\left(\begin{smallmatrix} 0\,0\,0\,0 \\ 0\,0\,0 \end{smallmatrix}\right)^1_1\Big\} \tag{A.1.9}$$

is 19-dimensional and is represented by symmetric rank-3 tensors, modulo arbitrary complex multiples of f_{abc}, $\mathrm{cok}(f) \approx \{\phi_{(abc)} \cong \phi_{(abc)} + \vartheta f_{abc}\}$.

The contribution $\left(\begin{smallmatrix}-2&0&0&0\\1&1&1\end{smallmatrix}\right)^{10}_{1}$ is represented by symmetric rank-2 symmetric tensors $\varphi_{(ab)} \in (-2\,0\,0\,0)$, multiplied by an overall factor of $\epsilon^{\alpha\beta\gamma}$, representing the $(1\,1\,1)$ tableau.

Similarly, tensoring by $\mathcal{T}_x = \left(\begin{smallmatrix}-1 & | & 0&0&1\\ 0 & | & 0&0\end{smallmatrix}\right)$, the resolution sequence (A.1.5) yields

$$
\begin{array}{c|c|c|c||l}
q & \left(\begin{smallmatrix}3 & | & 0&0&1\\ 3 & | & 0&0\end{smallmatrix}\right) & \begin{smallmatrix}\left(\begin{smallmatrix}0 & | & 0&0&1\\ 3 & | & 0&0\end{smallmatrix}\right)\\ \left(\begin{smallmatrix}2 & | & 0&0&1\\ 0 & | & 0&0\end{smallmatrix}\right)\end{smallmatrix} & \left(\begin{smallmatrix}-1 & | & 0&0&1\\ 0 & | & 0&0\end{smallmatrix}\right) & \mathcal{T}_x|_{\mathcal{M}} \\
\hline
0: & 0 & 0 & \left(\begin{smallmatrix}-1&0&0&1\\0&0&0\end{smallmatrix}\right)^{15}_{1} & \dashrightarrow H^0(\mathcal{M},\mathcal{T}_x) \\
1: & 0 & 0 & 0 & \dashrightarrow H^1(\mathcal{M},\mathcal{T}_x) \\
2: & 0 & \left(\begin{smallmatrix}0&0&0&1\\1&1&1\end{smallmatrix}\right)^{4}_{1} & 0 & \dashrightarrow H^2(\mathcal{M},\mathcal{T}_x) \\
3: & 0 & 0 & 0 & \Rightarrow 0 \\
4: & \left(\begin{smallmatrix}1&1&1&1\\1&1&1\end{smallmatrix}\right)^{1}_{1} & 0 & 0 & - \\
5: & 0 & 0 & 0 & -
\end{array}
\tag{A.1.10}
$$

There clearly are no effective differential maps, so $E_0^{\star,\star} = E_\infty^{\star,\star}$ and there is a non-zero contribution to $H^q(\mathcal{M},\mathcal{T}_x)$ for $q{=}0,1,2$, while $H^3(\mathcal{M},\mathcal{T}_x) = 0$.

✎ Compute $H^q(\mathcal{M},\mathcal{E}_g)$ and $H^q(\mathcal{M},\mathcal{T}_y)$. ❧

Through the *Künneth formula*$^{(L)}$, all the information for Seq. (9.1.2) is now ready; in with it. Charting the long exact cohomology sequence under Seq. (9.1.1), we have

$$
\begin{array}{lclcl}
\mathcal{T}_{\mathcal{M}} & \xrightarrow{i} & \mathcal{T}_X|_{\mathcal{M}} & \xrightarrow{f,g} & \mathcal{E}|_{\mathcal{M}} \\
\hline
H^0(\mathcal{M},\mathcal{T}_{\mathcal{M}}) & \xrightarrow{i_0} & \begin{smallmatrix}\left(\begin{smallmatrix}-1&0&0&1\\0&0&0\end{smallmatrix}\right)^{15}_{1}\\ \left(\begin{smallmatrix}0&0&0&0\\-1&0&1\end{smallmatrix}\right)^{1}_{8}\end{smallmatrix} & \begin{smallmatrix}\xrightarrow{f},\ \searrow g\\ \xrightarrow{g}\end{smallmatrix} & \begin{smallmatrix}\left[\left(\begin{smallmatrix}-3&0&0&0\\0&0&0\end{smallmatrix}\right)^{20}_{1} \Big/ f\left(\begin{smallmatrix}0&0&0&0\\0&0&0\end{smallmatrix}\right)^{1}_{1}\right]\\ \left[\left(\begin{smallmatrix}-1&0&0&0\\-3&0&1\end{smallmatrix}\right)^{4}_{10} \Big/ g\left(\begin{smallmatrix}0&0&0&0\\0&0&0\end{smallmatrix}\right)^{1}_{1}\right]\end{smallmatrix} \\
H^1(\mathcal{M},\mathcal{T}_{\mathcal{M}}) & \xrightarrow{i_1} & \left(\begin{smallmatrix}0&0&0&1\\1&1&1\end{smallmatrix}\right)^{4}_{1} & \xrightarrow{f} & \left(\begin{smallmatrix}-2&0&0&0\\1&1&1\end{smallmatrix}\right)^{10}_{1} \\
H^2(\mathcal{M},\mathcal{T}_{\mathcal{M}}) & \xrightarrow{i_2} & 2\left(\begin{smallmatrix}1&1&1&1\\1&1&1\end{smallmatrix}\right)^{1}_{1} & & 0 \\
H^3(\mathcal{M},\mathcal{T}_{\mathcal{M}}) & \xrightarrow{i_3} & 0 & & 0
\end{array}
\tag{A.1.11}
$$

Clearly, $H^3(\mathcal{M},\mathcal{T}_{\mathcal{M}})$ vanishes. On general grounds (§ 1.2), we also know that $H^0(\mathcal{M},\mathcal{T}_{\mathcal{M}}) = 0$, but this can be computed here explicitly.

Relying on the exactness of the long cohomology sequence, we see that it is sufficient to discern the action of the maps labeled by f and g to determine $H^q(\mathcal{M},\mathcal{T}_\mathcal{M})$ completely. In particular, denoting the combined triple map by j_0, and the f-map in $q\!=\!1$ by j_1, we have

$$H^0(\mathcal{M},\mathcal{T}_\mathcal{M}) \approx \ker(j_0)\ , \tag{A.1.12}$$
$$H^1(\mathcal{M},\mathcal{T}_\mathcal{M}) \approx \ker(j_1) + \mathrm{cok}(j_0)\ , \tag{A.1.13}$$
$$H^2(\mathcal{M},\mathcal{T}_\mathcal{M}) \approx 2\big({}^{1\,1\,1\,1}_{\,1\,1\,1}\big) + \mathrm{cok}(j_1)\ . \tag{A.1.14}$$

To be pedantic, recall that '+' denotes *extension*$^{(L)}$, not a holomorphic direct sum although this does not seem to have surfaced in any of the physically relevant computations.

Most notably, determining the action of maps in a long exact cohomology sequence (or more generally, in the spectral sequences) is *in general* a tough problem; consequently, so is the computation of $H^q(\mathcal{M},\mathcal{T}_\mathcal{M})$. However, if the Bott–Borel–Weil theorem 9.2 can be applied as we do it here and as can be done for all complete intersections in products of generalized flag spaces, the computation becomes an easy (albeit possibly lengthy) exercise in linear algebra. Moreover, as we will discuss at the end of this volume, it also links into some modern techniques in quantum field theories.

Now, the $q\!=\!1$ single map corresponds to

$$j_1:\qquad f_{abc}(\epsilon^{\alpha\beta\gamma}\,\lambda^c)\ \longmapsto\ (\epsilon^{\alpha\beta\gamma}\,\theta_{(ab)})\ . \tag{A.1.15}$$

Clearly, no component of the 4-vector λ^c is annihilated, so $\ker j_1\ =\ 0$. However,

$$\mathrm{cok}(j_1) \approx \{\,\epsilon^{\alpha\beta\gamma}\,\theta_{(ab)} \cong \epsilon^{\alpha\beta\gamma}\,\theta_{(ab)} + (f_{abc}\epsilon^{\alpha\beta\gamma}\,\lambda^c)\,\}\ , \tag{A.1.16}$$

and so $H^2(\mathcal{M},\mathcal{T}_\mathcal{M}) \approx H^{2,2}(\mathcal{M})$ is the extension

$$H^2(\mathcal{M},\mathcal{T}_\mathcal{M}) = \{\,(*J_x)\,,\,(*J_y)\,\}+\{\,\epsilon^{\alpha\beta\gamma}\,\theta_{(ab)} \cong \epsilon^{\alpha\beta\gamma}\,\theta_{(ab)}+(f_{abc}\epsilon^{\alpha\beta\gamma}\,\lambda^c)\,\}\ . \tag{A.1.17}$$

By $(*J_x)$ and $(*J_y)$ we of course mean the (2,2)-classes dual to the pullbacks to $\mathcal{M}$ of the Kähler (1,1)-classes of $\mathbb{P}^3$ and $\mathbb{P}^2$, respectively. Straightforwardly then,

$$H^1(\mathcal{M},\mathcal{T}^*_\mathcal{M}) = \{\,J_x\,,\,J_y\,\}+\{\,\theta^{(ab)}\,\epsilon^{cdef}\ :\ f_{abc}\theta^{(ab)}=0\,\}\ . \tag{A.1.18}$$

The dimensions add up to $1+1+(10-4) = 8$, and we have used here that

$$(a, b, \cdots, d) \xrightarrow{\text{Serre-duality}} (1-d, \cdots, 1-b, 1-a) \ , \tag{A.1.19}$$

because $\mathcal{K}^*_{\mathbb{P}^n} = (1|1, \cdots, 1)$. (See also *exact sequence*$^{(L)}$.)

At $q=0$, we have the combined map

$$j_0 : \quad \begin{array}{ccc} \left(\begin{smallmatrix} -1&0&0&1 \\ 0&0&0 \end{smallmatrix}\right)^{15}_{1} & \xrightarrow{\ f\ } & \left[\left(\begin{smallmatrix} -3&0&0&0 \\ 0&0&0 \end{smallmatrix}\right)^{20}_{1} \Big/ f\left(\begin{smallmatrix} 0&0&0&0 \\ 0&0&0 \end{smallmatrix}\right)^{1}_{1}\right] \\ & \searrow^{g} & \\ \left(\begin{smallmatrix} 0&0&0&0 \\ -1&0&1 \end{smallmatrix}\right)^{1}_{8} & \xrightarrow{\ g\ } & \left[\left(\begin{smallmatrix} 0&0&0&0 \\ -1&0&1 \end{smallmatrix}\right)^{1}_{8} \Big/ g\left(\begin{smallmatrix} 0&0&0&0 \\ 0&0&0 \end{smallmatrix}\right)^{1}_{1}\right] \end{array} \ , \tag{A.1.20}$$

represented by

$$\begin{array}{ccc} \{\lambda_a{}^b\} & \xrightarrow{\ f\ } & \{(\lambda_{(a}{}^d f_{bc)d}\} \\ & \searrow^{g} & \\ \{\lambda_\alpha{}^\beta\} & \xrightarrow{\ g\ } & \{(\lambda_a{}^d g_{d\,\alpha\beta\gamma}) + (g_{a\,\delta(\alpha\beta}\,\lambda_{\gamma)}{}^\delta)\} \end{array} \ . \tag{A.1.21}$$

✎ Show that $H^0(\mathcal{M}, \mathcal{T}_\mathcal{M}) = \ker(j_0) = 0$. (Caution: in addition to showing that no element of the domain is annihilated in contraction with f_{abc} and $g_{a\,\alpha\beta\gamma}$, it must be shown that only zero is mapped to the null element(s) of the target, which is a *quotient*.) ❧

The cokernel is simple to express upon noticing that, for example, $f\left(\begin{smallmatrix} 0&0&0&0 \\ 0&0&0 \end{smallmatrix}\right)$ and $f\left(\begin{smallmatrix} -1&0&0&1 \\ 0&0&0 \end{smallmatrix}\right)$ may be combined into f acting on an arbitrary (not traceless) 4×4 matrix. Since $\ker(j_1) = 0$, $H^1(\mathcal{M}, \mathcal{T}_\mathcal{M}) = \operatorname{cok}(j_0)$,

$$H^1(\mathcal{M}, \mathcal{T}_\mathcal{M}) \ = \ \left\{ \begin{array}{lcl} \{\phi_{(abc)} & \cong & \phi_{(abc)} + (\lambda_{(a}{}^d f_{bc)d})\} \\ \{\varphi_{a(\alpha\beta\gamma)} & \cong & \varphi_{a(\alpha\beta\gamma)} + (\lambda_a{}^d g_{d\,\alpha\beta\gamma}) + (g_{a\,\delta(\alpha\beta}\,\lambda_{\gamma)}{}^\delta)\} \end{array} \right\} , \tag{A.1.22}$$

where $\lambda_a{}^b$ and $\lambda_\alpha{}^\beta$ are now arbitrary invertible 4×4 and 3×3 complex matrices. The dimensions add up to $20{\cdot}1 + 4{\cdot}10 - 16 - 9 = 35 = h^{2,1}$.

A.2 Computing Cotangent Bundle Valued Forms

To compute $H^q(\mathcal{M}, \mathcal{T}^*_\mathcal{M})$, we consider the dual of Seq. (9.1.1),

$$0 \to \mathcal{E}^*|_\mathcal{M} \xrightarrow{\nabla\xi} \mathcal{T}^*_\mathcal{X}|_\mathcal{M} \xrightarrow{i} \mathcal{T}^*_\mathcal{M} \to 0 \ . \tag{A.2.1}$$

In view of *Serre duality*$^{(L)}$, this is of course an unnecessary exercise if $\mathcal{M}$ is Calabi–Yau, since then

$$H^q(\mathcal{M}, \mathcal{T}^*_\mathcal{M})^* \ \approx \ H^{n-q}(\mathcal{M}, \mathcal{T}_\mathcal{M} \otimes \mathcal{K}_\mathcal{M}) \ \approx \ H^{n-q}(\mathcal{M}, \mathcal{T}_\mathcal{M}) \ . \tag{A.2.2}$$

In the general case however, when $\mathcal{K}_{\mathcal{M}} \not\approx \mathbb{C}$, the computation of $H^q(\mathcal{M}, \mathcal{T}^*_{\mathcal{M}})$ provides the *Hodge numbers*$^{(L)}$ and the standard (p,q)-cohomology ring. On the other hand, $H^q(\mathcal{M}, \mathcal{T}_{\mathcal{M}})$ provides information on deformations of the complex structure: $H^1(\mathcal{M}, \mathcal{T}_{\mathcal{M}})$ spans the local deformations.

To determine $H^q(\mathcal{M}, \mathcal{T}^*_{\mathcal{M}})$ from the long exact cohomology sequence associated to Seq. (A.2.1), we need to fill in the blanks for $H^q(\mathcal{M}, \mathcal{T}^*_{X})$ and $H^q(\mathcal{M}, \mathcal{E}^*)$. This information (see Table A.1) is obtained from the resolution (A.1.5), upon tensoring with the irreducible components of

$$\mathcal{T}^*_X = \left(\begin{array}{c|c} 1 & -1\;0\;0 \\ 0 & 0\;0 \end{array}\right) \oplus \left(\begin{array}{c|c} 0 & 0\;0\;0 \\ 1 & -1\;0 \end{array}\right), \qquad \mathcal{E}^* = \left(\begin{array}{c|c} 3 & 0\;0\;0 \\ 0 & 0\;0 \end{array}\right) \oplus \left(\begin{array}{c|c} 1 & 0\;0\;0 \\ 3 & 0\;0 \end{array}\right). \tag{A.2.3}$$

✎ Verify the results in Table A.1. ❧

With this information in the long exact cohomology sequence accompanying Seq. (A.2.1), we have

$$\begin{array}{ccccc}
\mathcal{E}^*|_{\mathcal{M}} & \xrightarrow{f,g} & \mathcal{T}^*_X|_{\mathcal{M}} & \xrightarrow{i} & \mathcal{T}^*_{\mathcal{M}} \\
\hline
0 & & 0 & \xrightarrow{i_0} & H^0(\mathcal{M}, \mathcal{T}^*_{\mathcal{M}}) \\
0 & & 2\left(\begin{smallmatrix} 0\,0\,0\,0 \\ 0\,0\,0 \end{smallmatrix}\right)^1_1 & \xrightarrow{i_1} & H^1(\mathcal{M}, \mathcal{T}^*_{\mathcal{M}}) \\
\left(\begin{smallmatrix} 1\,1\,1\,3 \\ 0\,0\,0 \end{smallmatrix}\right)^{10}_1 & \xrightarrow{f} & \left(\begin{smallmatrix} 0\,1\,1\,1 \\ 0\,0\,0 \end{smallmatrix}\right)^4_1 & \xrightarrow{i_2} & H^2(\mathcal{M}, \mathcal{T}^*_{\mathcal{M}}) \\
\left[\ker \left(\begin{smallmatrix} 1\,1\,1\,4 \\ 1\,1\,1 \end{smallmatrix}\right)^{20}_1 \xrightarrow{f} \left(\begin{smallmatrix} 1\,1\,1\,1 \\ 1\,1\,1 \end{smallmatrix}\right)^1_1\right] & \xrightarrow{f} & \left(\begin{smallmatrix} 0\,1\,1\,2 \\ 1\,1\,1 \end{smallmatrix}\right)^{15}_1 & & \\
 & \nearrow_{g} & & \xrightarrow{i_0} & H^3(\mathcal{M}, \mathcal{T}^*_{\mathcal{M}}) \\
\left[\ker \left(\begin{smallmatrix} 1\,1\,1\,2 \\ 1\,1\,4 \end{smallmatrix}\right)^{1}_8 \xrightarrow{g} \left(\begin{smallmatrix} 1\,1\,1\,1 \\ 1\,1\,1 \end{smallmatrix}\right)^1_1\right] & \xrightarrow{g} & \left(\begin{smallmatrix} 1\,1\,1\,1 \\ 0\,1\,2 \end{smallmatrix}\right)^{1}_8 & &
\end{array} \tag{A.2.4}$$

Table A.1: The non-zero cohomology on $\mathcal{M}$, required for $H^q(\mathcal{M}, \mathcal{T}^*_{\mathcal{M}})$.

Bundle	**Non-zero cohomology on $\mathcal{M}$**
$\left(\begin{array}{c\|c} 1 & -1\;0\;0 \\ 0 & 0\;0 \end{array}\right)$	$H^1 = \left(\begin{smallmatrix} 0\,0\,0\,0 \\ 0\,0\,0 \end{smallmatrix}\right)^1_1, \quad H^2 = \left(\begin{smallmatrix} 0\,1\,1\,1 \\ 0\,0\,0 \end{smallmatrix}\right)^4_1, \quad H^3 = \left(\begin{smallmatrix} 0\,1\,1\,2 \\ 1\,1\,1 \end{smallmatrix}\right)^{15}_1$
$\left(\begin{array}{c\|c} 0 & 0\;0\;0 \\ 1 & -1\;0 \end{array}\right)$	$H^1 = \left(\begin{smallmatrix} 0\,0\,0\,0 \\ 0\,0\,0 \end{smallmatrix}\right)^1_1, \quad H^3 = \left(\begin{smallmatrix} 1\,1\,1\,1 \\ 0\,1\,2 \end{smallmatrix}\right)^1_8$
$\left(\begin{array}{c\|c} 3 & 0\;0\;0 \\ 0 & 0\;0 \end{array}\right)$	$H^2 = \left(\begin{smallmatrix} 1\,1\,1\,3 \\ 0\,0\,0 \end{smallmatrix}\right)^{10}_1, \quad H^3 = \left[\ker \left(\begin{smallmatrix} 1\,1\,1\,4 \\ 1\,1\,1 \end{smallmatrix}\right)^{20}_1 \xrightarrow{f} \left(\begin{smallmatrix} 1\,1\,1\,1 \\ 1\,1\,1 \end{smallmatrix}\right)^1_1\right]_{19}$
$\left(\begin{array}{c\|c} 1 & 0\;0\;0 \\ 3 & 0\;0 \end{array}\right)$	$H^3 = \left[\ker \left(\begin{smallmatrix} 1\,1\,1\,2 \\ 1\,1\,4 \end{smallmatrix}\right)^{4}_{10} \xrightarrow{g} \left(\begin{smallmatrix} 1\,1\,1\,1 \\ 1\,1\,1 \end{smallmatrix}\right)^1_1\right]_{39}$

Comparing with (A.1.11), we see Serre duality at work. We can again analyze the action of each of the indicated maps $j_q : H^q(\mathcal{M},\mathcal{E}^*) \xrightarrow{f,g} H^q(\mathcal{M},\mathcal{T}^*_X)$, just as we did for $H^q(\mathcal{M},\mathcal{T}_\mathcal{M})$.

It is easy to re-derive Eq. (A.1.18) from (A.2.4). Also,

$$H^2(\mathcal{M},\mathcal{T}^*_\mathcal{M}) \;=\; \left\{ \begin{matrix} \{\, \epsilon^{a_1a_2a_3a_4}\epsilon^{\alpha_1\alpha_2\alpha_3}\phi^{(bcd)} \;:\; f_{abc}\phi^{(bcd)} = 0\,\} \\ \{\, \epsilon^{a_1a_2a_3a_4}\epsilon^{\alpha_1\alpha_2\alpha_3}\varphi^{b(\beta\gamma\delta)} \;:\; g_{b\,\alpha\beta\gamma}\varphi^{b(\beta\gamma\delta)} = 0\,\} \end{matrix} \right\}, \tag{A.2.5}$$

where the map

$$\left[\ker \begin{pmatrix} 1&1&1&2 \\ 1&1&4 \end{pmatrix}^1_8 \xrightarrow{g} \begin{pmatrix} 1&1&1&1 \\ 1&1&1 \end{pmatrix}^1_1\right] \xrightarrow{g} \begin{pmatrix} 0&1&1&2 \\ 1&1&1 \end{pmatrix}^{15}_1 \tag{A.2.6}$$

was chosen ineffective while the other two are effective. This freedom is exactly the counterpart of the freedom we have in *choosing* representatives for the quotient (A.1.22). Clearly, there are many different but equivalent choices which must be borne in mind in calculations with cohomology groups.

A.3 Relation to Polynomial Deformation

Tracing the calculation in the preceding sections, we see that there is always a contribution to $H^1(\mathcal{M},\mathcal{T}_\mathcal{M})$ essentially of the form

$$\{H^0(\mathcal{X},\mathcal{E})/[H^1(\mathcal{X},\mathcal{E}) \oplus H^0(\mathcal{X},\mathcal{T}_X)]\}\ . \tag{A.3.1}$$

Now, elements of $H^0(\mathcal{X},\mathcal{E})$ are just the sections of $\mathcal{E}$, usually polynomials of the homogeneity of the defining constraints. As we have seen above, the quotient with respect to $H^1(\mathcal{X},\mathcal{E})$ means taking such sections modulo additive complex multiplets of the defining sections. The quotient with respect to $H^0(\mathcal{X},\mathcal{T}_X)$ is precisely the quotient with respect to linear reparametrizations.

In full generality, consider a configuration $[\mathcal{X}\|\mathcal{E}]$ where $\mathcal{X}$ is the embedding space and $\mathcal{E}$ is the bundle the sections of which are used to define $\mathcal{M} \in [\mathcal{X}\|\mathcal{E}]$ (not necessarily as a complete intersection). Then we define the space of 'polynomial deformations' to be

$$\mathfrak{d}[\mathcal{X}\|\mathcal{E}] \;\stackrel{\text{def}}{=}\; \left\{\frac{H^0(\mathcal{X},\mathcal{E})}{\mathrm{Aut}(\mathcal{X})\times \mathrm{Aut}(\mathcal{E})}\right\}, \tag{A.3.2}$$

where $\mathrm{Aut}(\mathcal{X})$ is the group of continuous *automorphisms*$^{(L)}$ (symmetries) of $\mathcal{X}$ and $\mathrm{Aut}(\mathcal{E})$ is the group of continuous automorphisms of $\mathcal{E}$. $\mathrm{Aut}(\mathcal{X})$ is usually just the group of coordinate reparametrizations of $\mathcal{X}$. $\mathrm{Aut}(\mathcal{E})$, on the other hand, may on occasion become pretty involved. Even in a simple example, $\mathcal{M} \in [5\|2\,4]$ where $\mathcal{E} = \mathcal{O}(2) \oplus \mathcal{O}(4)$, $\mathrm{Aut}(\mathcal{E})$ is generated by adding a complex multiple of the two defining polynomials to quadric and quartic monomials, respectively, but also by adding a quadric multiple of the first defining polynomial (the quadric) to quartic monomials. Without the spectral sequence to systematize all possible maps, accounting for all these can get out of hand rather quickly.

Depending on the relation between $\mathcal{E}$ and the normal bundle $\mathcal{N}_{X/\mathcal{M}}$ (A.1.1), the space of 'polynomial deformations' (A.3.2) is related to $H^1(\mathcal{M}, \mathcal{T}_\mathcal{M})$. For complete intersections, $\mathcal{E} = \mathcal{N}_{X/\mathcal{M}}$ and in favorable cases moreover, $\mathfrak{d}[\mathcal{X}\|\mathcal{E}]$ is *equal* to $H^1(\mathcal{M}, \mathcal{T}_\mathcal{M})$ (see below).

A.3.1 Polynomial deformations

Clearly, arbitrary polynomials (more generally, sections) of the homogeneity of the defining polynomials (sections)—modulo the above ideal—may be considered as effective deformations of the defining polynomials (sections). So, a subset (possibly all) of $H^1(\mathcal{M}, \mathcal{T}_\mathcal{M})$ may be spanned by effective deformations of the defining polynomials; this was derived in Ref. [84] using standard tensor calculus. Now, general theory [35] shows that $H^1(\mathcal{M}, \mathcal{T}_\mathcal{M})$—for any complex manifold—spans the local deformations of the complex structure on $\mathcal{M}$. On the other hand, effective deformations of the defining constraints span the local deformations of the embedding $\mathcal{M} \hookrightarrow \mathcal{X}$. In fortunate cases, these two different notions of deformation coincide and polynomial deformations in fact do span $H^1(\mathcal{M}, \mathcal{T}_\mathcal{M})$ completely and with no redundancy.

Having gone through a generally valid calculation of $H^1(\mathcal{M}, \mathcal{T}_\mathcal{M})$, we see that polynomial deformations may in general be both incomplete and redundant. More precisely, the above cohomology computation involves

$$\begin{array}{ccccccc}
\vdots & & \vdots & & & & \vdots \\
\downarrow & & \downarrow & & & & \downarrow \\
H^0(\mathcal{X}, \mathcal{T}_X) & & H^0(\mathcal{X}, \mathcal{E}) & & & & H^1(\mathcal{X}, \mathcal{T}_X) \\
\downarrow \varrho_0^{\mathcal{T}_X} & & \downarrow \varrho_0^{\mathcal{E}} & & & & \downarrow \varrho_1^{\mathcal{T}_X} \\
H^0(\mathcal{M}, \mathcal{T}_X) & \to & H^0(\mathcal{M}, \mathcal{E}) & \to & H^1(\mathcal{M}, \mathcal{T}_\mathcal{M}) & \xrightarrow{i_0} & H^1(\mathcal{M}, \mathcal{T}_X) \\
\downarrow & & \downarrow & & & & \downarrow \\
0 & & 0 & & & & \vdots
\end{array} \tag{A.3.3}$$

Clearly, the restrictions $\varrho_0^{\mathcal{T}_X}$ and $\varrho_0^{\mathcal{E}}$ are in general *not* 1–1, that is, there will in general be elements of $H^0(\mathcal{M}, \mathcal{T}_X)$ and $H^0(\mathcal{M}, \mathcal{E})$ which do not arise as restrictions of $H^0(\mathcal{X},\mathcal{T}_{\mathcal{X}})$ and $H^0(\mathcal{X},\mathcal{E})$, respectively. So, by approximating

$$\{H^0(\mathcal{M},\mathcal{E})/H^0(\mathcal{M},\mathcal{T}_{\mathcal{X}})\} \rightsquigarrow \{H^0(\mathcal{X},\mathcal{E})/H^0(\mathcal{X},\mathcal{T}_{\mathcal{X}})\}\ , \tag{A.3.4}$$

we may be undercounting the 'numerator' or the 'denominator,' or even both of them. Finally, *in principle*, there is no reason for the map i_0 to be null. So, some elements of $H^1(\mathcal{M},\mathcal{T}_{\mathcal{M}})$ may have nothing to do with elements of the quotients (A.3.4), but could instead be the preimages under i_0 of certain elements in $H^1(\mathcal{M},\mathcal{T}_{\mathcal{X}})$. (Caution, again: $H^1(\mathcal{M},\mathcal{T}_{\mathcal{X}})$ cannot be simply replaced by $H^1(\mathcal{X},\mathcal{T}_{\mathcal{X}})$, the restriction $\varrho_1^{\mathcal{T}_{\mathcal{X}}}$ need not be 1–1.)

It is, however, an *experimental* fact, observed upon a case by case computation, that for complete intersection Calabi–Yau 3-folds in products of complex projective spaces* i_0 does vanish [148]. It does not appear to be known why this is so and how generally this is true.

Even with $i_0 = 0$, however, much room remains for polynomial deformations to fail in providing a correct representation of $H^1(\mathcal{M},\mathcal{T}_{\mathcal{M}})$. On the other hand, if it can be ascertained that $\varrho_0^{\mathcal{T}_{\mathcal{X}}}$ and $\varrho_0^{\mathcal{E}}$ in fact *are injective maps*$^{(L)}$, they are also isomorphisms and the polynomial deformation interpretation gives us a handy computational tool. It is indeed a felicitous twist of Fate that polynomial deformations correctly represent $H^1(\mathcal{M},\mathcal{T}_{\mathcal{M}})$ for the three manifolds (8.0.1)—those which attracted the most physicists' attention.

For complete intersections in products of complete projective spaces, a simple criterion for redundancy of $\mathfrak{d}[\mathcal{X}\|\mathcal{E}]$ has been derived [146]:

1. If there is no decomposing 1-leg (see *1-leg-decomposable*$^{(L)}$) in the diagram of $[\mathcal{X}\|\mathcal{E}]$, then $E_0^{k+1,k}(\mathcal{T}_{\mathcal{X}}) = 0$ and $E_0^{q,k}(\mathcal{T}_{\mathcal{X}}) = E_\infty^{q,k}(\mathcal{T}_{\mathcal{X}})$. So,

$$H^0(\mathcal{M},\mathcal{T}_{\mathcal{X}}) = \sum_{k=0}^{K} E_0^{k,k}(\mathcal{T}_{\mathcal{X}})\ , \quad \text{and} \quad H^2(\mathcal{M},\mathcal{T}_{\mathcal{X}}) = \sum_{k=0}^{K} E_0^{k+2,k}(\mathcal{T}_{\mathcal{X}})\ , \tag{A.3.5}$$

   while $H^1(\mathcal{M},\mathcal{T}_{\mathcal{X}}) = 0 = H^3(\mathcal{M},\mathcal{T}_{\mathcal{X}})$.

2. If, moreover, there is a decomposing 0-leg, $H^0(\mathcal{M},\mathcal{T}_{\mathcal{X}})$ has a contribution other than $E_0^{0,0}(\mathcal{T}_{\mathcal{X}}) = H^0(\mathcal{X},\mathcal{T}_{\mathcal{X}})$ which cannot be canceled out since $E_0^{k+1,k}(\mathcal{T}_{\mathcal{X}}) = 0$ and so $\mathfrak{d}[\mathcal{X}\|\mathcal{E}]$ is ineffective. That is, there are equivalence relations (and non-trivial elements also, as it turns out) amongst elements of $H^1(\mathcal{M},\mathcal{T}_{\mathcal{M}})$ which are not accounted for by Eq. (A.3.2).

* Clearly, only a finite list was actually explicitly searched. In view of certain identities subsumed here in Lemma 2.2, a list of about 8000 *configurations*$^{(L)}$ contains all topological types [85] and this is the list which has been checked.

There is unfortunately no such simple "selection rule" for the $\mathcal{E}$-valued cohomology. Similarly, when the embedding space contains factors other than complex projective spaces, the above criteria become substantially more complicated.

For the manifold (A.1.2), the polynomial deformations are spanned by

$$\begin{bmatrix} \phi_{(abc)}x^a x^b x^c \\ \varphi_{a(\alpha\beta\gamma)}x^a y^\alpha y^\beta y^\gamma \end{bmatrix} \cong \begin{bmatrix} \phi_{(abc)}x^a x^b x^c \\ \varphi_{a(\alpha\beta\gamma)}x^a y^\alpha y^\beta y^\gamma \end{bmatrix} + (x^a \lambda_a{}^b \partial_b + y^\alpha \lambda_\alpha{}^\beta \partial_\beta) \begin{bmatrix} f(x) \\ g(x,y) \end{bmatrix}. \tag{A.3.6}$$

It is easy to identify this with the quotient (A.1.22), through contracting the latter with x's and y's. For example,

$$\phi_{(111)} \to (x^1)^3 \ , \quad \phi_{(123)} \to x^1 x^2 x^3 \ , \quad \text{and so on} \ldots \tag{A.3.7}$$

The relationship between the tensor coefficients such as $\phi_{(abc)}$ and monomials is another type of *duality*$^{(L)}$, in that for every component $\phi_{(abc)}$, for some fixed a, b, c, there is a unique monomial $x^a x^b x^c$ which contracts with $\phi_{(abc)}$ into $\phi(x)$.

Notation: We adhere to the standard relativity notation in which coordinates are contravariant vectors (their components having upper indices), at the risk of clumsiness when it comes to powers of coordinates. Keeping consistently to this convention essentially identifies the duality between (totally symmetric) tensor coefficients *vs.* coordinate monomials with the Einstein summation convention and clarifies much of the calculations as usual in tensor calculus.

This relation shows that the cohomology computation discussed above may be considered as a generalization of the polynomial deformations—if nothing else than at least in that all types of tensors can be discussed not only the totally symmetric ones which correspond to coordinate monomials. Also, through contraction with coordinates, we can translate the generally valid results of the above cohomology computations into the polynomial deformation type language which is better known to the physics audience and is closely related to certain recent results of quantum field theory, as we will see at the end of this volume.

A.3.2 A residue formula

Given the tensor coefficient *vs.* coordinate monomial duality, all results obtained for polynomial deformations have an immediate translation into the tensor language. For example, consider the spectral sequence accompanying the resolution (A.1.5). We have

q	$\left(\begin{smallmatrix}4 & \vert & 0\,0\,0\\ 3 & \vert & 0\,0\end{smallmatrix}\right)$	$\left(\begin{smallmatrix}1 & \vert & 0\,0\,0\\ 3 & \vert & 0\,0\end{smallmatrix}\right)$ $\left(\begin{smallmatrix}3 & \vert & 0\,0\,0\\ 0 & \vert & 0\,0\end{smallmatrix}\right)$	$\left(\begin{smallmatrix}0 & \vert & 0\,0\,0\\ 0 & \vert & 0\,0\end{smallmatrix}\right)$	$\mathcal{O}_{\mathcal{M}}$
0:	0	0	$\left(\begin{smallmatrix}0\,0\,0\,1\\ 0\,0\,0\end{smallmatrix}\right)^1_1$	$H^0(\mathcal{M},\mathbb{C})$
1:	0	0	0	$H^1(\mathcal{M},\mathbb{C})$
2:	0	0	0	$H^2(\mathcal{M},\mathbb{C})$
3:	0	0	0	$H^3(\mathcal{M},\mathbb{C})$
4:	0	0	0	—
5:	$\left(\begin{smallmatrix}1\,1\,1\,1\\ 1\,1\,1\end{smallmatrix}\right)^1_1$	0	0	—

(A.3.8)

This provides a single generator of $H^0(\mathcal{M},\mathbb{C})$, a scalar $\vartheta \in \left(\begin{smallmatrix}0\,0\,0\,0\\ 0\,0\,0\end{smallmatrix}\right)$. Similarly, there is a single generator for

$$H^3(\mathcal{M},\mathbb{C}) \;=\; H^{0,3}(\mathcal{M}) \;\ni\; \epsilon^{abcd}\epsilon^{\alpha\beta\gamma} \;\in\; \left(\begin{smallmatrix}1\,1\,1\,1\\ 1\,1\,1\end{smallmatrix}\right), \tag{A.3.9}$$

being contravariant themselves, these cannot be contracted with coordinates. However, $\epsilon^{abcd}\epsilon^{\alpha\beta\gamma}$ are clearly dual to $\epsilon_{abcd}\epsilon_{\alpha\beta\gamma}$, which in turn can be contracted with x's and y's. The ϵ-tensor densities being totally antisymmetric, there is only one possibility:

$$\epsilon^{abcd} \;\overset{\star}{\sim}\; \epsilon_{abcd} \;\overset{\star}{\sim}\; x^a\mathrm{d}x^b\mathrm{d}x^c\mathrm{d}x^d \;. \tag{A.3.10}$$

The double duality provides us with a natural identification

$$\epsilon^{abcd} \;\approx\; x^a\mathrm{d}x^b\mathrm{d}x^c\mathrm{d}x^d \;\overset{\text{def}}{=}\; (x\mathrm{d}^3x) \tag{A.3.11}$$

since they are both dual to ϵ_{abcd}. (Similarly for $\epsilon^{\alpha\beta\gamma}$.)

Recall now that standard tensor calculus methods yield [84]

$$H^{3,0}(\mathcal{M}) \;\ni\; \Omega \;\overset{\text{def}}{=}\; \oint_{\Gamma(f)}\oint_{\Gamma(g)} \frac{\epsilon_{abcd}x^a\mathrm{d}x^b\mathrm{d}x^c\mathrm{d}x^d\;\epsilon_{\alpha\beta\gamma}y^\alpha\mathrm{d}y^\beta\mathrm{d}y^\gamma}{f(x)g(x,y)} \;. \tag{A.3.12}$$

Here the contour integrals are along circuits $\Gamma(f)$ and $\Gamma(g)$ in $\mathbb{P}^3 \times \mathbb{P}^2$ and go around the hypersurfaces $f(x) = 0$ and $g(x,y) = 0$, respectively. The above double contour integral would vanish were it not for the division by $f(x)$ and $g(x,y)$. These create poles at the $f(x) = 0$ and $g(x,y) = 0$ hypersurfaces, rendering the double integral non-zero precisely on the common zero-set, that is on $\mathcal{M}$. Also, the double integral is needed to convert the holomorphic 5-form $[(x\mathrm{d}^3x)(y\mathrm{d}^2y)]$ into the holomorphic 3-form of Calabi–Yau manifolds[†]. The integrals are most easily evaluated using the residue formula.

Comparing Eq. (A.3.12) and (A.3.9), we see that the duality between $H^{3,0}$ and $H^{0,3}$ reflects itself in the contravariancy of the tensor coefficients

$$\epsilon^{abcd}\epsilon^{\alpha\beta\gamma} \qquad vs. \qquad \epsilon_{abcd}\epsilon_{\alpha\beta\gamma} \ , \tag{A.3.13}$$

respectively.

In general, therefore, there is an integral formula similar to Eq. (A.3.12) for all cohomology representatives obtained using the Koszul complex and spectral sequences. However, the tensor coefficients alone are usually sufficient to recover this integral expression completely.

One may ask if there are general formulae of the type (L.18) for arbitrary bundles over an n-fold $\mathcal{X}$. In general, the answer is 'no'. However, the situation is much more favorable for holomorphic *tensor* bundles, $\mathcal{T}^r_s \stackrel{\text{def}}{=} (\otimes^r\mathcal{T}_\mathcal{X} \otimes \otimes^s\mathcal{T}^*_\mathcal{X})$, because a local coordinate system x^μ at a point $p \in \mathcal{X}$ determines a natural frame ∂_μ for the holomorphic tangent bundle, $\mathcal{T}_\mathcal{X}$, and $\mathrm{d}z^\mu$ for the holomorphic cotangent bundle, $\mathcal{T}^*_\mathcal{X}$. At a point $p \in \mathcal{X}$, a holomorphic vector field $\mathcal{V}$ can then be written as

$$\mathcal{V}_p(x) \;=\; x^\mu\, v_\mu{}^\nu\, \partial_\nu \;+\; \text{higher order terms} \ , \tag{A.3.14}$$

and let $\boldsymbol{v}_p = [v_\mu{}^\nu]_p$. Under a local change of coordinates, $\boldsymbol{v}_p$ is determined up to conjugation. Therefore, an *invariant polynomial*[(L)] $P^k(\boldsymbol{v}_p)$ is an invariant of $\mathcal{V}$ and p and we expect this polynomial to carry global information. Indeed, we have

Theorem A.1 (The Bott Residue Formula). *Let* $\boldsymbol{\Theta}$ *be a curvature 2-form for the holomorphic tangent bundle on a complex manifold* $\mathcal{X}$ *and* $\mathcal{V}$ *a global vector field over* $\mathcal{X}$. *Then*

$$\int_\mathcal{M} P(\tfrac{i}{2\pi}\boldsymbol{\Theta}) \;=\; \sum_{\mathcal{V}(p)=0} \frac{P(\boldsymbol{v}_p)}{\det(\boldsymbol{v}_p)} \ , \tag{A.3.15}$$

for any invariant polynomial $P(\boldsymbol{v})$ *of degree* $\dim\mathcal{X}$.

Very much like the *Hopf Index Theorem*[(L)], this allows us to replace certain integrals over $\mathcal{X}$ with *finite* sums over the set of zeros of a global (holomorphic) vector field. In conjunction with the *Lefschetz fixed-point formulae*[(L)], it is often possible to further relate such integrals to traces of the action of certain symmetries of $\mathcal{X}$ over some cohomology on $\mathcal{X}$.

[†] So, the criteria for Ricci-flatness in § 9.5 are in fact tailored to construct the non-zero holomorphic 3-form Ω via an integral formula like Eq. (A.3.12).

A.4 Sequencing Sequences

In retrospect, the computation of $H^q(\mathcal{M}, \mathcal{T}_\mathcal{M})$ and also of $H^q(\mathcal{M}, \mathcal{T}^*_\mathcal{M})$ described above involves computing several intermediate cohomology groups which we do not really need in the end. One wonders if it is possible to avoid that, somewhat like collapsing the square of short exact sequences into a spectral sequence described in § 9.3.2 and § 9.3.3.

Consider a (not necessarily Calabi–Yau) hypersurface $\mathcal{M} \subset \mathcal{X}$, defined as the zero-set of a section $f(x)$ of a line bundle $\mathcal{E}$ over $\mathcal{X}$; $\mathcal{M} = f^{-1}(0)$. The computation of $H^q(\mathcal{M}, \mathcal{T}^*_\mathcal{M})$, as described above, relies on the diagram

$$
\begin{array}{ccccccc}
 & & 0 & & 0 & & \\
 & & \downarrow & & \downarrow & & \\
 & & (\mathcal{E}^*)^{\otimes 2} & & \mathcal{T}^*_\mathcal{X} \otimes \mathcal{E}^* & & \\
 & & \downarrow & & \downarrow & & \\
 & & \mathcal{E}^* & & \mathcal{T}^*_\mathcal{X} & & \\
 & & \downarrow & & \downarrow & & \\
0 \to & & \mathcal{E}^*|_\mathcal{M} & \to & \mathcal{T}^*_\mathcal{X}|_\mathcal{M} & \to & \mathcal{T}^*_\mathcal{M} \to 0 \\
 & & \downarrow & & \downarrow & & \\
 & & 0 & & 0 & &
\end{array}
\tag{A.4.1}
$$

in which the column-sequences are sheaf-exact over $\mathcal{X}$ and the row-sequence is exact on $\mathcal{M}$. Superficially, this looks almost like the diagram (9.3.14), in that there is a mesh of long exact cohomology sequences in which we input the cohomology of the bundles framed by the dotted square. From these, we determine the cohomology of the bundles directly beneath the dotted square and this then leads to the required cohomology of the bundle at the lower right. Given the formal similarity *in usage*, one expects that that the diagram (A.4.1) can be collapsed to a single (sheaf-)exact sequence, something like

$$
0 \to \mathcal{E}^{*\otimes 2} \to \begin{array}{c} \mathcal{T}^*_\mathcal{X} \otimes \mathcal{E}^* \\ \oplus \\ \mathcal{E}^* \end{array} \to \mathcal{T}^*_\mathcal{X} \xrightarrow{\varrho} \mathcal{T}^*_\mathcal{M} \to 0 . \tag{A.4.2}
$$

This would then induce a spectral sequence just like (9.3.14) is collapsed to the exact sequence (9.3.7), which then induces the wanted spectral sequence.

There is, however, a major difference between the diagram (9.3.14) and the (A.4.1) one. Namely, while the former is naturally completed into a 3×3

square of short exact sequences, the diagram (A.4.1) cannot be complete in such a way. To begin with, there is no natural map $\mathcal{E}^* \to \mathcal{T}_{\mathcal{X}}^*$. Of course, restricted to $\mathcal{M}$, the map $(\nabla f)(x) : \mathcal{E}^*|_{\mathcal{M}} \to \mathcal{T}_{\mathcal{X}}^*|_{\mathcal{M}}$ is natural, since it is independent of the choice of the connection

$$(\nabla f)(x) \;=\; (\mathrm{d}f)(x) + (\Gamma \cdot f)(x) \;=\; \mathrm{d}f(x) \tag{A.4.3}$$

on $\mathcal{M}$. *Away* from $\mathcal{M}$ in $\mathcal{X}$, however, ∇f indeed depends on the choice of the connection and so does not provide a well defined bundle map. Thus, there is no well defined map $\mathcal{E}^{*\otimes 2} \to \mathcal{T}_{\mathcal{X}}^* \otimes \mathcal{E}^*$ either and the sequence (A.4.2) fails to be exact in any sense.

However, when $\mathcal{X}$ is a Grassmannian $G_{(k,n)}$ of k-planes in $\mathbb{C}^n$ (including the simpler case of projective spaces, $\mathbb{P}^n = G_{(1,n+1)}$), there is a straightforward remedy for this. The two off-diagonal bundles in the dotted rectangle of diagram (A.4.1) are connected by a short exact sequence, known as the *tautological sequence*[(L)]; tensoring the dual of Seq. (A.4.1) by $\mathcal{E}^* \otimes \mathcal{S}$, we have that‡

$$0 \to \mathcal{T}_{\mathcal{X}}^* \otimes \mathcal{E}^* \to \mathcal{S}^{\otimes n} \otimes \mathcal{E}^* \to \mathcal{E}^* \to 0 \tag{A.4.4}$$

is exact. This fits in the diagram (A.4.1), in the NE–SW (2 o'clock–8 o'clock) direction and induces the NW–SE (10 o'clock–4 o'clock) sheaf-exact sequence threaded in as shown below

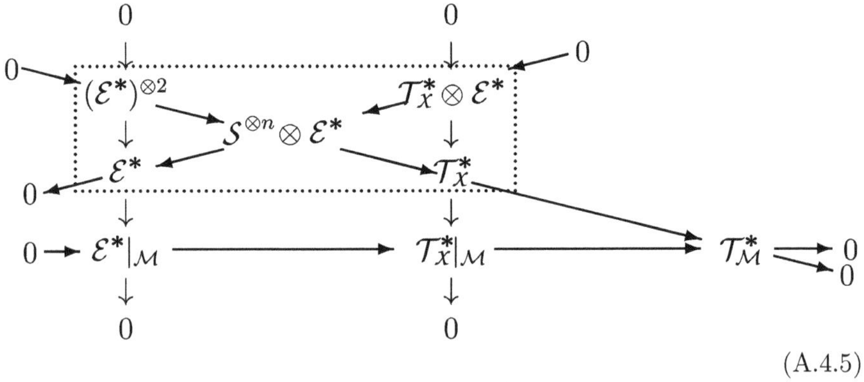

(A.4.5)

The exactness of the 'diagonal' (NW–SE) sequence relies on the general fact that, given the solid arrows as indicated in the diagram

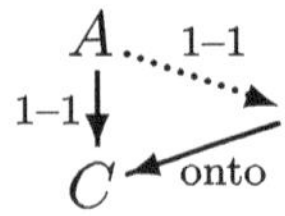

(A.4.6)

‡ For $G_{(1,n+1)} = \mathbb{P}^n$, $\mathcal{S} = \mathcal{O}_{\mathbb{P}^n}(-1)$.

we can always recover the dotted map $A \to B$ to have no kernel and such that the diagram be commutative[§]. To see this, recall that if $B \xrightarrow{\alpha} C$ is onto then $C = \{B/\ker(\alpha)\}$. Now, the map $A \to C$ naturally implies a map to the numerator B of the quotient $C = \{B/\ker(\alpha)\}$. It can have no kernel, for if some non-zero element of A would be annihilated in $A \to B$, it would be mapped to $0 \in C$ by $B \to C$. From commutivity, this would mean that the same non-zero element is also annihilated in $A \to C$, which contradicts the assumption that $A \to C$ is 1–1 (has no kernel).

This explains the map $\mathcal{E}^{*\otimes 2} \xrightarrow{1-1} \mathcal{S}^{\otimes n} \otimes \mathcal{E}^*$ in the diagram (A.4.5), for $f(x) \neq 0$, that is, away from $\mathcal{M}$ in $\mathcal{X}$. Along $\mathcal{M}$, the map $\mathcal{E}^{*\otimes 2} \to \mathcal{E}^*$ is zero and so is then $\mathcal{E}^{*\otimes 2} \xrightarrow{1-1} \mathcal{S}^{\otimes n} \otimes \mathcal{E}^*$. The rest of the 'diagonal' sequence follows by similar arguments and we obtain the sheaf-exact sequence

$$0 \longrightarrow \mathcal{E}^{*\otimes 2} \longrightarrow \mathcal{S}^{\otimes n} \otimes \mathcal{E}^* \longrightarrow \mathcal{T}_{G_{(k,n)}}{}^* \xrightarrow{\varrho} \mathcal{T}^*_{\mathcal{M}} \longrightarrow 0 \;, \qquad \text{(A.4.7)}$$

which induces a spectral sequence very much like Seq. (9.3.7) induces the spectral sequence with the 0th level (9.3.16).

This spectral sequence is then analyzed just as that in § 9.3.3 and, of course, yields the same result. The benefit lies in the fact that we need to compute the cohomology of only three bundles over $\mathcal{X}$, namely $\mathcal{E}^{*\otimes 2}$, $\mathcal{S}^{\otimes n} \otimes \mathcal{E}^*$ and $\mathcal{T}_{G_{(k,n)}}$, instead of four bundles over $\mathcal{X}$ and two auxiliary ones over $\mathcal{M}$, as required in the computation of § A.1. The generalization to complete intersections and/or flag spaces with more than two 'stripes,' $\mathbb{F}_{[n_1,\dots,n_f]}$ with $f > 2$, the result is less simple and we forego $\mathbb{F}_{[n_1,\dots,n_f]}$ with $f > 2$, the result is less simple and we forego an analysis here.

[§] A diagram is commutative if by starting at one end, from a particular element, we obtain the same element no matter which way we trace through the diagram.

Chapter B

Other Tangent Bundle Related Cohomology

There are many other cohomology groups, besides $H^q(\mathcal{M}, \mathcal{T}_\mathcal{M})$ and $H^q(\mathcal{M}, \mathcal{T}^*_\mathcal{M})$, which we can easily access using the Koszul complex and related spectral sequence. Recalling the immediate physical application (see Table 0.2), we describe here the computation of $H^q(\mathcal{M}, \text{End}\,\mathcal{T}_\mathcal{M})$, as a simple and physically relevant case of the more general $H^q(\mathcal{M}, \mathcal{T}_\mathcal{M}{}^{\otimes r} \otimes \mathcal{T}^*_\mathcal{M}{}^{\otimes s})$, for all of which the present techniques are valid.

B.1 Some General Facts

As noted in Chapter 0, by End $\mathcal{T}_\mathcal{M}$ (more generally, End $\mathcal{V}$) we mean the bundle of traceless *endomorphisms* [(L)] of the holomorphic tangent bundle $\mathcal{T}_\mathcal{M}$ (arbitrary vector bundle $\mathcal{V}$. Given a local frame for $\mathcal{T}_\mathcal{M}$, $\varphi^\mu(x)$, End$\mathcal{T}_\mathcal{M}$ acts on it by simple matrix multiplication,

$$\vartheta^\mu{}_\nu(x) \, : \, \varphi^\nu(x) \longmapsto \varphi^\mu(x) = (\vartheta^\mu{}_\nu \varphi^\nu)(x) \, , \tag{B.1.1}$$

where $\vartheta^\alpha{}_\beta(x)$ is local and (Hermitian) matrix-valued. Clearly, $\vartheta^\mu{}_\nu \in$ End $\mathcal{T}_\mathcal{M}$ belongs to the traceless part of $\mathcal{T}_\mathcal{M} \otimes \mathcal{T}^*_\mathcal{M}$, that is, End $\mathcal{T}_\mathcal{M}$ is a (holomorphic) sub-bundle of $\mathcal{T}_\mathcal{M} \otimes \mathcal{T}^*_\mathcal{M}$.

A little thought reveals that the short sequence of holomorphic bundles and appropriate bundle maps

$$0 \to \text{End}\,\mathcal{T}_\mathcal{M} \xrightarrow{i} (\mathcal{T}_\mathcal{M} \otimes \mathcal{T}^*_\mathcal{M}) \xrightarrow{\text{tr}} \text{Tr}[\mathcal{T}_\mathcal{M} \otimes \mathcal{T}^*_\mathcal{M}] \to 0 \tag{B.1.2}$$

is exact. Of course, the projection on the trace is onto and its *kernel*[(L)] consists precisely of traceless endomorphisms, while the embedding i has no kernel and $\mathrm{cok}(i) \stackrel{\text{def}}{=} \{(\mathcal{T}_{\mathcal{M}} \otimes \mathcal{T}^*_{\mathcal{M}})/\mathrm{End}\,\mathcal{T}_{\mathcal{M}}\}$ may clearly be identified with $\mathrm{Tr}[\mathcal{T}_{\mathcal{M}} \otimes \mathcal{T}^*_{\mathcal{M}}] \approx \mathbb{C}$. This suffices to express the $\mathrm{End}\,\mathcal{T}_{\mathcal{M}}$-valued cohomology in terms of the $(\mathcal{T}_{\mathcal{M}} \otimes \mathcal{T}^*_{\mathcal{M}})$-valued and the $\mathbb{C}$-valued cohomology on $\mathcal{M}$.

However, we also have a comfortable simplification in this case, since the bundle $\mathbb{C} \approx \mathrm{Tr}[\mathcal{T}_{\mathcal{M}} \otimes \mathcal{T}^*_{\mathcal{M}}]$ may also be regarded as a sub-bundle of $\mathcal{T}_{\mathcal{M}} \otimes \mathcal{T}^*_{\mathcal{M}}$. In other words, there exists a map $\tau\colon \mathrm{Tr}[\mathcal{T}_{\mathcal{M}} \otimes \mathcal{T}^*_{\mathcal{M}}] \to (\mathcal{T}_{\mathcal{M}} \otimes \mathcal{T}^*_{\mathcal{M}})$ such that $\tau \circ j = \mathbb{1}_{(\mathcal{T}_{\mathcal{M}} \otimes \mathcal{T}^*_{\mathcal{M}})}$ and $j \circ \tau = \mathbb{1}_{\mathrm{Tr}[\mathcal{T}_{\mathcal{M}} \otimes \mathcal{T}^*_{\mathcal{M}}]}$. τ is simply (fiber-wise) multiplication with the identity matrix. Because of this, we have that in fact

$$(\mathcal{T}_{\mathcal{M}} \otimes \mathcal{T}^*_{\mathcal{M}}) = \mathrm{End}\,\mathcal{T}_{\mathcal{M}} \oplus \mathrm{Tr}[\mathcal{T}_{\mathcal{M}} \otimes \mathcal{T}^*_{\mathcal{M}}] \tag{B.1.3}$$

holomorphically, not just in the C^∞ category. This implies that

$$H^q(\mathcal{M}, \mathcal{T}_{\mathcal{M}} \otimes \mathcal{T}^*_{\mathcal{M}}) = H^q(\mathcal{M}, \mathrm{End}\,\mathcal{T}_{\mathcal{M}}) \oplus H^q(\mathcal{M}, \mathbb{C}) . \tag{B.1.4}$$

A split exact sequence: Generally, let $0 \to \mathcal{A} \xrightarrow{\iota} \mathcal{B} \xrightarrow{j} \mathcal{C} \to 0$ be a short exact sequence of holomorphic bundles over some compact complex space $\mathcal{X}$ and let there also exists a bundle map $\mathcal{C} \xrightarrow{\gamma} \mathcal{B}$ such that $j \circ \gamma = \mathbb{1}_{\mathcal{C}}$ and $\gamma \circ j = \mathbb{1}_{\mathcal{B}}$. Then the short exact sequence is said to *split*, $\mathcal{B} = \mathcal{A} \oplus \mathcal{C}$ holomorphically and

$$H^q(\mathcal{X}, \mathcal{B}) = H^q(\mathcal{X}, \mathcal{A}) \oplus H^q(\mathcal{X}, \mathcal{C}) . \tag{B.1.5}$$

Furthermore, note that $(\mathcal{T}_{\mathcal{M}} \otimes \mathcal{T}^*_{\mathcal{M}})$ is self-adjoint; on a Calabi–Yau n-fold where $\mathcal{K}_{\mathcal{M}} \approx \mathbb{C}$, *Serre duality*[(L)] implies that $H^q(\mathcal{M}, \mathcal{T}_{\mathcal{M}} \otimes \mathcal{T}^*_{\mathcal{M}}) \approx H^{n-q}(\mathcal{M}, \mathcal{T}^*_{\mathcal{M}} \otimes \mathcal{T}_{\mathcal{M}})$.

Now we recall a theorem, which is proven in Ref. [43] for bundles over $\mathbb{P}^n$ but is in fact valid for any (para)compact complex manifold [226].

Definition B.1. *A holomorphic bundle $\mathcal{V}$ over a compact Kähler n-fold $\mathcal{X}$ (with Kähler class J) is stable if*

$$\frac{\int_X J^{n-1} \wedge c_1(\mathcal{V})}{\mathrm{rank}\,\mathcal{V}} > \frac{\int_X J^{n-1} \wedge c_1(\mathfrak{V}_i)}{\mathrm{rank}\,\mathfrak{V}_i} , \tag{B.1.6}$$

for all i, where $\mathfrak{V}_i$ are subsheaves of the sheaf corresponding to $\mathcal{V}$.

In fortunate cases, it suffices to test the condition over sub-bundles. Also, stability may be reformulated more directly, just in terms of curvature 2-forms of $\mathcal{V}$ and $\mathfrak{V}_i$. In fact, there may exist a more general and more robust definition of stability, but this will suffice for our present purposes.

Theorem B.1. *If* $\mathcal{V}$ *is a stable holomorphic bundle over a paracompact complex manifold* $\mathcal{X}$, *then* $H^0(\mathcal{X}, \mathrm{End}\,\mathcal{V}) = 0$.

In general, it is quite laborious to prove that a particular bundle is stable. However, Ref. [226] proves that the holomorphic tangent bundle of a regular Calabi–Yau n-fold ($h^{1,0} = 0$, see *Genus*$^{(L)}$) is stable, prompting the following

Corollary B.1. *For a regular Calabi–Yau n-fold* $\mathcal{M}$,

$$H^q(\mathcal{M}, \mathcal{T}_{\mathcal{M}} \otimes \mathcal{T}^*_{\mathcal{M}}) \approx \begin{cases} \mathbb{C}, & q = 0 \pmod{n}; \\ H^q(\mathcal{M}, \mathrm{End}\,\mathcal{T}_{\mathcal{M}}), & \textit{otherwise}. \end{cases} \tag{B.1.7}$$

For the currently interesting spaces, Calabi–Yau 3-folds, we can compute $H^1(\mathcal{M}, \mathcal{T}_{\mathcal{M}} \otimes \mathcal{T}^*_{\mathcal{M}})$ and then simply use that $H^1(\mathcal{M}, \mathcal{T}_{\mathcal{M}} \otimes \mathcal{T}^*_{\mathcal{M}}) = H^1(\mathcal{M}, \mathrm{End}\,\mathcal{T}_{\mathcal{M}})$. In practice, we compute all $H^q(\mathcal{M}, \mathcal{T}_{\mathcal{M}} \otimes \mathcal{T}^*_{\mathcal{M}})$ and check the calculations against Corollary B.1 and Serre duality.

B.2 Computing $H^q(\mathcal{M}, \mathrm{End}\mathcal{T}_{\mathcal{M}})$

Before we engage in the Koszul complex/spectral sequence computation, let's consider the deformation theoretic alternative.

B.2.1 Deformation theoretic results

Comparing the computation $H^q(\mathcal{M}, \mathrm{End}\,\mathcal{T}_{\mathcal{M}})$ with that of $H^q(\mathcal{M}, \mathcal{T}_{\mathcal{M}})$ in the previous chapter, one asks if the 'polynomial deformation' type of representation could be applied (at least in part) to represent $H^1(\mathcal{M}, \mathrm{End}\,\mathcal{T}_{\mathcal{M}})$. Roughly, in view of Eq. (B.1), if (some) $\mathcal{T}_{\mathcal{M}}$-valued 1-forms can be represented by deformations of the defining polynomial, can (some) $\mathcal{T}_{\mathcal{M}} \otimes \mathcal{T}^*_{\mathcal{M}}$-valued 1-forms can be represented by deformations of the *differentials* of the defining polynomial?

In other words, given the defining polynomials* $f(x)$ and $g(x,y)$ as specified in Eq. (A.1.3), we consider their differentials

$$\mathrm{d}x^a(f_{abc}x^b x^c)\,, \quad \mathrm{d}x^a(g_{a\,\alpha\beta\gamma}y^\alpha y^\beta x^\gamma)\,, \quad \mathrm{d}y^\alpha(g_{a\,\alpha\beta\gamma}x^a y^\beta x^\gamma)\,. \tag{B.2.1}$$

* We again consider the example presented in Chapter A rather than trying to formulate this in some general and probably bewildering notation. This will hopefully suffice to present the idea and the Reader will have no insurmountable problem in extending this to any other desired model.

The variations of each of these can be cast in two irreducible types; for example

$$\delta\,\mathrm{d}x^a(f_{abc}x^bx^c) \;=\; \mathrm{d}x^a(\phi_{(abc)}x^bx^c) \;+\mathrm{d}x^a(\varphi_{a(bc)}x^bx^c)\;. \tag{B.2.2}$$

The former, involving the totally symmetric tensor coefficient $\phi_{(abc)}$ is a total differential $\mathrm{d}x^a(\phi_{(abc)}x^bx^c) = \frac{1}{3}\mathrm{d}(\phi_{(abc)}x^ax^bx^c)$ and will not qualify for representing a harmonic form. On the other hand, $\mathrm{d}x^a(\varphi_{a(bc)}x^bx^c)$ involves the mixed-symmetric tensor $\varphi_{a(bc)} \sim (-2,-1,0,0)$, which is symmetric in $b \leftrightarrow c$, but satisfies

$$\varphi_{a(bc)} \;+\; \varphi_{b(ca)} \;+\; \varphi_{c(ab)} \;=\; 0\;. \tag{B.2.3}$$

Its components are obtained from

$$\phi_{a(bc)} \;\stackrel{\text{def}}{=}\; \phi_{a|(bc)} \;-\; \phi_{(abc)}\;, \tag{B.2.4}$$

and

$$\phi_{a(bc)} \;=\; \tfrac{1}{3}\big(\phi_{a|(bc)} - \phi_{b|(ac)}\big) \;+\; \tfrac{1}{3}\big(\phi_{a|(bc)} - \phi_{c|(ba)}\big)\;, \tag{B.2.5}$$

where $\phi_{a|(bc)} \stackrel{\text{def}}{=} \frac{1}{2}(\phi_{a|b|c} + \phi_{a|c|b})$.

Notation: In general, we use parentheses to denote symmetrization, square brackets to denote skew-symmetrization. The indices which do not participate in a symmetrization are separated off by vertical pipes.

Proceeding in this fashion with the available differentials of the defining polynomials $f(x)$ and $g(x,y)$, we obtain 20+32 deformations

$$\mathrm{d}x^a(\varphi_{a(bc)}x^bx^c)\;, \qquad \mathrm{d}y^\alpha(\varphi_{a\,\alpha(\beta\gamma)}x^ay^\beta x^\gamma)\;. \tag{B.2.6}$$

This easily saturates the lower bound $\dim H^1(\mathcal{M},\mathrm{End}\,\mathcal{T}_\mathcal{M}) \geq \dim H^1(\mathcal{M},\mathcal{T}_\mathcal{M})$ derived in Ref. [168]†.

B.2.2 The general method

By tensoring Seq. (9.1.1) with $\mathcal{T}^*_\mathcal{M}$ and the dual of Seq. (9.1.1) with $\mathcal{E}$ and with $\mathcal{T}_\mathcal{X}$ respectively, we obtain three short exact sequences that fit together into the diagram with exact rows and columns, shown in Figure B.1.

† In fact, using the *Mirror map* $^{(L)}$, a stronger bound would follow, namely that $\dim H^1(\mathcal{M},\mathrm{End}\,\mathcal{T}_\mathcal{M}) \geq \dim H^1(\mathcal{M},\mathcal{T}_\mathcal{M}) + \dim H^1(\mathcal{M},\mathcal{T}^*_\mathcal{M})$.

$$\begin{array}{ccccccccc}
 & & & & 0 & & 0 & & \\
 & & & & \downarrow & & \downarrow & & \\
 & & & & (\mathcal{T}_{\mathcal{X}}\otimes\mathcal{E}^*)|_{\mathcal{M}} & & (\mathcal{E}\otimes\mathcal{E}^*)|_{\mathcal{M}} & & \\
 & & & & \downarrow & & \downarrow & & \\
 & & & & (\mathcal{T}_{\mathcal{X}}\otimes\mathcal{T}_{\mathcal{X}}^*)|_{\mathcal{M}} & & (\mathcal{E}\otimes\mathcal{T}_{\mathcal{X}}^*)|_{\mathcal{M}} & & \\
 & & & & \downarrow & & \downarrow & & \\
0 & \to & \mathcal{T}_{\mathcal{M}}\otimes\mathcal{T}_{\mathcal{M}}^* & \to & \mathcal{T}_{\mathcal{X}}|_{\mathcal{M}}\otimes\mathcal{T}_{\mathcal{M}}^* & \to & \mathcal{E}|_{\mathcal{M}}\otimes\mathcal{T}_{\mathcal{M}}^* & \to & 0\\
 & & & & \downarrow & & \downarrow & & \\
 & & & & 0 & & 0 & &
\end{array}$$

Figure B.1: The schematics of computing $H^*(\mathcal{M}, \mathcal{T}_{\mathcal{M}} \otimes \mathcal{T}_{\mathcal{M}}^*)$.

It relates the vector bundle $\mathcal{T}_{\mathcal{M}} \otimes \mathcal{T}_{\mathcal{M}}^*$ over $\mathcal{M}$—intrinsic to $\mathcal{M}$—to the restriction to $\mathcal{M}$ of various vector bundles over the (rather better known) embedding space $\mathcal{X}$.

All three short exact sequences appearing in the diagram (B.1) induce long exact cohomology sequences. These are tangled as dictated by the original diagram (B.1): one may imagine the horizontal one being crossed from above by the vertical ones once for each q at $H^q(\mathcal{M}, \mathcal{T}_{\mathcal{X}} \otimes \mathcal{T}_{\mathcal{M}}^*)$ and at $H^q(\mathcal{M}, \mathcal{E} \otimes \mathcal{T}_{\mathcal{M}}^*)$. Through these latter two sets of cohomology groups, we relate $H^\star(\mathcal{M}, \mathcal{T}_{\mathcal{M}} \otimes \mathcal{T}_{\mathcal{M}}^*)$ to

$$H^q(\mathcal{M}, \mathcal{E} \otimes \mathcal{E}^*)\ , \quad H^q(\mathcal{M}, \mathcal{E} \otimes \mathcal{T}_{\mathcal{X}}^*)\ , \quad H^q(\mathcal{M}, \mathcal{T}_{\mathcal{X}} \otimes \mathcal{T}_{\mathcal{X}}^*)\ . \tag{B.2.7}$$

These three sets of cohomology groups are determined, in terms of cohomological data entirely on $\mathcal{X}$, from the spectral sequences induced by Seq. (9.3.7) tensored by $\mathcal{E} \otimes \mathcal{E}^*$, $\mathcal{E} \otimes \mathcal{T}_{\mathcal{X}}^*$ and $\mathcal{T}_{\mathcal{X}} \otimes \mathcal{T}_{\mathcal{X}}^*$, respectively.

To complete the plan, we need to fill in the various cohomology groups—entirely over $\mathcal{X}$—into the lower left quadrant of these spectral sequences [see the chart (9.3.16)]. Iterating the *Künneth formula*$^{(L)}$

$$H^q(\mathcal{X} \times \mathcal{Y}, \mathcal{V}) = \bigoplus_{r+s=q} H^r(\mathcal{X}, \mathcal{V}|_{\mathcal{X}}) \otimes H^s(\mathcal{Y}, \mathcal{V}|_{\mathcal{Y}})\ , \tag{B.2.8}$$

these are given in terms of cohomology groups on the factors of the embedding space. For the flag space type factors, all cohomology is determined by the Bott–Borel–Weil theorem 9.2; more generally, one need not know *all* the cohomology of a factor in $\mathcal{X}$—only those needed in the Koszul spectral sequences for (B.2.7), that is

$$\begin{gathered} H^q(\mathcal{M}, \wedge^k\mathcal{E}^* \otimes \mathcal{E} \otimes \mathcal{E}^*)\ , \quad H^q(\mathcal{M}, \wedge^k\mathcal{E}^* \otimes \mathcal{E} \otimes \mathcal{T}_{\mathcal{X}}^*)\ , \\ H^q(\mathcal{M}, \wedge^k\mathcal{E}^* \otimes \mathcal{T}_{\mathcal{X}} \otimes \mathcal{T}_{\mathcal{X}}^*)\ , \end{gathered} \tag{B.2.9}$$

for $k = 0, \ldots, K$. Of course, to work the Koszul spectral sequence machinery, we also need to know the 'differential' maps d_r acting in (9.3.16). While this is the hardest part in a general application of spectral sequences, for embeddings in products of flag spaces $\{G/P\}$—this is elementary linear algebra of G-tensors (see *flag spaces*$^{(L)}$ and the Bott–Borel–Weil theorem 9.2).

By iterating the construction which resulted in diagram (B.1), we can straightforwardly obtain any ${\mathcal{T}_{\mathcal{M}}}^{\otimes r} \otimes {\mathcal{T}^*_{\mathcal{M}}}^{\otimes s}$ tensor bundle and whence the correspondingly valued cohomology on $\mathcal{M}$.

B.3 Simple Examples

To illustrate the foregoing description, we follow Ref. [121] and consider two simple configurations, both chosen from the list (8.0.1) in view of their physical application.

B.3.1 A bi-cubic Calabi–Yau 3-fold

To begin with, consider $\mathcal{M} \in \left[\begin{smallmatrix}2\\2\end{smallmatrix}\middle\|\begin{smallmatrix}3\\3\end{smallmatrix}\right]^{2;\,83}_{-162}$, that is, the space of solutions of a bi-cubic constraint

$$f(x,y) \stackrel{\text{def}}{=} f_{abc\,\alpha\beta\gamma}\, x^a x^b x^c\, y^\alpha y^\beta y^\gamma \;=\; 0\,. \tag{B.3.1}$$

The defining polynomial $f(x,y)$ is a section of $\mathcal{E} = \left(\begin{smallmatrix}-3\\-3\end{smallmatrix}\middle|\begin{smallmatrix}0&0\\0&0\end{smallmatrix}\right)$, in the obvious notation for the tensor product $(-3|00)_x \otimes (-3|00)_y$. The tangent bundle is $\mathcal{T}_X = \mathcal{T}_x \oplus \mathcal{T}_y$, where $\mathcal{T}_x = \left(\begin{smallmatrix}-1\\0\end{smallmatrix}\middle|\begin{smallmatrix}0&1\\0&0\end{smallmatrix}\right)$ and $\mathcal{T}_y = \left(\begin{smallmatrix}0\\-1\end{smallmatrix}\middle|\begin{smallmatrix}0&0\\0&1\end{smallmatrix}\right)$. Therefore:

$$\mathcal{E} \otimes \mathcal{E}^* = \left(\begin{smallmatrix}0\\0\end{smallmatrix}\middle|\begin{smallmatrix}0&0\\0&0\end{smallmatrix}\right)\,,$$

$$\mathcal{E} \otimes \mathcal{T}^*_X = \left(\begin{smallmatrix}-2\\-3\end{smallmatrix}\middle|\begin{smallmatrix}-1&0\\0&0\end{smallmatrix}\right) \oplus \left(\begin{smallmatrix}-3\\-2\end{smallmatrix}\middle|\begin{smallmatrix}0&0\\-1&0\end{smallmatrix}\right)\,,$$

$$\mathcal{T}_X \otimes \mathcal{E}^* = \left(\begin{smallmatrix}2\\3\end{smallmatrix}\middle|\begin{smallmatrix}0&1\\0&0\end{smallmatrix}\right) \oplus \left(\begin{smallmatrix}3\\2\end{smallmatrix}\middle|\begin{smallmatrix}0&0\\0&1\end{smallmatrix}\right)\,,$$

$$\mathcal{T}_X \otimes \mathcal{T}^*_X = 2\left(\begin{smallmatrix}0\\0\end{smallmatrix}\middle|\begin{smallmatrix}0&0\\0&0\end{smallmatrix}\right) \oplus \left(\begin{smallmatrix}-1\\1\end{smallmatrix}\middle|\begin{smallmatrix}0&1\\-1&0\end{smallmatrix}\right) \oplus \left(\begin{smallmatrix}1\\-1\end{smallmatrix}\middle|\begin{smallmatrix}-1&0\\0&1\end{smallmatrix}\right) \oplus \left(\begin{smallmatrix}0\\0\end{smallmatrix}\middle|\begin{smallmatrix}-1&1\\0&0\end{smallmatrix}\right) \oplus \left(\begin{smallmatrix}0\\0\end{smallmatrix}\middle|\begin{smallmatrix}0&0\\-1&1\end{smallmatrix}\right)\,.$$

Restricting the ambient cohomology to the submanifold: For each of these ten irreducible bundles, we employ the spectral sequence separately. As there is only one defining function, the Koszul resolution (9.3.7) collapses to a simple short exact sequence

$$0 \to \left(\begin{smallmatrix}3\\3\end{smallmatrix}\middle|\begin{smallmatrix}0&0\\0&0\end{smallmatrix}\right) \stackrel{f}{\to} \left(\begin{smallmatrix}0\\0\end{smallmatrix}\middle|\begin{smallmatrix}0&0\\0&0\end{smallmatrix}\right) \to \left(\begin{smallmatrix}0\\0\end{smallmatrix}\middle|\begin{smallmatrix}0&0\\0&0\end{smallmatrix}\right)_{\mathcal{M}} \to 0 \tag{B.3.2}$$

Table B.1: The cohomology input required for $H^q(\mathcal{M}, \mathrm{End}\, \mathcal{T}_{\mathcal{M}})$, for the bi-cubic Calabi–Yau 3-fold in $\mathbb{P}^2 \times \mathbb{P}^2$.

Bundle	Cohomology on $\mathcal{M}$	Bundle	Cohomology on $\mathcal{M}$
$\left(\begin{array}{c\|cc}0&0&0\\0&0&0\end{array}\right)$	$H^0=\left(\begin{smallmatrix}0&0&0\\0&0&0\end{smallmatrix}\right)^1_1,\ H^3=\left(\begin{smallmatrix}1&1&1\\1&1&1\end{smallmatrix}\right)^1_1$		
$\left(\begin{array}{c\|cc}-2&-1&0\\-3&0&0\end{array}\right)$	$H^0=\left(\begin{smallmatrix}0&0&0\\0&0&0\end{smallmatrix}\right)^1_1+\left(\begin{smallmatrix}-2&-1&0\\-3&0&0\end{smallmatrix}\right)^8_{10}$	$\left(\begin{array}{c\|cc}-3&0&0\\-2&-1&0\end{array}\right)$	$H^0=\left(\begin{smallmatrix}0&0&0\\0&0&0\end{smallmatrix}\right)^1_1+\left(\begin{smallmatrix}-3&0&0\\-2&-1&0\end{smallmatrix}\right)^{10}_8$
$\left(\begin{array}{c\|cc}2&0&1\\3&0&0\end{array}\right)$	$H^3=\left(\begin{smallmatrix}1&2&3\\1&1&4\end{smallmatrix}\right)^8_{10}+\left(\begin{smallmatrix}1&1&1\\1&1&1\end{smallmatrix}\right)^1_1$	$\left(\begin{array}{c\|cc}3&0&0\\2&0&1\end{array}\right)$	$H^3=\left(\begin{smallmatrix}1&1&4\\1&2&3\end{smallmatrix}\right)^{10}_8+\left(\begin{smallmatrix}1&1&1\\1&1&1\end{smallmatrix}\right)^1_1$
$\left(\begin{array}{c\|cc}-1&0&1\\1&-1&0\end{array}\right)$	$H^1=\left(\begin{smallmatrix}-1&0&1\\0&0&0\end{smallmatrix}\right)^8_1,\ H^2=\left(\begin{smallmatrix}1&1&1\\0&1&2\end{smallmatrix}\right)^1_8$	$\left(\begin{array}{c\|cc}1&-1&0\\-1&0&1\end{array}\right)$	$H^1=\left(\begin{smallmatrix}0&0&0\\-1&0&1\end{smallmatrix}\right)^1_8,\ H^2=\left(\begin{smallmatrix}0&1&2\\1&1&1\end{smallmatrix}\right)^8_1$
$\left(\begin{array}{c\|cc}0&0&0\\0&-1&1\end{array}\right)$	All cohomology vanishes	$\left(\begin{array}{c\|cc}0&-1&1\\0&0&0\end{array}\right)$	All cohomology vanishes

and the spectral sequence is equivalent to the corresponding long exact cohomology sequence. Using the Bott–Borel–Weil theorem, we fill in the SW (lower left) quadrant of $E_0^{q,k}(\mathcal{V})$, setting $\mathcal{V}$ equal, one by one, to each of the ten bundles above. Just by elementary $U(3)$ transformation properties, all possible actions of f in this spectral sequence vanish. That means that $E_0^{q,k}(\mathcal{V}) = E_\infty^{q,k}(\mathcal{V})$ and the results in Table B.1 are easily obtained. Of course, the superscripts and subscripts are the dimensions of the $\mathbb{P}^2_x$- and $\mathbb{P}^2_y$-factor cohomology groups and the total dimension is therefore the product. The symmetry obtained by interchanging the factors $\mathbb{P}^2_x \leftrightarrow \mathbb{P}^2_y$ corresponds to swapping the columns in Table B.1. Also there is *Serre duality*$^{(L)}$:

$$H^q(\mathcal{M}, \mathcal{V}) = H^{3-q}(\mathcal{M}, \mathcal{V}^*)^* \otimes \left(\begin{smallmatrix}1&1&1\\1&1&1\end{smallmatrix}\right) . \tag{B.3.3}$$

Using the information in Table B.1, the long exact cohomology sequence accompanying the right most column sequence of the diagram (B.1) implies that $H^1(\mathcal{M}, \mathcal{E} \otimes \mathcal{T}^*_{\mathcal{M}})$ and $H^3(\mathcal{M}, \mathcal{E} \otimes \mathcal{T}^*_{\mathcal{M}})$ vanish, that $H^2(\mathcal{M}, \mathcal{E} \otimes \mathcal{T}^*_{\mathcal{M}}) = \left(\begin{smallmatrix}1&1&1\\1&1&1\end{smallmatrix}\right)^1_1$ and that $H^0(\mathcal{M}, \mathcal{E} \otimes \mathcal{T}^*_{\mathcal{M}})$ equals the extension

$$\left(\begin{smallmatrix}0&0&0\\0&0&0\end{smallmatrix}\right)^1_1 + \left[\left(\begin{smallmatrix}-2&-1&0\\-3&0&0\end{smallmatrix}\right)^8_{10} \oplus \left(\begin{smallmatrix}-3&0&0\\-2&-1&0\end{smallmatrix}\right)^{10}_8\right] , \tag{B.3.4}$$

in total, a vector space of dimension 161.

From the long exact cohomology sequence accompanying the central column sequence of the diagram (B.1), we obtain:

$$H^0(\mathcal{M}, \mathcal{T}_X \otimes \mathcal{T}^*_{\mathcal{M}}) = 2\left(\begin{smallmatrix}0&0&0\\0&0&0\end{smallmatrix}\right)^1_1 , \tag{B.3.5}$$

$$H^1(\mathcal{M}, \mathcal{T}_X \otimes \mathcal{T}^*_{\mathcal{M}}) = \left(\begin{smallmatrix}-1&0&1\\0&0&0\end{smallmatrix}\right)^8_1 \oplus \left(\begin{smallmatrix}0&0&0\\-1&0&1\end{smallmatrix}\right)^1_8 , \tag{B.3.6}$$

and that

$$0\to \begin{matrix}\left(\begin{smallmatrix}1&1&1\\0&1&2\end{smallmatrix}\right)^{1}_{8}\\ \left(\begin{smallmatrix}0&1&2\\1&1&1\end{smallmatrix}\right)^{8}_{1}\end{matrix} \to H^2(\mathcal{M},\mathcal{T}_X\otimes\mathcal{T}^*_{\mathcal{M}}) \to \left[\begin{matrix}\left(\begin{smallmatrix}1&2&3\\1&1&4\end{smallmatrix}\right)^{8}_{10}\\ \left(\begin{smallmatrix}1&1&4\\1&1&1\end{smallmatrix}\right)^{10}_{8}\end{matrix} + 2\left(\begin{smallmatrix}1&1&1\\1&1&1\end{smallmatrix}\right)^{1}_{1}\right] \xrightarrow{j}$$
$$\xrightarrow{j} 2\left(\begin{smallmatrix}1&1&1\\1&1&1\end{smallmatrix}\right)^{1}_{1} \to H^3(\mathcal{M},\mathcal{T}_X\otimes\mathcal{T}^*_{\mathcal{M}}) \to 0\,. \tag{B.3.7}$$

As before, vertical stacks represent direct sums. It is not hard to verify (see Appendix A of Ref. [121] for the details) that the map j here acts in what is essentially the only possible way consistent with $U(3)\times U(3)$ tensor algebra—as the identity map $\left(\begin{smallmatrix}1&1&1\\1&1&1\end{smallmatrix}\right)\to\left(\begin{smallmatrix}1&1&1\\1&1&1\end{smallmatrix}\right)$. These contributions are therefore canceled out, whence $H^3(\mathcal{M},\mathcal{T}_X\otimes\mathcal{T}^*_{\mathcal{M}})$ vanishes and

$$H^2(\mathcal{M},\mathcal{T}_X\otimes\mathcal{T}^*_{\mathcal{M}}) = \begin{matrix}\left(\begin{smallmatrix}1&2&3\\1&1&4\end{smallmatrix}\right)^{8}_{10}\\ \left(\begin{smallmatrix}1&1&4\\1&2&3\end{smallmatrix}\right)^{10}_{8}\end{matrix} + \begin{matrix}\left(\begin{smallmatrix}1&1&1\\0&1&2\end{smallmatrix}\right)^{1}_{8}\\ \left(\begin{smallmatrix}0&1&2\\1&1&1\end{smallmatrix}\right)^{8}_{1}\end{matrix}\;, \tag{B.3.8}$$

a vector space of dimension 176.

If there were a non-vanishing map $\left(\begin{smallmatrix}1&2&3\\1&1&4\end{smallmatrix}\right)\xrightarrow{j}\left(\begin{smallmatrix}1&1&1\\1&1&1\end{smallmatrix}\right)$, it would essentially be induced from the map $\mathcal{T}^*_X|_{\mathcal{M}}\to\mathcal{E}|_{\mathcal{M}}$, that is from a contraction with $\mathrm{d}f(x,y)$. It follows that it would therefore be given as (up to overall factors $\epsilon^{abc}\epsilon^{\alpha\beta\gamma}$)

$$f_{abc\,\alpha\beta\gamma}\cdot\vartheta^{i(jk)\,(\mu\nu\rho)} \quad\longmapsto\quad \lambda\,. \tag{B.3.9}$$

By $U(3)\times U(3)$ covariance, there can be no such map (no way to contract the tensors on the left-hand side of the map (B.3.9) into a scalar).

The End$\mathcal{T}_{\mathcal{M}}$-valued cohomology on the submanifold: With the information obtained so far, we are ready for the base row of the diagram (B.1). The accompanying long exact cohomology sequence breaks into two exact sequences (dual to each other), the first of which is

$$0\to H^0(\mathcal{M},\mathcal{T}_{\mathcal{M}}\otimes\mathcal{T}^*_{\mathcal{M}}) \xrightarrow{\iota} 2\left(\begin{smallmatrix}0&0&0\\0&0&0\end{smallmatrix}\right) \xrightarrow{j} \left[\left(\begin{smallmatrix}0&0&0\\0&0&0\end{smallmatrix}\right) + \begin{matrix}\left(\begin{smallmatrix}-4&-1&0\\-3&0&0\end{smallmatrix}\right)\\ \left(\begin{smallmatrix}-3&0&0\\-4&-1&0\end{smallmatrix}\right)\end{matrix}\right] \xrightarrow{\delta}$$
$$\xrightarrow{\delta} H^1(\mathcal{M},\mathcal{T}_{\mathcal{M}}\otimes\mathcal{T}^*_{\mathcal{M}}) \xrightarrow{\iota} \begin{matrix}\left(\begin{smallmatrix}0&0&0\\-1&0&1\end{smallmatrix}\right)\\ \left(\begin{smallmatrix}-1&0&1\\0&0&0\end{smallmatrix}\right)\end{matrix} \to 0\,. \tag{B.3.10}$$

To specify $H^0(\mathcal{M},\mathcal{T}_{\mathcal{M}}\otimes\mathcal{T}^*_{\mathcal{M}})$ and $H^1(\mathcal{M},\mathcal{T}_{\mathcal{M}}\otimes\mathcal{T}^*_{\mathcal{M}})$, all we need is to determine the action of the map j. Considering its domain and the target, the only possible action of j here—as in Seq. (B.3.7) too—is the identity. This identifies (and thus cancels out) two $\left(\begin{smallmatrix} 0&0&0\\ 0&0&0 \end{smallmatrix}\right)$ terms, leaving

$$H^0(\mathcal{M}, \mathcal{T}_{\mathcal{M}} \otimes \mathcal{T}^*_{\mathcal{M}}) = \begin{pmatrix} 0&0&0\\ 0&0&0 \end{pmatrix}^1_1 , \tag{B.3.11}$$

in agreement with Corollary (B.1). This leaves

$$H^1(\mathcal{M}, \mathcal{T}_{\mathcal{M}} \otimes \mathcal{T}^*_{\mathcal{M}}) = \begin{matrix} \begin{pmatrix} -1&0&1\\ 0&0&0 \end{pmatrix}^8_1 \\ \begin{pmatrix} 0&0&0\\ -1&0&1 \end{pmatrix}^1_8 \end{matrix} + \begin{matrix} \begin{pmatrix} -2&-1&0\\ -3&0&0 \end{pmatrix}^8_{10} \\ \begin{pmatrix} -3&0&0\\ -2&-1&0 \end{pmatrix}^{10}_8 \end{matrix} , \tag{B.3.12}$$

a vector space of dimension 176, as found in Ref. [115]. The remaining part of the exact sequence yields

$$H^2(\mathcal{M}, \mathcal{T}_{\mathcal{M}} \otimes \mathcal{T}^*_{\mathcal{M}}) = \begin{matrix} \begin{pmatrix} 1&2&3\\ 1&1&4 \end{pmatrix}^8_{10} \\ \begin{pmatrix} 1&1&4\\ 1&2&3 \end{pmatrix}^{10}_8 \end{matrix} + \begin{matrix} \begin{pmatrix} 1&1&1\\ 0&1&2 \end{pmatrix}^1_8 \\ \begin{pmatrix} 0&1&2\\ 1&1&1 \end{pmatrix}^8_1 \end{matrix} , \tag{B.3.13}$$

in agreement with (B.3.3).

Beyond deformation theory: In this simple case, we used the $U(3) \times U(3)$ transformation properties in a rather elementary way, merely to obtain independently what can also be ascertained using Corollary B.1. The computations become more complicated in the next example and the transformation properties delivered by the Bott–Borel–Weil theorem will be absolutely essential. Now, the result (B.3.12) reveals more than the deformation theoretic discussion in § B.2.1.

The 8×10 dimensional contribution $\left(\begin{smallmatrix} -2&-1&0\\ -3&0&0 \end{smallmatrix}\right)$ is represented by $\vartheta_{a(bc)\,(\alpha\beta\gamma)}$; it is totally symmetric in α, β, γ, symmetric in b, c, but vanishes when a, b, c are symmetrized. Contracting with a $\mathrm{d}x$, two x's and three y's, we obtain

$$\mathrm{d}x^a \vartheta_{a(bc)\,(\alpha\beta\gamma)}\, x^b x^c\, y^\alpha y^\beta y^\gamma , \tag{B.3.14}$$

precisely as foretold in § B.2.1. These are non-trivial variations of $\mathrm{d}_x f(x, y)$. The other 80-dimensional contribution is obtained analogously, after $x \leftrightarrow y$.

However, the two 8-dimensional contributions do not have such an interpretation. They are simply two 3×3, traceless Hermitian matrices $\lambda_a{}^b$ and $\lambda_\alpha{}^\beta$. Tracing back through the calculations (see Table B.1), we see that the first one arises from the product $\mathcal{T}_{\mathbb{P}^2_x} \otimes \mathcal{T}^*_{\mathbb{P}^2_y}$. So, it acquires a natural interpretation as a type of linear reparametrization which maps $\mathcal{T}_{\mathbb{P}^2_y}$ to $\mathcal{T}_{\mathbb{P}^2_x}$. It is not clear from a strictly deformation theory point of view why such contributions should arise, whereas the foregoing computation based on the Koszul complex and spectral sequence provides it readily. In general, there are many more contributions, the computation becomes rather more tedious and may easily get out of (human) hand even though it is completely straightforward; perhaps a Reader adept at programming might try to develop an automated version of this.

B.3.2 Another example

Recall the Calabi–Yau 3-fold $\mathcal{M} \in \left[\begin{smallmatrix} 3 \\ 2 \end{smallmatrix} \middle\| \begin{smallmatrix} 3 & 1 \\ 0 & 3 \end{smallmatrix}\right]^{8;35}_{-54}$, defined as the zero set of the defining polynomials $f(x)$ and $g(x,y)$, as specified in Eq. (A.1.3),

$$f(x) \overset{\text{def}}{=} f_{abc}\, x^a x^b x^c \,, \qquad g(x,y) \overset{\text{def}}{=} g_{a\alpha\beta\gamma}\, x^a y^\alpha y^\beta y^\gamma \,, \tag{B.3.15}$$

with $x \in \mathbb{P}^3$ and $y \in \mathbb{P}^2$. As noted before, f and g are sections of $\left(\begin{array}{r|r} -3 & 0\,0\,0 \\ 0 & 0\,0 \end{array}\right)$ and $\left(\begin{array}{r|r} -1 & 0\,0\,0 \\ -3 & 0\,0 \end{array}\right)$, respectively and $\mathcal{E}$ is now the rank-two direct sum of these two line bundles. The tangent bundle on $\mathcal{X}$ is given similarly as in the previous case,

$$\mathcal{T}_{\mathcal{X}} = \left(\begin{array}{r|r} -1 & 0\,0\,1 \\ 0 & 0\,0 \end{array}\right) \oplus \left(\begin{array}{r|r} 0 & 0\,0\,0 \\ -1 & 0\,1 \end{array}\right) . \tag{B.3.16}$$

We therefore have

$$\mathcal{E} \otimes \mathcal{E}^* = 2\left(\begin{array}{r|r} 0 & 0\,0\,0 \\ 0 & 0\,0 \end{array}\right) \oplus \left(\begin{array}{r|r} -2 & 0\,0\,0 \\ 3 & 0\,0 \end{array}\right) \oplus \left(\begin{array}{r|r} 2 & 0\,0\,0 \\ -3 & 0\,0 \end{array}\right) ,$$

$$\mathcal{E} \otimes \mathcal{T}^*_{\mathcal{X}} = \left(\begin{array}{r|r} -2 & -1\,0\,0 \\ 0 & 0\,0 \end{array}\right) \oplus \left(\begin{array}{r|r} -3 & 0\,0\,0 \\ 1 & -1\,0 \end{array}\right) \oplus \left(\begin{array}{r|r} 0 & -1\,0\,0 \\ -3 & 0\,0 \end{array}\right) \oplus \left(\begin{array}{r|r} -1 & 0\,0\,0 \\ -2 & -1\,0 \end{array}\right) ,$$

$$\mathcal{T}_{\mathcal{X}} \otimes \mathcal{E}^* = \left(\begin{array}{r|r} 2 & 0\,0\,1 \\ 0 & 0\,0 \end{array}\right) \oplus \left(\begin{array}{r|r} 3 & 0\,0\,0 \\ -1 & 0\,1 \end{array}\right) \oplus \left(\begin{array}{r|r} 0 & 0\,0\,1 \\ 3 & 0\,0 \end{array}\right) \oplus \left(\begin{array}{r|r} 1 & 0\,0\,0 \\ 2 & 0\,1 \end{array}\right) ,$$

$$\mathcal{T}_{\mathcal{X}} \otimes \mathcal{T}^*_{\mathcal{X}} = 2\left(\begin{array}{r|r} 0 & 0\,0\,0 \\ 0 & 0\,0 \end{array}\right) \oplus \left(\begin{array}{r|r} -1 & 0\,0\,1 \\ 1 & -1\,0 \end{array}\right) \oplus \left(\begin{array}{r|r} 1 & -1\,0\,0 \\ -1 & 0\,1 \end{array}\right) \oplus \left(\begin{array}{r|r} 0 & -1\,0\,1 \\ 0 & 0\,0 \end{array}\right) \oplus \left(\begin{array}{r|r} 0 & 0\,0\,0 \\ 0 & -1\,1 \end{array}\right) .$$

B.3.3 Restricting the ambient cohomology to the submanifold

As in the previous example, we compute the cohomology groups on $\mathcal{M}$ valued in each of these irreducible bundles by using the corresponding spectral sequence. Recall that the resolution sequence (9.3.7), for this complete intersection, becomes Seq. (A.1.5)

$$0 \to \left(\begin{array}{c|ccc}4 & 0&0&0\\ 3 & 0&0&\end{array}\right) \begin{array}{c} \overset{f}{\nearrow} \left(\begin{array}{c|ccc}1 & 0&0&0\\ 3 & 0&0&\end{array}\right) \overset{g}{\searrow} \\ \underset{-g}{\searrow} \left(\begin{array}{c|ccc}3 & 0&0&0\\ 0 & 0&0&\end{array}\right) \underset{f}{\nearrow} \end{array} \left(\begin{array}{c|ccc}0 & 0&0&0\\ 0 & 0&0&\end{array}\right) \to \left(\begin{array}{c|ccc}0 & 0&0&0\\ 0 & 0&0&\end{array}\right)_{\mathcal{M}} \to 0\,. \qquad (A.1.5')$$

The arrows labeled by f and g denote maps induced by contraction with the respective polynomials, i.e., tensor coefficients f_{abc} and $g_{a\,\alpha\beta\gamma}$.

Unlike the previous example, not all spectral sequences now converge at the first level. The action of the spectral sequence differentials, all induced by f and g, must be determined. To show that this is nothing complicated, consider for example the spectral sequence for $\mathcal{E}_f \otimes \mathcal{T}_y{}^* = \left(\begin{array}{c|ccc}-3 & 0&0&0\\ 1 & -1&0&\end{array}\right)$:

q: $\left(\begin{array}{c\|ccc}1 & 0&0&0\\ 4 & -1&0&\end{array}\right)$	$\left(\begin{array}{c\|ccc}-2 & 0&0&0\\ 4 & -1&0&\end{array}\right)$ $\left(\begin{array}{c\|ccc}0 & 0&0&0\\ 1 & -1&0&\end{array}\right)$		$\left(\begin{array}{c\|ccc}-3 & 0&0&0\\ 1 & -1&0&\end{array}\right)$	$\overset{\varrho}{\to}$	$(\mathcal{E}_1 \otimes \mathcal{T}_y{}^*)\|_{\mathcal{M}}$
0: 0	0		0		$\Rightarrow 0$
1: 0	$\left(\begin{smallmatrix}0&0&0&0\\0&0&0\end{smallmatrix}\right)^1_1$	$\overset{f}{\to}$	$\left(\begin{smallmatrix}-3&0&0&0\\0&0&0\end{smallmatrix}\right)^{20}_1$	⋯⋯▸	see below
2: 0	$\left(\begin{smallmatrix}-2&0&0&0\\0&1&2\end{smallmatrix}\right)^{10}_8$ ⋯▸ (to row 1)		0		$\Rightarrow 0$
3: 0	0		0		$\Rightarrow 0$
4: 0	0		0		—
5: 0	0		0		—

(B.3.17)

The map labeled by f acts by taking

$$f_{abc}\,:\; \left[\lambda \in \left(\begin{smallmatrix}0&0&0&0\\0&0&0\end{smallmatrix}\right)\right] \longmapsto \left[(\lambda f_{abc}) \in \left(\begin{smallmatrix}-3&0&0&0\\0&0&0\end{smallmatrix}\right)\right], \qquad (B.3.18)$$

is clearly 1–1 and of maximum rank (=1), and yields the only non-vanishing differential in this spectral sequence. The corresponding chart of $E_2^{q,k}(\mathcal{E}_1 \otimes \mathcal{T}_y{}^*)$, the next and final level, is therefore obtained by replacing $E_1^{1,1} = \left(\begin{smallmatrix}0&0&0&0\\0&0&0\end{smallmatrix}\right)$ with 0 and $E_1^{1,0} = \left(\begin{smallmatrix}-3&0&0&0\\0&0&0\end{smallmatrix}\right)$ by its quotient with $f\left(\begin{smallmatrix}0&0&0&0\\0&0&0\end{smallmatrix}\right)$. Therefore, the only non-zero $\mathcal{E}_f \otimes \mathcal{T}_y{}^*$-valued cohomology on $\mathcal{M}$ is

$$H^1(\mathcal{M}, \mathcal{E}_f \otimes \mathcal{T}_y{}^*) = \left(\begin{smallmatrix}-2&0&0&0\\0&1&2\end{smallmatrix}\right)^{10}_8 + \left[\text{cok}\;\left(\begin{smallmatrix}0&0&0&0\\0&0&0\end{smallmatrix}\right)^1_1 \overset{f}{\to} \left(\begin{smallmatrix}-3&0&0&0\\0&0&0\end{smallmatrix}\right)^{20}_1\right]_{19}. \qquad (B.3.19)$$

In this way we obtain all the cohomology on $\mathcal{M}$, valued in the various bundles listed above. The results are:

Bundle	**Non-zero cohomology**
$\left(\begin{array}{c\|ccc} 0 & 0 & 0 & 0 \\ 0 & 0 & 0 & \end{array}\right)$	$H^0 = \left(\begin{smallmatrix} 0&0&0&0 \\ 0&0&0 \end{smallmatrix}\right)^1_1 \; H^3 = \left(\begin{smallmatrix} 1&1&1&1 \\ 1&1&1 \end{smallmatrix}\right)^1_1$
$\left(\begin{array}{c\|ccc} -2 & 0 & 0 & 0 \\ 3 & 0 & 0 & \end{array}\right)$	$H^1 = \ker\left[\left(\begin{smallmatrix} -1&0&0&0 \\ 1&1&4 \end{smallmatrix}\right)^4_{10} \xrightarrow{g} \left(\begin{smallmatrix} -2&0&0&0 \\ 1&1&1 \end{smallmatrix}\right)^{10}_1\right]$
$\left(\begin{array}{c\|ccc} -2 & -1 & 0 & 0 \\ 0 & 0 & 0 & \end{array}\right)$	$H^0 = \left(\begin{smallmatrix} 0&0&0&0 \\ 0&0&0 \end{smallmatrix}\right)^1_1 + \left(\begin{smallmatrix} -2&-1&0&0 \\ 0&0&0 \end{smallmatrix}\right)^{20}_1, \quad H^1 = \left(\begin{smallmatrix} -1&-1&0&0 \\ 1&1&1 \end{smallmatrix}\right)^6_1$
$\left(\begin{array}{c\|ccc} -3 & 0 & 0 & 0 \\ 1 & -1 & 0 & \end{array}\right)$	$H^1 = \left(\begin{smallmatrix} -2&0&0&0 \\ 0&1&2 \end{smallmatrix}\right)^{10}_8 + \text{cok}\left[\left(\begin{smallmatrix} 0&0&0&0 \\ 0&0&0 \end{smallmatrix}\right)^1_1 \xrightarrow{f} \left(\begin{smallmatrix} -3&0&0&0 \\ 0&0&0 \end{smallmatrix}\right)^{20}_1\right]$
$\left(\begin{array}{c\|ccc} 0 & -1 & 0 & 0 \\ -3 & 0 & 0 & \end{array}\right)$	$H^0 = \left(\begin{smallmatrix} 0&0&0&0 \\ 0&0&0 \end{smallmatrix}\right)^1_1 \; H^1 = \left(\begin{smallmatrix} 0&1&1&1 \\ 0&0&0 \end{smallmatrix}\right)^4_1$
$\left(\begin{array}{c\|ccc} -1 & 0 & 0 & 0 \\ -2 & -1 & 0 & \end{array}\right)$	$H^0 = \left(\begin{smallmatrix} 0&0&0&0 \\ 0&0&0 \end{smallmatrix}\right)^1_1 + \left(\begin{smallmatrix} -1&0&0&0 \\ -2&-1&0 \end{smallmatrix}\right)^4_8$
$\left(\begin{array}{c\|ccc} 0 & -1 & 0 & 1 \\ 0 & 0 & 0 & \end{array}\right)$	$H^1 = \left(\begin{smallmatrix} 0&1&1&1 \\ 0&0&0 \end{smallmatrix}\right)^4_1 \; H^2 = \left(\begin{smallmatrix} 0&0&0&1 \\ 1&1&1 \end{smallmatrix}\right)^4_1$
$\left(\begin{array}{c\|ccc} 0 & 0 & 0 & 0 \\ 0 & -1 & 1 & \end{array}\right)$	All cohomology vanishes
$\left(\begin{array}{c\|ccc} -1 & 0 & 0 & 1 \\ 1 & -1 & 0 & \end{array}\right)$	$H^1 = \left(\begin{smallmatrix} 0&0&0&1 \\ 0&1&2 \end{smallmatrix}\right)^4_8 + \left(\begin{smallmatrix} -1&0&0&1 \\ 0&0&0 \end{smallmatrix}\right)^{15}_1, \quad H^2 = \left(\begin{smallmatrix} 1&1&1&1 \\ 0&1&2 \end{smallmatrix}\right)^1_8$

(B.3.20)

where we have omitted entries that can be obtained using *Serre duality*$^{(L)}$; in the present case, it yields

$$H^q(\mathcal{M}, \mathcal{V}) = H^{3-q}(\mathcal{M}, \mathcal{V}^*)^* \otimes \left(\begin{smallmatrix} 1&1&1&1 \\ 1&1&1 \end{smallmatrix}\right). \tag{B.3.21}$$

To proceed, the sequence (A.2.1) is tensored with each of $\mathcal{E}_f$, $\mathcal{E}_g$, $\mathcal{T}_x$ and $\mathcal{T}_y$. This yields four separate short exact sequences, refining the two vertical exact sequences in the diagram (B.1). From the accompanying long exact cohomology sequences we then determine the $\mathcal{E}_i \otimes \mathcal{T}^*_{\mathcal{M}}$-valued and the $\mathcal{T}_i \otimes \mathcal{T}^*_{\mathcal{M}}$-valued cohomology on $\mathcal{M}$. Let's denote generically by j all maps in the Koszul spectral sequence and also all the maps induced from $\nabla\xi$ in Seq. (A.2.1) which are represented as various possible contractions with (possibly products of) the defining tensors f_{abc} and $g_{a\,\alpha\beta\gamma}$. This representation is again sufficient to locate and determine all non-vanishing maps between the cohomology groups. The resulting cohomology data is then plugged into the long exact cohomology sequence accompanying the short exact row sequence in the diagram (B.1).

✎ In the next couple of paragraphs, only the last few steps—the most complicated ones—are presented. Fill in the rest. ❧

B.3.4 Endomorphism valued cohomology on the submanifold

In the base row short exact sequence in the diagram (B.1), we study the action of j for each q separately,

$$\cdots \to H^q(\mathcal{M}, \mathcal{T}_{\mathcal{M}} \otimes \mathcal{T}^*_{\mathcal{M}}) \to H^q(\mathcal{M}, \mathcal{T}_{\mathcal{X}} \otimes \mathcal{T}^*_{\mathcal{M}}) \xrightarrow{j} H^q(\mathcal{M}, \mathcal{E} \otimes \mathcal{T}^*_{\mathcal{M}}) \to \cdots \tag{B.3.22}$$

and determine thereby $H^q(\mathcal{M}, \mathcal{T}_{\mathcal{M}} \otimes \mathcal{T}^*_{\mathcal{M}})$.

For $q = 0$, we have

$$2\left(\begin{smallmatrix} 0\,0\,0\,0 \\ 0\,0\,0 \end{smallmatrix}\right) \xrightarrow{j} \left[\left(\begin{smallmatrix} -2\,-1\,0\,0 \\ 0\;\;0\,0 \end{smallmatrix}\right) \oplus \left(\begin{smallmatrix} -1\;\;0\,0\,0 \\ -2\,-1\,0 \end{smallmatrix}\right) \oplus \left(\begin{smallmatrix} 0\,0\,0\,0 \\ 0\,0\,0 \end{smallmatrix}\right)\right] . \tag{B.3.23}$$

As before, with Seq. (B.3.7), j here identifies the copy of $\left(\begin{smallmatrix} 0\,0\,0\,0 \\ 0\,0\,0 \end{smallmatrix}\right)$ in the domain with the one in the target. This leaves us with $H^0(\mathcal{M}, \mathcal{T}_{\mathcal{M}} \otimes \mathcal{T}^*_{\mathcal{M}}) = \left(\begin{smallmatrix} 0\,0\,0\,0 \\ 0\,0\,0 \end{smallmatrix}\right)$, in agreement with Corollary B.1.

$$\operatorname{cok} j = \left[\left(\begin{smallmatrix} -2\,-1\,0\,0 \\ 0\;\;0\,0 \end{smallmatrix}\right) \oplus \left(\begin{smallmatrix} -1\;\;0\,0\,0 \\ -2\,-1\,0 \end{smallmatrix}\right)\right] \tag{B.3.24}$$

is mapped via the connecting homomorphism to $H^1(\mathcal{M}, \mathcal{T}_{\mathcal{M}} \otimes \mathcal{T}^*_{\mathcal{M}})$.

For $q = 1$, the stack of cohomology groups on both sides of the map j is rather longish (see Figure B.2). The computation itself is however less complicated than the wiring diagram which Figure B.2 resembles at first—we determine all mappings essentially by $U(4) \times U(3)$ covariance. In Figure B.2, we have written any mapping derived from f or g as just f or g.

For example, to obtain the *cokernel*$^{(L)}$

$$\left[\begin{array}{lcl} & \overset{g}{\nearrow} & \left(\begin{smallmatrix} -1\,-1\,0\,0 \\ 1\;\;1\,1 \end{smallmatrix}\right)^{6}_{1} \\ \operatorname{cok}\left(\begin{smallmatrix} -1\,0\,0\,0 \\ 1\,1\,4 \end{smallmatrix}\right)^{4}_{10} & \xrightarrow{g} & \left(\begin{smallmatrix} -2\,0\,0\,0 \\ 1\,1\,1 \end{smallmatrix}\right)^{10}_{1} \\ & \overset{g}{\searrow} & \left(\begin{smallmatrix} -2\,0\,0\,0 \\ 0\,1\,2 \end{smallmatrix}\right)^{10}_{8} \end{array}\right]_{56}$$

note that the linear transformation

$$g : \left(\begin{smallmatrix} -1\,0\,0\,0 \\ 1\,1\,4 \end{smallmatrix}\right)^{4}_{10} \begin{array}{l} \nearrow \left(\begin{smallmatrix} -1\,-1\,0\,0 \\ 1\;\;1\,1 \end{smallmatrix}\right)^{6}_{1} \\ \to \left(\begin{smallmatrix} -2\,0\,0\,0 \\ 1\,1\,1 \end{smallmatrix}\right)^{10}_{1} \\ \searrow \left(\begin{smallmatrix} -2\,0\,0\,0 \\ 0\,1\,2 \end{smallmatrix}\right)^{10}_{8} \end{array} \tag{B.3.25}$$

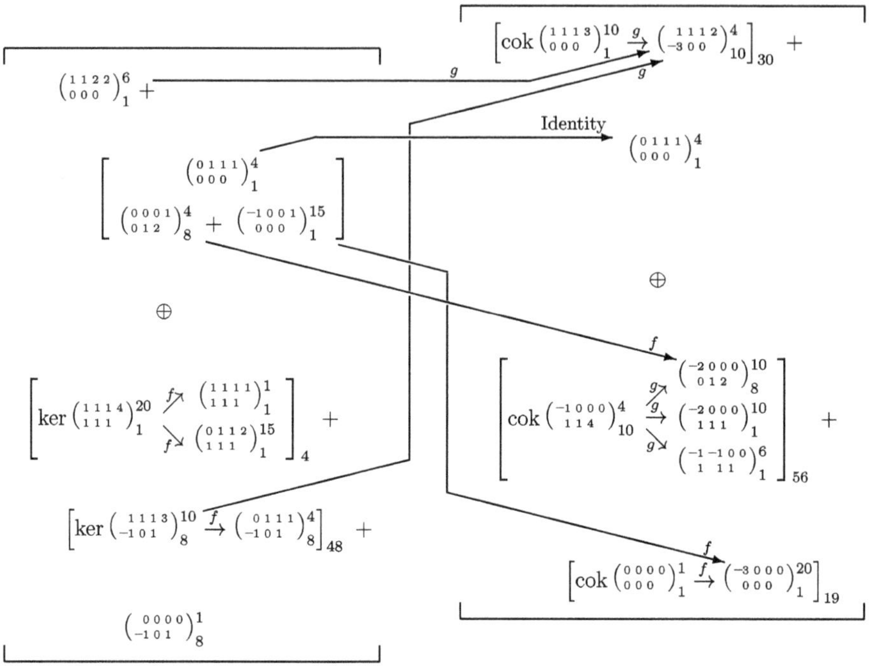

Figure B.2: The mapping $H^1(\mathcal{M}, \mathcal{T}_{\mathcal{X}} \otimes \mathcal{T}^*_{\mathcal{M}}) \longrightarrow H^1(\mathcal{M}, \mathcal{E} \otimes \mathcal{T}^*_{\mathcal{M}})$.

is 1–1 (is injective, has no kernel) for generic choice of $g(x,y)$, that is, for generic $g_{a\,\alpha\beta\gamma} = g_{a(\alpha\beta\gamma)}$. We verify this here explicitly, but leave other such calculations as an exercise.

Firstly, the $\left(\begin{smallmatrix}-1&0&0&0\\ &1&1&4\end{smallmatrix}\right) = \left(\begin{smallmatrix}-1&0&0&0\\ &0&0&3\end{smallmatrix}\right) \otimes \left(\begin{smallmatrix}0&0&0&0\\ &1&1&1\end{smallmatrix}\right)$ Young tableau corresponds to a tensor $\lambda_a{}^{\alpha\beta\gamma} = \lambda_a{}^{(\alpha\beta\gamma)}$. The combined map can be represented as

$$
\begin{aligned}
& \nearrow \theta_{[ab]} &&= \lambda_{[a}{}^{(\alpha\beta\gamma)} g_{b]\alpha\beta\gamma} \,, \\
g:\ \lambda_a{}^{(\alpha\beta\gamma)} & \rightarrow \theta_{(ab)} &&= \lambda_{(a}{}^{(\alpha\beta\gamma)} g_{b)\alpha\beta\gamma} \,, \\
& \searrow \theta_{(ab)\,\delta}{}^{\gamma} &&= (\delta^{\gamma\mu}_{\nu\delta} - \tfrac{1}{3}\delta^{\gamma\mu}_{\delta\nu}) \lambda_{(a}{}^{(\alpha\beta\nu)} g_{b)\alpha\beta\mu} \,.
\end{aligned}
\tag{B.3.26}
$$

The combined kernel of these three maps then consists of tensors $\lambda_a{}^{(\alpha\beta\gamma)}$ which are annihilated by all three maps, that is, those non-zero components of $\lambda_a{}^{(\alpha\beta\gamma)}$ which satisfy the system of three *linear* tensorial equations

$$\lambda_{[a}{}^{\alpha\beta\gamma} g_{b]\alpha\beta\gamma} = 0 \qquad \text{6 equations ,} \tag{B.3.27}$$

$$\lambda_{(a}{}^{\alpha\beta\gamma}\, g_{b)\alpha\beta\gamma} = 0 \qquad 10 \text{ equations}\ , \tag{B.3.28}$$

$$(\delta^{\gamma\mu}_{\nu\delta} - \tfrac{1}{3}\delta^{\gamma\mu}_{\delta\nu})\lambda_{(a}{}^{(\alpha\beta\nu)} g_{b)\alpha\beta\mu} = 0 \qquad 80 \text{ equations}\ , \tag{B.3.29}$$

corresponding to the three maps, respectively. (The $(\delta^{\gamma\mu}_{\nu\delta} - \frac{1}{3}\delta^{\gamma\mu}_{\delta\nu})$ prefactor projects the trace-free part.)

Now, these are 96 equations in only 40 unknowns and it is only reasonable to expect that all components of $\lambda_a{}^{\alpha\beta\gamma}$ are forced to vanish by these equations—*for a generic choice* of $g_{a\,\alpha\beta\gamma}$. Making a specific choice of $g_{a\,\alpha\beta\gamma}$ such as

$$g_{1\,111} = g_{2\,222} = g_{3\,333} = 1\ , \qquad g_{0\,123} = \kappa \neq 0\ , \tag{B.3.30}$$

where all other $g_{a\,\alpha\beta\gamma} = 0$, it can be shown explicitly that the choice is 'generic enough,' that is, all $\lambda_a{}^{\alpha\beta\gamma} = 0$ and there is no kernel.

Consider now the sub-generic case, when $g_{0\,123} = \kappa = 0$ [210]. It is not hard to show that the complete intersection is still smooth. However, the combined mapping (B.3.26) changes drastically. Since all non-zero components of $g_{a\,\alpha\beta\gamma}$ have $\alpha = \beta = \gamma$, the four components $\lambda_a{}^{(123)}$ cannot possibly occur non-trivially in the map (B.3.26), will not be set to zero and therefore span (part of) the kernel.

✎ Find the full kernel of the combined map (B.3.26), with $g_{a\,\alpha\beta\gamma}$ as in Eq. (B.3.30), and when $\kappa = 0$. ❧

On the other hand, the cokernel of the combined (generic) map is the space spanned by the 96 components $\{\theta_{[ab]}, \theta_{(ab)}, \theta_{(ab)\,\delta}{}^{\gamma}\}$, taken however modulo the combined additive image of the 40 components of $\lambda_a{}^{\alpha\beta\gamma}$. That is—for a 'generic enough' choice of $f(x)$ and $g(x,y)$—we can use the 40 independent components of the three different $(\lambda \cdot g)$'s to gauge away forty components of the θ's. As usual, representatives of this quotient are not uniquely defined and there are various (but equivalent) choices. Although it may become cumbersome to keep track of this, it will eventually be useful to toggle between such equivalent choices when it comes to compute some Yukawa couplings.

✎ Find the full cokernel of the combined map (B.3.26), where $g_{a\,\alpha\beta\gamma}$ is given in Eq. (B.3.30) with $\kappa = 0$. ❧

Finally, after completing several similar computations, $H^1(\mathcal{M}, \operatorname{End}\mathcal{T}_{\mathcal{M}})$ is obtained as the double extension $A + B + C$, where

$$A \;=\; \ker\left[\begin{pmatrix}1&1&1&4\\1&1&1&\end{pmatrix}^{20}_{1} \begin{matrix}\overset{f}{\nearrow}\\ \underset{f}{\searrow}\end{matrix} \begin{matrix}\begin{pmatrix}1&1&1&1\\1&1&1&\end{pmatrix}^{1}_{1}\\[4pt] \begin{pmatrix}0&1&1&2\\1&1&1&\end{pmatrix}^{15}_{1}\end{matrix}\right]_4\ , \tag{B.3.31}$$

$$B = \ker \left[\begin{array}{ccc} \left(\begin{smallmatrix} 1&1&2&2 \\ 0&0&0& \end{smallmatrix}\right)^{6}_{1} & \overset{g}{\searrow} & \\ \left(\begin{smallmatrix} 1&1&1&3 \\ 0&0&0& \end{smallmatrix}\right)^{10}_{1} & \overset{g}{\rightarrow} & \left(\begin{smallmatrix} 1&1&1&2 \\ -3&0&0& \end{smallmatrix}\right)^{4}_{10} \\ \left(\begin{smallmatrix} 1&1&1&3 \\ -1&0&1& \end{smallmatrix}\right)^{10}_{8} & \overset{g}{\nearrow} \overset{f}{\rightarrow} & \left(\begin{smallmatrix} 0&1&1&1 \\ -1&0&1& \end{smallmatrix}\right)^{4}_{8} \end{array} \right]_{24} , \tag{B.3.32}$$

$$C = \left(\begin{smallmatrix} 0&0&0&0 \\ -1&0&1& \end{smallmatrix}\right)^{1}_{8} + \left[\left(\begin{smallmatrix} -2&-1&0&0 \\ 0&0&0& \end{smallmatrix}\right)^{20}_{1} \oplus \left(\begin{smallmatrix} -1&0&0&0 \\ -2&-1&0& \end{smallmatrix}\right)^{4}_{8} \right] . \tag{B.3.33}$$

$H^2(\mathcal{M}, \mathrm{End}\, \mathcal{T}_{\mathcal{M}})$ is given dually.

Explicit tensorial representation: The contribution A in (B.3.31) is rather simple. It consists of elements of

$$\left(\begin{smallmatrix} 1&1&1&4 \\ 1&1&1& \end{smallmatrix}\right) \sim \theta^{(abc)} \epsilon^{ijkl}\, \epsilon^{\alpha\beta\gamma} , \tag{B.3.34}$$

which are annihilated in the mapping, by f, into $\left(\begin{smallmatrix} 1&1&1&1 \\ 1&1&1& \end{smallmatrix}\right) \oplus \left(\begin{smallmatrix} 0&1&1&2 \\ 1&1&1& \end{smallmatrix}\right)$. Up to overall factors of $\epsilon^{ijkl}\epsilon^{\alpha\beta\gamma}$, these Young tableaux correspond to scalars and traceless Hermitian matrices. The restrictions defining the kernel are easily enforced by

$$\begin{aligned} f_{abc}\, \theta^{(abc)} &= 0 , \\ (\delta^{ai}_{jd} - \tfrac{1}{4}\delta^{ai}_{dj}) f_{abc}\, \theta^{(bcd)} &= 0 . \end{aligned} \tag{B.3.35}$$

Combining these two conditions in the obvious way, we have that

$$A \sim \{ \, (\theta^{(abc)}\, \epsilon^{ijkl}\, \epsilon^{\alpha\beta\gamma}) \; : \; f_{abc}\, \theta^{(bcd)} = 0 \, \} . \tag{B.3.36}$$

Now, the contribution B (B.3.32) is mostly the dual of (B.3.26), which we have considered before. Omitting overall factors of ϵ^{ijkl}, we introduce irreducible $U(4) \times U(3)$ tensor variables $\theta^{[ab]}$, $\theta^{(ab)}$ and $\theta^{(ab)\,\alpha}{}_{\beta}$ for the three Young tableaux on the left hand side of B, respectively. The kernel will be described by a subset of these variables which vanish upon the mapping in B. The last of these, $\theta^{(ab)\,\alpha}{}_{\beta}$, must also be trace-free in α, β and is subject to the additional restriction

$$f_{abc}\, \theta^{(ab)\,\alpha}{}_{\beta} = 0 , \tag{B.3.37}$$

which describes the kernel of the bottom mapping in B. With a generically chosen f_{abc}, this produces 32 independent equations in 80 variables in $\theta^{(ab)\alpha}{}_{\beta}$, leaving 48 of them free to parametrize the kernel. For specific choices, this has to be verified explicitly; the Fermat-type choice

$$f(x) = \sum_{a=0}^{3} (x^a)^3 , \qquad \Longleftrightarrow \qquad f_{abc} = \begin{cases} 1 & a = b = c, \\ 0 & \text{otherwise} \end{cases} \tag{B.3.38}$$

is 'generic enough' in this sense, that is, it provides a map of maximum rank.

The kernel of the combined mapping into $\left(\begin{smallmatrix} 0 & 0 & 0 & 1 \\ -3 & 0 & 0 & \end{smallmatrix}\right)$ is described by

$$\phi_1\, g_{b(\alpha\beta\gamma)}\, \theta^{[ab]} + \phi_2\, g_{b(\alpha\beta\gamma)}\, \theta^{(ab)} + \phi_3\, g_{b(\alpha\delta\gamma}\, \theta^{(ab)\delta}{}_{\beta)} = 0 \ . \tag{B.3.39}$$

The parameters ϕ_i describe the triple mapping more precisely and are non-zero but otherwise arbitrary. Choosing $g_{a\,\alpha\beta\gamma}$ generically, this produces 40 independent relations among $6+10+48$ components of our tensor variables, leaving 24 to span the combined kernel representing the contribution B.

Finally, the three Young tableaux in (B.3.33) correspond to

$$\theta^{\alpha}{}_{\beta}\ , \qquad \theta_{a(bc)}\ , \qquad \theta_{a\,\alpha(\beta\gamma)}\ . \tag{B.3.40}$$

The last two of these occur as non-trivial variations of

$$\mathrm{d}_x f(x) = dx^a\, f_{abc}\, x^b x^c\ , \qquad \text{and} \qquad \mathrm{d}_y g(x,y) = dy^{\alpha}\, g_{a\,\alpha\beta\gamma}\, x^a\, y^{\beta} y^{\gamma}\ , \tag{B.3.41}$$

and cover the result of the polynomial deformation method; the remaining 36 components have been missed.

Special defining polynomials: Consider again the smooth but subgeneric case, when $g_{0\,123} = \kappa = 0$ [210]. Just as the combined mapping (B.3.26) suffers drastic changes, so does the final result for $H^1(\mathcal{M}, \mathrm{End}\,\mathcal{T}_{\mathcal{M}})$. In practice, we have to go back through the computations and carefully reconsider all instances of contraction with $g_{a\,\alpha\beta\gamma}$. As in (B.3.26), there will be cases where the kernel, and so also the cokernel of a map induced by contraction with $g_{a\,\alpha\beta\gamma}$ becomes larger in the limit $g_{0\,123} = \kappa \to 0$, so we expect new elements of $H^1(\mathcal{M}, \mathrm{End}\,\mathcal{T}_{\mathcal{M}})$.

There is however another, subtler change: the representatives themselves change and in fact discontinuously so in the limit $g_{0\,123} = \kappa \to 0$. In particular, choose [210]

$$f_{abc} = \begin{cases} 1,\ a=b=c, \\ 0,\ \text{otherwise}; \end{cases} \qquad g_{a\,\alpha\beta\gamma} = \begin{cases} \kappa,\ (a,\alpha,\beta,\gamma) = (0,1,2,3), \\ 0,\ \text{otherwise.} \end{cases} \tag{B.3.42}$$

The contribution A in (B.3.36) now becomes

$$\{\ \theta^{012}, \theta^{013}, \theta^{023}, \theta^{123}\ \}\ . \tag{$B.3.36'$}$$

These and the contributions C, presented in (B.3.40), are unchanged in the limit $g_{0\,123} = \kappa \to 0$.

For the contribution B, specified in (B.3.32), a convenient basis is, in the limit $g_{0\,123} = \kappa \to 0$, provided by

$$\{\ (\theta^{(ab)\gamma}{}_{\delta}\epsilon^{ijkl})\ :\ a \neq b,\ \alpha = \gamma,\ \theta^{(ab)\gamma}{}_{\gamma} = 0\ \}_{12}\ , \tag{B.3.43}$$

$$\{\ (\theta^{(\hat{a}\hat{b})\gamma}{}_{\delta}\epsilon^{ijkl})\ :\ \delta = \hat{a},\ \hat{a}, \hat{b}, \gamma \text{ different }\}_{6}\ , \tag{B.3.44}$$

$$\{\ (\theta^{(0\hat{b})\gamma}{}_{\delta}\epsilon^{ijkl})\ :\ \hat{b}, \gamma, \delta \text{ different }\}_{6}\ , \tag{B.3.45}$$

where the caret on an index denotes that the index does not take the value 0. When we deform the 3-fold by switching on $g_{0\,123} = \kappa \neq 0$, the first 18 of these remain unchanged. However, the last six (B.3.45) become replaced with certain linear combinations, such as

$$\theta^{(01)3}{}_{2} \to \left(\theta^{(01)3}{}_{2} - \left(\frac{g_{2\,222}}{g_{0\,123}}\right)\theta^{(21)2}{}_{1}\right)\,. \tag{B.3.46}$$

Finally, there are also 20 elements of $H^1(\mathcal{M}, \mathrm{End}\,\mathcal{T}_{\mathcal{M}})$ and of course 20 matching elements of $H^2(\mathcal{M}, \mathrm{End}\,\mathcal{T}_{\mathcal{M}})$ which cannot be canceled out with the special choice of $g_{a\,\alpha\beta\gamma}$ given in (B.3.42). However, as soon as $g_{0\,123} = \kappa \neq 0$, these 20 elements of $H^1(\mathcal{M}, \mathrm{End}\,\mathcal{T}_{\mathcal{M}})$ are mapped to the matching 20 from $H^2(\mathcal{M}, \mathrm{End}\,\mathcal{T}_{\mathcal{M}})$; the former 20 are no longer closed and the latter 20 become exact and so drop out of the $\mathrm{End}\,\mathcal{T}_{\mathcal{M}}$-valued cohomology. A suitable basis for these 20 elements of $H^1(\mathcal{M}, \mathrm{End}\,\mathcal{T}_{\mathcal{M}})$ is

$$\begin{array}{ll} \{\ \theta^{(0b)}\epsilon^{ijkl}\ \}_{4}\ , & \{\ \theta^{(0\hat{b})\gamma}{}_{\delta}\epsilon^{ijkl}\ :\ \hat{b} = \delta \neq \gamma\ \}_{6}\ , \\ \{\ \theta_{a}{}^{(123)}\epsilon^{\mu\nu\rho}\ \}_{4}\ , & \{\ \theta_{\hat{a}}{}^{(\alpha\beta\gamma)}\epsilon^{\mu\nu\rho}\ :\ \hat{a} = \alpha = \beta \neq \gamma\ \}_{6}. \end{array} \tag{B.3.47}$$

B.4 Cohomology and Moduli

The General deformation theory of Kodaira and Spencer (see Ref. [35]) ensures that $H^1(\mathcal{M}, \mathcal{T}_{\mathcal{M}})$ spans the deformations of the complex structure of any compact complex manifold $\mathcal{M}$—*locally* near the original choice of the complex structure for $\mathcal{M}$. Consider now the manifold $\mathcal{M}$ embedded in some compact complex manifold $\mathcal{X}$ as the vanishing set of a (system of) section(s) of a bundle $\mathcal{E}$ over $\mathcal{X}$; that is, $\mathcal{M} \in [\mathcal{X} \| \mathcal{E}]$.

By varying the defining equations of the manifold $\mathcal{M}$, embedded in $\mathcal{X}$, we span the 'polynomial deformation' space $\mathfrak{d}[\mathcal{X} \| \mathcal{E}]$ of the embedding $\mathcal{M} \hookrightarrow \mathcal{X}$. The image of this shows up in $H^1(\mathcal{M}, \mathcal{T}_{\mathcal{M}})$ as the 'polynomial deformation' of the complex structure, as discussed above.

The results of the preceding Chapter show that $\dim H^q(\mathcal{M}, \mathcal{T}_{\mathcal{M}})$ and also $\dim H^q(\mathcal{M}, \mathcal{T}^*_{\mathcal{M}}) = h^{1,q}$ are independent of the choice of the complex structure, and in particular, the defining polynomials of $\mathcal{M}$—as long as $\mathcal{M}$ is smooth. By contrast, we have seen that $H^q(\mathcal{M}, \mathrm{End}\,\mathcal{T}_{\mathcal{M}})$ exhibits a more interesting behavior which is known as *semi-continuity*: $\dim H^1(\mathcal{M}, \mathrm{End}\,\mathcal{T}_{\mathcal{M}})$ is minimal for generic choices of the complex structure, but increases in jumps as we specialize to less and less generic choices. These changes are accurately computed using the Koszul spectral sequence as described here.

So far, we have considered varying the complex structure of the (Calabi–Yau) 3-fold $\mathcal{M}$. The choice of the Kähler class J is another important datum on $\mathcal{M}$, relevant for (super)string compactification. There is no reason why $H^1(\mathcal{M}, \mathrm{End}\,\mathcal{T}_{\mathcal{M}})$ should vary with the choice of the complex structure but be inert under variations of the choice of the Kähler class. In fact, the *mirror map* $^{(L)}$ strongly suggests that $H^1(\mathcal{M}, \mathrm{End}\,\mathcal{T}_{\mathcal{M}})$ depends on the choice $J \in H^{1,1}(\mathcal{M})$ in precisely the same way as it depends on the choice of the complex structure. Unfortunately, this dependence does not show up in the Koszul spectral sequence computation and cannot be controlled in this way.

Finally, we note that $H^1(\mathcal{M}, \mathrm{End}\,\mathcal{T}_{\mathcal{M}})$ spans the deformations of the complex structure of the total space of $\mathcal{T}_{\mathcal{M}}$, the holomorphic tangent bundle of $\mathcal{M}$. Variations of this choice are also beyond the Koszul spectral sequence computation as described here.

Comparing the results of the Koszul spectral sequence with other, more stringy methods (some of which will be discussed at the end of this volume) suggests that the Koszul spectral sequence computation refers to (1) a specified choice of the complex structure (as controlled by the choice of the defining equations), but (2) a generic choice of the Kähler class and the complex structure of $\mathcal{T}_{\mathcal{M}}$. That is, certain other methods involve special choices of the Kähler class (metric) and also yield what appears to be a bigger dimensional $H^1(\mathcal{M}, \mathrm{End}\,\mathcal{T}_{\mathcal{M}})$. Similarly, certain properties of the ring structures generated by $H^1(\mathcal{M}, \mathcal{T}_{\mathcal{M}})$, $H^1(\mathcal{M}, \mathcal{T}^*_{\mathcal{M}})$ and $H^1(\mathcal{M}, \mathrm{End}\,\mathcal{T}_{\mathcal{M}})$ seem to support this possibility. Clearly, however, the problem of describing the variation of these cohomology groups and the ring structures which they generate remains a rather new and largely uncharted territory in the jungle.

Chapter C

The (2, 1) Triple Couplings and Generalization

Continuing the discussion in Chapter 8, we now turn to the task of computing the $\mathbf{27}^3$ Yukawa couplings, as given in Eq. (0.5.7):

$$\kappa_{\alpha\beta\gamma} \stackrel{\text{def}}{=} \int_{\mathcal{M}} \left(\Omega^{\overline{\mu}\,\overline{\nu}\,\overline{\rho}}\ \varphi_{(\alpha)\,\overline{\mu}}{}^{\sigma}\ \varphi_{(\beta)\,\overline{\nu}}{}^{\tau}\ \varphi_{(\gamma)\,\overline{\rho}}{}^{\lambda}\ \Omega_{\sigma\tau\lambda} \right) . \qquad (0.5.7')$$

Of course, as it stands, this expression has little practical use, since we have next to no knowledge about the actual functional form of the integrand. In a way, we need to reformulate this integral formula into something more tractable.

This is not at all unlike the computation described in Chapter 8, where an integral was related to intersection numbers which were obtained by counting, or the spectral sequence computations presented in Chapters 9–B, where the solutions of some partial differential equations are described without ever solving them*. In particular examples, there may be more than one way of doing this and not infrequently, some of these "translations" of the original problem will offer a solution which is not quite complete in some aspect. Usually, however, different methods can be combined to complete the desired computations.

---

* One might say that most of the tricks of the trade boil down to integration without integrating. To this end, the general theory of residues comes in handy. We have seen one of its results in Eq. (A.3.12) and many of the subsequent facts about residue class (quotient) rings follow from this general theory; see Chapter 5 of Ref. [23] for a general discussion.

C.1 The Deformation Theoretic (2, 1) Triple Couplings

In a good number of complete intersection type embeddings, there is a useful computational tool which turns the integral (0.5.7) into a product and which can often be evaluated by inspection [84].

It relies on two facts: (1) On a complex manifold $\mathcal{M}$, harmonic (2,1)-forms, or equivalently, elements of $H^1(\mathcal{M}, \mathcal{T}_\mathcal{M})$ parametrize locally the space of complex structures of $\mathcal{M}$. (2) For many complete intersections, the deformation space of the embedding $\mathcal{M} \hookrightarrow \mathcal{X}$, as defined in Eq. (A.3.2) is in fact isomorphic to the space of complex structure locally around any particular choice. The precise conditions when this happens were discussed in Chapter A. We now assume that these conditions are met.

The degree-3 cohomology group

$$H^3 \quad = \quad H^{3,0} \oplus H^{2,1} \oplus H^{1,2} \oplus H^{0,3} \tag{C.1.1}$$

is then completely parametrized and given the structure of a *residue class ring*,

$$\mathcal{R}_{[\mathcal{X}\|\mathcal{E}]} \stackrel{\text{def}}{=} \{\,\mathbb{C}[(\mathcal{X}\,)] \,/\, \Im\big[\mathrm{Aut}[\mathcal{X}\|\mathcal{E}]\big]\,\}\ . \tag{C.1.2}$$

$\mathbb{C}[(\mathcal{X}\,)]$ is the ring of homogeneous polynomials over $\mathcal{X}$ with complex coefficients, generated from polynomials of the degree of the defining constraints, $f_a(x) = 0$, so that there is a natural grading in multiples of $\deg(f_a)$. $\Im\Big[\mathrm{Aut}[\mathcal{X}\|\mathcal{E}]\Big]$ is the ideal generated by various symmetries of the configuration, such as linear reparametrizations of the embedding space, suitable rescalings of the defining constraints and so on. When polynomial deformations represent $H^{2,1}$ faithfully, the grading of $\mathcal{R}$ corresponds to the decomposition (C.1.1) and so there is a unique equivalence class, sometimes a single monomial, $\mathcal{Q} \in \mathcal{R}$, corresponding to elements of $H^{0,3}$. If $\mathcal{m}_\alpha$ are elements of $\mathcal{R}$ corresponding to $H^{2,1}$, then

$$\mathcal{m}_\alpha{\cdot}\mathcal{m}_\beta{\cdot}\mathcal{m}_\gamma \quad \equiv \quad \kappa_{\alpha\beta\gamma}{\cdot}\mathcal{Q}\ (\mathrm{mod}\Im)\ , \tag{C.1.3}$$

where $\Im$ is standing for $\Im\Big[\mathrm{Aut}[\mathcal{X}\|\mathcal{E}]\Big]$. The so defined constant $\kappa_{\alpha\beta\gamma}$ must then be equal to the Yukawa coupling (0.5.7). Some numerical characteristics of $\mathcal{R}$ can be obtained easily using the standard results given in § 7.4.

On *algebraic varieties*, which are generally defined as the zero-set of a polynomial constraint system, the residue class ring is one of the fundamental objects [27, 38]. So much so, in fact, that the variety itself can be practically disregarded and one may proceed simply by manipulating $\mathcal{R}$ and various related structures. The beautiful simplicity† of complete intersection spaces where the polynomial deformation technology does faithfully represent $H^{2,1}$ and the related Yukawa couplings implies that in these cases the residue class ring $\mathcal{R}$ which is explicitly constructed from the space of 'polynomial deformations' coincides with the abstract such ring of the corresponding abstract algebraic variety.

— ❦ —

To illustrate this technique, we use the result (A.3.6). The $m_\alpha \in \mathcal{R}$ which correspond to $H^{2,1}(\mathcal{M})$ of the complete intersection manifold (A.1.2) are

$$\begin{bmatrix} \phi_{(abc)} x^a x^b x^c \\ \varphi_{a(\alpha\beta\gamma)} x^a y^\alpha y^\beta y^\gamma \end{bmatrix} \cong \begin{bmatrix} \phi_{(abc)} x^a x^b x^c \\ \varphi_{a(\alpha\beta\gamma)} x^a y^\alpha y^\beta y^\gamma \end{bmatrix} + (x^a \lambda_a{}^b \partial_b + y^\alpha \lambda_\alpha{}^\beta \partial_\beta) \begin{bmatrix} f(x) \\ g(x,y) \end{bmatrix}, \tag{A.3.6'}$$

where $f(x)$ and $g(x,y)$ are the defining equations of bi-degree (3,0) and (1,3), respectively, in coordinates $x \in \mathbb{P}^3$ and $y \in \mathbb{P}^2$. Note in fact how obvious the relationship between the monomial representatives and the tensorial representatives appears in Eq. (A.3.6′): we may regard $x^a x^b x^c$ as the relevant objects and $\phi_{(abc)}$ merely as coefficients, or we may think of the monomials as basis-elements and of $\phi_{(abc)}$ as the tensor components. Calculating with $\phi_{(abc)}$ is then just like representing a vector by its components: v^μ.

Roughly, the equivalence relation in (A.3.6) is like gauging monomials equivalent to zero if they are proportional to the gradients of the defining equations. However, one has to be careful in doing so. Since the equivalence relations stem from the action of the linear reparametrization groups, the number of relations imposed using these generators must be precisely equal to the dimension of the reparametrization group. $\dim \mathrm{Aut}[\mathcal{X} \| \mathcal{E}] = 25$ in the above case ($\dim(PGL(n,\mathbb{C})) = n^2 - 1$, plus the overall rescaling of the defining equations). This is precisely accounted for by the two matrices $[\lambda_a{}^b]$ and $[\lambda_\alpha{}^\beta]$.

The various components of $[\lambda_a{}^b]$ and $[\lambda_\alpha{}^\beta]$ can then be chosen so as to eliminate one or another monomial from the list of candidates for a basis. This ambiguity leads to many different choices of bases, which are all of course equivalent. This is completely described by the Koszul complex.

† ... in principle, although not always also in practice ...

The example at hand corresponds to the mapping

$$\begin{array}{ccc} \begin{pmatrix} -1\,0\,0\,1 \\ 0\,0\,0 \end{pmatrix}^{15}_{1} & \xrightarrow{\ f\ } & \left[\begin{pmatrix} -3\,0\,0\,0 \\ 0\,0\,0 \end{pmatrix}^{20}_{1} \Big/ f \begin{pmatrix} 0\,0\,0\,0 \\ 0\,0\,0 \end{pmatrix}^{1}_{1} \right] \\ & \searrow^{g} & \\ \begin{pmatrix} 0\,0\,0\,0 \\ -1\,0\,1 \end{pmatrix}^{1}_{8} & \xrightarrow{\ g\ } & \left[\begin{pmatrix} 0\,0\,0\,0 \\ -1\,0\,1 \end{pmatrix}^{1}_{8} \Big/ g \begin{pmatrix} 0\,0\,0\,0 \\ 0\,0\,0 \end{pmatrix}^{1}_{1} \right] \end{array} , \tag{A.1.20$'$}$$

the cokernel of which is identified as $H^1(\mathcal{M}, \mathcal{T}_\mathcal{M})$ and presented in (A.1.22). The horizontal maps correspond to using the $\lambda_a{}^b$ and the $\lambda_\alpha{}^\beta$ matrix in 'gauging away' monomials from $\{\phi_{(abc)}x^a x^b x^c\}$ and $\{\phi_{a(\alpha\beta\gamma)}x^a y^\alpha y^\beta y^\gamma\}$, respectively. The slanted map represents using the $\lambda_a{}^b$ matrix to 'gauge away' monomials from the second set. Therefore, the true ideal is generated by the horizontal maps modulo the effect of the slanted map. More explicitly, we find that the ideal is not generated by all the three non-zero gradients

$$[\partial_a f(x)] \ , \quad [\partial_a g(x,y)] \ , \quad [\partial_\alpha g(x,y)] \ , \tag{C.1.4}$$

independently, but rather by

$$\left\{ \frac{[\partial_a f(x)] \oplus [\partial_a g(x,y)] \oplus [\partial_\alpha g(x,y)]}{[\partial_a f(x) - \partial_a g(x,y)]} \right\} , \tag{C.1.5}$$

where the condition on the generators of the ideal (in the denominator) represents the action of the slanted map since it allows us to replace a relation generated by $\partial_a f(x)$ with another relation generated by $\partial_a g(x,y)$ in the action of the ideal on the tensors to the right of the combined map (A.1.20). This quotient ideal is equivalent to

$$\Im_{\rm K}[(\partial_a f + \partial_a g),\, (\partial_\alpha g)] \ , \tag{C.1.6}$$

and was called the *Koszul ideal* in (9.3.15).

To describe this residue class ring structure more succinctly, we should require the relations to be homogeneous and have $\deg(\partial_a f(x)) = \deg(\partial_a g(x,y))$. This implies that $2\deg(x^a) = 3\deg(y^\alpha)$, i.e., $\deg(x^a) = 3$, $\deg(y^\alpha) = 2$ and clearly also that $\deg(f(x)) = \deg(g(x,y)) = 9$. We record $\deg((\partial_a f - \partial_a g)) = 6$ and $\deg(\partial_\alpha g) = 7$. With this, the cohomology has just been given the structure of a *graded* quotient ring, for which the results in § 7.4 apply straightforwardly and we have

$$P_t(\mathbb{C}[x^a, y^\alpha]/\Im_{\rm K}) \ = \ \frac{(1-t^6)^4 \cdot (1-t^7)^3}{(1-t^3)^4 \cdot (1-t^2)^3} \ , \tag{C.1.7}$$

which gives $P_t = \ldots + 35t^9 + \ldots$, corresponding to $\dim H^1(\mathcal{M}, \mathcal{T}_{\mathcal{M}}) = 35$. In general, the result will appear as a sum of such contributions, each of which can be assigned a grade. This makes the full cohomology ring into a bi-graded ring, where one grading comes from the different degrees of the homogeneous coordinates in order for the Koszul ideal to be generated by homogeneous relations and the other grading labels the different contributions in formulae of the type (9.3.20).

— ❦ —

With the simple choice $f(x) \stackrel{\text{def}}{=} \sum_{a=0}^{3}(x^a)^3$ and $g(x,y) \stackrel{\text{def}}{=} \sum_{\alpha=1}^{3} x^\alpha (y^\alpha)^3$, we have that

$$\partial_a f(x) = 3(x^a)^2 \ , \quad \partial_a g(x,y) = (y^\alpha)^3 \quad \text{and} \quad \partial_\alpha g(x,y) = 3x^\alpha (y^\alpha)^2 \ . \tag{C.1.8}$$

It is then not hard to show that

$$\mathcal{Q} \ \propto \ (x^0 x^1 x^2 x^3)(x^1 y^1\, x^2 y^2\, x^3 y^3) \ , \tag{C.1.9}$$

or, more generally,

$$\mathcal{Q} \ \propto \ \big(\det[f''(x)]\big)\big(\det[\ddot{g}(x,y)]\big) \ , \tag{C.1.10}$$

where primes denotes derivative with respect to the x's and dots are derivatives with respect to the y's[‡].

C.2 A General Trace Formula

We now wish to generalize the polynomial deformation formula for the Yukawa coupling [84] in a way that would let us compute Yukawa coupling when polynomial deformations do not represent $H^{2,1}(\mathcal{M})$ faithfully.

To that end, we will (1) translate the polynomial deformation formula into the tensorial representations' language and then also (2) provide a dual representation of the formula, partly for ease of computation but also because that will prove to have the most general form[§].

---

‡ The other contribution is proportional to $\det[\ddot{f}] = 0$; see Ref. [84], Eq. (4.34).

§ ⌈*I am grateful to Per Berglund for partly unpublished collaboration on these topics, some of which became published ([75, 318] in particular) after the initial printing of the Bestiary [272].*⌋

C.2.1 The Yukawa coupling synopsis

On writing Eq. (C.1.10) explicitly out, we obtain the rather bewildering expression

$$\mathcal{Q} = d_{a_1a_2a_3a_4}\, x^{a_1}\, x^{a_2}\, x^{a_3}\, x^{a_4}\, d_{b_1b_2b_3\,\beta_1\beta_2\beta_3}\, x^{b_1}\, x^{b_2}\, x^{b_3}\, y^{\beta_1}\, y^{\beta_2}\, y^{\beta_3}\ , \tag{C.2.1}$$

where we have abbreviated

$$d_{a_1a_2a_3a_4} \stackrel{\text{def}}{=} \varepsilon^{b_1b_2b_3b_4}\, \varepsilon^{c_1c_2c_3c_4}\, f_{a_1b_1c_1}\, f_{a_2b_2c_2}\, f_{a_3b_3c_3}\, f_{a_4b_4c_4}\ , \tag{C.2.2}$$

$$d_{b_1b_2b_3\,\beta_1\beta_2\beta_3} \stackrel{\text{def}}{=} \varepsilon^{\alpha_1\alpha_2\alpha_3}\, \varepsilon^{\gamma_1\gamma_2\gamma_3}\, g_{b_1\alpha_1\beta_1\gamma_1}\, g_{b_2\alpha_2\beta_2\gamma_2}\, g_{b_3\alpha_3\beta_3\gamma_3}\ , \tag{C.2.3}$$

up to some uninteresting numerical factors.

Note that $\mathcal{Q}$ contains precisely two of both the ε^{abcd}- and the $\varepsilon^{\alpha\beta\gamma}$-tensors. Recall the residue formula (A.3.12),

$$H^{3,0}(\mathcal{M}) \ni \Omega \stackrel{\text{def}}{=} \oint_{\Gamma(f)} \oint_{\Gamma(g)} \frac{\epsilon_{abcd}\, x^a \mathrm{d}x^b \mathrm{d}x^c \mathrm{d}x^d\ \epsilon_{\alpha\beta\gamma}\, y^\alpha \mathrm{d}y^\beta \mathrm{d}y^\gamma}{f(x) g(x,y)}\ . \tag{A.3.12$'$}$$

Its dual, $\overline{\Omega} \in H^{0,3}(\mathcal{M})$, is therefore represented by $\epsilon^{abcd}\epsilon^{\alpha\beta\gamma}$, as was also obtained by the Koszul computation (A.3.9). So, $\mathcal{Q}$ in fact represents the *square* of the dual of the holomorphic 3-form, Ω. Since the Yukawa coupling definition (0.5.7) contains precisely two Ω's,

$$\Big(\Omega{\cdot}\mathfrak{m}_\alpha{\cdot}\mathfrak{m}_\beta{\cdot}\mathfrak{m}_\gamma{\cdot}\Omega\Big) \longmapsto \Big(\Omega{\cdot}(\overline{\Omega}\,\kappa_{\alpha\beta\gamma}\,\overline{\Omega}){\cdot}\Omega\Big) \longmapsto \kappa_{\alpha\beta\gamma} \tag{C.2.4}$$

is the formal synopsis for the Yukawa coupling computation.

— ❦ —

An alternative and both more general and more rigorous description of this can be given as follows (see also Ref. [222, 442]). One observes that elements of $H^1(\mathcal{M}, \mathcal{T}_\mathcal{M})$ act on elements of $H^{n-1,1}(\mathcal{M}) = H^1(\mathcal{M}, \wedge^{n-1}\mathcal{T}^*_\mathcal{M})$ essentially by the natural contraction between $\mathcal{T}_\mathcal{M}$ and $\mathcal{T}^*_\mathcal{M}$ and by the wedge product:

$$H^1(\mathcal{M}, \mathcal{T}_\mathcal{M}) \ni \varphi\ :\ H^q(\mathcal{M}, \wedge^p\mathcal{T}^*_\mathcal{M}) \longrightarrow H^{q+1}(\mathcal{M}, \wedge^{p-1}\mathcal{T}^*_\mathcal{M})\ . \tag{C.2.5}$$

In meticulous tensorial notation, this reads

$$\begin{aligned}(\mathrm{d}z^{\overline{\mu}}\varphi_{\overline{\mu}}{}^\alpha\partial_a)\cdot(\mathrm{d}z^{\nu_1}\cdots\mathrm{d}z^{\nu_p}\,\mathrm{d}z^{\overline{\rho}_1}\cdots\mathrm{d}z^{\overline{\rho}_q}\,\omega_{\nu_1\cdots\nu_p\overline{\rho}_1\cdots\overline{\rho}_q}) &\longmapsto \\ \longmapsto \textstyle\sum_i(-)^{i+p}\delta_a^{\nu_i}\,\mathrm{d}z^{\nu_1}\cdots\mathrm{d}z^{\nu_{i-1}}\mathrm{d}z^{\nu_{i+1}}\cdots\mathrm{d}z^{\nu_p}\wedge& \\ \mathrm{d}z^{\overline{\mu}}\mathrm{d}z^{\overline{\rho}_1}\cdots\mathrm{d}z^{\overline{\rho}_q}\,(\varphi_{\overline{\mu}}{}^\alpha\,\omega_{\nu_1\cdots\nu_{i_1}\,\alpha\,\nu_{i+1}\cdots\nu_p\overline{\rho}_1\cdots\overline{\rho}_q})&\ ,\end{aligned} \tag{C.2.6}$$

where we have omitted the wedge products for brevity. This provides a map

$$H^1(\mathcal{M},\mathcal{T}_\mathcal{M}) \longrightarrow \mathrm{Hom}\big(H^q(\mathcal{M},\wedge^p\mathcal{T}^*_\mathcal{M}),\, H^{q+1}(\mathcal{M},\wedge^{p-1}\mathcal{T}^*_\mathcal{M})\big)\ , \qquad (C.2.7)$$

and recall that generally $\mathrm{Hom}(A,B) = B\otimes A^*$. Quite straightforwardly then,

$$\begin{aligned}\mathrm{Sym}^n H^1(\mathcal{M},\mathcal{T}_\mathcal{M}) &\longrightarrow \mathrm{Hom}\big(H^0(\mathcal{M},\wedge^n\mathcal{T}^*_\mathcal{M}),\, H^n(\mathcal{M},\mathbb{C})\big)\\ &= \mathrm{Hom}\big(H^{n,0}(\mathcal{M}),\, H^{0,n}(\mathcal{M})\big)\ .\end{aligned} \qquad (C.2.8)$$

Since $H^{n,0}(\mathcal{M})$ is naturally the dual of $H^{0,n}(\mathcal{M})$,

$$\mathrm{Sym}^n H^1(\mathcal{M},\mathcal{T}_\mathcal{M}) \longrightarrow \big(H^{0,n}(\mathcal{M})\big)^{\otimes 2}\ . \qquad (C.2.9)$$

This map is the Yukawa coupling (0.5.7).

Note: Should $H^{n-k,k} = 0$ for $k < m$ (and by conjugation also for $k > n-m$), the map

$$\mathrm{Sym}^{(n-m)} H^1(\mathcal{M},\mathcal{T}_\mathcal{M}) \longrightarrow \big(H^{0,n-m}(\mathcal{M})\big)^{\otimes 2}\ . \qquad (C.2.10)$$

still defines a (generalized) Yukawa coupling.

— ❦ —

C.2.2 Tensorial translation

Following the gist of Eq. (C.1.3), we form a triple product of the tensor representatives, for our sample model, $\varphi_{(abc)}$ and $\varphi_{a(\alpha\beta\gamma)}$. Then multiply this with the square of the tensor representative of $\Omega \in H^{3,0}$, in the present case, $\epsilon^{abcd}\,\epsilon^{\alpha\beta\gamma}$. Finally, we massage the so obtained tensor using the relations from the ideal until the product is brought into the form of $\mathcal{Q}$.

As the reduction modulo the ideal may become nerve-wrecking and time-consuming (without a clever computer program), we note that the same result may be obtained also in a *dual* computation, sometimes with more efficiency. To that end, consider the defining equations such as $f(x) = f_{abc}x^a x^b x^c$ and isolate their tensor coefficients, which we might call the defining tensors. Then simply swap their indices to obtain the dual defining tensors¶, such as

$$f^{(abc)} \overset{*}{\sim} f_{abc}\ , \qquad g^{a(\alpha\beta\gamma)} \overset{*}{\sim} g_{a(\alpha\beta g)}\ . \qquad (C.2.11)$$

¶ The acronym DDT readily comes to mind, but we should prefer to avoid the pun.

If $f_a(x)$ denote some generic defining polynomials, the dual defining tensors will be denoted by f^a; note the similarity to the variables used in § 9.3.1 to represent the duals of the line bundles of which the defining polynomials are sections.

Back to Eq. (C.2.3): the Yukawa coupling can now be written, with the aid of the dual defining tensors as

$$\kappa_{\alpha\beta\gamma} \quad = \quad \mathrm{Tr}_{\mathcal{X},f^a}\Big[\boldsymbol{\epsilon}^2 \cdot \boldsymbol{t}_\alpha \cdot \boldsymbol{t}_\beta \cdot \boldsymbol{t}_\gamma\Big] , \tag{C.2.12}$$

where $\boldsymbol{\epsilon}$ is the tensor representative of $\Omega \in H^{3,0}(\mathcal{M})$ and $\boldsymbol{t}_\alpha$ are the tensor representatives of $H^{2,1}(\mathcal{M})$ obtained by the Koszul spectral sequence computation. By "$\mathrm{Tr}_{\mathcal{X},f^a}$" we mean taking traces using the invariant tensors of the tensor algebra on $\mathcal{X}$ and the dual defining tensors. For example, the product

$$(\epsilon_{abcd}\epsilon_{\alpha\beta\gamma})\,(\epsilon_{efgh}\epsilon_{\delta\lambda\kappa})\;\phi_{(ijk)}\;\phi_{(lmn)}\;\varphi_{p(\mu\nu\rho)} \tag{C.2.13}$$

can only be contracted (completely!) with four $f^{(\cdots)}$'s and three $g^{\cdot(\cdots)}$'s. Of course, there are several ways to do so, but they are related by the symmetry properties of the tensors involved; this contributes to the numerical factor which we have so far neglected. However, when comparing two or more Yukawa couplings, the relative numerical factor is of course the measure of the relative strength of the two couplings.

The above formalism clearly generalizes to essentially arbitrary tensors the monomials of the polynomial deformations, which correspond to totally symmetric tensors of the same degree as the defining equations. Also, the formula depends only on the defining equations and as usual in tensor calculations and being a trace, it is coordinate independent. Note in particular that the number of the ϵ-tensors is precisely given in the definition of the Yukawa coupling and that their presence (and skew symmetry) restricts the possible contractions rather severely.

Of course, in models where polynomial deformations do faithfully represent $H^{2,1}$, the $\boldsymbol{t}$'s (that is, the tensor coefficients such as $\phi_{(abc)}$ and $\varphi_{a(\alpha\beta\gamma)}$) and the coordinate monomials $\boldsymbol{m}$'s may be used interchangeably. More generally, however, only the tensorial representation is going to represent $H^{2,1}$ faithfully and yield the Yukawa coupling for arbitrary complete intersections in products of flag spaces☿. For embeddings in more general spaces, the Koszul computation is of course valid, but it expresses the cohomology

☿ *While coordinate monomials are dual precisely to totally symmetric tensor coefficients, tensor coefficients with partial symmetry are determined iteratively, such as in (B.2.3)–(B.2.5), and so are dual to relations between "source" and "target" monomials. In 2018,*

on the complete intersection in terms of the cohomology on the embedding space. At that point, of course, it all hinges upon how well we know the latter one, which is the primary reason for embedding in projective spaces and some of their simple generalizations—their cohomology is best known.

— ❦ —

At first, Eq. (C.2.12) may seem *ad hoc*, but actually it is the only expression that has an invariant meaning and is constructed entirely from "invariant tensors". Since $f_a(x)$ vanish on the manifold, their tensor coefficients are special in the sense that they characterize the complete intersection manifold. Just as the tensor algebra on $\mathbb{P}^n$ has the Kronecker tensor as the only invariant tensor and the Levi-Civita ϵ-tensor as an "almost invariant," when we pass to a subspace $\mathcal{M}$, the tensor algebra on $\mathcal{M}$ will be inherited from that of the embedding space, modified however by the inclusion of the new "invariant tensors"—those corresponding to the defining equations.

In the example which we used throughout the Chapter, we consider the Calabi–Yau manifold

$$\mathcal{M} \in \left[\begin{matrix}3\\2\end{matrix}\middle\|\begin{matrix}3 & 1\\0 & 3\end{matrix}\right] , \tag{C.2.14}$$

that is, $\mathcal{M} \hookrightarrow \mathbb{P}^3 \times \mathbb{P}^2$ as the space of common solutions to a cubic constraint $f(x) = 0$ over $\mathbb{P}^3$ and a constraint $g(x, y) = 0$ of bi-degree (1,3) in $\mathbb{P}^3 \times \mathbb{P}^2$. While the tensor algebra over $\mathbb{P}^n$ is the $U(n+1)$-tensor algebra since $\mathbb{P}^n = \left\{\frac{U(n+1)}{U(1)\times U(n)}\right\}$, the only tensorial objects that carry information about the subspace $\mathcal{M} \subset \mathbb{P}^3 \times \mathbb{P}^2$ are the tensor coefficients of the defining polynomials $f(x)$ and $g(x, y)$.

C.3 The $(1,1)^3$ and Other Yukawa Couplings

Since the formula (C.2.12) is so general in view of the above explanation, we expect something similar to be valid also for other Yukawa couplings:

$$\mathring{\kappa}_{ABC} = \int_{\mathcal{M}} e_A \wedge e_B \wedge e_C , \tag{C.3.1}$$

Andre Lukas showed me a working Mathematica program that effectively computes the cohomology groups discussed herein in such a dual fashion. Similar, mechanized computation of the most frequently needed cohomology groups has been developed even for the much larger pool of toric embeddings [331].⌋

$$\mathring{\kappa}_{ijk} = \int_{\mathcal{M}} \Omega \wedge \mathrm{Tr}\Big(s_i \wedge s_j \wedge s_k\Big) \ , \tag{C.3.2}$$

$$\mathring{\kappa}_{\alpha jC} = \int_{\mathcal{M}} \Omega \wedge \mathrm{Tr}\Big(\varphi_\alpha \wedge s_j \wedge e_C\Big) \ . \tag{C.3.3}$$

Indeed, this is almost true. However, in $\mathring{\kappa}_{ABC}$, which we have already discussed in Chapter 8, and in $\mathring{\kappa}_{\alpha jC}$ there is some degree of freedom. Namely, the pull-backs of the Kähler forms appear as simple scalars in the Koszul computation. When our complete intersection is embedded in a single projective space, there is a single Kähler form, which is by the Lefschetz hyperplane theorem inherited by the complete intersection and this degree of freedom is simply an overall multiplicative factor. However, when the embedding space has more than one factor, this becomes more involved.

This is seen by translating into the tensorial language the $\mathring{\kappa}_{ABC}$ Yukawa coupling as computed in Chapter 8. For the model (C.2.14), the Koszul computation provides a basis

$$H^1(\mathcal{M}, \mathcal{T}^*_{\mathcal{M}}) \ = \ \{\, J_x,\ J_y,\ \epsilon^{abcd}\vartheta^{(ef)} \,\} \ , \tag{C.3.4}$$

where J_x and J_y are the (pull-backs of the) Kähler forms of $\mathbb{P}^3$ and $\mathbb{P}^2$ respectively. The trace-formula would allow the following couplings

$$(J_r, J_s, J_t) \ , \qquad (J_r, \epsilon^{0123}\vartheta^{(ab)}, \epsilon^{0123}\vartheta^{(cd)}) \ , \tag{C.3.5}$$

where $r, s, t = x, y$ unrestrictedly and a, b, c, d must all be different. The latter couplings are, in detail,

$$J_r \cdot f_{a_1b_1c_1} f_{a_2b_2c_2} f_{a_3b_3c_3} f_{a_4b_4c_4} (\epsilon^{a_1a_2a_3a_4}\,\vartheta^{(c_1c_2)})(\epsilon^{b_1b_2b_3b_4}\,\vartheta^{(c_3c_4)}) \tag{C.3.6}$$

for specific choices of the superscripts c_i—as dictated by the chosen form of the defining polynomial $f(x)$—and $r = x, y$. The various couplings for $r, s, t = x, y$ are all unrelated by any of the Koszul spectral sequence machinery as understood in the above Chapters.

On the other hand, the eight tensorial representatives of $H^1(\mathcal{M}, \mathcal{T}^*_{\mathcal{M}})$ are clearly in 1–1 correspondence with the 8-dimensional basis that we can define along the results in Eq. (8.5.10). In fact those results can be used directly, identifying [77]:

$$J_x \ = \ 3E_0 - \sum_{i=1}^{6} E_i \ , \qquad J_y \ = \ 3E_0' \ , \tag{C.3.7}$$

and

$$\begin{aligned}
\epsilon\,\phi^{(01)} &\longleftrightarrow -(E_1+\omega^2E_2+\omega E_3)\ ,\\
\epsilon\,\phi^{(02)} &\longleftrightarrow \quad (\omega^2E_4+\omega E_5+E_6)\ ,\\
\epsilon\,\phi^{(03)} &\longleftrightarrow \overline{\gamma}\left[H+\omega^2(E_1+E_2+E_3)+\omega(E_4+E_5+E_6)\right]\ ,\\
\epsilon\,\phi^{(12)} &\longleftrightarrow \gamma\left[H+\omega(E_1+E_2+E_3)+\omega^2(E_4+E_5+E_6)\right]\ ,\\
\epsilon\,\phi^{(13)} &\longleftrightarrow \quad (\omega E_4+\omega^2E_5+E_6)\ ,\\
\epsilon\,\phi^{(23)} &\longleftrightarrow -(E_1+\omega E_2+\omega^2E_3)\ .
\end{aligned} \tag{C.3.8}$$

Upon this identification, half of the couplings in (C.3.5) vanish and the remaining ones are all related in value to a single overall scale.

The vanishing of half of the couplings in (C.3.5) can be caused simply by *including* an additional requirement to the trace-formula (C.2.12). Note that J_x is in fact a double pull-back. First, it is inherited by the cubic surface in $\mathbb{P}^3$ and then by the Calabi–Yau 3-fold. Because of the fact that it is a pull-back of a (1,1)-form on the cubic 2-fold in $\mathbb{P}^3$, we have that J_x^3 vanishes identically. Moreover, the $\epsilon^{0123}\vartheta^{(ab)}$'s also correspond to (1,1)-forms inherited from this 2-fold, so that—upon this consideration—the only non-vanishing Yukawa couplings are

$$(J_x,J_x,J_y)\ ,\qquad (J_x,J_y,J_y)\ ,\qquad (J_y,\epsilon^{0123}\vartheta^{(ab)},\epsilon^{0123}\vartheta^{(cd)})\ . \tag{C.3.9}$$

Only the relative values within the third group of couplings can be determined by the structure so far revealed; the relative values with respect to the first two remain undetermined. Exactly the same freedom will occur in any other Yukawa coupling computation which involves a tensorial representative which in fact is a scalar. Generally, such representatives correspond to (pull-backs of) Kähler forms or other generators of the (r,r)-cohomology of the embedding space (flag spaces may contain higher such forms which are independent from the Kähler form).

Using also the representatives (B.3.36), (B.3.39) and (B.3.40), the couplings $\mathring{\kappa}_{ijk}$ in Eq. (C.3.2) and $\mathring{\kappa}_{\alpha jC}$ in (C.3.3) can be evaluated also, up to the same freedom in the $(1,1)$-cohomology (for a similar calculation, see Ref. [77]).

In this respect, the Koszul spectral sequence computation as presented above manifests the dependence on the choice of complex structure, through the dependence on the defining polynomials. In addition, it does feature some degree of freedom controlled by a choice in the $(1,1)$-cohomology (variations of the Kähler class). Also, it provides an evaluation of Yukawa couplings in the same tensorial language and therefore it allows a comparison of all four types of Yukawa couplings (0.5.7),

(0.5.5), (0.5.9) and (0.5.11), which is much more than any other available technique. Finally, a complete and detailed identification between these calculations and the Landau–Ginzburg orbifold counterparts [159, 227] seems to be possible [72].

C.4 Using Symmetries

Needless to say, when the space $\mathcal{M}$ exhibits symmetries, these can be employed to simplify the computation of the Yukawa couplings.

Given the tensorial representation of $H^1(\mathcal{M}, \mathcal{T}_{\mathcal{M}})$, $H^1(\mathcal{M}, \mathcal{T}^*_{\mathcal{M}})$ and also of $H^1(\mathcal{M}, \mathrm{End}\,\mathcal{T}_{\mathcal{M}})$, it is straightforward to deduce the lifted action of a symmetry $\varpi : \mathcal{M} \to \mathcal{M}$ to the cohomology. The Yukawa couplings must of course be invariant under any reparametrization of $\mathcal{M}$ and so also with respect to ϖ. This usually annihilates a good deal of the Yukawa couplings and relates many of them. In a way, this is just an application of the Wigner–Eckart theorem, where several different physically measurable quantities $X^{(I)}$ can be "reduced" to a common quantity

$$X^{(I)} = (\text{Clebsch–Gordan})^{(I)} \cdot X_{\text{reduced}} , \tag{C.4.1}$$

and the Clebsch–Gordan coefficients can be computed purely by symmetry considerations.

Also, one may consider constructing the complete intersection 3-fold one hypersurface at a time and computing the cohomology for every intermediate step. The advantage of doing so becomes apparent when we recall, once again, the example (C.2.14). Thinking of the Calabi–Yau 3-fold $\mathcal{M}$ as a hypersurface in $\Sigma \times \mathbb{P}^2$, where Σ is the cubic surface in $\mathbb{P}^3$, allows us to completely determine the required cohomology on Σ. Most of the properties of representatives of the $H^{\star,\star}(\Sigma)$ cohomology are retained as these representatives are inherited by $H^{\star,\star}(\mathcal{M})$. This can be traced by following the representatives carefully through the *Künneth formula*$^{(L)}$ for $\Sigma \times \mathbb{P}^2$ and using the *Lefschetz Hyperplane Theorem*$^{(L)}$ to deduce precisely which of these representatives are inherited by the 3-fold $\mathcal{M}$.

Finally, the *Lefschetz fixed point formulae*$^{(L)}$ can be used, for one of the generators of the various symmetries at a time, to decompose the cohomology groups of curves and 2-folds according to representations of the symmetry groups. Combining these results, the same decomposition can then be achieved for any complete intersection 3-fold which can be represented as a complete intersection in products of curves and surfaces. One of the main parts of Ref. [153] is precisely such a computation, pertaining to a particularly simple and very symmetric member of the so-called Tian–Yau family of Calabi–Yau 3-folds [223], represented by the configuration

$$\left[\begin{array}{c||ccc} 3 & 3 & 0 & 1 \\ 3 & 0 & 3 & 1 \end{array}\right] . \tag{C.4.2}$$

We refer the interested Reader to consult Ref. [153] for details of this calculation. In view of the rather more general method of Koszul spectral sequences, one may regard this as an independent verification of computations. Alternatively, of course, one may not be interested in the complete answer, so a simpler computational method may be preferable.

Part III
Changelings

Chapter D

Parameter Spaces: From Afar

In most parts of this volume, we have been discussing ways of constructing Calabi–Yau 3-folds (and in fact also spaces of other dimensions and types) and some of the most basic and general methods of computing the various numerical characteristics, guided in this latter task by a prospective physics application. In the process, it has become obvious that the number of topologically distinct Calabi–Yau 3-folds is rather immense. By their construction, they typically come in families and groups, and a general classification is beginning to take shape, albeit a very fuzzy one thus far.

D.1 The Programme

Before trying to describe what is known of these preliminary concepts of classification, it is only fair to say that the topic seems to be in an early stage of assembling a puzzle, without a templet picture and without a clue as to how many pieces there are. Some of the pieces however do fit together, rather nicely.

D.1.1 Identification numbers

Consulting any algebraic geometry text, one finds that curves are essentially classified by their genus (number of handles), which is equal to $(1-\frac{1}{2}\chi_E)$. For surfaces, the situation becomes more complicated and one is advised first of all to compute the Kodaira dimension* [23, 37]:

---

* The name "dimension" is in fact descriptive. There does exist an abstractly defined space, called however the Iitaka variety, the dimension of which equals the Kodaira dimension; for details, see Ref. [37, pp. 247–248].

Definition D.1. *For a compact complex n-fold $\mathcal{X}$ with plurigenera*$^{(L)}$ $\mathfrak{P}_m$, *the Kodaira dimension is defined to be:*

$-\infty$ *(or -1), if $\mathfrak{P}_m = 0$ for all $m > 0$;*
$\mathbf{0}$, *if $0 < \mathfrak{P}_m < \dim \mathcal{X}$;*
$\boldsymbol{k}$, *if $m^{-k}\mathfrak{P}_m$ is bounded and $0 < k < n$ is the smallest such integer;*
$\boldsymbol{n}$, *in all other cases.*

There are no other types (✎ Prove this. ❧).
All Calabi–Yau n-folds have vanishing Kodaira dimension.

— ❦ —

Next, (for vanishing Kodaira dimension) we are advised to check *regularity*, measured by $h^{1,0}$; this too vanishes for Calabi–Yau 3-folds (except if we agree to include complex 3-tori, the product of a complex torus and a K3 and the generalizations thereof, so called *abelian varieties*).

Finally, there are several numerical characteristics such as the Hodge numbers, which feature in Wall's classification Theorem 8.1, wherein a broad class of 3-folds, including Calabi–Yau ones, are classified up to diffeomorphism. We must caution here that a serious misunderstanding might arise between Calabi–Yau practitioners of mathematical *vs.* physics persuasion. A collection of valuable numerical characteristics with which to identify a Calabi–Yau model is supplied with the Yukawa couplings (see Chapters 0, 8 and C). However, there are "classical" and "quantum" Yukawa couplings, which are supposed to agree in a certain limit, but might be drastically different otherwise. In particular, the "classical" $(1,1)^3$ Yukawa couplings appear in Wall's theorem and are purely topological. The "quantum" $(1,1)^3$ Yukawa couplings include also a quantum correction term which depends on the mesh of rational curves in the manifold and the choice of the Kähler class and are therefore far from topological in the usual sense.

This being a bestiary *for physicists*, we will tip the balance in that direction when in a dilemma. Fortunately, I believe, there will always exist a limit in which the two points of view will agree and a bridge is possible.

D.1.2 Cartography or genealogy?

By simply naming examples and attaching a string of numerical characteristics not much more is done, however, than assembling a telephone directory.

Even such a simple-minded task is far from completion: computer lists of models have been compiled but it is unclear how to unambiguously tell one model from another since it is not in general known precisely what numerical characteristics identify a model.

A more unifying description would probably be universally agreed to be preferable. Amongst mathematicians, there are in general rather different points of view, ranging from "teleological" [216] to "biological" [202]. Physicists have of course an altogether different point of view, in that Calabi–Yau 3-folds occur only as a sector and are fitted into superstring theory models. Yet, perhaps surprisingly, some conclusions and conjectures will jibe.

— ❦ —

First and foremost, Calabi–Yau spaces come bundled not only with a string of numerical characteristics (not all of which do we always know how to obtain), but also with a number of continuous parameters. Clearly, one generally tries to model these in the image of the far better understood *moduli spaces* of algebraic curves and surfaces.

As it has perhaps by now become obvious, by a Calabi–Yau 3-fold we *usually* mean a compact real 6-dimensional space endowed with a complex structure and a *complexified* Kähler class. Some examples in string theory suggest that the requirement of Kählerness might be relaxed although it is not clear what is precisely the broader class which is to replace this; the *Hodge decomposition*$^{(L)}$ of cohomology is however supposed to persist in some suitable sense. Unless otherwise stated, we will assume that the Calabi–Yau 3-fold is Kähler. In addition, most constructions and analyses never really leave the holomorphic category and one is actually talking about algebraic varieties over the field of complex numbers.

Finally, the complexification of the Kähler class happens owing to requirements of supersymmetry; the action (0.1.2) which describes the propagation of superstrings contains not only the hermitian metric $G_{\mu\overline{\nu}}$ on our 3-fold but also an antihermitian 2-form $B_{\mu\overline{\nu}}$. Although they occur in the antihermitian combination $(B+iG)$ and the conjugate thereof, for certain purposes, the combinations $(B\pm G)$ might be used [94]. If $G_{\mu\overline{\nu}}$ is a Kähler metric, $iG_{\mu\overline{\nu}}\mathrm{d}z^{\mu}\,\mathrm{d}z^{\overline{\nu}}$ represents the corresponding Kähler class, denoted J and so $\not{G} \stackrel{\mathrm{def}}{=} (B-G) = (B+iJ)$, which is the complex variable usually taken to span the complexified Kähler cone.

In any case, therefore, there are two parameter spaces associated to a (topological class of) Calabi–Yau 3-fold(s) $\mathcal{M}$, the space of complex structures, which we denote $\mathfrak{M}$ and the complexified space of Kähler classes, which we denote by $\mathfrak{W}$. Now, quite generally, one first chooses a complex structure and is only then able to talk about (1,1)-cohomology classes. This would imply that the two parameter spaces are related non-trivially, that is that even the definition of $\mathfrak{W}$ depends on where in $\mathfrak{M}$ one is. However, because there are no non-trivial (2,0)- and (0,2)-cohomology classes, all of the degree-2 cohomology, $H^2(\mathcal{M}, \Bbbk)$, can be identified with the (1,1)-cohomology. This makes $\mathfrak{W}$ fairly independent of $\mathfrak{M}$, although not absolutely so; at certain special regions of $\mathfrak{M}$, the space $\mathfrak{W}$ does undergo changes, tied to certain subtle change in the 3-fold parametrized [235]. Also, more drastically, at special regions where the parametrized 3-fold undergoes certain degenerations, $\mathfrak{W}$ changes rather drastically. We will give examples of this behavior below. In view of this, by the parameter space of a (topological class of a) Calabi–Yau 3-fold we will mean the pair $(\mathfrak{M}, \mathfrak{W})$, with their dependence implied.

It becomes obvious then that the parameter space $(\mathfrak{M}, \mathfrak{W})$ is a smooth complex space at its generic points but we expect some degree of roughness in lower-dimensional subsets. In particular, we certainly expect $\mathcal{M}$ to have finite quotient singularities since there certainly exist Calabi–Yau 3-folds in every topological class which have rather special (finite) symmetries. This is very much alike the $\mathbb{Z}_2$- and $\mathbb{Z}_3$-symmetric points in the moduli space of the torus.

— ❦ —

One striking result is that the parameter spaces of topologically distinct Calabi–Yau spaces do touch along certain special regions, which we loosely speaking call "boundary" although the dimension of these regions is at least one (complex!) dimension smaller than the generic part. Such 'interface' regions correspond to certain singular limits of the respective Calabi–Yau 3-folds. By including these limit points, the parameter spaces of a large number of simply connected Calabi–Yau spaces are explicitly shown to join together to form a connected 'web' [149, 90, 92]. This resonates strongly with Reid's conjecture [203] †, which in its full span still remains a conjecture; see however also [305, 430, 450], and more recently [381].

We will presently describe some of this *global* picture and will turn to some preliminary description of the local (and metric) properties of $(\mathfrak{M}, \mathfrak{W})$ in the next Chapter. The Reader should bare in mind that this is still a very

† Reid calls the so morphing Calabi–Yau 3-folds "rakshasa," which named this part of the book.

rapidly developing field and the present account is only meant as a rough sketch. It will transpire, I hope, that the conglomeration of Calabi–Yau 3-folds, tentatively and partially aligned by the web of Ref. [149], do seem to be organizable into an orderly chart, which moreover appears to take a hierarchical (genealogical?) structure. Finally, I believe that this structure will play a rôle not only in the mathematical classification but will also have a physical interpretation, perhaps not too far from the ideas in Ref. [149, 90, 92].

D.2 Splitting: New Manifolds for Old

As described so far, Calabi–Yau spaces may be constructed in a variety of ways. The moduli spaces of complete intersections in products of complex projective spaces have been explicitly shown to be linked together [149, 92]. This was done utilizing the process of *splitting* [85, 149] and we now turn to describe this.

We first consider a simple example to illustrate how it is possible to move continuously from the moduli space of one Calabi–Yau manifold to the moduli space of another.

$$\check{\mathcal{M}} \in \begin{bmatrix} 5 \\ 1 \end{bmatrix}\!\!\begin{Vmatrix} 4 & 1 & 1 \\ 0 & 1 & 1 \end{Vmatrix}^{2,86}_{-168}, \qquad \begin{cases} h^{1,1}=2,\ h^{2,1}=86, \\ \chi_{E}=-168, \end{cases} \tag{D.2.1}$$

for which the defining equations, encoded by the columns of the matrix (D.2.1), may without loss of generality be written as

$$\begin{aligned} Q(x) &= 0 \ , \\ x^1y^1 + x^2y^2 &= 0 \ , \\ x^3y^1 + x^4y^2 &= 0 \ . \end{aligned} \tag{D.2.2}$$

A relatively simple choice of transverse polynomials is obtained by choosing, in addition to the bi-linear constraints, $Q(x) = \sum_{a=1}^{6}(x^a)^4$. We may also without loss of generality think of this example as described by the configuration

$$\check{\mathcal{M}} \in \begin{bmatrix} \mathfrak{X} \\ \mathbb{P}^1 \end{bmatrix}\!\!\begin{Vmatrix} 1 & 1 \\ 1 & 1 \end{Vmatrix} \ , \qquad \mathfrak{X} \in [5\|4] \ , \tag{D.2.3}$$

that is, $\mathfrak{X}$ is the quartic 4-fold in $\mathbb{P}^5$ defined by $Q(x) = 0$.

For any particular $x \in \mathfrak{X}$, we are left with a system of two equations in two variables, y^i. Being homogeneous coordinates of a $\mathbb{P}^1$, y^1 and y^2

cannot both vanish, it follows that the determinant of the coefficients must vanish:

$$\mathcal{C}(x) \stackrel{\text{def}}{=} x^1x^4 - x^2x^3 = 0 \ . \tag{D.2.4}$$

The quadric hypersurface $\mathcal{M}^\sharp \subset \mathfrak{X}$, defined by $\mathcal{C}(x) = 0$, is clearly a Calabi–Yau 3-fold and may be regarded as the projection $\mathcal{M}^\sharp = \wp(\check{\mathcal{M}})$ along the $\mathbb{P}^1$ in (D.2.2). Alternatively, we may identify $\check{\mathcal{M}} = \lim_{R\to 0}\check{\mathcal{M}}$, where R is the average radius of the $\mathbb{P}^1$. Clearly, being an intersection of the quartic $Q(x) = 0$ and the quadric $\mathcal{C}(x) = 0$ in $\mathbb{P}^5$, $\mathcal{M}^\sharp$ is singular at $x^1, x^2, x^3, x^4 = 0$ as there $Q(x)$ and $\mathcal{C}(x)$, but also $\mathrm{d}\mathcal{C}(x)$ vanish. Since $Q(x)$ is a (generic) quadric, this happens at four points. For the simple choice $Q(x) = \sum_{i=1}^{6}(x^i)^4$, the singular points are at $(0{:}0{:}0{:}0{:}\omega{:}1)$, with $\omega^4 = -1$. Note that singularities induced from such a projection occur at points and are nodes (recall § 4.4.1 and 4.4.2); this follows from the smoothness of $\check{\mathcal{M}}$ and Bertini's theorem, and can also easily be checked.

To smooth out $\mathcal{M}^\sharp$, we perturb the defining equation (D.2.4) of $\mathcal{M}^\sharp \subset \mathfrak{X}$ and obtain

$$\mathcal{C}(x) = r^2(x) \ . \tag{D.2.5}$$

Here, $r^2(x)$ is a quadric perturbation, non-zero at the nodes of $\mathcal{M}^\sharp$, such as $r^2(x) = t\big(\sum_{a=1}^{6}(x^a)^2\big)$. For any small $t \neq 0$, this defines a non-singular quadric $\mathcal{M}^\flat{}_t$ in $\mathfrak{X}$ in $\mathbb{P}^5$.

Rather explicitly, the conifold[‡] $\mathcal{M}^\sharp$ occurs at a common point in the boundary of the relevant parameter spaces of non-singular 3-folds such as $\check{\mathcal{M}}$ and smooth 3-folds such as $\mathcal{M}^\flat{}_t \subset \mathfrak{X}$. Alternatively, the conifold $\mathcal{M}^\sharp$ can be smoothed in two topologically distinct ways: by a *small resolution* into $\check{\mathcal{M}}$, or by a *deformation* into $\mathcal{M}^\flat$. That is,

$$\begin{array}{ccc} & & \check{\mathcal{M}} \\ & & \big\downarrow \wp \\ \mathcal{M}^\flat & \xrightarrow{r^2\to 0} & \mathcal{M}^\sharp \end{array} \ , \qquad \begin{array}{ccc} & & \check{\mathcal{M}} \\ & & \big\uparrow \text{small resol.} \\ \mathcal{M}^\flat & \xleftarrow{\text{deform.}} & \mathcal{M}^\sharp \end{array} \tag{D.2.6}$$

Superficially, it may appear that the present situation is an isolated case. Not so; it exemplifies an ubiquitous process [149] which connects

‡ Following Ref. [92], we use this name for a space which has only isolated conical singularities. For all of these, the square-integrable cohomology (in, say, the projectively induced metric) is dual to intersection homology, which simplifies explicit calculations; see however also [250].

parameter spaces of a great many Calabi–Yau spaces and which—suitably generalized—may in fact link the moduli spaces of all simply connected Calabi–Yau manifolds. In this large class (described below) there always exist conifolds which occur as limit points (in the two different senses) in the two respective parameter spaces. These conifolds have isolated nodes, each described locally by an equation of the type (D.2.4):

$$XY - UV = 0 \; . \tag{D.2.7}$$

The quantities X, Y, U, V (generally polynomials in the homogeneous coordinates on the embedding $\mathbb{P}^n$'s) may be regarded as local coordinates, whence they describe a cone with apex at the origin, as discussed in § 4.4. Note that the conifold $\mathcal{M}^\sharp$ is in fact defined by such a determinantal equation also *globally*, inside $\mathfrak{X}$.

D.2.1 Nodes: locally and what to do with them

To analyze the local properties of nodes, we consider an open neighborhood of $p^\sharp \in \mathcal{M}^\sharp \subset \mathfrak{X}$ around each node. The neighborhood is locally like $\mathbb{C}^4$ and the four polynomials X, Y, U, V serve as local coordinates. In § 4.4 we saw that this cone can be viewed either as a complex cone over $\mathbb{P}^1 \times \mathbb{P}^1$, or a real cone over $S^2 \times S^3$.

Next, we compare the two distinct ways of smoothing a node, by following the two processes $\mathcal{M}^\sharp \to \mathcal{M}^\flat$ and $\mathcal{M}^\sharp \to \check{\mathcal{M}}$, respectively.

To begin with, consider the deformation introduced in Eq. (D.2.5). Locally at the nodes, we can treat r^2 as a parameter independent of U, V, X, Y. This is easily justified by considering the above simple choices: For $r^2(x) = t\big(\sum_{a=1}^{6}(x^a)^5\big)$ at the 4 nodes $(0, 0, e^{2i\pi(2k+1)/8}, 1)$ and with $k = 0, 1, 2, 3$, we have $r^2 = -t(1 \pm i)$, clearly non-zero for $t \neq 0$. A node $p^\sharp \in \mathcal{M}^\sharp$ then corresponds to $r = 0$ in

$$\vec{\xi}^2 = \vec{\eta}^2 = r^2 \; , \qquad \vec{\xi} \cdot \vec{\eta} = 0 \; , \tag{4.4.2$'$}$$

where a local and holomorphic change of variables has been performed. In the $\mathcal{M}^\sharp \to \mathcal{M}^\flat$ process, each node is being "inflated" into an S^3 described by

$$\vec{\xi}^2 = r^2 \; ; \quad \vec{\eta} = 0 \; . \tag{D.2.8}$$

In § 4.4, we have described a different way of smoothing, when each node $p^\sharp$ becomes replaced by a $\mathbb{P}^1$, aligned with one or the other $\mathbb{P}^1$ in the base of the complex cone. That this is indeed what happens when replacing

$\mathcal{M}^\sharp$ with $\check{\mathcal{M}}$, is easily seen by analyzing what happens to the y's from Eq. (D.2.2). Since $x^1x^4 - x^2x^3 = 0$, the matrix

$$\begin{bmatrix} x^1 & x^2 \\ x^3 & x^4 \end{bmatrix} \tag{D.2.9}$$

has rank 1, except at a node where $x^1, x^2, x^3, x^4 = 0$ and the rank is 0. Generically, therefore, the equations fix y^1/y^2 uniquely and hence specify a point in $\mathbb{P}^1$. In contrast, y^1 and y^2 remain undetermined when $x^1, x^2, x^3, x^4 = 0$ and $\check{\mathcal{M}}$ contains a copy of the embedding $\mathbb{P}^1$ there. When we project along the embedding $\mathbb{P}^1$, this entire copy of $\mathbb{P}^1$ is collapsed to a node of $\mathcal{M}^\sharp$. This is depicted in Figure 6.3.

So, in the $\mathcal{M}^\sharp \to \check{\mathcal{M}}$ process, each node $p^\sharp$ is resolved into a copy of $\mathbb{P}^1 = S^2$ generated by $y^1/y^2 = -x^2/x^1 = -x^4/x^3$. This indeed is the small resolution, the present simplicity of which is special to three dimensions [2]. These two processes are depicted schematically in Figure D.1.

D.2.2 The flop

The careful Reader may have complained to first solving $\vec{\xi}^2 = r^2$ from Eq. (4.4.2) which produced the "$S^3_{\vec{\xi}}$," spanned by the $\vec{\xi}$. Starting from $\vec{\eta}^2 = r^2$ instead, we find an "$S^3_{\vec{\eta}}$". So, with which of these two S^3's do we "align" the S^3 introduced in the deformation $\mathcal{M}^\sharp \to \mathcal{M}^\flat$? In fact, there is a continuous set of choices, parametrized by an angle:

$$\vec{\xi}(\theta) \stackrel{\text{def}}{=} \vec{\xi}\cos\theta + \vec{\eta}\sin\theta\ , \quad \vec{\eta}(\theta) \stackrel{\text{def}}{=} -\vec{\xi}\sin\theta + \vec{\eta}\cos\theta\ . \tag{D.2.10}$$

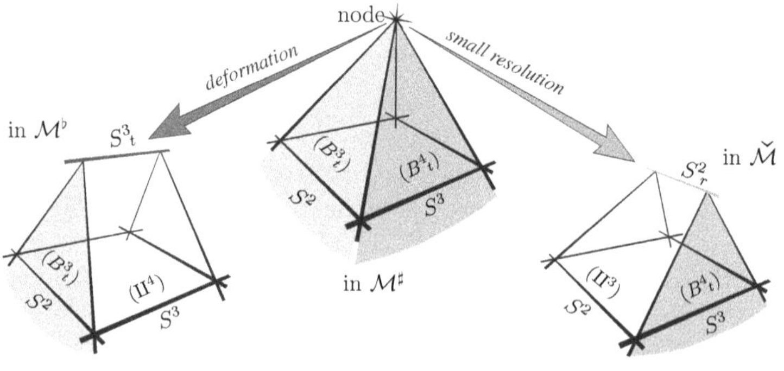

Figure D.1: Local transformation of cycles near a node under a smoothing deformation *vs.* a small resolution.

Since $\vec{\xi}(\theta)$ and $\vec{\eta}(\theta)$ satisfy Eq. (4.4.2), Eq. (D.2.10) establish a (homotopy) equivalence between all these choices, including "S^3_ξ" ($\theta = 0$) and "S^3_η" ($\theta = \pi$).

As for small resolutions, however, no such equivalence exists and a topological ambiguity does persist, as described in general in § 4.4. Here, this ambiguity is seen explicitly. Each node is resolved into a $\mathbb{P}^1$ generated by $y^1/y^2 = -x^2/x^1 = -x^4/x^3$. Instead, a node could be resolved into a $\mathbb{P}^1$ parametrized by $y^1/y^2 = -x^3/x^1 = -x^4/x^2$ and this obviously requires the transposition

$$\begin{bmatrix} x^1 & x^2 \\ x^3 & x^4 \end{bmatrix} \to \begin{bmatrix} x^1 & x^3 \\ x^2 & x^4 \end{bmatrix} . \tag{D.2.11}$$

This could be done at each of the nodes independently. In a way, the *global* construction $\wp\colon \check{\mathcal{M}} \to \mathcal{M}^\sharp$ provides a "ready made" Kähler small resolution of the conifold $\mathcal{M}^\sharp$, answering positively the otherwise (in general) rather difficult question, whether any of the 2^N small resolutions are Kähler.

In the present example, we can flop all four exceptional $\mathbb{P}^1$'s, by relabeling $x^2 \leftrightarrow x^3$ *globally*. In principle, the result of the above transposition, if possible (especially in other examples, when the elements are other than linear in the coordinates), yields a new 3-fold: the global *flop* of $\check{\mathcal{M}}$, which we might denote by $\widehat{\mathcal{M}}$. In the present case, however, we can also relate the two by deformation: Bertini's theorem tells us that the parameter space of smooth quartics is connected, so that, whatever our choice of the quartic $Q(x)$, $Q(x)|_{x^2 \leftrightarrow x^3}$ can be deformed back into $Q(x)$ smoothly.

D.2.3 Transvestite cycles

Let us now follow the change of some numerical characteristics, using the above analysis. In the process $\check{\mathcal{M}} \to \mathcal{M}^\flat$, we replace an S^2 by and S^3 for each of the N nodes in $\mathcal{M}^\sharp$. Knowing that $\chi_{_E}(S^3) = 0$ and $\chi_{_E}(S^2) = 2$, we have

$$\chi_{_E}(\check{\mathcal{M}}) = \chi_{_E}(\mathcal{M}^\flat) + 2N . \tag{D.2.12}$$

Consider now the diagram in Figure D.1. It represents the local transformation occurring at the neighborhood containing a minimal $S^3{}_t \subset \mathcal{M}^\flat$, which is being crushed into a node of $\mathcal{M}^\sharp$ by deformation and then replaced by an $S^2 \subset \mathcal{M}$ by small resolution. In view of the general discussion of small resolutions in § 4.4, for all A, D, E type singularities the transformation is merely an iteration of the small resolution part.

Now, in $\mathcal{M}^\flat$, the ($\mathbb{R}_+$-like) generators of the cone form a copy of $B^3{}_t$ by passing through the points of the S^2 in the base at every fixed point of the minimal $S^3{}_t$, drawn in Figure D.1 as the crest of the "tent". This is suggested by the shaded area. This $B^3{}_t$ is then completed, possibly into a cycle, elsewhere in $\mathcal{M}^\flat$, outside of this neighborhood. Note that the analogous construction obtained by collecting all ($\mathbb{R}_+$-like) generators passing through the points of the S^3 in the base while keeping a point of the S^2 in the base fixed yields a "4-hoop" $\amalg^4 \approx S^3 \times \mathbb{R}_+$, the boundary of which is the minimal $S^3{}_t$. This clearly is not a cycle for it has a boundary, but will become important as we progress towards the right hand side of Figure D.1. Similarly, the S^2 in the boundary is not a cycle because it can be contracted (along the $\mathbb{R}_+$-like generators of the cone) to a point (on $S^3{}_t$).

It should be obvious that precisely if $B^3{}_t$ becomes completed (away from $S^3{}_t$ in $\mathcal{M}^\flat$) into a 3-cycle, $C^{(3)}$, $S^3{}_t$ is homotopically non-trivial and the cycle that $S^3{}_t$ represents is dual to $C^{(3)}$. Each can be "propagated" along the other one and they meet in a single point although, of course, $S^3{}_t$ may develop into some different 3-space as it is propagated along a representative of $C^{(3)}$. This is very much like in the case of Riemann surfaces.

Now deform $\mathcal{M}^\flat$ into $\mathcal{M}^\sharp$: both S^3 and S^2 in the base can be contracted to the vertex of the cone and therefore fail to represent non-trivial 3- and 2-cycles respectively. However, nothing has happened to the cycle represented by the 3-space the portion of which we see in Figure D.1 as $B^3{}_t$; if this was a non-trivial cycle in $\mathcal{M}^\flat$, it remains to be so in a 'smooth homology' of $\mathcal{M}^\sharp$. However, using Theorem 7.3, we see that it fails to be a non-trivial element of $IH_3(\mathcal{M}^\sharp)$, since it becomes an open interval when the singular point is removed and is therefore contractible (outside of this neighborhood). Now, as for the "4-hoop" $\amalg^4$, we see that its boundary, $S^3{}_t$, has shrank to a point; this was the only local obstruction for it to be a portion of a representative of a 4-cycle. If there is no obstruction outside of this neighborhood, we have just obtained a new 4-cycle. Note that, also by Theorem 7.3, this 4-cycle survives as an element of $IH_4(\mathcal{M}^\sharp)$ because of the different treatment of $IH_3(\mathcal{M}^\sharp)$ and $IH_k(\mathcal{M}^\sharp)$ for $k \neq 3$. Also, there is a dual element of $IH_2(\mathcal{M}^\sharp)$ even though there is no contribution to the 'smooth' $H_2(\mathcal{M}^\sharp)$ from this neighborhood.

Finally, we pass to the small resolution, replacing the node with the minimal $S^2{}_r$, drawn as the crest of the "tent"-like neighborhood in the right hand side in Figure D.1. Now $B^3{}_t$ develops a hole—$S^2{}_r$ itself—and becomes a "3-hoop" $\amalg^3 \approx S^2 \times \mathbb{R}_+$, representing therefore the trivial 3-cycle. Each

S^3 in the base is homologous to a point on $S^2{}_r$ and hence homologically trivial also; there is no local 3-homology left. On the other hand, each S^2 in the base (forming a family parametrized by the S^3) is homologous to the $S^2{}_r$ in the crest and is locally non-trivial as an element of $H_2(\mathcal{M})$. Similarly, the "4-hoop" $\amalg^4$, the boundary of which, $S^3{}_t$, has been contracted into a point is possibly a portion of a representative of a non-trivial 4-cycle, $C^{(4)}$. Indeed, precisely if it is non-trivial—then $S^2{}_r$ is a local representative of a non-trivial 2-cycle and these two are dual to each other.

D.2.4 Three deformations of four nodes

When running the process Eq. (D.2.2)–(D.2.4)–(D.2.5) backwards, $\mathcal{M}^\flat \xrightarrow{t\to 0} \mathcal{M}^\sharp \xrightarrow{\wp^{-1}} \check{\mathcal{M}}$, we easily count four nodes in $\mathcal{M}^\sharp$ as four solutions for (x_5/x_6) of $Q(x) = 0$ at $x_1 = x_2 = x_3 = x_4 = 0$. Therefore, there must be four $S^3{}_{t^\alpha}$'s in $\mathcal{M}^\flat$ which are being crushed into nodes during the $\mathcal{M}^\flat \to \mathcal{M}^\sharp$ deformation. It is, however, an elementary fact that there are only three independent quadric polynomials in two variables. It follows that there must be precisely one relation between the four $S^3{}_{t^\alpha}$'s and can be represented as a 4-chain, Z^4 (completing the $\amalg^4$-hoops in $\mathcal{M}^\flat$), the boundary of which is precisely the union of these four 3-spheres. Also, it cannot possibly be a boundary of something else, since then the boundary of a boundary would not be empty. So, the four minimal 3-spheres $S^3{}_{t^\alpha}$ are *the only obstruction* for the "4-chain" Z^4 to become a 4-cycle!

As we pass to $\mathcal{M}^\sharp$, it is obvious that $\amalg^4$ becomes a 4-cycle and the Theorem 7.3 guarantees that there is also a dual element in $IH_2(\mathcal{M}^\sharp)$, although an explicit representative is hard to find. When we pass to the small resolution, $\check{\mathcal{M}}$, essentially nothing happens to this 4-cycle. The introduction of the four $S^2{}_{t^\alpha}$, however provides explicit representatives for the 2-cycle dual to the 4-cycle represented by $\amalg^4$ (with the $S^3{}_{t^\alpha}$ boundaries contracted to points). Moreover, the four S^2's are in fact homologous to each other by "propagation" along $\amalg^4$. Figure D.2 might perhaps clarify the situation.

Table D.1 displays the effect on the cohomology groups. In $k = 3$, the $3+3$ denote the three independent $S^3{}_{t^\alpha}$'s plus their three dual cycles, so that $0+3$ means that while the three $S^3{}_{t^\alpha}$'s have become trivial, their dual cycles are still non-trivial in the 'smooth homology' group $H_k(\mathcal{M}^\sharp)$. In $IH_k(\mathcal{M}^\sharp)$, however, neither of these 3-spaces represents a non-trivial cycle, whence the "$0+0$" entry. Throughout, $x = 1$ and $y = 174$ count cycles that do not depend on the four described neighborhoods.

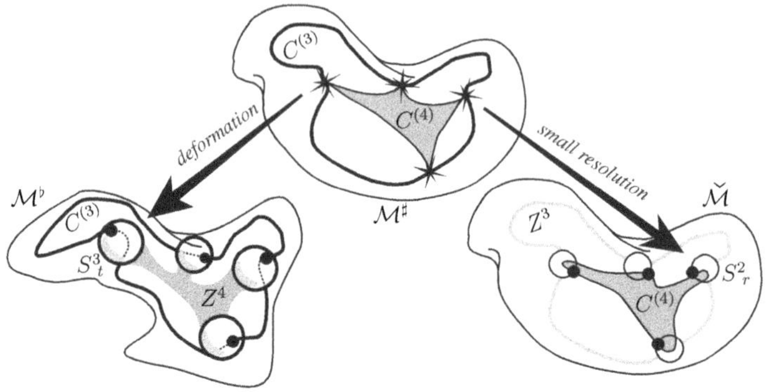

Figure D.2: Global transformation of cycles in a nodal (singular) deformation which is then followed by a small resolution.

Table D.1: The effect of "splitting" on homology.

k	$H_k(\mathcal{M}^\flat)$	$H_k(\mathcal{M}^\sharp)$	$IH_k(\mathcal{M}^\sharp)$	$H_k(\mathcal{M})$	2024 $\lceil SH_k(\mathcal{M})^\ddagger \rfloor$
0	1	1	1	1	1
1	0	0	0	0	0
2	x	x	$x+1$	$x+1$	$x+1$
3	$y+3+3$	$y+0+3$	$y+0+0$	$y+0+0$	$y+3+3$
4	x	$x+1$	$x+1$	$x+1$	$x+1$
5	0	0	0	0	0
6	1	1	1	1	1

‡ 2024 ⌈*The right-most column is the combined homology [250], complementing* IH_*⌋

The Reader must have noted the fact that $IH_k(\mathcal{M}^\sharp) = H_k(\mathcal{M})$. Indeed, this is just an explicitly computed consequence of Theorem 7.4. Finally, because the L^2-cohomology and the intersection homology are dual§ for the case of conical singularities, the same relation will hold for cohomology: $H^{p,q}_{(2)}(\mathcal{M}^\sharp) = H^{p,q}(\check{\mathcal{M}})$. In view of Theorem 7.4, the same will be true if instead of nodes, other A, D, E-singularities had appeared.

D.2.5 101 deformations (of 125 nodes)

Consider now the Calabi–Yau manifold embedded in $\mathbb{P}^4$ by means of

$$\mathcal{Q}(x;\psi) \stackrel{\text{def}}{=} \sum_{i=1}^{5} {x_i}^5 - 5\psi \prod_{i=1}^{5} x_i = 0\,. \tag{D.2.13}$$

§ With a suitable class of norms (metrics) for the L^2-cohomology.

At $\psi=1$, this quintic has 125 nodes of the type $(x_1:x_2:x_3:x_4:x_5) = (\alpha:\beta:\gamma:\delta:\omega)$, where $\alpha^5=\beta^5=\gamma^5=\delta^5=\omega^5=1$. Using projectivity, we may set $\omega=1$; for the gradients of $\mathcal{Q}$ to vanish (condition for singularity), it suffices to set $\alpha\beta\gamma\delta=1$, which fixes, e.g., δ. We are thus left with three fifth roots of unity, α, β, γ, chosen freely—hence $5^3=125$ nodes. One of these is (1:1:1:1:1).

Now, similarly to the case above, we can deform the singular quintic $\mathcal{Q}(x;1)$ by adding arbitrary quintic monomials. The trouble is, there are only 101 independent quintic monomials (modulo linear reparametrizations). On the other hand, there are 125 nodes, each being deformed into an ${S^3}_{t^\alpha}$. These 125 S^3's however can represent only 101 independent cycles and there must exist precisely 24 relations between them, each represented by a 4-chain Z^4_a, $a=1,\ldots,24$. At the conifold, $\psi = 1$, these 4-chains become 4-cycles, $C^{(4)}_a$, and give rise to 24 new elements of IH_4 and also IH_2, by duality. Indeed, Ref. [211] reports $\dim H^{1,1} = 25$ for the small resolution of this conifold with 125 nodes (1 generator for the Kähler class and 24 associated to $C^{(4)}_a$-type cycles).

D.3 A View into Moduli Spaces

Towards our promise in the first section of this Chapter: we will reinterpret the processes (D.2.6) as operations on the respective parameter spaces. To that end, recall that as discussed above, Calabi–Yau compactification is specified by giving a Calabi–Yau manifold $\mathcal{M}$ with a complex structure and (complexified) Kähler class. Thus, there are two distinct types of deformation of this structure; we first study, in turn, the parameter spaces for these.

D.3.1 Complex structure moduli

From the general discussion in sections 4.3 and 4.4, we know that $(\check{\mathcal{M}}-E) = (\mathcal{M}^\sharp - \{p^\sharp\})$, where E is the collection of 2-spheres into which the nodes $p^\sharp$ have been resolved. As there are only finitely many nodes and small resolutions and since the defining polynomial(s) of $\mathcal{M}^\sharp$ depend(s) holomorphically on those defining $\check{\mathcal{M}}$, we can locally identify the parameters of $\check{\mathcal{M}}$ with those of $\mathcal{M}^\sharp$, that is, $\check{\mathfrak{M}} \sim \mathfrak{M}^\sharp$. In fact, we can identify

$$\check{\mathfrak{M}} \xrightarrow{\sim} \mathfrak{M}^\sharp = \lim_{t\to 0} \mathfrak{M}^\flat \subset {\mathfrak{M}^\flat}_+ \,, \tag{D.3.1}$$

where $\check{\mathfrak{M}}$ and $\mathfrak{M}^\sharp$ parametrize the complex structure for $\check{\mathcal{M}}$ and $\mathcal{M}^\sharp$ respectively and ${\mathfrak{M}^\flat}_+$ parametrizes the complex structure of $\mathcal{M}^\flat$, including (in the 'boundary') conifold singularizations thereof. To orient the Reader, we had $\dim \check{\mathfrak{M}} = \dim \mathfrak{M}^\sharp = 86$ while $\dim \mathfrak{M}^\flat = 89$ in (D.2.6). Locally, the map $\xrightarrow{\sim}$ may be treated as an identification. Whenever polynomial deformations represent $H^1(\mathcal{M}, \mathcal{T}_\mathcal{M})$ faithfully, these parameter spaces and maps can be constructed explicitly; although this is clearly not the case in general, the above relations can be generalized appropriately [149].

D.3.2 Kähler class moduli

At any complex structure (point $t \in \mathfrak{M}$), the Kähler $(1,1)$-class $[J]$ determines, by Yau's theorem, a positive Kähler metric $J = ig_{\mu\bar{\nu}} dz^\mu \wedge dz^{\bar{\nu}}$. As J is a $(1,1)$-form, it is determined as a linear combination of $(1,1)$-forms $J = v^A e_A$ from a complete set and with real coefficients v^A.

For J to determine a positive metric (such a $(1,1)$-form is said to be positive) the allowed values of v^A must be restricted and we may, in general, end up with an empty set. $h^{1,1} \geq 0$ is only a necessary, not a sufficient condition for a manifold to be Kähler; for a $(1,1)$-form to qualify for a Kähler form, it must satisfy

$$\int_\mathcal{C} J > 0\ , \qquad \int_\mathcal{S} J^2 > 0\ , \qquad \int_\mathcal{M} J^3 > 0\ . \tag{4.3.16$'$}$$

It is a simple fact that if J_x and J_y are both positive, so is $v^x J_x + v^y J_y$ for v^x, v^y real and positive. The parameter space of the Kähler structure $\mathfrak{Re}[\mathfrak{W}]$ is a then cone in $\mathbb{R}^{h^{1,1}}$, called the Kähler cone. As discussed above, we include the antihermitian 2-form B by expanding $B = u^A e_A$, writing $\not{\mathcal{G}} \stackrel{\text{def}}{=} (B+iJ) = (u^A+iv^A)e_A$ for the complexified Kähler form and $\mathfrak{W}$ for the so complexified Kähler cone.

These characterizations of (the real part of) $\mathfrak{W}$ depend in general on the choice of $t \in \mathfrak{M}$. We expect therefore the combined parameter space of Calabi–Yau manifolds to be a space fibered over the space of complex structures, with Kähler cones as fibers. 2024[*The apparent asymmetry $\mathfrak{W} \leftrightarrow \mathfrak{M}$ should be suspect: a Kähler manifold as discussed throughout this book within the complex algebraic framework may also be described within a symplectic geometry framework*¶, *the latter of which is well adapted to address the geometry of the (complexified) completely enlarged space of*

¶ A Kähler manifold, $\mathcal{X}$, simultaneously admits both a complex structure, $\mathfrak{J}$, and a symplectic structure, ω, which are mutually compatible. With these, $g(u,v) := \omega(u, \mathfrak{J}v)$ is symmetric and positive definite, and so a Riemannian metric—the corresponding Kähler metric on $\mathcal{X}$, such as $G_{\mu\bar{\nu}}$, implicitly present throughout this volume, since (0.1.3).

Kähler classes but is beyond the scope of this volume; we return to this in § *F.8.*⌋

For many complete intersection (Calabi–Yau) manifolds $\mathcal{M}$ embedded in products of flag spaces $\mathcal{X} = \mathbb{F}_{[\mathbf{n}_1]} \times \cdots \times \mathbb{F}_{[\mathbf{n}_m]}$, at least a subgroup of $H^{1,1}(\mathcal{M})$ will equal (an isomorphic image of) $H^{1,1}(\mathcal{X})$ (see Chapters 2 and 3). The latter cohomology group has a basis consisting of the generators $J_{(r)} \in H^2(\mathbb{F}_{[\mathbf{n}_r]}, \mathbb{Z})$. Let $\mathring{e}_s$ denote the remaining, 'vanishing generators' of $H^{1,1}(\mathcal{M})$. Positive one forms are then given as $v^r J_{(r)} + v_0^s \mathring{e}_s$, with suitable coefficients v_0^s and positive and large enough v^r. When $J_{(r)}$ span all of $H^{1,1}(\mathcal{M})$, the combined parameter space is *locally* and at a generic choice a direct product^[◊].

That the above emphasis on locality and genericity is not pedantic nit-picking is seen from a simple example tailored from the surfaces studied in § 3.1.2 and 8.6. Consider the Calabi–Yau 3-fold defined as

$$\mathcal{M}_\varepsilon \in \begin{bmatrix} 3 \\ 1 \\ 1 \\ 1 \end{bmatrix}\!\!\begin{Vmatrix} 1 & 1 & 2 \\ 1 & 1 & 0 \\ 0 & 0 & 2 \\ 0 & 0 & 2 \end{Vmatrix} \quad : \quad \left\{ \begin{array}{lr} z_0\, w_0 & +\, z_1\, w_1 = 0 \;, \\ z_2\, w_0 & +\, [\sum_{i=0}^{2} a_i z_i + \varepsilon z_3]\, w_1 = 0 \;, \\ & Q(z,x,y) = 0 \;. \end{array} \right. \tag{D.3.2}$$

Here z, w, x, y are, respectively, homogeneous coordinates for $\mathbb{P}^3 \times \mathbb{P}^1 \times \mathbb{P}^1 \times \mathbb{P}^1$ and $Q(z,x,y)$ is a smooth degree-$(2,2,2)$ polynomial in z, x, y. The Lefschetz hyperplane theorem guarantees that the (pull-backs of the) Kähler forms, J_z, J_w, J_x, J_y generate $H^{1,1}(\mathcal{M}_\varepsilon, \mathbb{Z})$ for all ε, including $\varepsilon = 0$. Also, $J(v^z, v^w, v^x, v^y) \stackrel{\text{def}}{=} v^z J_z + v^w J_w + v^x J_x + v^y J_y$ is Kähler if $v^z, v^w, v^x, v^y > 0$.

For $\varepsilon \neq 0$, we may use Eq. (8.6.3) to replace $\{J_z, J_w\}$ with $\{a, b\}$ since $\left[\begin{smallmatrix} 3 \\ 1 \end{smallmatrix} \middle\| \begin{smallmatrix} 1 & 1 \\ 1 & 1 \end{smallmatrix}\right]_{\varepsilon \neq 0}$ is $\mathbb{P}^1 \times \mathbb{P}^1$. In terms of $\{a, b, J_x, J_y\}$, the real Kähler cone is given by the wedge $v^a, v^b, v^x, v^y > 0$, which implies $v^z, v^x, v^y > 0$ and $v^w > -v^z$! So, the $0 > v^w > -v^z$ region of $\Re\mathfrak{e}[\mathfrak{W}_{\varepsilon \neq 0}]$ is not allowed in the $\varepsilon = 0$ case. The reason for this is that $\mathcal{M}_{\varepsilon=0}$ contains an elliptic surface (related to the line of self-intersection -2 in the surface $\left[\begin{smallmatrix} 3 \\ 1 \end{smallmatrix} \middle\| \begin{smallmatrix} 1 & 1 \\ 1 & 1 \end{smallmatrix}\right]_{\varepsilon = 0} = \mathbf{F}_2$) and $J(v^z, v^w, v^x, v^y)$, with $v^z, v^x, v^y > 0$ and $v^w > -v^z$, fails to be positive over this exceptional set unless $v^w > 0$. In Ref. [235], it was proven that

[◊] In the general case and at a generic point in the parameter space, the integral cohomology still gives a canonical *local* product structure to the fibration over the space of complex structures and with the *real* (1,1)-cohomology as the fiber. However, the location of the (positive) Kähler cone within this fiber might depend on the complex structure.

this typifies the only process in which smooth but special deformations of the complex structure affect the (complexified) Kähler cone.

— ❦ —

Roughly, the v^r's may be identified with the sizes of the factors $\mathbb{F}_{[\mathbf{n}_r]}$ in $\mathcal{X}$. In Eq. (D.2.1), we had $J(\mathcal{M}^\flat) = v^x J_x$ and $\mathfrak{W}^\flat$ is the positive semi-axis $\{0 < v^x < \infty\}$ while $J(\check{\mathcal{M}}) = v^x J_x + v^y J_y$ so that $\check{\mathfrak{W}}$ is the first quadrant $\{0 < v^x, v^y < \infty\}$.

Note that $\mathcal{M}^\sharp$ is singular and does not admit a Kähler class in the usual sense; however, we define a real cohomology class on $\mathcal{M}^\sharp$ to be positive if its restriction to the complement of the nodes is represented by a positive $(1,1)$-form the cube of which has a finite integral. We can therefore identify both $\mathfrak{W}^\sharp$ and $\mathfrak{W}^\flat$with the positive multiples of J_x and have

$$\check{\mathfrak{W}}_+ \supset (\lim_{\mathbb{P}^1\to\cdot} \check{\mathfrak{W}}) \xleftarrow{\sim} \mathfrak{W}^\sharp = \mathfrak{W}^\flat\ . \tag{D.3.3}$$

Here $\check{\mathfrak{W}}_+$ is the Kähler cone of $\check{\mathcal{M}}$ together with its boundaries. Also, the map "$\xleftarrow{\sim}$" is an identification between $\mathfrak{W}^\sharp$ and the (corresponding subset of the) boundary $\lim_{\mathbb{P}^1\to\cdot}\check{\mathfrak{W}}$.

— ❦ —

The combined parameter space of $\mathcal{M}^\sharp$ can be identified therefore with an interfacing region in the boundary of both the parameter space for $\check{\mathcal{M}}$ and that for $\mathcal{M}^\flat$. This fact motivated the suggestion [149, 90, 92] that the process (D.2.6) might acquire a physical interpretation as a phase transition: certain paths, passing through the parameter space of $\mathcal{M}^\sharp$, connect points of distinct parameter spaces which represent non-singular Calabi–Yau 3-folds of distinct homotopy types.

D.3.3 An odd example

To illustrate some properties of nodal singularizations, we turn to another example of a deformation class of smooth CICYs and some of its nodal limits for which no Kähler small resolution is known.

Let $\mathcal{M}$ now be the quintic hypersurface in $\mathbb{P}^4$:

$$I = {x_5}^3\Big(\sum_{i=1}^4 {x_i}^2\Big) + \sum_{i=1}^5 a_i {x_i}^5\ , \tag{D.3.4}$$

where x_i are the homogeneous coordinates of $\mathbb{P}^4$. On the $x_5 = 0$ hyperplane $\mathbb{P}^3 \subset \mathbb{P}^4$

$$I = \sum_{i=1}^{4} a_i {x_i}^5 , \qquad \mathrm{d}I = 5 \sum_{i=1}^{4} dx_i \cdot a_i {x_i}^4 . \tag{D.3.5}$$

For generic a_i, $I = 0 = \mathrm{d}I$ only at $x_i = 0$, which is not in $\mathbb{P}^4$, so the singular points of $\mathcal{M}$ are all in the $x_5 \neq 0$ coordinate patch and we set $x_5 = 1$.

0. For a generic choice of a_i, $I = 0$ and $dI = 0$ have no common solution, so that the quintic hypersurface $I = 0$ in $\mathbb{P}^4$ is a manifold.

Let now $a_5 \to 0$. Then, $dI = 0$ implies $x_i(5a_i{x_i}^3 + 2) = 0$ and candidate singular points ($p^\sharp$) are parametrized as

$$x_i = -\alpha_i \sqrt[3]{2/5a_i} \cdot \omega^{k_i} ,$$

where $\omega = e^{2i\pi/3}$, $k_i = 0, 1, 2$, $\alpha_i = 0, 1$, $i = 1, 2, 3, 4$. At these points

$$I(p^\sharp) = \tfrac{3}{5} \sqrt[3]{\tfrac{4}{25}} \Big(\sum_{i=1}^{4} \frac{\alpha_i \omega^{2k_i}}{{a_i}^{2/3}} \Big)$$

and we have three cases:

1. $(0, 0, 0, 0, 1)$, i.e., $\alpha_i = 0$, $i = 1, 2, 3, 4$ satisfies both $I = 0$ and $dI = 0$ regardless of the choice of the a_i's. There $\det I'' = 16 \neq 0$ and this singular point is a node.

Since $\omega^3 = 1$, other singularities (solutions of both $I = 0$ and $dI = 0$) are possible only if at least three of the four ${a_i}^{2/3}$ are equal. Thus, we have:

2. If $a_1 = a_2 = a_3 = a \neq a_4$, simultaneous solutions to $I = 0$ and $dI = 0$ are of the form $(\omega^{2k_1}, \omega^{2k_2}, \omega^{2k_3}, 0, -\sqrt[3]{5a/2})$, where k_1, k_2, k_3 are all different; there are 6 such points. All these are nodes since $\det I'' = -432 \neq 0$.

3. If $a_i = a$, $i = 1, 2, 3, 4$, singular points are found as in case **2.**, with obvious permutations, and now there are 24 nodes.

Starting from a non-singular quintic (case **0.**) with $h^{2,1} = 101$, we have $\dim \mathfrak{M}^\flat = 101$. Sending $a_5 \to 0$, while keeping no three of the other four a_i equal, produces a nodal quintic in $\mathbb{P}^4$ which features a single node. Clearly, the space of complex structures for this conifold satisfies $\dim \mathfrak{M}^\sharp{}_{(1)} = 100$. Moreover, for every point $t^\sharp \in \mathfrak{M}^\sharp{}_{(1)}$, there is a $\mathbb{C}$-like disk in $\mathfrak{M}^\flat$, with the origin at $t^\sharp$. In this sense $\mathfrak{M}^\sharp{}_{(1)}$ lies *inside* the boundary of $\mathfrak{M}^\flat$.

Let $\mathfrak{M}^\sharp{}_{(N)}$ denote the space of complex structures for quintic conifolds in $\mathbb{P}^4$ with at least N nodes (such as the examples above and those in Ref. [149, 90, 91]), and which is the N-fold self-intersection of $\mathfrak{M}^\sharp{}_{(1)}$. From the fact that the number of available smoothing deformations is less than or equal to the number of nodes, we conclude

Lemma D.1. *Let $\mathfrak{M}$ denote the space of complex structures for a Calabi–Yau 3-fold and let $\mathfrak{M}^\sharp$ denote the singularity divisor in $\mathfrak{M}$. Then, (1) the self-intersections of $\mathfrak{M}^\sharp$ are not all transversal.*

Let $\mathfrak{M}^{\sharp\sharp}_{(\delta)}$ be the set where $\mathfrak{M}^\sharp$ self-intersects non-transversely, with $\delta = N - c$, N the number of components self-intersecting at a generic point of $\mathfrak{M}^{\sharp\sharp}_{(\delta)}$, and c the actual codimension of $\mathfrak{M}^{\sharp\sharp}_{(\delta)} \subset \mathfrak{M}^\sharp$. (2) Only the conifolds parametrized by $\mathfrak{M}^{\sharp\sharp}_{(\delta)}$ may have a Kähler small resolution and δ new (1,1)-cohomology classes.

Sketch of proof: Point (1) is clear, although often ignored.

The codimension c can be easily calculated by counting the smoothing deformations locally at each point t_0 in $\mathfrak{M}^{\sharp\sharp}_{(\delta)}$, as has been done in the above example. N counts the number of nodes in the 3-fold corresponding to t_0. As discussed above, their difference $\delta \geq 0$ implies that there are δ relations between the corresponding 3-cycles, which in turn implies the existence of δ 'vanishing 4-cycles' upon small resolution. These are isomorphic to δ new (1,1)-cohomology classes.

For conifolds which are parametrized by points of $\mathfrak{M}^\sharp$, but not in $\mathfrak{M}^{\sharp\sharp}_{(\delta)}$, there are no new (1,1)-cohomology classes, the candidate Kähler form is a linear combination of "old" (1,1)-classes and must be exact on the exceptional sets of the small resolution, hence null. As described above, however, when $\delta > 0$, there are new (1,1)-cohomology classes to allow positivity for the candidate Kähler class. ☑

Although nodes are produced by (local) deformations, for the conifold to be a complete intersection, the global nature of the defining equations may enforce more than one node to occur. In the example just above, the nodes occur in odd-numbered groups of 1, 7 and 25.

D.4 Many Calabi–Yau Manifolds are Connected

Generalizations of the simple process (D.2.6) have been proven to link the parameter spaces of many large families of Calabi–Yau 3-folds [149]. We now present some of these generalizations.

D.4.1 Determinantal splitting

Complete intersection Calabi–Yau (CICY) manifolds were described in Chapter 2; each of them is defined by a system of transverse polynomial constraints $\{f_a = 0\}_{a=1\ldots K}$ in $\mathcal{X} = \mathbb{P}_1^{n_1} \times \cdots \times \mathbb{P}_m^{n_m}$. Recall the *configuration matrix*, an integer-valued matrix of the general form

$$[\mathcal{X} \,\|\, \mathcal{E}] = \begin{bmatrix} n_1 \\ \vdots \\ n_m \end{bmatrix}\!\!\left\|\begin{matrix} q_1^1 & \cdots & q_K^1 \\ \vdots & \ddots & \vdots \\ q_1^m & \cdots & q_K^m \end{matrix}\right], \qquad q_a^r \geq 0, \tag{D.4.1}$$

where n_r is the dimension of $\mathbb{P}_r^{n_r}$, the rth factor in $\mathcal{X}$, and q_a^r is the degree of homogeneity of the ath polynomial with respect to the homogeneous coordinates of $\mathbb{P}_r^{n_r}$. $\mathcal{E}$ simply denotes the matrix of the degrees q_a^r.

A straightforward generalization of the relation

$$\begin{bmatrix} 5 \\ 1 \end{bmatrix}\!\!\left\|\begin{matrix} 4 & 1 & 1 \\ 0 & 1 & 1 \end{matrix}\right] \;\leftrightarrow\; [5\|4,2] \tag{D.4.2}$$

is provided by any two configurations of the form

$$\left[\begin{matrix} \mathcal{X} \\ n \end{matrix}\right\|\left.\begin{matrix} \mathcal{E} & a_1 & a_2 & \ldots & a_{n+1} \\ 0 & 1 & 1 & \ldots & 1 \end{matrix}\right] \longleftrightarrow \left[\mathcal{X} \,\middle\|\, \mathcal{E}\,,\ \sum_{k=1}^{n+1} a_k\right]. \tag{D.4.3}$$

The $\mathcal{X}$ and $\mathcal{E}$ may be substituted as in Eq. (D.4.1) and a_i are column vectors. Owing to Eq. (2.1.17), if the l.h.s. configuration represents a Calabi–Yau 3-fold, so does the r.h.s. one. The relation "$\longleftrightarrow$" specifies that the parameter spaces of the two configurations have common limit points (parametrizing conifolds) and that the one on the l.h.s. has a strictly smaller dimension. The relation is self-evident upon writing out the last $n+1$ equations which are linear in the $n+1$ coordinates of $\mathbb{P}^n$ and results in a determinant equation of degree $\sum_{k=1}^{n+1} a_k$. With the constraints represented by $\mathcal{E}$, this equation generically describes a conifold which may be smoothed to give a manifold belonging to the configuration on the right. We refer to this as a *determinantal contraction*, while passing from the r.h.s. of Eq. (D.4.3) to the l.h.s. was called a *determinantal splitting* [149]. We also use these terms for the corresponding replacement of one moduli space by another.

By straightforward iteration of (D.4.3) we have, for example,

$$[5\|4\ 2] \leftrightarrow \left[\begin{smallmatrix}5\\1\end{smallmatrix}\middle\|\begin{smallmatrix}4&1&1\\0&1&1\end{smallmatrix}\right] \leftrightarrow \left[\begin{matrix}5\\1\\1\end{matrix}\middle\|\begin{matrix}3&1&1&1\\1&1&0&0\\0&0&1&1\end{matrix}\right] \leftrightarrow \left[\begin{matrix}5\\1\\1\\1\end{matrix}\middle\|\begin{matrix}2&1&1&1&1\\1&1&0&0&0\\0&1&1&0&0\\0&0&0&1&1\end{matrix}\right]$$

$$\leftrightarrow \left[\begin{matrix}5\\1\\1\\1\\1\end{matrix}\middle\|\begin{matrix}1&1&1&1&1&1\\1&1&0&0&0&0\\0&1&1&0&0&0\\0&0&1&1&0&0\\0&0&0&0&1&1\end{matrix}\right] \leftrightarrow \left[\begin{matrix}1\\1\\1\\1\end{matrix}\middle\|\begin{matrix}2\\2\\2\\2\end{matrix}\right]. \tag{D.4.4}$$

Indeed, we have [149]

Theorem D.1. *The parameter spaces of all Calabi–Yau complete intersections in products of complex projective spaces are connected.*

Proof. Starting any Calabi–Yau complete intersection with a configuration matrix $[\mathcal{X}\|\mathcal{E}]$ as in (2.1.5), perform determinantal splittings iteratively, introducing only $\mathbb{P}^1$'s into the embedding space, in a way that finally leaves each constraint at most linear in at most one $\mathbb{P}^{n_r}_r$ factor in $\mathcal{X}$ with $n_r \neq 1$. Perform next determinantal contractions of all $\mathbb{P}^{n_r}_r$'s, $n_r \neq 1$. This leaves a Calabi–Yau complete intersection which is embedded in a product of only $\mathbb{P}^1$'s. Finish with all possible determinantal contractions of $\mathbb{P}^1$'s: $\left[\begin{matrix}1\\1\\1\\1\end{matrix}\middle\|\begin{matrix}2\\2\\2\\2\end{matrix}\right]$ is the result of the above algorithm, regardless of the initial Calabi–Yau complete intersection. Thus, every Calabi–Yau complete intersection in products of complex projective spaces connects to this one, and the statement follows. The moduli spaces of all such Calabi–Yau complete intersections therefore piece together, much the same as they did in the simple example above, and form a *web* of Calabi–Yau complete intersections. Figure D.3 illustrates this procedure. ☑

— ❦ —

Suffice it here to note that a similar systematic procedure connects into this same web all multiple branched covers of almost Fano 3-folds and fibered products presented in Chapter 6 [149, 92]. At the local level, all these procedures involve exactly the same process, a singularizing deformation followed by a small resolution (or the other way around). Compare this with Theorem 7.3 and also the well-understood notion of the "universal curve," whereby moduli spaces of Riemann surfaces of different genera are pieced together in a hierarchical fashion.

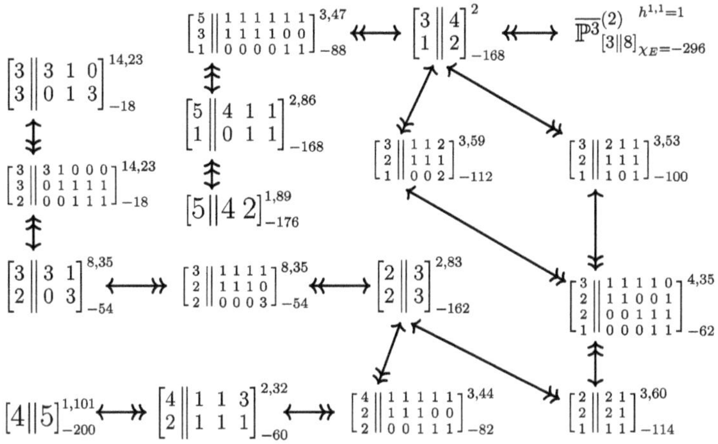

Figure D.3: A minute region of the web of connected Calabi–Yau manifolds.

2024 ⌈*In the three decades since the original publication of the Bestiary [272], the "conifold transitions" of § D.2, generating the connected web of Calabi–Yau 3-folds [149, 92] a minuscule portion of which is shown in Figure D.3, have been studied extensively and also generalized; see Refs. [430, 305, 450, 465, 405, 311, 368, 415] to start with.*

The physical mechanism and characteristics of these transitions were soon understood within the context of string compactifications with $N=2$ target-space supersymmetry [467], and they were quickly extended to M- and F-theory [445, 477]; see also [329, 419, 406] and [297]. The physics aspects of such **geometric transitions** *among the $N=1$-supersymmetric heterotic compactifications are still actively studied and not fully understood; see for example [297].*⌋

Chapter E

Parameter Spaces: A Closer Look

E.1 Metric? What Metric?

The main result of the preceding Chapter was that the relevant parameter spaces of a large number of Calabi–Yau 3-folds piece together into a web, so that there exist (large) variations of parameters which connect (ultimately) all such 3-folds with one another. The typical situation is sketched in Figure E.1. One inquires next if the (shortest) distances between two topologically distinct 3-folds in this web are finite. To answer this, some appropriate metric must be chosen on the parameter space and we face several seemingly different options. Fortunately, these choices end up resulting in the same structure, indicating that a general theory, not unlike the Teichmüller theory on the moduli space of Riemann surfaces, is probably in store. 2024⌈*The process sketched in Figure E.1 has been rigorously shown to be continuous (in the Gromov–Hausdorff topology) even when* $h^{1,1}\to 0$ *[381], and so indeed connects standard Ansatz superstring compactifications as discussed herein*. In turn, processes with more general compactification Ansätze have also been suggested, wherein passing through the conifold leaves the physics spectra unchanged [297], raising the question of whether these can then be deformed back to the standard Ansatz, as needed and "on the same side" of the conifold.*⌋

* Calabi–Yau spaces with $h^{1,1}=0$ cannot be Kähler, but may well mirror rigid ($h^{2,1}=0$) such spaces.

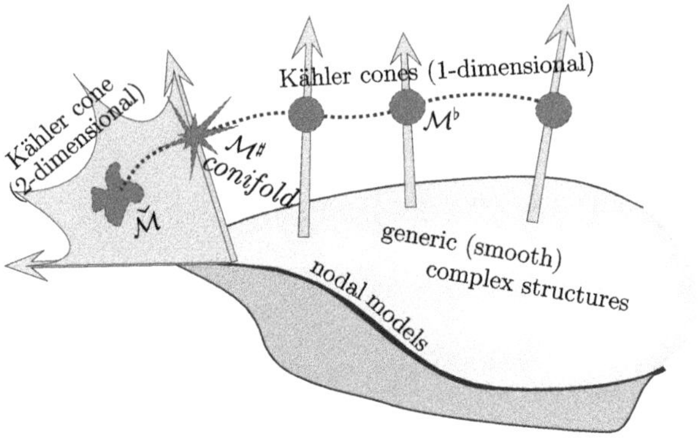

Figure E.1: A sketch of the parameter spaces of two Calabi–Yau 3-folds of different topology type, which connect at an interface, corresponding to Calabi–Yau conifolds.

Recall that the combined parameter space is a pair, $(\mathfrak{M}, \mathfrak{W})$. Points in $\mathfrak{M}$ parametrize the complex structure for $\mathcal{M}$, or equivalently, specify one harmonic 3-form to be the *holomorphic* 3-form Ω (determined up to a multiplicative constant). Points in $\mathfrak{W}$ specify a complexified Kähler class $[\not{G}]$, and by Yau's theorem a unique Ricci-flat Kähler metric corresponding to the imaginary part of $[\not{G}]$.

E.1.1 Effective kinetic terms

As we have seen in Chapter 0, Calabi–Yau compactification of the Heterotic (super)string features $h^{2,1}$ massless (spacetime $N=1$) superfields $\Phi^\alpha(x)$, each containing a family of Standard Model particles, together with $h^{2,1}$ gauge-singlet "moduli superfields" (the spin-$(1/2,0)$ superfields in the first part of Table 0.2). These are associated with metric variations of the pure type,

$$\varphi_{\overline{\mu}}{}^{\rho} \stackrel{\text{def}}{=} (g^{\rho\overline{\nu}}\,\delta g_{\overline{\mu}\,\overline{\nu}}) \quad\sim\quad H^1(\mathcal{M}, \mathcal{T}_{\mathcal{M}}) \;. \tag{E.1.1}$$

There also occur $h^{1,1}$ massless families $\Psi^A(x)$ of mirror-particles likewise accompanied by gauge-singlet moduli superfields and associated with metric variations of mixed type

$$e_{\overline{\mu}\,\rho} \stackrel{\text{def}}{=} \delta g_{\rho\overline{\mu}} \quad\sim\quad H^1(\mathcal{M}, \mathcal{T}^*_{\mathcal{M}}) \;. \tag{E.1.2}$$

Recall, $H^1(\mathcal{M}, \mathcal{T}^*_{\mathcal{M}}) \approx H^{1,1}(\mathcal{M})$ and $H^1(\mathcal{M}, \mathcal{T}_{\mathcal{M}}) \approx H^{2,1}(\mathcal{M})$; see (1.2.3).

To lowest order, we obtain the effective Lagrangian density upon superspace integration of

$$\mathcal{L}_{\text{eff.}} = \int d^2\theta\, d^2\bar{\theta}\, \Big[N_{\alpha\bar{\beta}}\Phi^{\alpha}(\Phi^{\beta})^{\dagger} + N_{A\bar{B}}\Psi^{A}(\Psi^{B})^{\dagger}\Big] + \int d^2\theta \Big[\kappa_{\alpha\beta\gamma}\Phi^{\alpha}\Phi^{\beta}\Phi^{\gamma} + \mathring{\kappa}_{ABC}\Psi^{A}\Psi^{B}\Psi^{C}\Big] + \text{ c.c.} \tag{E.1.3}$$

Following Ref. [218], we find explicit expressions for these terms with respect to a basis $\{\varphi_\alpha\}_{\alpha=1\dots h^{2,1}}$ for $H^{(2,1)}(\mathcal{M})$ and $\{e_A\}_{A=1\dots h^{1,1}}$ for $H^{(1,1)}(\mathcal{M})$:

$$N_{\alpha\bar{\beta}} = i\int_{\mathcal{M}} \varphi_\alpha \wedge \overline{\varphi_\beta}\ , \qquad N_{A\bar{B}} = \int_{\mathcal{M}} e_A \wedge *e_B\ . \tag{E.1.4}$$

For discussions including also the gauge non-invariant massless superfields see Ref. [123, 120].

E.1.2 The natural line element

From a relativist's point of view, we have a complex manifold $\mathcal{M}$ with a (Ricci-flat or not) Hermitian metric, $g_{\mu\overline{\nu}}$ and we wish to describe the metric properties of the space of metrics. In addition, of course, we have also the anti-Hermitian 2-form B, the variation of which is also considered. Furthermore, we know from the analysis that resulted in Table 0.2, that we are dealing with the variations

$$\delta g_{\mu\nu}\ ,\quad \delta g_{\overline{\mu}\,\overline{\nu}}\ ,\qquad \delta g_{\mu\overline{\nu}}\ ,\quad \delta B_{\mu\overline{\nu}}\ . \tag{E.1.5}$$

On the space locally spanned by these variables, we must choose a metric. To do so, we shall neglect: (1) derivative terms (these are systematically introduced by σ-model corrections), (2) a possible term $(g^{mn}\delta g_{mn})^2$, since $g^{mn}\delta B_{mn}$ vanishes, this term would spoil the symmetry between g_{mn} and B_{mn}, and (3) terms involving the $E_8 \times E_8$ Yang–Mills sector (for the inclusion of this see Ref. [123]). Within these restrictions, the most general metric is

$$\begin{aligned} ds^2 &= \frac{1}{V}\int_{\mathcal{M}} \|\delta g_{\mu\nu}\|^2 + \|\delta g_{\mu\overline{\nu}} \pm \delta B_{\mu\overline{\nu}}\|^2\ ,\\ &= \frac{2}{V}\int_{\mathcal{M}} g^{\kappa\overline{\mu}}g^{\lambda\overline{\nu}}\left(\delta g_{\kappa\lambda}\delta g_{\overline{\mu}\,\overline{\nu}} + (\delta g_{\kappa\overline{\nu}}\delta g_{\lambda\overline{\mu}} + \delta B_{\kappa\overline{\nu}}\delta B_{\overline{\mu}\lambda})\right) g^{\frac{1}{2}} d^6x\ . \end{aligned} \tag{E.1.6}$$

Somewhat surprisingly perhaps, this "metric on the space of metrics" is block-diagonal, with a separate block corresponding to variations of the complex structure and of the Kähler class, respectively. Note, however, that this is not a direct product space metric: while the (1,1)- and the (2,1)-sectors are *orthogonal*, they are not independent. The factor $g^{\kappa\bar{\mu}}g^{\lambda\bar{\nu}}$ depends on both parameters.

— ❦ —

On general grounds, one would expect that the expressions (E.1.4) and (E.1.6) agree and we now turn to verifying that.

E.2 The (2, 1)-Forms

Consider first the deformations of the complex structure (we follow Ref. [222]). Define

$$\varphi_{\alpha\,\kappa\lambda\bar{\mu}} \stackrel{\text{def}}{=} -\tfrac{1}{2}\Omega_{\kappa\lambda}{}^{\bar{\nu}}\frac{\partial g_{\bar{\mu}\bar{\nu}}}{\partial t^\alpha}\,, \qquad \varphi_\alpha \stackrel{\text{def}}{=} \tfrac{1}{2}\varphi_{\alpha\,\kappa\lambda\bar{\mu}}\,dx^\kappa\wedge dx^\lambda\wedge dx^{\bar{\mu}}\,. \tag{E.2.1}$$

The parameters t^α, $\alpha = 1,\dots,h^{2,1}$ are chosen to be local coordinates in an open neighborhood of $\mathfrak{M}$. Each φ_α is a harmonic (2,1)-form. Inverting, we have

$$\delta g_{\bar{\mu}\bar{\nu}} = -\frac{1}{\|\Omega\|^2}\overline{\Omega}_{\bar{\mu}}{}^{\rho\sigma}\varphi_{\alpha\,\rho\sigma\bar{\nu}}\,\delta t^\alpha\,, \tag{E.2.2}$$

where $\|\Omega\|^2 \stackrel{\text{def}}{=} \frac{1}{3!}\Omega_{\mu\nu\rho}\overline{\Omega}^{\mu\nu\rho}$ and is constant on the Calabi–Yau 3-fold, but varies over $\mathfrak{M}$.

✎ Derive

$$\mathbf{G}_{\alpha\bar{\beta}}\,\delta t^\alpha\,\delta t^{\bar{\beta}} = -\left(\frac{i\int_{\mathcal{M}}\varphi_\alpha\wedge\overline{\varphi}_{\bar{\beta}}}{i\int_{\mathcal{M}}\Omega\wedge\overline{\Omega}}\right)\delta t^\alpha\,\delta t^{\bar{\beta}}\,. \tag{E.2.3}$$

❧

This is called the Weil–Petersson metric [222] and coincides, up to the normalization by the (nonstandard) volume $i\int_{\mathcal{M}}\Omega\overline{\Omega}$, with the matrix $N_{\alpha\bar{\beta}}$ in Eq. (E.1.4).

— ❦ —

We however continue with an observation which is attributed to K. Kodaira (see Ref. [222])

$$\frac{\partial\Omega}{\partial t^\alpha} = K_\alpha\Omega + \varphi_\alpha\,, \tag{E.2.4}$$

where the K_α may depend on t^α but not on the coordinates of $\mathcal{M}$. Essentially, this can be seen as follows. Consider a pure (3,0)-form change under a variation of the complex structure, that is, varying the z's into a linear combination of z's and $\overline{z}$'s. To lowest order, only one of its three indices might change from holomorphic to antiholomorphic and the result is partly a (3,0)-form and partly a (2,1)-form. With a little care, it can also be shown that these terms are harmonic.

From Eq. (E.2.3) and (E.2.4), it follows immediately that

$$\mathbf{G}_{\alpha\bar{\beta}} = -\frac{\partial}{\partial t^\alpha}\frac{\partial}{\partial t^{\bar{\beta}}} \log\left(i\int_{\mathcal{M}} \Omega \wedge \overline{\Omega}\right) \tag{E.2.5}$$

and the space of complex structures is itself Kähler, with the logarithm of the nonstandard volume form, $i\int_{\mathcal{M}} \Omega\overline{\Omega}$, being the Kähler potential.

— ❦ —

The holomorphic 3-form has so far shown up in several different ways as an essential marking, so we will describe this in a little more detail, following Ref. [92, 87]. Recall that, after all, Ω is just an element of the real degree-3 cohomology. By choosing which of the available harmonic 3-forms is regarded as the purely holomorphic one, we have specified the complex structure. The available choices are then parametrized by expanding Ω over a basis of the dual 3-homology.

Let (A^a, B_b) , $a, b = 0, \ldots, h^{2,1}$ be a canonical homology basis for $H_3(\mathcal{M}; \mathbb{Z})$ and let (α_a, β^b) be the dual cohomology basis so that

$$\int_{A^b} \alpha_a = \int_{\mathcal{M}} \alpha_a \wedge \beta^b = \delta_a{}^b \,, \qquad \int_{B_a} \beta^b = \int_{\mathcal{M}} \beta^b \wedge \alpha_a = -\delta_a{}^b \,. \tag{E.2.6}$$

Also,

$$z^a \stackrel{\text{def}}{=} \int_{A^a} \Omega \,, \qquad \mathcal{G}_a \stackrel{\text{def}}{=} \int_{B_a} \Omega \tag{E.2.7}$$

are called the *periods* of Ω. Bryant and Griffiths [82] show that, locally in the moduli (effective parameter) space, the complex structure of $\mathcal{M}$ is entirely determined by a choice of the z^a, so that $\mathcal{G}_a$ may be regarded (locally!) as a function of z. The z^a cannot vanish simultaneously. Since Ω is defined only up to an overall scaling $\Omega(z) \to \lambda\Omega(z)$, the same must be required for $z^a \to \lambda z^a$. We may therefore regard the z^a as projective coordinates for the complex structure and Ω as being homogeneous of degree 1 in these coordinates:

$$z^a \in \mathbb{P}^{h^{2,1}} \,, \qquad \Omega(\lambda z) = \lambda\Omega(z) \,. \tag{E.2.8}$$

Finally, we may now expand Ω as

$$\Omega = z^a \alpha_a - \mathcal{G}_a(z)\beta^a \,. \tag{E.2.9}$$

It easily follows that

$$e^{-K_{(2,1)}} = i\int_{\mathcal{M}} \Omega\wedge\overline{\Omega} = -2\Im m\,(\overline{z}^a\,\mathcal{G}_a)\;, \tag{E.2.10}$$

where $K_{(2,1)} = -\log(i\int_{\mathcal{M}}\Omega\overline{\Omega})$ is the Kähler potential on $\mathfrak{M}$. We see that the $\mathcal{G}_a(z)$ completely determine this metric. On the other hand, the $\mathcal{G}_a(z)$ themselves are determined by integration from the Yukawa couplings (see below) up to and additive term $\frac{1}{2}\kappa_{ab}z^a z^b$ where κ_{ab} is a symmetric (locally) constant matrix. Recalling now that these Yukawa couplings receive no quantum correction, It follows that this metric structure on $\mathfrak{M}$ is exact up to this κ_{ab}.

— ❦ —

Eq. (E.2.4) induces a number of vanishing relations such as

$$\varOmega_{(1)} \stackrel{\text{def}}{=} i\mathrm{d}z^a\int_{\mathcal{M}}\Big(\Omega\wedge\frac{\partial\Omega}{\partial z^a}\Big) = 0\;. \tag{E.2.11}$$

This implies that $2\mathcal{G}_a = \frac{\partial}{\partial z^a}(z^c\mathcal{G}_c)$ and so $\mathcal{G}_a$ is (locally!) the gradient of a degree-2 homogeneous function

$$\mathcal{G}_a = \frac{\partial\mathcal{G}}{\partial z^a}\;,\quad \mathcal{G}(\lambda z) = \lambda^2\mathcal{G}(z)\;. \tag{E.2.12}$$

Several other identities of similar form may be derived from $\mathrm{d}\varOmega_{(1)} = 0$ and

$$\varOmega_{(2)} \stackrel{\text{def}}{=} i\mathrm{d}z^a\otimes\mathrm{d}z^b\int_{\mathcal{M}}\Big(\Omega\wedge\frac{\partial^2\Omega}{\partial z^a\partial z^b}\Big) = 0\;, \tag{E.2.13}$$

but they do not restrict the z's and the $\mathcal{G}(z)$ any further.

The third derivative of $\int_{\mathcal{M}}\Omega\wedge\Omega$ is the lowest non-zero one and relates to the Yukawa coupling between $(2,1)$-forms [84]. Using Eq. (E.2.9) and Eq. (E.2.12):

$$\frac{\partial^3}{\partial z^a\partial z^b\partial z^c}\Big(i\int_{\mathcal{M}}\Omega\wedge\Omega\Big) = i\int_{\mathcal{M}}\Big(\Omega\wedge\frac{\partial^3\Omega}{\partial z^a\partial z^b\partial z^c}\Big) = i\frac{\partial^3\mathcal{G}}{\partial z^a\partial z^b\partial z^c}\;, \tag{E.2.14}$$

up to lower order terms (and vanishing at least in these coordinates). In general, therefore, this expression appears ill defined, as it seems to depend on the basis chosen to express it. Note, however, that we can easily introduce a covariant derivative from Eq. (E.2.4):

$$D_\alpha \stackrel{\text{def}}{=} \frac{\partial}{\partial t^\alpha} - K_\alpha\;,\qquad K_\alpha \stackrel{\text{def}}{=} -\frac{\partial}{\partial t^\alpha}\log\Big(i\int_{\mathcal{M}}\Omega\wedge\Omega\Big)\;, \tag{E.2.15}$$

which also defines φ_α as a proper vector. In the left hand side of formula (E.2.14), we may freely replace all partial derivatives with D_α, the end result will not change [219, 87]. A little thought reveals that the Yukawa coupling is independent of any connection because of the vanishing relations of the type (E.2.11) and (E.2.13). This is very much like the fact that gradients of a section are independent of any connection on the zero-locus of the section, which we have noted in the first Chapters.

Clearly not all values of z^a are allowed since we must have

$$\left(i\int_{\mathcal{M}} \Omega\wedge\overline{\Omega}\right) = -2\Im m(\overline{z}^a\mathcal{G}_a) > 0 \ . \tag{E.2.16}$$

— ❦ —

The vector space $H^{(3,0)}\oplus H^{(2,1)}$ is a linear subspace of $H^3(\mathcal{M})$ and varies holomorphically with z^a. A basis for $H^{(3,0)}\oplus H^{(2,1)}$ is then

$$\omega_a \stackrel{\text{def}}{=} \frac{\partial\Omega}{\partial z^a} = \alpha_a - \frac{\partial^2\mathcal{G}}{\partial z^a\partial z^b}\beta^b \ , \tag{E.2.17}$$

in which the period matrix $\varpi \stackrel{\text{def}}{=} \left(\int_{A^b}\omega_a, \int_{B_b}\omega_a\right)$ simplifies:

$$\varpi = (1,\Xi) \ , \qquad \Xi_{ab} = \frac{\partial^2\mathcal{G}}{\partial z^a\partial z^b} \ . \tag{E.2.18}$$

Since $i\int_{\mathcal{M}}\omega_a\wedge\overline{\omega}_b = -2\Im m\left(\frac{\partial^2\mathcal{G}}{\partial z^a\partial z^b}\right)$, we learn that $\Im m\,\Xi$ has signature $(h^{2,1},1)$ ($h^{2,1}$ plus signs and one minus). This is analogous to lying in the Siegel upper half plane, just as is the action of the symplectic group on Ξ to the action on period matrices of Riemann surfaces. With

$$\begin{bmatrix}\tilde\alpha\\ \tilde\beta\end{bmatrix} = \begin{bmatrix}A & B\\ C & D\end{bmatrix}\begin{bmatrix}\alpha\\ \beta\end{bmatrix} \ , \qquad \begin{bmatrix}A & B\\ C & D\end{bmatrix} \in \mathrm{Sp}(2h^{2,1}+2;\mathbb{Z}) \tag{E.2.19}$$

being the action on a basis for H_3, we have

$$\begin{bmatrix}\widetilde{\partial\mathcal{G}}\\ \widetilde{z}\end{bmatrix} = \begin{bmatrix}A & B\\ C & D\end{bmatrix}\begin{bmatrix}\partial\mathcal{G}\\ z\end{bmatrix} \ , \tag{E.2.20}$$

from which it easily follows that

$$\widetilde{\Xi} = \left[A\Xi + B\right]\left[C\Xi + D\right]^{-1} \ . \tag{E.2.21}$$

E.3 The $(1,1)$-Forms

Corresponding to Eq. (E.1.6), we have on $H^{(1,1)}$

$$\mathbf{G}(\varrho,\sigma) = \frac{1}{2V}\int_{\mathcal{M}} \varrho_{\mu\bar\nu}\,\sigma_{\rho\bar\sigma}\,g^{\mu\bar\sigma}g^{\rho\bar\nu}\,g^{\frac{1}{2}}\mathrm{d}^6x = \frac{1}{2V}\int_{\mathcal{M}} \varrho\wedge *\sigma \ , \tag{E.3.1}$$

for real $(1,1)$-forms, ϱ and σ. Owing to the positivity of $g^{\mu\bar\nu}$, $G(\ ,\ )$ is positive. Up to normalization by the (now standard) volume $V = \frac{1}{n!}\int_{\mathcal{M}} J^n$, this is precisely the metric adopted in point-field theory [218, 97]. Also, $G(\varrho,\sigma)$ may be rewritten entirely in terms of the (topological) cubic form $\mathring{\kappa}(\varrho,\sigma,\tau) \stackrel{\text{def}}{=} \int \varrho\wedge\sigma\wedge\tau$,

$$\mathbf{G}(\varrho,\sigma) = -\frac{3}{2}\left[\frac{\mathring{\kappa}(\varrho,\sigma,J)}{\mathring{\kappa}(J,J,J)} - \frac{3}{2}\frac{\mathring{\kappa}(\varrho,J,J)\mathring{\kappa}(\sigma,J,J)}{\mathring{\kappa}^2(J,J,J)}\right] , \tag{E.3.2}$$

where we have used that

$$*\sigma = -J\wedge\sigma + \frac{3}{2}\frac{\mathring{\kappa}(\sigma,J,J)}{\mathring{\kappa}(J,J,J)}\, J\wedge J\ . \tag{E.3.3}$$

Since $H^{2,0} = H^{0,2} = 0$, we may choose the basis used in Eq. (E.1.4) to be integral and expand

$$\not{G} \ \stackrel{\text{def}}{=}\ (B+iJ) = w^A e_A\ , \qquad w^A = u^A + iv^A \ \in \mathfrak{W}. \tag{E.3.4}$$

Straightforwardly,

$$\mathbf{G}_{A\bar{B}} \stackrel{\text{def}}{=} \mathbf{G}(e_A,e_B) = \frac{\partial}{\partial w^A}\frac{\partial}{\partial w^{\bar{B}}} \log \mathring{\kappa}(J,J,J)\ . \tag{E.3.5}$$

Therefore, $\mathcal{W}$, the space of the w^A is also a Kähler manifold and the Kähler potential is the logarithm of the (now standard) volume of the Calabi–Yau manifold. If we define [87]

$$\mathcal{F}(w)\ \ \stackrel{\text{def}}{=}\ \ -\frac{1}{3!}\frac{\mathring{\kappa}_{ABC}w^Aw^Bw^C}{w^0}\ , \tag{E.3.6}$$

then

$$e^{-K_{(1,1)}}\ \ =\ \ \frac{4}{3}\int_{\mathcal{M}} J^3\ \ =\ \ -2\Im m\,(\overline{w}^A\,\mathcal{F}_A)\ , \tag{E.3.7}$$

the similarity of which to Eq. (E.2.10) may appear quite startling.

Again, not all values of v^A are allowed, since we must ensure that

$$V = \mathring{\kappa}_{ABC}\, v^Av^Bv^C > 0\ , \tag{E.3.8}$$

this being the mildest of the three conditions (4.3.16), $\int_{\mathcal{M}} J^3 > 0$. The other two, $\int_S J^2 > 0$ and $\int_C J > 0$ for every surface $S\subset\mathcal{M}$ and every curve $C\subset\mathcal{M}$ are harder to verify.

We can however obtain another restriction since Eq. (E.3.2) must be a positive definite metric. Upon rewriting,

$$\mathbf{G}(\varrho,\sigma) = -\frac{3}{2}\left\{\frac{\mathring{\kappa}(\varrho,\sigma,J)}{\mathring{\kappa}(J,J,J)} - \frac{\mathring{\kappa}(\varrho,J,J)\mathring{\kappa}(\sigma,J,J)}{\mathring{\kappa}^2(J,J,J)}\right\} + \frac{3}{4}\frac{\mathring{\kappa}(\varrho,J,J)\mathring{\kappa}(\sigma,J,J)}{\mathring{\kappa}^2(J,J,J)}\,, \tag{E.3.9}$$

we have just decomposed this metric into a the "J-parallel" and "J-orthogonal" part. That is, every (1, 1)-form may be written as

$$\varrho = \varrho^{\perp} + \varrho^{\|} = \frac{1}{6}(\varrho_{mn}J^{mn})J + \left(\varrho - \frac{1}{6}\varrho_{mn}J^{mn}J\right)\,, \tag{E.3.10}$$

where $\varrho^{\|}$ is covariantly constant if ϱ is harmonic. This is the Lefschetz decomposition $H^2 = (H^2_{\|} \approx H^0) \oplus H^2_{\perp}$, where $H^2_{\perp}$ is the group of primitive two-forms.

In this notation, Eq. (E.3.9) decomposes the metric into $\mathbf{G}(\varrho^{\perp},\sigma^{\perp})$ and $\mathbf{G}(\varrho^{\|},\sigma^{\|})$ both of which must be positive definite. $\mathbf{G}(\varrho^{\|},\sigma^{\|})$ clearly is, but

$$\mathbf{G}(\varrho^{\perp},\sigma^{\perp}) = -\frac{3}{2}\frac{\mathring{\kappa}(\varrho^{\perp},\sigma^{\perp},J)}{\mathring{\kappa}(J,J,J)} > 0 \tag{E.3.11}$$

imposes a restriction. That is, $\mathring{\kappa}_{AB}(v) \stackrel{\text{def}}{=} \mathring{\kappa}_{ABC}v^A$ ought to be negative definite on $H^2_{\perp}$ but positive definite on $H^2_{\|}$. Thus it must have signature $(1, h^{1,1}{-}1)$ (one plus and the rest minus), which can change only if one of the eigenvalues passes through zero. The boundary of the allowed region is therefore given by

$$\det \mathring{\kappa}_{AB}(v) = 0\ . \tag{E.3.12}$$

The fundamental restriction on the parameters is of course that the Kähler form J be positive and this implies both positive volume and positive $\mathbf{G}_{A\bar{B}}$.

E.3.1 Coordinate-free description of the metric

The foregoing can be summarized in more mathematical terms. We have observed that for a Calabi–Yau manifold $\mathcal{M}$

$$H^{0,3}(\mathcal{M}) \subseteq H^3(\mathcal{M},\mathbb{C}) = H^3(\mathcal{M},\mathbb{Z}) \otimes_{\mathbb{Z}} \mathbb{C} \tag{E.3.13}$$

and that the integral periods of Ω determine (at least locally) the holomorphic structure on $\mathcal{M}$.

Whereas positivity of J depends in principle on the choice of a complex structure for $\mathcal{M}$, the reality of J does not, as it can be characterized in terms of the reality of the integral periods, via the identity

$$H^2(\mathcal{M},\mathbb{Z}) \otimes_{\mathbb{Z}} \mathbb{R} = H^2(\mathcal{M},\mathbb{R}) \subseteq H^2(\mathcal{M},\mathbb{C}) = H^{1,1}(\mathcal{M}) \tag{E.3.14}$$

independently of the holomorphic structure on $\mathcal{M}$. For similar reasons, $\int_{\mathcal{M}} J^3$ depends only on the cohomology class $[J]$ and is independent of the holomorphic structure on $\mathcal{M}$. We note that if $[J]$ is positive, so is $\int_{\mathcal{M}} J^3$, but that the converse does not hold in general. However, the set of positive classes is an open subset of the set of classes with positive cubes.

The exponentials of the Kähler potentials on $\mathfrak{M}$ and $\mathfrak{W}$, $e^{-K_{(2,1)}} = i\int_{\mathcal{M}} \Omega\wedge\overline{\Omega}$ and $e^{-K_{(1,1)}} = \frac{3}{4}\int_{\mathcal{M}} J^3$, respectively, can both be interpreted as volumes for $\mathcal{M}$ (the latter one being standard). However, for the purpose of defining a metric on the metric moduli space we impose no normalization but allow Ω and J to vary independently. Moreover we can fix the integral periods of either Ω or J, thereby making it meaningful to vary one while fixing the other (locally, at a generic choice). The two subspaces so defined become thereby orthogonal and we define a metric on each.

The integral periods of J may be regarded as coordinates for the subspaces defined by holding Ω constant, and take the metric on such a subspace to be the Hessian of $\log(\int_{\mathcal{M}} J^3)$. For the subspaces of constant J, we choose any moduli on which the periods of Ω can be chosen to depend holomorphically, and we define the metric to be the Kähler metric with the Kähler form $\partial\overline{\partial}\log(i\int_{\mathcal{M}} \Omega\wedge\overline{\Omega})$. This is known as the Weil–Petersson metric for the (complex structure) moduli space of $\mathcal{M}$, and is independent of all the choices involved. Note that Ω is undetermined up to multiplication by a holomorphic function $\Omega \to f(t)\Omega$ but this does not affect the metric basically because $\partial\overline{\partial}\log(f\overline{f}) = 0$.

E.4 Finiteness of Distances between Distinct Manifolds

In the foregoing, we have seen that the Weil–Petersson type metric is favored both from the mathematical and from the (point-field) physical vantage point of view. Section F.2 will reveal that it also appears in the full string theory, at the lowest non-zero order. We then wish to estimate the distances in the relevant parameter spaces for the process

$$\begin{array}{ccc} & & \check{\mathcal{M}} \\ & & \big\downarrow{\scriptstyle \wp} \\ \mathcal{M}^\flat & \xrightarrow{r^2\to 0} & \mathcal{M}^\sharp \end{array} \quad . \tag{E.4.1}$$

In the space of complex structures of $\mathcal{M}^\flat$, the $\xrightarrow{r^2\to 0}$ process traces a "horizontal path," while a "vertical path" in the space of (complexified)

Kähler classes of $\check{\mathcal{M}}$ is traced by $\xrightarrow{\wp}$; see Figure E.1. We now discuss these in turn.

We need to show that the (exponentials of the) Kähler potentials on $\mathfrak{M}$ and $\mathfrak{W}$, $(i\int_{\mathcal{M}}\Omega\overline{\Omega})$ and $\int_{\mathcal{M}}J^3$, respectively, may degenerate but mildly enough so that the distance to the conifold converges. This is automatic for $\int_{\mathcal{M}}J^3$, as the dependence on the parameters is completely polynomial, but we also expect significant contributions from world-sheet instanton corrections. As discussed above, these dominate if their sizes are small—which is precisely the case when the Calabi–Yau 3-fold singularizes (the rational curves can tag onto the collapsing $\mathbb{P}^1$'s in the $\check{\mathcal{M}} \xrightarrow{\wp} \mathcal{M}^\sharp$ process). Theorem 7.4 ensures that the canonical divisor of $\check{\mathcal{M}}$ is being inherited by $\mathcal{M}^\sharp$, so the only uncertainty is indeed the behavior of $\int_{\mathcal{M}}J^3$. This remains an open problem although one expects, in view of the *mirror map* $^{(L)}$, that this is going to be as well behaved as $(i\int_{\mathcal{M}}\Omega\overline{\Omega})$.

E.4.1 Finiteness in the spaces of complex structure

Following Ref. [92], we present here a proof based on the monodromy properties of the A^a and B_a cycles.

The singularities which we wish to describe are isolated hypersurface singular points and we can consider a $\mathbb{C}^1$-family of manifolds (see § D.3.3) $\mathcal{M}^\flat{}_t$ given by hypersurfaces in some 4-fold and defined by

$$q + tp = 0\ , \qquad t \in \mathbb{C}\ , \quad |t| < \rho\ , \tag{E.4.2}$$

where $q = 0$ is a conifold and p does not vanish at the nodes of q. For any $0 \neq t < \rho$, $\mathcal{M}^\flat{}_t$ is smooth and a symplectic basis $\{A^a, B_a\}$ for $H_3(\mathcal{M}^\flat{}_t, \mathbb{Z})$ may be chosen such that $A^1, \ldots, A^k$ span the subgroup of $H_3(\mathcal{M}^\flat{}_t, \mathbb{Z})$ generated by the 3-spheres discussed in § D.2.3. This is possible since none of these 3-spheres meets any of the others. Then $B_{k+1}, \ldots, B_{h^{2,1}+1}$ do not meet these 3-spheres either and can be represented by cycles well removed from the nodes. $B_1, \ldots, B_k$ on the other hand, do meet these 3-spheres and each will include some of the 3-hoops, II^3, also discussed in § D.2.3.

Since $\{A^a, B_a\}$ is an integral basis it is locally constant as a function of t. However, if we make a complete circuit in the t-disk around the origin (which parametrizes the conifold), the basis will in general change. The $B_a(t)$ being simply integral cycles dual to the $A^a(t)$ and since $A^a(t) \cap A^b(t) = 0$, there is nothing to prevent the $B_a(t)$ from acquiring multiples of the $A^a(t)$. If we denote the monodromy operator by Σ, that is $\Sigma(A^a(t))$

and $\Sigma(B_a(t))$ are the results of transporting $A^a(t)$ and $B_a(t)$ around $t = 0$ in the t-disk, then we see that

$$\Sigma(A^a) = A^a \ , \qquad \Sigma(B_a) = B_a + \sigma_{ab} A^b \ , \tag{E.4.3}$$

where σ_{ab} is an integral matrix the entries of which vanish for a or $b > k$. From the fact that Σ respects the symplectic structure it follows that $\sigma_{ab} = \sigma_{ba}$. The erudite Reader will note the similarity with the Dehn twists of Riemann surfaces.

From § E.2 we recall that the periods $z^a(t) = \int_{A^a} \Omega_t$ and $\xi_a(t) = \int_{B_a} \Omega_t$ are holomorphic if $t \neq 0$ and if Ω_t depends holomorphically on t and has the local representation as in § D.2.3. We now need to estimate these periods.

Changing coordinates suitably, the equation $\mathcal{C}(x) = r^2(x)$ from Eq. (D.2.5) which defines $\mathcal{M}^\flat$ near the node of $\mathcal{M}^\sharp$ becomes

$$x^2 + y^2 + uv = t^2 \ , \qquad t^2 > 0 \ . \tag{E.4.4}$$

The quadratic expression on the left may be regarded as the ratio q/p, comparing to Eq. (E.4.2). S^3_t is then given by x, y real and $v = \overline{u}$, while the 3-cap B^3_t is given by $x > 0$, y imaginary and $v = -\overline{u}$.

The holomorphic 3-form now becomes

$$\Omega = f \frac{dx \wedge dy \wedge du}{u} \ , \tag{E.4.5}$$

where f does depends on p, but is non-zero and bounded at each of the nodes. The integral

$$\int_{S^3_t} \Omega \;=\; \int_{x^2+y^2 \leq t^2} dxdy \int_{u\bar{u}=t^2-x^2-y^2} f\,\frac{du}{u} \;\sim\; t^2 \tag{E.4.6}$$

vanishes quadratically as $t^2 \to 0$, since its absolute value is bounded by $2\pi^2 t^2 F$ where $|f| \leq F$.

✎ Derive this. ❧

On the other hand, by writing $y = i\gamma$ and R for the radius of the boundary of the neighborhood, we have

$$\int_{B^3_t} \Omega = \int_0^{\sqrt{(R^2+t^2)/2}} dx \int_0^{\sqrt{x^2-t^2}} d\gamma \int_{u\bar{u}=x^2-\gamma^2-t^2} f \frac{du}{u} \ . \tag{E.4.7}$$

Just as in the previous case, the last integral reduces to $\int_0^{2\pi} i f d\theta$, and the full expression manifestly remains bounded as $t^2 \to 0$. It is perhaps amusing to write out the first few terms:

$$\int_{B_t^3} \Omega \sim \Big| \int_0^{\sqrt{(R^2+t^2)/2}} \mathrm{d}x \int_0^{\sqrt{x^2-t^2}} \mathrm{d}\gamma \int_{u\bar{u}=x^2-\gamma^2-t^2} f\frac{\mathrm{d}u}{u} \Big| , \tag{E.4.8}$$

$$\sim f_0 \Big| \Big[x\sqrt{x^2-t^2} - t^2 \log\Big(x + \sqrt{x^2-t^2} \Big) \Big]_0^{\sqrt{(R^2+t^2)/2}} \Big| ,$$

$$\sim \Big\{ \tfrac{1}{2}R^2 + \Big| \log\Big(-i\sqrt{2}(\tfrac{t}{R}) \Big) \Big| t^2 - \frac{1}{4R^2}t^4 + \frac{1}{8R^4}t^6 + \dots \Big\} ;$$

only the t^2 term has a non-constant coefficient.

This calculation shows that $\lim_{t\to 0} z^a(t) = 0$ for $1 \leq a \leq k$, and can be extended to show that all the periods are bounded. Now, $\mathcal{G}_a(t)$ is not single valued for $a \leq k$, because of the monodromy. However,

$$\widehat{\mathcal{G}}_a(t) \stackrel{\text{def}}{=} \mathcal{G}_a(t) - \tfrac{1}{2\pi i}\sigma_{ab} z^b \log(t) \tag{E.4.9}$$

is single-valued since action of the monodromy cancels between the two terms on the right hand side. From the single-valuedness and the boundedness, it follows that the singularity at $t = 0$ is removable so that $\widehat{\mathcal{G}}_a(t)$ is also holomorphic at 0.

With these,

$$\begin{aligned} i\int_{\mathcal{M}^\flat{}_t} \Omega \wedge \overline{\Omega} &= -2\Im m \left(\overline{z}^a(t)\, \xi_a(t)\right) , \\ &= -2\Im m \Big(\overline{z}^a(t)\, \widehat{\xi}_a(t) + \tfrac{1}{2i\pi}\overline{z}^a(t)\sigma_{ab} z^b(t)\, \log(t) \Big) . \end{aligned} \tag{E.4.10}$$

Since $z^a(0) = 0 = z^b(0)$ for $\sigma_{ab} \neq 0$, this has the form

$$a(t) + b(t)|t|^2 \log|t| \ , \tag{E.4.11}$$

where $a(t)$ and $b(t)$ are C^∞ functions and $a(0) \neq 0$. This being the asymptotic behavior of the Kähler potential for the Weil–Petersson metric (along the disk in the moduli space which is parametrized by t), it is easy to see that the metric diverges at worst logarithmically. Therefore, the distance to the origin remains finite even though the metric becomes singular at $t = 0$.†

† The Reader unfamiliar with the type of computation presented here may benefit from consulting Ref. [92], where the result (E.4.11) has also been derived by straightforward but less rigorous computation, largely following Ref. [91]; for a fully rigorous proof, see [474].

In fact, much worse asymptotic behavior of $i\int_{\mathcal{M}}\Omega\wedge\overline{\Omega}$ would still lead to finite distance: even $i\int_{\mathcal{M}}\Omega\wedge\overline{\Omega}\sim|F(t)|^2\{a(t)+b(t)|t|^2\log|t|\}$ with $F(t)$ analytic in the neighborhood of $t=0$ but singular at $t=0$. It is therefore natural to expect certain singularizations of Calabi–Yau manifolds, with singularities worse than nodes also to be at a finite distance from the smooth manifolds. Indeed, Ref. [144] proves finite distance for singularizations of Calabi–Yau manifolds which have at worst isolated singular points with conical neighborhoods defined by

$$x^{e/a}+y^{e/b}+u^{e/c}+v^{e/d}=0\ , \tag{E.4.12}$$

with all exponents integral, (x,y,u,v) local coordinates of weights (a,b,c,d) and $e<a+b+c+d$.

Eq. (E.4.12), in particular, straightforwardly subsumes the cases of A_k, E_6 and E_8 singularities and the D_k and E_7 singularities can be shown to be orbifolds of (E.4.12). Therefore, indeed all A,D,E-singularities are at finite distance in the moduli space, which is related to the fact that the canonical divisor behaves relatively tamely in this singularizations.

Note also that we have used cohomology theory to describe the moduli spaces locally—which is indeed the general technique. In view of the § 7.3 discussion, the L^2-cohomology seems strongly motivated by recalling that the Calabi–Yau manifolds we study herein, $\mathcal{M}$, occur as *ground states* in the *underlying* world-sheet quantum field theory. Also, the Standard Model we aim to derive must stem from a *target-space* quantum field theory over the 9+1-dimensional space that includes $\mathcal{M}$ (see § 0.1). Since string theory *does* involve singularizations at both levels, the L^2-integrability standardly assumed in quantum field theories must adapt correspondingly. However, as the range of singularizations and defects that are admissible in string theory is not yet settled (see however [441]), neither can the ultimate criterion for a "true" stringy (co)homology theory. It then feels encouraging that the combined (co)homology theory of Ref. [250] is complementary to the intersection (co)homology (see Table D.1), and so is better suited for both conifold transitions (Chapter D) and the mirror map. However, it remains an open question whether this suffices in describing all members of the connected web of Calabi–Yau 3-folds Ref. [203], such as CICYs [149], their toric generalizations [305, 430, 431, 295], and related phenomena [445, 477, 419, 310, 297, 423]; see also § F.8.

2024 *Both the Yukawa couplings (E.2.14) and the geometry of the moduli space (E.2.7)–(E.2.21) are controlled by a single holomorphic function, $\mathcal{G}(z)$, of suitable moduli coordinates, z. The analogous is true (E.3.6)–(E.3.7)—as* *it must, by mirror symmetry—on the Kähler side. This property is known as* special geometry *[219, 378] and has been identified as a hallmark of string theory [356]. This exhibits the conceptual* symmetry $\mathfrak{W} \leftrightarrow \mathfrak{M}$ *in the geometry of these moduli spaces, mending the apparent asymmetry noted on p. 144, in the § D.3 discussion of these moduli spaces. As noted in the Remark containing Eq. (1.2.16), Calabi–Yau n-folds come in* mirror pairs $(\mathcal{M}, \mathcal{W})$ *such that, completing (L.43):*

$$\begin{aligned} H^q(\mathcal{W}, \wedge^p \mathcal{T}^*_{\mathcal{W}}) = H^{p,q}(\mathcal{W}) \;&\cong_Q\; H^{n-p,q}(\mathcal{M}) \overset{(1.2.3)}{=\!=} H^q(\mathcal{M}, \wedge^p \mathcal{T}_{\mathcal{M}}) \;, \\ H^q(\mathcal{M}, \wedge^p \mathcal{T}^*_{\mathcal{M}}) = H^{p,q}(\mathcal{M}) \;&\cong_Q\; H^{n-p,q}(\mathcal{W}) \overset{(1.2.3)}{=\!=} H^q(\mathcal{W}, \wedge^p \mathcal{T}_{\mathcal{W}}) \;. \end{aligned} \tag{E.4.13}$$

This "mirror symmetry" is known to persist [68] throughout most known constructions [309] and has been proven for $n = 1, 2$ by M. Kontsevich (the first of Refs. [279]) and P. Seidel [461], respectively.

Comparing the far-left and far-right valuing bundles in (E.4.13) suggests that the mirror mapping may be thought of as the formal swapping of tangent and cotangent bundles,

$$(\mathcal{M}; \mathcal{T}_{\mathcal{M}}, \mathcal{T}^*_{\mathcal{M}}) \;\leftrightarrow\; (\mathcal{W}; \mathcal{T}^*_{\mathcal{W}}, \mathcal{T}_{\mathcal{W}}) \;. \tag{E.4.14}$$

Instead of the holomorphic tangent bundle $\mathcal{T}_{\mathcal{M}} := \bigcup_{p \in \mathcal{M}} T_p$ separately from the cotangent bundle $\mathcal{T}^*_{\mathcal{M}} := \bigcup_{p \in \mathcal{M}} T^*_p$, the mirror mapping (E.4.14) suggests considering the combined bundle, $\bigcup_{p \in \mathcal{M}} (T_p \oplus T^*_p)$. This is indeed the key starting point of the "generalized geometry" framework [404], originally motivated by analyzing constrained systems [366]. This framework was soon found to correspond to the quantum field theory framework of "double field theory" [471]. A key motivation is also provided by the fact that each Calabi–Yau n-fold varies over both *its complex structure moduli space, $\mathfrak{M}$, as well as over its complexified Kähler structure space, $\mathfrak{W}$, which gives rise to a "universal geometry" of heterotic string vacua [351]. Since* End $\mathcal{T}_{\mathcal{M}}$ *is the traceless part of $\mathcal{T}_{\mathcal{M}} \otimes \mathcal{T}^*_{\mathcal{M}}$, both generalized and universal geometry then also provide a natural home to the "E_6-singlets," corresponding to the elements of $H^1(\mathcal{M}, \operatorname{End} \mathcal{T}_{\mathcal{M}})$; see § 0.8 and Chapter B. We will return to these notions in § F.8.*

Part IV
Concordance

Chapter F

A Prelude to Quantum Geometry

Having spent essentially all of this volume on various techniques and methods in (mostly algebraic) geometry, as adapted for dealing with Calabi–Yau 3-folds, we are now ready to look back at our initial aim—superstring compactification. At this point then a disclaimer is in order (if not already overdue). It is simply physically not possible to collect all the vigorously industrious study in this area, compress it into a volume that could be held in a single hand and fairly represent all results. Instead, a modest sample is presented, with the hopes that the Reader will find them inviting and motivating. Meanwhile, developments march on, and the Reader is strongly advised to catch up with Refs. [241, 174], [100, 101, 102], [178], [57, 59, 60, 75, 71, 88] and the references therein and thereafter.

F.1 Some Generalities

To describe the propagation of the heterotic superstring, we need to deal with local (1,0)-supersymmetry on the world-sheet and we adopt the superspace formalism (for a general reference, see Ref. [21]).

F.1.1 Flat spacetime

Certainly the simplest example of some relevance is the case where the heterotic string propagates through flat spacetime, that is, 9+1-dimensional

Minkowski spacetime [160]. The action for the underlying 2-dimensional field theory has two main parts, $S^{D=10}_{String}$ and S_{Gauge}. The former describes string propagation through spacetime

$$S^{D=10}_{String} = \int_{\Sigma} \mathrm{d}^2\sigma\, \mathrm{d}\varsigma^-\; E^{-1}\left[i\tfrac{1}{2}\, \eta_{\underline{m}\,\underline{n}}\, (\nabla_+ \boldsymbol{X}^{\underline{m}})\, (\nabla_= \boldsymbol{X}^{\underline{n}}) \right], \tag{F.1.1}$$

where E is the superdeterminant of the super-Zweibein. The off-shell string coordinate superfields are given by

$$\boldsymbol{X}^{\underline{m}}(\varsigma^+, \sigma, \tau) = X^{\underline{m}}(\sigma, \tau) \;+\; \varsigma^+ \psi_+{}^{\underline{m}}(\sigma, \tau)\;. \tag{F.1.2}$$

Here, $X^{\underline{m}}$ map the world-sheet into the flat Minkowski spacetime, the latter being specified through $\eta_{\underline{m}\,\underline{n}}$. Imposing the equations of motion,

$$\boldsymbol{X}^{\underline{m}}(\varsigma^+, \sigma, \tau) = X^{\underline{m}}(\sigma^{\ddagger}) + X^{\underline{m}}(\sigma^{=}) \;+\; \varsigma^+ \psi_+{}^{\underline{m}}(\sigma^{\ddagger})\;. \tag{F.1.3}$$

Thus, on-shell $X^{\underline{m}}$ contains both left- and right-moving modes while the equation of motion for each $\psi_+{}^{\underline{m}}$ implies that it is left-moving, i.e., independent of $\sigma^{=}$.

To propagate in a curved spacetime instead, the replacement

$$\eta_{\underline{m}\,\underline{n}} \;\to\; G_{(\underline{m}\,\underline{n})}(\boldsymbol{X}) \;+\; B_{[\underline{m}\,\underline{n}]}(\boldsymbol{X}) \tag{F.1.4}$$

would be necessary, introducing interactions among the $X^{\underline{m}}$ and $\psi_+{}^{\underline{m}}$, non-linear in $X^{\underline{m}}$. In the effective (target spacetime) theory, the background (vacuum expectation) value of these $X(\sigma)$-dependent coupling tensors determine the geometry of the spacetime in which the string propagates. Their fluctuations represent the various states stemming from the $D{=}10$ supergravity sector.

As classical field theories, the above action is perfectly consistent. For quantum field theories, several of the symmetries will in general be violated by anomalies. The standard field theory trick is of course to find collections of fields such that all anomaly contributions cancel out. To give ourselves more freedom in doing so, we add another type of terms to the action.

F.1.2 The Yang–Mills sector

The second part, S_{Gauge}, introduces the degrees of freedom usually referred to as 'internal' and thereby allows us to shift the burden of anomaly cancellation into something less conspicuous than, say, 26-dimensional

Minkowski spacetime. Luckily, perhaps, this summand is (by far) not unique. Consider the fermionic representation

$$S^{(32)}_{Gauge} = -\tfrac{1}{2}\int_\Sigma \mathrm{d}^2\sigma\,\mathrm{d}\varsigma^-\; E^{-1}\left[\boldsymbol{\eta}_-{}^{\hat I}\,\delta_{\hat I\hat J}\nabla_+\,\boldsymbol{\eta}_-{}^{\hat J}\right], \tag{F.1.5}$$

with $\hat I = 1,\ldots,32$, where only the lowest component field of each $\boldsymbol{\eta}_-{}^{\hat I}$ propagates and which is enforced by the equations of motion to be right-moving, i.e., independent of $\sigma^{\mp}$. A simple calculation tells that, together with $S^{D=10}_{String}$, this provides a consistent model, the *Spin*$(32)/\mathbb{Z}_2$-heterotic string.

For the *Spin*$(32)/\mathbb{Z}_2$ model, these 32 fermionic fields $\eta_-{}^{\hat I}$ form the vector representation of *Spin*(32). The 496 fermionic bilinear combinations

$$\eta_-{}^{[\hat I}\eta_-{}^{\hat J]} \tag{F.1.6}$$

span the *complete* adjoint representation and generate the full group, so that σ-model coupling tensors can be added through minimal coupling

$$\delta_{\hat I\hat J}\nabla_+ \;\to\; \delta_{\hat I\hat J}\nabla_+ \;+\; A_{+\,[\hat I\hat J]}(\boldsymbol{X})\;, \tag{F.1.7}$$

and generate *Spin*(32). The world-sheet gauge (super)field $A_{+\,[\hat I\hat J]}$ is the so-called "pull-back" of the target space gauge field:

$$A_{+\,[\hat I\hat J]}(\boldsymbol{X}) \;=\; (\nabla_+\boldsymbol{X}^{\underline{m}})\,A_{\underline{m}\,[\hat I\hat J]}(\boldsymbol{X}) \;\;. \tag{F.1.8}$$

Finally, the passage $\eta_{\underline{m}\,\underline{n}} \to G_{(\underline{m}\,\underline{n})}(\boldsymbol{X}) + B_{[\underline{m}\,\underline{n}]}(\boldsymbol{X})$ need not be done for the full span of 9+1-spacetime dimensions, rather, we will want to leave 3+1-spacetime dimensions flat and curve only the others. Clearly, many other possibilities exist—from the 2-dimensional point of view—involving the usual palette of fields: scalars, spinors, and so on.

The partitioning of the ten spacetime coordinate fields into 3+1 flat ones and 6 curved ones suspiciously looks like compactification, but leaves two essential questions. (1) How do we know that the curved space is compact? (2) What other criteria are there to be satisfied? The answer to the latter one is that the criteria are essentially translations of the various conditions known from point field theory (with the known exception of modular invariance). As to the compactness, the actions so far described in fact do not know anything about it. However, recall that the fundamental object in quantum field theory is the partition function(al), which involves (path-)integration over the various fields used. The integration range is where the quantum field theory registers compactness and also other global information. In general, this will not be easily accessible. See Ref. [135] for more on this.

F.2 Calabi–Yau σ-Models in General

To simplify somewhat the notation, we temporarily restrict to the bosonic sector*. Consider the action functional governing the propagation of strings in an "internal" space:

$$\mathcal{A} = -\frac{1}{2\pi\alpha'}\int_\Sigma \Big\{(G_{mn} + iB_{mn})(\bar{\partial}\boldsymbol{X}^m(\sigma))(\partial\boldsymbol{X}^n(\sigma)) \\ + (G_{mn} - iB_{mn})(\partial\boldsymbol{X}^m(\sigma))(\bar{\partial}\boldsymbol{X}^n(\sigma))\Big\}\ .$$

On Calabi–Yau spaces, it is standard to choose a background Kähler metric $G_{\mu\bar{\nu}}$ and set $B = 0$ in vacuum. The relevant σ-model describing a Calabi–Yau background action is then

$$\mathcal{A}_0 = -\frac{1}{\pi\alpha'}\int_\Sigma G_{\mu\bar{\nu}}(Z,\bar{Z})\Big[(\bar{\partial}Z^\mu(\sigma))(\partial Z^{\bar{\nu}}(\sigma)) + (\partial Z^\mu(\sigma))(\bar{\partial}Z^{\bar{\nu}}(\sigma))\Big]\ . \tag{F.2.1}$$

To qualify for describing superstring vacua, the above action needs to be conformally invariant which is, up to four loop corrections, equivalent to the Ricci-flatness of $G_{\mu\bar{\nu}}(Z,\bar{Z})$. At four loops, the condition acquires a correction proportional to the double derivative of the Euler density, but it is possible to redefine G and B suitably; this does not change their respective cohomology classes [196]. This being so we may as well concentrate on the Ricci-flat metric for simplicity.

Any choice of such a $G_{\mu\bar{\nu}}(Z,\bar{Z})$ simply specifies a point in the relevant moduli space $(\mathfrak{M},\mathfrak{W})$. To determine the metric on the moduli space at that point, we shall need to consider deformations† of this choice and evaluate second order effects. We skim through this here, following Ref. [94]; a more complete and careful field theoretic analysis is found in Ref. [120].

* In this Euclidean world-sheet formulation, $\bar{\partial} \leftrightarrow \partial_{\ddagger}$ and $\partial \leftrightarrow \partial_{=}$.

† It is important to note that the deformations of the σ-model action (F.2.1) describing Calabi–Yau vacua of superstring theories may be parametrized through deformations of the complex structure and of the Kähler class.

Defining the following variations‡ of $\mathcal{A}_0$,

$$\mathcal{A}_{(1,1)} \stackrel{\text{def}}{=} -\frac{1}{\pi\alpha'}\int_\Sigma \Big[\mathcal{G}^+_{\mu\bar\nu}(\bar\partial Z^\mu(\sigma))(\partial Z^{\bar\nu}(\sigma)) + \mathcal{G}^-_{\mu\bar\nu}(\partial Z^\mu(\sigma))(\bar\partial Z^{\bar\nu}(\sigma))\Big]\,, \tag{F.2.2}$$

$$\mathcal{A}_{(2,1)} \stackrel{\text{def}}{=} -\frac{1}{\pi\alpha'}\int_\Sigma \Big[\mathcal{G}_{\mu\nu}(\bar\partial Z^\mu(\sigma))(\partial Z^{\nu}(\sigma)) + \mathcal{G}_{\overline{\mu\nu}}(\partial Z^{\overline{\mu}}(\sigma))(\bar\partial Z^{\bar\nu}(\sigma))\Big]\,, \tag{F.2.3}$$

$\mathcal{A}_0 + \mathcal{A}_{(1,1)} + \mathcal{A}_{(2,1)}$ is the most general expression for an action that still has the geometrical meaning of embedding the world-sheet into the target (Calabi–Yau) manifold.

For a conformally invariant theory, Zamolodchikov's metric [243] is the (fourth order) residue of the irreducible two-point function $\langle\Phi_I(\sigma)\Phi_K(0)\rangle$ where the set $\{\Phi_I(\sigma)\}$ parametrizes the variations of the background action density. As always,

$$\lim_{\sigma\to 0}|\sigma|^4\frac{\delta}{\delta J^I(\sigma)}\frac{\delta}{\delta J^K(0)}\,, \tag{F.2.4}$$

where $J^I(\sigma)$ are suitable sources probing the variation $\Phi_I(\sigma)$. For $\mathcal{A}_0 + \mathcal{A}_{(1,1)} + \mathcal{A}_{(2,1)}$, we introduce the world-sheet sources through

$$\begin{aligned} \mathcal{G}^+_{\mu\bar\nu} &= J^{A+}(\sigma)\,\mathcal{G}^+_{A\mu\bar\nu}(Z,\bar Z)\,, \qquad & \mathcal{G}^-_{\mu\bar\nu} &= J^{A-}(\sigma)\,\mathcal{G}^-_{A\mu\bar\nu}(Z,\bar Z)\,,\\ \mathcal{G}_{\mu\nu} &= T^{\alpha}(\sigma)\,\mathcal{G}_{\alpha\mu\nu}(Z,\bar Z)\,, & \mathcal{G}_{\bar\mu\bar\nu} &= T^{\bar\alpha}(\sigma)\,\mathcal{G}_{\bar\alpha\bar\mu\bar\nu}(Z,\bar Z)\,. \end{aligned} \tag{F.2.5}$$

Of course, we have

$$\mathcal{G}^\pm_{\mu\bar\nu} \;=\; \delta G_{\mu\bar\nu} \pm i\delta B_{\mu\bar\nu}\,. \tag{F.2.6}$$

We can evaluate the Zamolodchikov metric perturbatively, by expanding $Z^\mu = z^\mu + \widetilde{Z}^\mu$, where z^μ is constant over the world-sheet and $\widetilde{Z}^\mu$ are fluctuations around z^μ. Similarly, we expand all functions of Z^μ, for example, $G_{\mu\bar\nu}(Z) = g_{\mu\bar\nu}(z) + \ldots$ Rather easily, this proves that the lowest non-zero contribution to the Zamolodchikov metric is the Weil–Petersson type metric in Eq. (E.1.6) [94], up to a basis-change [120].

The alert Reader might have noticed that $\overline{\mathcal{G}^\pm_{\mu\bar\nu}\,\bar\partial Z^\mu\,\partial Z^{\bar\nu}} = \mathcal{G}^\mp_{\mu\bar\nu}\,\bar\partial Z^\mu\,\partial Z^{\bar\nu}$. It therefore appears that another parametrization may be more appropriate for the complexified Kähler cone, one in which the two types of variations

‡ The correspondence identifies $\mathrm{d}\bar\sigma\,\bar\partial Z^{\underline{m}} \leftrightarrow \mathrm{d}z^{\underline{m}}$, so that $\partial Z^\mu\,\bar\partial Z^{\bar\nu} \leftrightarrow \mathrm{d}z^\mu \mathrm{d}z^{\bar\nu} \leftrightarrow H^1(\mathcal{M}, \mathcal{T}^*_\mathcal{M})$ and $\partial Z^{\bar\mu}\,\bar\partial Z^{\bar\nu} \leftrightarrow g^{\rho\bar\mu}\partial_\rho \mathrm{d}z^{\bar\nu} \leftrightarrow H^1(\mathcal{M},\mathcal{T}_\mathcal{M})$ in the second terms, the first being their conjugates: world-sheet $\bar\partial$-derivatives map to the Dolbeault cohomology on $\mathcal{M}$, with ∂-defined valuation.

will be one another's complex conjugate. Indeed, since

$$\Big[\delta G_{\mu\overline{\nu}} \pm \delta B_{\mu\overline{\nu}}\Big] \equiv \tfrac{1}{\pm 2i}(1\pm i)\Big[\mathcal{G}^{+}_{\mu\overline{\nu}} \pm i\,\mathcal{G}^{-}_{\mu\overline{\nu}}\Big]\,, \tag{F.2.7}$$

we have $W^A\,\mathcal{G}_{A\,\mu\overline{\nu}} = [\delta G_{\mu\overline{\nu}} - \delta B_{\mu\overline{\nu}}]$ and $W^{\overline{A}}\,\overline{\mathcal{G}}_{\overline{A}\,\mu\overline{\nu}} = [\delta G_{\mu\overline{\nu}} + \delta B_{\mu\overline{\nu}}]$. Then, the constant mode of $W^A(\sigma)$ is exactly the parameter w^A introduced in Eq. (E.3.4). Furthermore, let us just state here that marginal operators as in $\mathcal{A}_{(1,1)}$ correspond to harmonic (1,1)-forms and become twisted-chiral superfields when supersymmetrized. This then provides the field-theoretic version of the complexified Kähler cone and space of complex structures (see also Ref. [97]).

On the other hand, the constant mode of $T^\alpha(\sigma)$ may be identified with the parameter t^α introduced in (E.2.1) as it is. Note also that only a non-holomorphic redefinition of coordinate fields can possibly bring $\mathcal{A}_0 + \mathcal{A}_{(2,1)}$ in the form of $\mathcal{A}_0$. This clearly corresponds to deformations of the complex structure. Hence the correspondence with (2,1)-forms which in turn correspond to marginal modes in $\mathcal{G}_{\overline{\mu}\overline{\nu}}$. These become chiral superfields when supersymmetrized. It transpires that the parameter space relevant to the above σ-model coincides with the combined space of complex structures and the Kähler cone (complexified through the inclusion of δB). In addition, it is precisely these geometrical data on the Calabi–Yau manifold that determine the essential physical observables, such as the Yukawa couplings, in the low-energy limit.

F.3 The Projective σ-Model

Just as Calabi–Yau spaces are being constructed from certain better known spaces, such as $\mathbb{P}^n$'s, world-sheet σ-models can also be treated in a similar fashion.

F.3.1 Three projective actions

We start with the well known supersymmetric $\mathbb{P}^n$ model in 2-dimensional spacetime with (2,2)-supersymmetry.[§] The classical action is

$$\begin{aligned}\mathcal{A}[\mathbb{P}^n] &\stackrel{\text{def}}{=} \textstyle\int_\Sigma \mathrm{d}^2\sigma\ \mathrm{d}^2\varsigma\,\mathrm{d}^2\overline{\varsigma}\,\Big(\,\|\boldsymbol{X}\|^2 e^{-\mathbf{V}} + \tfrac{n+1}{2f}\boldsymbol{V}\Big)\,,\\ \|\boldsymbol{X}\|^2 &\stackrel{\text{def}}{=} \textstyle\sum_{\mu=0}^{n} \boldsymbol{X}^\dagger{}_\mu\,\boldsymbol{X}^\mu\ .\end{aligned} \tag{F.3.1}$$

§ Only a quarter of this supersymmetry was needed by local coordinate reparametrization independence on the world-sheet. The *additional* 3/4 of the (2,2)-supersymmetry may well be relaxed, which is known to provide many physically interesting generalizations; see [371, 459, 463].

It involves $n{+}1$ chiral superfields, $\boldsymbol{X}^\mu$, and a gauge superfield, $\boldsymbol{V}$ for which no kinetic term is introduced. $\boldsymbol{V}$ gauges a complexified and supersymmetric version of $U(1)$, which we denote by $U(1;\mathbb{C})$ and which includes $U(1)$, dilation and their supersymmertic counterpart, a chiral symmetry acting on the fermion component fields in $\boldsymbol{X}^\mu$. Indeed, the body of the classical configuration space (spanned by the unconstrained bosonic fields) is $[\mathbb{C}^{n+1} \smallsetminus \{0\}]/U(1,\mathbb{C}) = \mathbb{P}^n$. The $U(1;\mathbb{C})$ gauge transformation, also known as the Kähler symmetry, acts by

$$\boldsymbol{X} \;\to\; e^{\Theta}\boldsymbol{X}\,, \qquad \boldsymbol{V} \;\to\; \boldsymbol{V} + \tfrac{1}{2i}(\Theta - \Theta^\dagger)\,, \tag{F.3.2}$$

where Θ is an arbitrary chiral superfield and $\Theta^\dagger$ its antichiral Hermitian conjugate.

The equation of motion for $\boldsymbol{V}$ reads $\boldsymbol{V} = \log\big(\frac{2f}{n+1}\|\boldsymbol{X}\|^2\big)$. Using this to eliminate $\boldsymbol{V}$ is equivalent to path-integration over $\boldsymbol{V}$ and produces, up to an uninteresting additive constant,

$$\begin{aligned} \mathcal{A}'[\mathbb{P}^n] &= \int_\Sigma \mathrm{d}^2\sigma\, \mathrm{d}^2\varsigma\, \mathrm{d}^2\bar{\varsigma}\; K_{(FS)}(\boldsymbol{X}^\dagger, \boldsymbol{X})\,, \\ K_{(FS)}(\boldsymbol{X}^\dagger, \boldsymbol{X}) &= \frac{n{+}12f}{\log}\|\boldsymbol{X}\|^2\,, \end{aligned} \tag{F.3.3}$$

where $\|\boldsymbol{X}\|^2 {=} \sum_{\mu=0}^{n} \boldsymbol{X}^\dagger_\mu \boldsymbol{X}^\mu$ and $K_{(FS)}$ is the Fubini–Study Kähler potential (5.1.8). Since the action (F.3.1) is quadratic in $\boldsymbol{X}$'s, it is also possible to integrate out the $\boldsymbol{X}$'s, which results in [107]:

$$\mathcal{A}''[\mathbb{P}^n] \;\stackrel{\text{def}}{=}\; \frac{n{+}1}{4\pi}\int_\Sigma \mathrm{d}^2\sigma \Big(\int \mathrm{d}\varsigma^+\, \mathrm{d}\bar{\varsigma}^-\; \boldsymbol{S}[\log(\boldsymbol{S}/\mu) - 1] \;+\; \text{h.c.}\Big) \;+\; \cdots \tag{F.3.4}$$

where the higher terms all involve $\int \mathrm{d}^2\varsigma\, \mathrm{d}^2\bar{\varsigma}$ -integrals.

F.3.2 Marginal operators

The superfields $\boldsymbol{S} \stackrel{\text{def}}{=} D_-\bar{D}_+\boldsymbol{V}$ and its Hermitian conjugate are both twisted-chiral superfields [136]. Since the leading terms in $\mathcal{A}''[\mathbb{P}^n]$ involve integration over only "half" of the superspace, these terms are protected by the usual non-renormalization theorems regarding (twisted) F-terms. So, while many D-terms supply only irrelevant operators [99], there do exist D-terms such as in Eq. (F.3.1) which have a marginal residue, represented in Eq. (F.3.4). That the kinetic terms considered here are not in the universality class represented by $K_0(\boldsymbol{X}^\dagger, \boldsymbol{X}) = \boldsymbol{X}^\dagger\boldsymbol{X}$ is most easily seen from the distinct behavior of the partition functionals. When defined with the kinetic term (F.3.3), a partition function clearly has an essential singularity at $\boldsymbol{X}^\mu = 0$. Removing the origin from the field-space, however, changes its topology and is seen to distinguish the two universality classes.

A few remarks are in order before carrying on to related models. Firstly, the continuous gauge symmetry (F.3.2) is anomalous; while the classical actions (F.3.1) and (F.3.3) are invariant, the action (F.3.4) is not, except for the leading term $\boldsymbol{S}$ and its conjugate [107]. The "leading log" term $\boldsymbol{S}\log(\boldsymbol{S}/\mu)$ and its conjugate carry the $U(1;\mathbb{C})$ gauge-anomaly of the $\mathbb{P}^n$ model; the numerical value is $(n{+}1)$ and equals the first Chern class of $\mathbb{P}^n$; the $1/4\pi$ factor turns out to be merely a suitable normalization. Also, because of the $(n{+}1)$ prefactor, however, a $\mathbb{Z}_{n+1}$ subgroup of (F.3.2) does remain a symmetry and its action has to be divided out. This will turn out to be the Lagrangian root of the GSO-type projection in the Landau–Ginzburg models.

Secondly, note that the equation of motion for $\boldsymbol{V}$ and the definition of the superfield $\boldsymbol{S}$ lets us re-interpret Eq. (F.3.4) in terms of the $\boldsymbol{X}$'s. To that end we record

$$\boldsymbol{S} \sim (D_-\boldsymbol{X}^\mu)\, G^{(FS)}_{\mu\bar\nu}\, (\bar D_+\boldsymbol{X}^{\dagger\bar\nu})\;, \qquad G^{(FS)}_{\mu\bar\nu} \stackrel{\text{def}}{=} \partial_\mu\, \partial_{\bar\nu} K_{(FS)}(\boldsymbol{X},\boldsymbol{X}^\dagger)\;, \tag{F.3.5}$$

where $G^{(FS)}_{\mu\bar\nu}$ is the Fubini–Study metric. Clearly, the (anticommuting) fermionic derivatives act as the field theory generalization of the dz's and d$\bar z$'s¶. Hence, the quantity

$$(\boldsymbol{S}\,\text{“}{+}\text{”}\,\boldsymbol{S}^\dagger) \stackrel{\text{def}}{=} \Big(\int \mathrm{d}\varsigma^+\,\mathrm{d}\bar\varsigma^-\; \boldsymbol{S} + \int \mathrm{d}\varsigma^-\,\mathrm{d}\bar\varsigma^+\; \boldsymbol{S}^\dagger\Big) \tag{F.3.6}$$

has a natural interpretation as the (super)field theory generalization of the Fubini–Study Kähler (1,1)-form; this supports the identification of the $U(1;\mathbb{C})$-anomaly with the first Chern class. Note that the term (F.3.6), with the interpretation (F.3.5), also appears when the "twisted half" of the fermionic integral is explicitly performed in Eq. (F.3.3).

Finally, the coupling constant f from $\mathcal{A}[\mathbb{P}^n]$ has undergone a "dynamical transmutation": in Eq. (F.3.4), it occurs through the renormalization 'mass' scale♁ $\mu{=}const.\,\varepsilon\, \mathrm{e}^{2\pi/f}$, where ε is the dimensional regularization parameter. When describing a σ-model with $\mathbb{P}^{n_1}\times\cdots\times\mathbb{P}^{n_m}$ target space, the effective action (F.3.4) is replaced by a linear combination of such

¶ Indeed, the lowest components of superfields such as $(D_\pm\boldsymbol{X}^\mu){=}\psi^\mu_\pm + \ldots$ have been identified as formal analogues of dz ever since the early work on σ-models [98]; see footnote ‡ on p. 313.

♁ In d-dimensional spacetime, μ has dimension $(d{-}2)/2$ and is actually dimensionless in $d=2$.

actions, one for each $\mathbb{P}^{n_i}$ factor, which contains the linear combination $\sum_{i=1}^{m}(const.+(2\pi/f_i))(\boldsymbol{S}_i“+”\boldsymbol{S}_i^\dagger)$. Indeed, the Kähler class of a product of $\mathbb{P}^n$'s is a linear superposition of the individual Kähler classes.

F.4 Constrained Models

Many Kähler manifolds are constructed by embedding in $\mathbb{P}^n$'s and rather important examples are obtained as the (sub)space of common solutions to a system of constraints in a product of complex projective spaces. Most naturally, one imposes the constraints $P^a(\boldsymbol{X}) = 0$ by means of Lagrange multiplier fields on a system of $\mathbb{P}^n$ models [185, 135]. The corresponding action is $\mathcal{A}_{\rm kin.} + \mathcal{A}_{\rm con.}$, where

$$\mathcal{A}_{\rm kin.} = \sum_{i=1}^{m} w^i \int_\Sigma {\rm d}^2\sigma\, {\rm d}^2\varsigma\, {\rm d}^2\bar\varsigma\, \Big(\|\boldsymbol{X}_i^{\mu_i}\|^2 {\rm e}^{-\mathbf{V}_i} + \frac{n_i{+}1}{2f_i}\boldsymbol{V}_i \Big) , \tag{F.4.1}$$

$$\mathcal{A}_{\rm con.} = \sum_{a=1}^{K} \int_\Sigma {\rm d}^2\sigma\, {\rm d}^2\varsigma\, \Big(\boldsymbol{\Lambda}_a\, P^a(\boldsymbol{X}) + {\rm h.c.} \Big) \tag{F.4.2}$$

and where $\boldsymbol{X}_i^{\mu_i}$ are coordinate superfields on $\mathbb{P}^{n_i}$. While this provides the exact field theory parallel of the construction of such Kähler spaces, its use in practical computations is limited, because of the non-linear couplings in (F.4.1) and since the Lagrange multiplier fields effectively introduce infinitely strong coupling (see however Ref. [169, 170] for some applications).

Since the terms in $\mathcal{A}_{\rm con.}$ are generally not quadratic in the $\boldsymbol{X}$'s, we cannot in general perform the $\int {\sf D}[\boldsymbol{X}]$ path-integration and obtain the constrained space analogues of Eq. (F.3.4) and (F.3.5). On general grounds, we know that analogous relations must exist although an explicit evaluation eludes us. Clearly, the analogue of $\boldsymbol{S}$ should then again be of the form (F.3.5), featuring, however the physical metric on the constrained space $\mathcal{M}$,

$$\boldsymbol{S}_\mathcal{M} \sim (D_-\boldsymbol{X}^\mu)\, G^{(\mathcal{M})}_{\mu\bar\nu}\, (\bar D_+\boldsymbol{X}^{\dagger\bar\nu}) , \tag{F.4.3}$$

where the indices μ and $\bar\nu$ have to be restricted so as to label directions locally (co)tangent to $\mathcal{M} \subset \mathbb{P}^{n_1}\times\cdots\times\mathbb{P}^{n_m}$. This is most easily performed by inserting appropriate local projection operators of the form $\mathbb{1}-[\partial P]$, where the matrix of gradients $\partial_M P^a(\boldsymbol{X})$ and its conjugate serve as projections (locally on $\mathcal{M}$) from $\mathbb{P}^{n_1}\times\cdots\times\mathbb{P}^{n_m}$ to the directions transversal to $\mathcal{M}$.

Having identified the $U(1;\mathbb{C})$-anomaly $(n{+}1)\,\boldsymbol{S}\log(\boldsymbol{S}/\mu)$ with the first Chern class of $\mathbb{P}^n$, one expects that integration over the $\boldsymbol{X}$'s, for a degree-q

polynomial constraint would reduce this anomaly to $(n+1-q)\,\boldsymbol{S}\log(\boldsymbol{S}/\mu)$. For example, if the constraint polynomial were linear, it would simply require a linear combination of $\boldsymbol{X}^{\mu}$'s to vanish. Integrating the remaining n $\boldsymbol{X}^{\mu}$'s would yield the decreased anomaly $n\,\boldsymbol{S}\log(\boldsymbol{S}/\mu)$, as expected.

In the general case, the explicit path-integration eludes us, but the anomaly contribution can be discerned by considering the transformation of the path-integral measure [130]. Alternatively, in particle physics parlance, integrating out the superfields $\boldsymbol{X}_i^{\mu_i}$ allows us to interpret the $\boldsymbol{\Lambda}_a$ as bound states of the charge-conjugates of those $\boldsymbol{X}$'s which appear in $P^a(\boldsymbol{X})$ (see Ref. [208] for a situation where such an interpretation is experimentally verified), so that the so "dressed" $\boldsymbol{\Lambda}_a$'s will have charge* $-q_a^i$ with respect to $\boldsymbol{V}_i$, where q_a^i is the degree of $P^a(\boldsymbol{X})$ with respect to the coordinate superfields of $\mathbb{P}^{n_i}$. There will also appear effective propagators for $\boldsymbol{\Lambda}_a$'s and $\boldsymbol{V}_i$'s ($\boldsymbol{S}_i$'s), whence the anomaly 1-loop diagrams yield the coefficients $(n_i{+}1 - \sum_{a=1}^{K} q_a^i)$ in place of $(n_i{+}1)$ in Eq. (F.3.4). This is in perfect agreement with our identification of the chiral ("Kähler") gauge anomaly with the first Chern class of the constrained subspace $\mathcal{M}$.

F.5 The Calabi–Yau Case

For $\mathcal{M}$ to be a Calabi–Yau space, the first Chern class of $\mathcal{M}$ has to vanish, so

$$\sum_{a=1}^{K} \deg_{\mathbb{CP}^{n_i}}\big(P^a(\boldsymbol{X})\big) \;=\; n_i{+}1\;, \qquad i=1,\dots,m\;. \tag{2.1.17}$$

Therefore, the "Kähler" $U(1)$ gauge anomaly of each $\boldsymbol{V}_i$ is completely canceled. However, the numerical pre-factor in the effective action analogous to $\mathcal{A}''[\mathbb{P}^n]$, Eq. (F.3.4), appears also to have been annihilated. Can it be that the anomaly cancellation efforts have been counter-productive, in that Eq. (2.1.17) not only ensures cancellation of the chiral gauge anomaly, but also kills the marginal terms in an effective action which represent the Kähler variations?

The answer clearly ought to be negative. In fact, it is not hard to see how that comes about. In any realistic situation, the couplings to (super)gravity

---

* Note that, before integrating the $\boldsymbol{X}^{\mu}$ out, the Lagrange superfields $\boldsymbol{\Lambda}_a$ in $\mathcal{A}_{\text{con.}}$ must have no charge with respect to the gauge fields $\boldsymbol{V}_i$; for, if they had, they would have to interact with the $\boldsymbol{V}_i$. In supersymmetric theories this would be possible only through kinetic terms for $\boldsymbol{\Lambda}_a$'s which would contradict their rôle as Lagrange multiplier superfields.

should be included. When considering 2-dimensional field theories to describe Calabi–Yau compactification of heterotic strings, for example, the constrained (product of) $\mathbb{P}^n$ model(s), Eq. (F.4.1) and (F.4.2), should be coupled to (1,0)-supergravity [135]. Indeed, the (super)gravitational anomaly [56] is nothing but the well-known trace-anomaly, i.e., the central charge of the Virasoro algebra. But then, the locally (1,0)-supersymmetric version of $\mathcal{A}_{\text{kin.}}+\mathcal{A}_{\text{con.}}$ in fact must be anomalous and precisely so as to cancel the contributions of a flat σ-model representing the 4-dimensional spacetime and of the super-reparametrization ghosts. In 2-dimensional field theories, the chiral gauge-, spin-3/2 and gravitational anomalies are all carried by the same operator [56]. It follows that the effective action for Calabi–Yau compactification of heterotic strings is bound to contain the marginal terms

$$\frac{c}{4\pi}\int_{\Sigma} \mathrm{d}^2\sigma \left(\int \mathrm{d}\varsigma^+\, \mathrm{d}\bar{\varsigma}^- \; \boldsymbol{S}_{CY}[\log(\boldsymbol{S}_{CY}/\mu) - 1] \; + \; \text{h.c.} \right) , \tag{F.5.1}$$

where c is the central charge, $c = 9$ for (2,2)-supersymmetric σ-models with a complex 3-dimensional target space. The $\boldsymbol{S}_{CY}$ may be thought of, in analogy to Eq. (F.4.3), as

$$\boldsymbol{S}_{CY} \;\sim\; (D_-\boldsymbol{X}^\mu)\, G^{(CY)}_{\mu\bar{\nu}}\, (\bar{D}_+\boldsymbol{X}^{\dagger\bar{\nu}}) \;\leftrightarrow\; \mathrm{d}x^\mu\, G^{(CY)}_{\mu\bar{\nu}}\, \mathrm{d}x^{\bar{\nu}} + \cdots , \tag{F.5.2}$$

where the same remark applies as for Eq. (F.4.3) and where $G^{(CY)}_{\mu\bar{\nu}}$ is the "repaired" combination of Fubini–Study metrics: from the linear combination

$$\sum_{i=1}^{m} (const.{+}(2\pi/f_i))(\boldsymbol{S}_i\,\text{"+"}\,\boldsymbol{S}_i^\dagger) , \tag{F.5.3}$$

one constructs first the Ricci-flat metric following Yau [54], and then corrects this along the lines in Ref. [196] to restore (super)conformal invariance of the complete model order by order. Thus, $\boldsymbol{S}_{CY}$ is the correspondingly "repaired" linear combination of the $\boldsymbol{S}_i$ from the $\mathbb{P}^n$ models. This again perfectly agrees with the identification of $(\boldsymbol{S}_{CY}\,\text{"+"}\,\boldsymbol{S}_{CY}^\dagger)$ with the Kähler class on the Calabi–Yau space and may be thought of as the result of integrating the true, superconformal D-term $\int \mathrm{d}^2\varsigma\, \mathrm{d}^2\bar{\varsigma}\, K_{CY}$ over the "twisted half" of the fermionic space. Note that even after dropping the anomalous $\boldsymbol{S}\log\boldsymbol{S}$ terms (after all, the anomaly eventually cancels out), the dependence on the f_i's remains through the $\log\mu$ factor; the f_i's in fact control the dependence of $\boldsymbol{S}_{CY}$ on the $\boldsymbol{S}_i$.

That the trace-anomaly had to appear through a marginal operator should have been obvious since it is a physical observable and can not be phased away through renormalization. That the gravitational and the gauge anomalies appear through the same marginal operator is a special feature of 2-dimensional spacetime; Eq. (F.5.1) then appears merely as a "twisted-chiral" re-write of the more standard expressions.

Finally, note that world-sheet instantons are expected to contribute to the terms (F.5.1). Determining these corrections appears to be a rather interesting problem for future study, because of the following intriguing possibility: Assume that the fully corrected terms will, instead of just $\boldsymbol{S}_{CY}$, contain a polynomial $Q(\boldsymbol{S})$ and that we may ignore the anomalous "$\boldsymbol{S}\log\boldsymbol{S}$" terms. The resulting action would be strikingly similar to a Landau–Ginzburg model in which the variations of $Q(\boldsymbol{S})$ are the marginal (twisted) F-terms describing the Kähler variations of our Calabi–Yau σ-model with target space $\mathcal{M}$. Alternatively, it could be interpreted as the Landau–Ginzburg model for describing the complex structure variations of the *mirror Calabi–Yau σ-model*, one which has target space $\mathcal{W}$, such that

$$H^q(\mathcal{M},\mathcal{T}_{\mathcal{M}}) \;\cong_Q\; H^q(\mathcal{W},\mathcal{T}_{\mathcal{W}}{}^*) \tag{F.5.4}$$

is an equivalence of ring structures (exact quantum Yukawa couplings), not only the equality of the dimensions of the respective spaces.

F.6 Landau–Ginzburg Mean-Field Theories

Now, by choosing the 'physical' gauge in which $\boldsymbol{V}_i = 0$ (gauge fields in 2-dimensions are unphysical) the kinetic term becomes

$$\mathcal{A}_{\text{kin.}} = \sum_{i=1}^{m} w^i \int_\Sigma \mathrm{d}^2\sigma\, \mathrm{d}^2\varsigma\, \mathrm{d}^2\bar\varsigma\; \|\boldsymbol{X}_i\|^2 \,, \tag{F.6.1}$$

which is just the standard Wess–Zumino type, in the generic universality class and irrelevant. In addition, we set $\boldsymbol{\Lambda}_a = \lambda_a$ (*const.*). This second step is not really any loss of generality—if we send $\lambda_a \to \infty$ at the end of the calculation. With some finite and big λ_a, we simply have a very steep potential, so that the fields are predominantly hovering about the minimum of the potential, $P^a(\boldsymbol{X}) = 0$. Taking the limit $\lambda_a \to \infty$ then corresponds to narrowing the potential until it becomes a delta function(al), constraining the fields at $P^a(\boldsymbol{X}) = 0$. In fact, if we agree to focus on the IR sector, then it does not really matter whether the higher frequency modes of the fields are constrained or not; as long as the low frequency modes are restricted to the subspace $P^a(\boldsymbol{X}) = 0$, global properties of the model will be successfully computed.

— ❦ —

With this in mind, we set $\lambda_a \to 1$ and the action reduces to

$$\int_\Sigma \mathrm{d}^2\sigma\, \mathrm{d}^2\varsigma\, \mathrm{d}^2\bar{\varsigma}\;\; K(\boldsymbol{X}, \boldsymbol{X}^\dagger) \quad + \quad \int_\Sigma \mathrm{d}^2\sigma\, W(\boldsymbol{X}) + h.c.\ , \tag{F.6.2}$$

where $K(\boldsymbol{X}, \boldsymbol{X}^\dagger)$ is the Kähler potential into which the kinetic term finally renormalizes and

$$W[\boldsymbol{X}] \;=\; \sum_{a=1}^{K} P^a(\boldsymbol{X}) \tag{F.6.3}$$

is the superpotential. In favorable circumstances, the so obtained model will have all the right properties with respect to the various anomalies and one proceeds to work with this.

However, there still exist superficial fields since without the non-trivial kinetic term we have lost the projectivity relation. For example, instead of having four local coordinates on $\mathbb{P}^4$, there would be the five homogeneous coordinates amongst which we cannot distinguish. Also, we have "lost" the Lagrange multipliers. However, one can perform a holomorphic change of variables, which will bring out a monomial in front of every P^a. These monomials can then be thought of as a single variable, provided that the total Jacobian is field-independent.☆ The condition for that turns out to be the Ricci-flatness condition (2.1.17). The change of variables, however, is not single-valued, so that one needs to pass to an orbifold and it is that Landau–Ginzburg Field Theory which corresponds to the Calabi–Yau space that we have started from.

The cyclic symmetry group by the action of which the Field Theory had to be divided is nothing else than the maximal subgroup of the (product of) $U(1)$ symmetry(ies)—one for each $\mathbb{P}^n$ factor in the embedding space—such that the constrained $\mathbb{P}^n$ action is left invariant. (Since the Kähler potential is strictly invariant, so must the superpotential be.) Being a remnant of a gauge symmetry, the necessity to divide out its action is obvious. Some powerful techniques have recently been advanced [229, 159, 227, 174], by which the Field Theory analogue of the (p, q)-cohomology can be

☆ Supersymmetry of course helps a lot, by canceling out the contributions from every Bose–Fermi pair. Zero modes, however are not paired and *do* contribute non-trivially.

determined, together with its ring structure and so the Yukawa couplings. The so obtained results strikingly resemble the Koszul spectral sequence. In particular, it is possible to show in case-by-case analysis that in both cases all the cohomology (untwisted and twisted, (c,c)- and (a,c)- sectors) is represented by elements of certain quotient rings, and where both the polynomial ring and the quotient ideal are *identical* [72]. The two methods are nevertheless rather different [241]; also, the Koszul spectral sequence is much more widely applicable while the Landau–Ginzburg orbifold analysis provides more detail.

F.7 How Singular a Space Can Superstrings Thread?

Consider again a Calabi–Yau constrained $\mathbb{P}^n$ model with the action (F.4.1), (F.4.2), where

$$\boldsymbol{X}^\mu \stackrel{\text{def}}{=} X^\mu + \varsigma^+ \xi_+{}^\mu + \varsigma^- \xi_-{}^\mu + \varsigma^+\varsigma^- \mathsf{X}^\mu \tag{F.7.1}$$

are the coordinate superfields and

$$\boldsymbol{\Lambda}_a \stackrel{\text{def}}{=} \Lambda_a + \varsigma^+ \lambda_{+a} + \varsigma^- \lambda_{-a} + \varsigma^+\varsigma^- \mathsf{L}_a \tag{F.7.2}$$

are the Lagrange superfields.

Path-integration over $\Lambda_a, \lambda_{\pm a}, \mathsf{L}_a$ and X^μ yield delta-functionals which enforce algebraic field equations. Upon the fermionic integration, these are:

$$\int \mathsf{D}[\Lambda_a] \Rightarrow 0 = P^a{}_{,\mu\nu}\, \xi_+{}^\mu \xi_-{}^\nu - P^a{}_{,\mu}\, \mathsf{X}^\mu , \tag{F.7.3}$$

$$\int \mathsf{D}[\lambda_{\pm a}] \Rightarrow 0 = P^a{}_{,\mu}\, \xi_\pm{}^\mu , \tag{F.7.4}$$

$$\int \mathsf{D}[\mathsf{L}_a] \Rightarrow 0 = P^a , \tag{F.7.5}$$

$$\int \mathsf{D}[\mathsf{X}^\mu] \Rightarrow 0 = G_{\mu\overline{\nu}}\, \mathsf{X}^{\overline{\nu}} - \Gamma_{\mu\overline{\nu}\,\overline{\rho}}\, \xi_+{}^{\overline{\nu}} \xi_-{}^{\overline{\rho}} - \Lambda_a\, P^a{}_{,\mu} , \tag{F.7.6}$$

with $G_{\mu\overline{\nu}} = \mathbf{K}_{,\mu\overline{\nu}}$ and $\Gamma_{\mu\overline{\nu}\,\overline{\rho}} = \mathbf{K}_{,\mu\overline{\nu}\,\overline{\rho}}$. Contracting Eq. (F.7.6) with $G^{\mu\overline{\rho}}\, \overline{P}^{\overline{a}}{}_{,\overline{\rho}}$, we obtain

$$\Lambda_a\, M^{a\overline{a}} = \overline{P}^{\overline{a}}{}_{,\overline{\mu}}\, \mathsf{X}^{\overline{\mu}} - \Gamma^{\overline{\mu}}{}_{\overline{\rho}\,\overline{\sigma}}\, \overline{P}^{\overline{a}}{}_{,\overline{\mu}}\, \xi_+{}^{\overline{\rho}} \xi_-{}^{\overline{\sigma}} , \tag{F.7.7}$$

$$M^{a\overline{a}} \stackrel{\text{def}}{=} (\, P^a{}_{,\mu}\, G^{\mu\overline{\mu}}\, \overline{P}^{\overline{a}}{}_{,\overline{\mu}} \,) , \tag{F.7.8}$$

here and hereafter, $G^{\mu\overline{\mu}}$ is used to raise indices.

F.7.1 Smooth target space

$M^{a\bar{a}}$ is easily seen to be invertible and actually positive definite on the constrained subspace, provided the matrix of gradients $[P^a{}_{,\mu}(\boldsymbol{X})]$ is of rank K where $P^a(\boldsymbol{X}) = 0$, i.e., provided the $\{P^a(\boldsymbol{X}) = 0\}$ subspace is smooth. Then,

$$\Lambda_a = M_{a\bar{a}}\,\overline{P}^{\bar{a}}{}_{,\bar{\mu}}\,\mathsf{X}^{\bar{\mu}} - M_{a\bar{a}}\,\overline{P}^{\bar{a}}{}_{,\bar{\mu}}\,\Gamma^{\bar{\mu}}{}_{\bar{\rho}\bar{\sigma}}\,\xi_+{}^{\bar{\rho}}\,\xi_-{}^{\bar{\sigma}}\,. \tag{F.7.9}$$

Path-integration over X^μ results in a delta-function that enforces Eq. (F.7.9).

As always, the matrix of gradients $[P^a{}_{,\mu}]$ is a proper tensor and independent of any connection Γ_μ on the constrained subspace where $P^a = 0$ since $P^a{}_{;\mu} \stackrel{\text{def}}{=} P^a{}_{,\mu} + \Gamma_\mu \cdot P^a = P^a{}_{,\mu}$. Furthermore, since the polynomials $P^a(\boldsymbol{X})$ are locally transverse to the subspace $\mathcal{M} \hookrightarrow \mathcal{X}$, the matrix $[P^a{}_{,\mu}(\boldsymbol{X})]$ projects onto the normal space to $\mathcal{M}$ at X^μ; this is the $\nabla_X f$ map of Eq. (1.3.6).

Upon elimination of the X's and $\overline{\mathsf{X}}$'s by means of their new equations of motion, the Lagrangian in (F.4.1) contains the four-fermion term

$$\Big[R_{\mu\bar{\nu}\rho\bar{\sigma}} + P^a{}_{,\mu;\rho}\,M^{a\bar{a}}\,\overline{P}^{\bar{a}}{}_{,\bar{\nu};\bar{\sigma}}\Big]\,\xi_+{}^{\mu}\,\xi_-{}^{\rho}\,\xi_+{}^{\bar{\nu}}\,\xi_-{}^{\bar{\sigma}}\,, \tag{F.7.10}$$

where

$$R_{\mu\bar{\nu}\rho\bar{\sigma}} \stackrel{\text{def}}{=} \mathbf{K}_{,\mu\rho\bar{\nu}\bar{\sigma}} - \Gamma_{\mu\rho}{}^{\kappa}\,\Gamma_{\kappa\bar{\nu}\bar{\sigma}}\,, \qquad P^a{}_{,\mu;\nu} \stackrel{\text{def}}{=} P^a{}_{,\mu\nu} + \Gamma_{\mu\nu}{}^{\rho}\,P^a{}_{,\rho} \tag{F.7.11}$$

are the standard Riemann tensor on $\mathcal{X}$ and the extrinsic curvature of the constrained subspace, respectively. The sum of the two contributions in Eq. (F.7.10) is indeed the induced Riemann tensor on the constrained subspace, to which the four-fermion term should couple.

F.7.2 Singular target space

We now prove the following:

Lemma F.1. *For a constrained σ-model to inherit the $(2,2)$-supersymmetry of the ambient σ-model and the possible singularities to be innocuous, it suffices that the constraints have non-zero Taylor series up to and including second order.*

Proof. Consider the case where $\text{rank}[P^a{}_{,\mu}] < K$ at a finite number of points of the $\{P^a = 0\}$ subspace $\mathcal{M}$, which is therefore singular at those points. Also, for ease of notation, we take $K = 1$. At a singular point, both P^1 and $P^1{}_{,\mu}$ vanish and there Eq. (F.7.3) becomes

$$\int \mathsf{D}[\Lambda_1] \;\Rightarrow\; 0 = P^1{}_{,\mu\nu}\, \xi_+{}^{\mu}\, \xi_-{}^{\nu}\ , \tag{F.7.12}$$

Eq. (F.7.5) is unchanged and restricts the string coordinates X^μ from $\mathcal{X}$ to the 3-dimensional constrained hypersurface $\mathcal{M}$ while Eq. (F.7.4) becomes trivial, $0 = 0$. However, the $\xi_\pm{}^\mu$ are now constrained by Eq. (F.7.12)—as long as the matrix $[P^1{}_{,\mu\nu}]$ does not vanish. It is easy to see that the $\xi_\pm{}^\mu$ now span a 3-dimensional cone with the tip at the singularity, as they should, tangential to the constrained hypersurface spanned by the X^μ subject to $P^1(X) = 0$.

Note that, for each fixed a, the matrix of second derivatives $[P^a{}_{,\mu\nu}]$, the so-called Hessian, is independent of any connection at a singularity since both the polynomials P^a and the gradients $P^a{}_{,\mu}$ vanish there and so

$$P^a{}_{;\mu\nu} = P^a{}_{,\mu\nu} + (\Gamma_{\mu,\nu} + \Gamma^\rho_{\mu\nu}\,\Gamma_\rho)P^a + \Gamma_{\mu\nu}{}^\rho P^a{}_{,\rho} = P^a{}_{,\mu\nu}\ . \tag{F.7.13}$$

The rank of $[P^a{}_{,\mu\nu}]$ has therefore an invariant meaning. In case of more constraints, the matrices of second derivatives $P^a{}_{,\mu\nu}$ form a direct sum and the analogous conclusion follows immediately.

Note also that, at a singularity, Eq. (F.7.6) becomes

$$\int \mathsf{D}[\mathsf{X}^\mu] \quad\Rightarrow\quad 0 = G_{\mu\overline{\nu}}\,\mathsf{X}^{\overline{\nu}} - \Gamma_{\mu\overline{\nu}\,\overline{\sigma}}\,\xi_+^{\overline{\nu}}\,\xi_-^{\overline{\sigma}} \tag{F.7.14}$$

and decouples from Eqs. (F.7.4), (F.7.5) and (F.7.12). Using these to eliminate all of the X^μ's at the singularity, the four-fermion term there is reduced to $R_{\mu\overline{\nu}\rho\overline{\sigma}}\,\xi_+{}^\mu\,\xi_-{}^\rho\,\xi_+{}^{\overline{\nu}}\,\xi_-{}^{\overline{\sigma}}$, with the ambient space Riemann tensor defined in (F.7.11) and the fermions subject to Eq. (F.7.12). At the singularity, the σ-model decouples from the extrinsic curvature term which occurred in Eq. (F.7.10) and would have been divergent since $M_{a\overline{a}}$ is the inverse of $M^{a\overline{a}}$ and the latter vanishes at any singularity. Writing $P^a = \phi^a + t\varphi^a$ with ϕ^a singular and φ^a smooth, divergent terms include t^{-1} and are seen to decouple in view of Eq. (F.7.12). ☑

We remark that all simple (A, D, E), modality-1 and modality-2 singular polynomials (see Ref. [1], p. 158–160 of the first book) satisfy this criterion. To see this, take for example the A_k polynomial germ, $f(x) = x^{k+1}$. A germ represents all constraints in a $\mathbb{C}^n$-like neighborhood which can be brought, through a holomorphic change of local coordinates, to the form

$$f(x_1) + {x_2}^2 + \cdots + {x_n}^2 = 0\ . \tag{F.7.15}$$

They all exhibit the same type of singularity and can be smoothed in the same way. The inclusion of squares of all the local coordinates not involved in the germ is called Morsification. Since the germs for all simple, modality-1 and modality-2 singularities involve less than four local coordinates, their Morsification always contains at least one coordinate occurring as a square and the matrix of second derivatives has at least rank one. It is important to note that the omission of even one local coordinate in Eq. (F.7.15) represents a *much worse* singularity than is indicated by the germ.

With the above result about supersymmetry and the fact that the σ-model decouples from the diverging extrinsic curvature, we conclude that isolated A, D, E type singularities are just as innocuous as orbifold singularities were well known to be [119].

F.8 Hindsighted Outlook

Fate is hindsight misnamed.

With three decades of intensive topical research in the rear-view mirror, the updating paragraphs on pp. 22, 62, 82, 149 and 164 can be expanded herein only in a rather telegraphic fashion. Even the references provided herein cannot possibly aim for anywhere close to completeness, but offer a partial (and tattered) roadmap to starting a more thorough and systematic literature search.

F.8.1 GLSM

As mentioned in the discussion closing § F.4, each Lagrange multiplier $\boldsymbol{\Lambda}_a$ may well be replaced by a dynamical superfield, with standard gauge-covariant kinetic terms and charge $-q_a^i$ with respect to $\boldsymbol{V}_i$. Additionally, rewriting all gauge-dependent contributions to the action in terms of $\boldsymbol{V}_i \to \boldsymbol{S}_i \stackrel{\text{def}}{=} D_-\bar{D}_+\boldsymbol{V}_i$ affords separating the (non-propagating) gauge component-fields from the dynamical "transversal" ones.❖ Finally, each "coordinate superfield," $\boldsymbol{X}^\mu$, in (F.4.1)–(F.4.2) may well interact with more than one of the $U(1;\mathbb{C})$-symmetries and so have an array of charges.

❖ Only in 1+1-dimensional spacetime of the world-sheet can this be done in a Lorentz-invariant way, since $Spin(1,1)$ is abelian and so has only 1-dimensional irreducible representations.

This sketches the key changes leading to the substantially more general framework of "gauged linear σ-models" (GLSM) [241], the quantum field theory of which is both technically better understood and computationally more effective than the framework of § F.4, as well as directly related to the much more general *toric* geometry [258, 42, 260, 255, 361, 257]. Its hallmark toric $(\mathbb{C}^*)^n$-action is provided by the multiple, $U(1;\mathbb{C})^n \approx (\mathbb{C}^*)^n$-gauge symmetry of the GLSM and naturally relates the $(\mathbb{C}^*)^n$-equivariant cohomology of toric spaces with the BRST cohomology of GLSMs, which explains their marvelous alignment.

Indeed, the cumulative world-sheet instanton-induced correction to the topological Yukawa coupling (0.5.5), the quantum cohomology ring in (1.2.16) and the short-distance geometry [60] of the (complexified, completely enlarged) Kähler class moduli space—all *essential* to mirror symmetry—were quickly determined [443] and found to have a rigorous mathematical foundation [408, 420, 264, 469]. This (completely) enlarged space of Kähler classes in the GLSM, probed by the so-called Fayet–Illiopoulos parameters, exhibits a fan-like partitioning ("phase diagram" [241, 393]) into sections that correspond to distinct low-energy appearances. The complexification made manifest by supersymmetry however renders these phases continuously connected: locations where the Kähler class and metric would undergo discontinuous changes (by flops or otherwise) may be circumnavigated by varying $B_{\mu\bar{\nu}}$; see (0.1.2) and § D.3.2. The "phase diagram" partitioning thus characterizes the distinct (limiting, low-energy) descriptions of the same underlying (high-energy) world-sheet quantum field theory.

In turn, this framework change has also made relaxing to the even more general world-sheet $(0,2)$-supersymmetry straightforward [371, 459]; for more recent developments, see, e.g., [391, 423] and the survey [463]. Suffice it here to say that $(0,2)$-supersymmetric GLSMs provide a natural framework for constructing and exploring the "non-standard" embedding $(\mathcal{T}_{\mathcal{M}} \rightsquigarrow \mathcal{V}) \subset \mathcal{V}_{E_8 \times E_8}$ or $\subset \mathcal{V}_{D_{16}}$, with rank $\mathcal{V} \geq 3$; see Refs. [476, 373] and also [449]. The ultimate relaxation to only the (world-sheet local) $(0,1)$-supersymmetry [134, 135] remains much less explored but is known to be insufficient to imply target-spacetime supersymmetry and so should be germane to describing its much-needed breaking; see also footnote § on p. 314.

The GLSM framework also gives rise to a rigorously formulated quantum cohomology [455] on the mathematical side and a particularly effective application of topological quantum field theories[✧] (TQFT) [478], and more recently (and with defects) [384]. This framework also enabled defining and computing a novel slew of invariants for Calabi–Yau manifolds [478, 427, 271], as well as developing methods for computing exact results in world-sheet gauge theories; see [332, 313]. Applications to GLSMs with Calabi–Yau target spaces and their moduli spaces led to an iterative formulation of "quantum geometry"; see [89, 88, 390].

This GLSM framework affords various generalizations, such as non-abelian gauge groups [375, 463] and more complex constraint systems [308, 307]. The former of these correspond to Grasmannians, more general flag spaces and C-spaces discussed in Chapter 3, while the latter relate to residue representations of the cohomology computed in Chapters 9–B [318], as well as the "double field theory" [294], mentioned in the updating close of § E.4, on p. 164.

— ❦ —

As concluded in § F.2, the metric deformation operators (F.2.2)–(F.2.3) become, respectively, twisted-chiral and chiral superfields, such that

$$\mathcal{A}_{(1,1)} \to \Theta : \quad \bar{D}_+\Theta = 0 = D_-\Theta, \qquad \text{twisted-chiral;} \tag{F.8.1}$$

$$\mathcal{A}_{(2,1)} \to \Phi : \quad \bar{D}_+\Phi = 0 = \bar{D}_-\Phi, \qquad \text{chiral.} \tag{F.8.2}$$

It is immediate that the products of (twisted-)chiral superfields are also (twisted-)chiral, so they form a *ring*$^{(L)}$, limiting to their more uniquely defined analogue in super-*conformal* theories [229]. The super-derivatives, $D_\pm, \bar{D}_\pm$, used in §§ F.3–F.4 satisfy the world-sheet $(2,2)$-supersymmetry algebra,

$$\{\bar{D}_+, D_+\} = 2i(\partial_{\mp}{:=}\partial_0{+}\partial_1), \qquad \{\bar{D}_-, D_-\} = 2i(\partial_{=}{:=}\partial_0{-}\partial_1), \tag{F.8.3}$$

reducing Lorentz-invariantly to separate left-moving ("+") and right-moving ("−") sectors [412]. Supersymmetric vacua in $(2,2)$-supersymmetric quantum field theories are annihilated by these super-derivatives, $\bar{D}_\pm\,|0\rangle = 0 = D_\pm\,|0\rangle$, and have vanishing energy $(H = i\hbar\partial_0)$ and linear

✧ As sketched in the following the topological ("A" and "B") "twists" of a supersymmetric world-sheet quantum field theory project on its quasi-topological correlation functions as observables [99].

momentum ($p = -i\hbar\partial_1$), owing to (F.8.3). Thus,

$$2i\partial_{\mp}\langle 0|\,\Phi\,\Phi'\,|0\rangle = 2i\,\langle 0|\,[\partial_{\mp},\Phi\,\Phi']\,|0\rangle = \langle 0|\,\{\bar{D}_+,[D_+,\Phi\,\Phi']\}\,|0\rangle\,, \qquad \text{(F.8.4)}$$

$$= \underline{\langle 0|\,\bar{D}_+}\,[D_+,\Phi\,\Phi']\,|0\rangle + \langle 0|\,[D_+,\Phi\,\Phi']\,\underline{\bar{D}_+\,|0\rangle} = 0, \qquad \text{(F.8.5)}$$

where the underlined factors vanish for supersymmetric vacua: purely chiral correlation functions are constant over the world-sheet and can be computed at any convenient location [358]. The analogous is true for purely twisted-chiral correlation functions and importantly simplifies the analysis in $(2,2)$-supersymmetric theories. Building on Witten's Morse-theoretic analysis of supersymmetric models [236] and reducibility (F.8.3), this provides for independent and quasi-topological "purely chiral" and "purely twisted-chiral" (sub)rings of observables [358].

Then, (**1**) twisting the considered superfields by appropriate complementary powers of the canonical bundle, (**2**) halving the $(2,2)$-supersymmetry algebra to the (nilpotent) BRST subalgebras $\{D_+,\bar{D}_-;\partial_{\mp},\partial_{=}\}$ and $\{D_+,D_-;\partial_{\mp},\partial_{=}\}$, and (**3**) restricting to the appropriate subring of observables produces the "topological" models "A" for $(\mathbf{27^*})^3$-couplings, and "B" for $\mathbf{27}^3$-couplings. For details, see the third-listed reference in [478], which also includes a "half-twisted" variant that is well suited to study the remaining **1**-involving dynamics (0.5.9) and (0.5.11).

F.8.2 Mirrors

Recall that the quantities $\mathcal{A}_{(1,1)}$ correspond to elements of $H^1(\mathcal{M},\mathcal{T}^*_{\mathcal{M}})$, and $\mathcal{A}_{(2,1)}$ to $H^1(\mathcal{M},\mathcal{T}_{\mathcal{M}})$. A glance at (F.8.1) and (F.8.2) reveals that, within the underlying world-sheet $(2,2)$-supersymmetric framework, the difference between them is merely a complex conjugation—*amongst the right-movers only,* corresponding to the cohomology valuation in $\mathcal{T}^*_{\mathcal{M}} \leftrightarrow \mathcal{T}_{\mathcal{M}}$.** This operation is thus identified as the field-theoretic root-source of mirror symmetry [388, 284, 444, 412, 407].

The mathematical translation, that Calabi–Yau 3-folds exist in pairs, $(\mathcal{M},\mathcal{W})$ satisfying (in full generality) (E.4.13), was first recognized—*and realized by explicit construction*—in Ref. [156]. The construction itself was soon generalized in two a priori distinct ways: by "transposition" [74] and by the toric "polar" operation [68], to be fully unified in a generalization [301]; see also [350], which contains the by now iconic

** For the bosonic degrees of freedom in (F.2.2)–(F.2.3), this corresponds to target-space conjugating the ∂-derivatives, $\partial Z^\mu \leftrightarrow \partial Z^{\bar{\mu}}$, but not the $\bar{\partial}$-derivatives.

left-right mirror-symmetric "shield" plot of $(h^{1,1}, h^{2,1})$ pairs for desingularized Calabi–Yau hypersurfaces in $\mathbb{P}^4_{\mathbf{w}}$. Being applicable to the largest database of Calabi–Yau 3-folds [431], the toric geometry framework (*abelian* GLSMs [241]) started a combinatorially expanding industry [255], which includes a generalization to toric CICYs [309], but also to CICYs in Grassmannians [310] and other (partial) flag spaces and related *non-abelian* GLSMs [375].

In turn, the Strominger–Yau–Zalow (SYZ) framework for understanding mirror symmetry [468, 470, 416, 396] as a T-duality, the homological mirror symmetry framework [279, 461, 392] and its *arithmetic* version [434, 464] all correlate the algebraic geometry on the "complex structure side" (as discussed herein) with the symplectic geometry on the "Kähler structure side" [263]; see also [414, 395]. Such a mirror correlation is indeed expected given the inherent alignment between the complex and symplectic structures in a Kähler manifold; see footnote ¶ on p. 144. SYZ mirror symmetry is also studied using Morse theory [360], an integral feature of supersymmetric theories [236]. There is also a non-Kähler version of SYZ mirror symmetry [433], which covers the non-Kähler Calabi–Yau solutions to the "Hull–Strominger system" [409, 312, 300]; see also [451] and references therein, as well as [383] for additional background.

A priori, mirror symmetry is an intrinsic feature of Calabi–Yau spaces but has been studied almost exclusively via various embedding constructions. In the simplest of these, both Calabi–Yau n-folds of a mirror pair, $(\mathcal{M}, \mathcal{W})$, may be constructed as hypersurfaces in a suitable pair of Fano $(n{+}1)$-folds, $(\mathcal{X}, \mathcal{Y})$, as zero loci of anticanonical sections, $f \in H^0(\mathcal{X}, \mathcal{K}^*)$ and $g \in H^0(\mathcal{Y}, \mathcal{K}^*)$. Finally, the *total space* of the anticanonical bundle of a variety is itself a non-compact Calabi–Yau space: by definition, the first Chern class along the fiber of $\mathcal{K}^*_{\mathcal{X}}$ cancels that along the base, $\mathcal{X}$. Thus, these embedding/construction-based studies of mirror symmetry generally follow a framework that fits into the following *commutative diagram*[††] [322]:

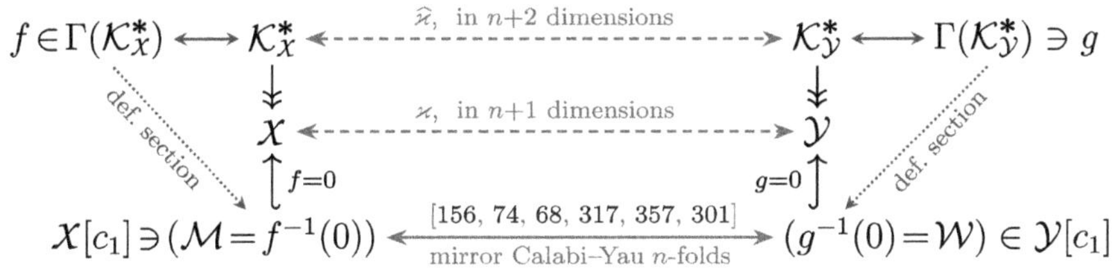

(F.8.6)

†† Distinct "A-to-B" paths along the mapping arrows give the same result; see footnote § on p. 11.

Within toric geometry, $\mathcal{X}$ and $\mathcal{Y}$ are specified by combinatorial data forming *reflexive polytopes*,[♯] and $\Delta_{\mathcal{X}}=(\Delta_{\mathcal{Y}})^{\circ}$, where $\circ$ denotes the lattice *polar* involution, so also $\Delta_{\mathcal{Y}}=(\Delta_{\mathcal{X}})^{\circ}$: Induced by the mirror map in n dimensions ("ground floor"), this defines the $\varkappa$-mapping in $(n{+}1)$ dimensions and for Fano toric varieties; the same Calabi–Yau n-fold can be a hypersurface in distinct embedding $(n{+}1)$-folds, but each will have a specific $\varkappa$-image. For $\mathcal{X}=\mathbb{P}^{n+1}$, $\mathcal{Y}=\underset{\text{MPCP}}{\mathrm{Bl}}[(\mathbb{P}^{n+1}/(\mathbb{Z}_{n+2})^n)]$, the *maximal projective crepant partial* (MPCP) desingularization [68] of the indicated quotient. The zero locus $g^{-1}(0)$ here automatically includes the requisite desingularization in the original quotient construction of mirror pairs [156]. One dimension higher yet, this mirror-symmetric framework also induces the $\widehat{\varkappa}$-mapping among the total spaces of the respective anticanonical bundles, themselves *non-compact* Calabi–Yau $(n{+}2)$-dimensional spaces, hinting at *local mirror symmetry* and *localization* techniques; see Refs. [407, 422, 473, 313] for details.

One would expect this general mirror-symmetric framework (F.8.6) to persist not only throughout toric hypersurfaces [430] but, with suitable adaptation, also beyond, to complete intersections of hypersurfaces, in products of toric varieties and then also Grassmannians and flag manifolds (§ 3.3): just as abelian GLSMs with abelian $U(1;\mathbb{C})^n \approx (\mathbb{C}^*)^n$-gauge symmetry have toric variety vacua, subspaces of Grassmannians and flag manifolds are vacua of GLSMs with non-abelian gauge symmetry [309, 310, 375]. In fact, there is no a priori reason for the mirror-symmetric framework (F.8.6) not to extend also to non-Fano and non-algebraic[**] embedding spaces $(\mathcal{X},\mathcal{Y})$; for the toric case, see [320, 321, 322]. (After all, Calabi–Yau 0-folds, pairs of points, are trivial to embed in Riemann surfaces of arbitrary genus.) The latter of these extension options generalizes both reflexive polytopes (spanning fans) into VEX multitopes (spanning multifans [439]), and the polar toric involution into its *stratified*$^{(L)}$ iterative *transpolar* generalization. Furthermore, such generalizations turn up *within the deformation families* [321] of generalized CICYs [296, 319], which are by now on rigorous mathematical footing [386]. The self-crossing, flip-folded and

---

♯ A lattice polytope is reflexive if it contains a single lattice point on its inside, from which each facet is at unit lattice distance. The polar of a reflexive polytope is also reflexive.

** Owing to their rigidity among spaces admitting maximal toric $(\mathbb{C}^*)^n$-action, toric varieties have been likened to "frigid crystals" [258], and it is virtually certain that string theory readily admits constructions that are much more general than that—calling on adventurous explorers!

otherwise multilayered of the VEX multitopes, which still have every facet at unit lattice distance from a "center" have turned up in the study of pre-symplectic spaces with toric action [424]. This seems to resonate with the above-observed correlation between the complex algebraic and symplectic structures provided by both homological and SYZ mirror symmetry: A symplectic structure degeneration at a certain (lower-dimensional) locus in a *pre-symplectic* space, say $\mathcal{Y}$ in (F.8.6), should then $\varkappa$-correspond to a degeneration of the complex structure at a $\varkappa$-corresponding locus in $\mathcal{X}$, making it *pre-complex.*

The Calabi–Yau n-folds of the desired mirror pair in (F.8.6), $(\mathcal{X}\supset\mathcal{M},\mathcal{W}\subset\mathcal{Y})$, may well have minimal (string-theory repairable☆☆, if any) intersections with these degeneration loci in $\mathcal{X}$ and $\mathcal{Y}$. Since target-space complex structure corresponds to at least $(0,2)$-supersymmetry on the world sheet, a degeneration of the target-space complex structure (if any is in fact incurred in the so-constructed Calabi–Yau n-fold) should correspond to a world-sheet $(p,2)\to(0,1)$-supersymmetry breaking, presumably indicating the inclusion of adequate supersymmetry-breaking defects.☆☆ The mirror-reflected consequences of a $\varkappa$-corresponding degeneration of the symplectic structure (if any are incurred in the so-constructed mirror Calabi–Yau n-fold) are not as clearly known. Such spaces will also fail to be Kähler, and require a more general approach; see [383], for example.

Finally, in the by now well-accepted fashion, the mirror-symmetric framework (F.8.6) must vary robustly and consistently over the combined pair of moduli spaces, $(\mathfrak{M},\mathfrak{W})_{\mathcal{X}}=(\mathfrak{W},\mathfrak{M})_{\mathcal{Y}}$, which is only locally a product, and a *foliation*$^{(L)}$ in general—and certainly also including moduli of the generalized geometry (§ E.4), and $H^1(\mathcal{M},\operatorname{End}\mathcal{T}_{\mathcal{M}})$-parametrized deformations in the "standard" Ansatz. These sweeping generalizations seem to be waiting for the more adventurous Reader.

F.8.3 . . . and perspectives

As generic phenomena among Calabi–Yau spaces, both mirror symmetry and conifold (and other) transitions (Chapters D and E) also serve to study and refine the characterization of Calabi–Yau spaces and their complex and Kähler structure moduli spaces, $(\mathfrak{M},\mathfrak{W})$. Inter alia, they helped establish the combined cohomology theory [250] as better

☆☆ Perhaps, by inclusion of some 'branes or other kinds of defects admissible in string theory, and a corresponding cohomology theory such as the combined cohomology of Ref. [250].

suited to string theory than the L^2 and intersection (co)homology [15, 26]; see footnote ✱ on p. 37 and Table D.1. They also serve in the study of the so-called *moduli stabilization*: In the target-space quantum field theory, $(\mathfrak{M}, \mathfrak{W})$ are spanned by vevs of moduli (super)fields, the (nonperturbative, complete) dynamics of which is expected to stabilize them at certain preferred values—hopefully, where a so-compactified string theory reproduces the Standard Model. The literature on this subject has been growing over the past two decades, where Refs. [418, 417, 332] should help start a systematic literature search.

The wish list of physically applicable expectations has grown from § 0.8 with cosmic aspirations: Throughout the first two decades of physically realistic string theory (since 1984 [143, 160, 93]), it was implicitly assumed that the observed 3+1-dimensional spacetime is flat, Minkowski $\mathbb{R}^{1,3}$—and was taken for granted in that wish list. Amidst the second string revolution (1994–2003), this assumption was proven false [454]: the expansion of the observed Universe is accelerating, consistent with having a minuscule but positive cosmological constant, Λ_{cc}, indicating an (asymptotically) de Sitter rather than Minkowski geometry of the vacuum; for a review and a first roadmap through the relevant literature, see Ref. [324].

It turns out that Calabi–Yau 3-fold string compactifications may well provide for a Minkowski ($\Lambda_{cc}=0$) to de Sitter ($\Lambda_{cc}>0$) *uplift,* provided the Calabi–Yau 3-fold is a barely and intricately smoothed conifold (see § D.2) [314]. In turn, more radical constructions have also been identified as a priori being able to provide for such cosmic configurations [411, 306, 369, 324]. However, the ability of such models to also reproduce the particle physics Standard Model remains to be determined.

Yet another possibility stems from a foundational reconsideration of the observable target-space physics stemming from superstrings [379]. This analysis indicates that spacetime in string theory is actually non-commutatively and chirally doubled,❖❖ (x^a_L, x^a_R), since closed strings support independent left- and right-moving world-sheet fields: recall the discussion of *oriented* loop-space on p. 2. The usual, by now *undergraduate* textbook [293] analysis of string theory identifies $x^a := (x^a_L + x^a_R)$ as the spacetime coordinates, ignoring the complementary degrees of freedom, $y_a := (x^a_L - x^a_R)$. Jointly, x^a and y_a span a $2\cdot(9+1)$-dimensional *almost symplectic* and *para-Hermitian* space, $\mathbb{X}$, with a compatible *foliation* [380]:

❖❖ The variables (x^a_L, x^a_R) are zero-modes of world-sheet chiral bosons and so contribute half of the anomaly of a conventional bosonic field.

At every point, $p \in \mathbb{X}$, the $\mathbb{X}_p = X_p \times Y_p$, where the quantum counterparts of the local coordinates fail to commute: $[\hat{x}^a, \hat{y}_b] = 2\pi i \lambda^2 \delta^a_b$, with λ the fundamental string length scale. Local diffeomorphisms are then integrable to finite translations only over various *polarizations* to some (9+1)-dimensional *maximal Lagrangian* (self-commuting) sub-spacetime—identifiable as the observable spacetime. This is just as quantum mechanics restricts to the coordinate, or momentum, or some other "half-dimensional" subspace of the classical phase-space. In this formalism, Calabi–Yau-compactifying $X_p \to \mathbb{R}^{1,3} \times \mathcal{M}$ at every point $p \in \mathbb{X}$ forces the complementary factor in the foliation to automatically include the mirror-Calabi–Yau space, $Y_p \to \mathbb{R}^{1,3} \times \mathcal{W}$ [325],[⬦⬦] which ends up having nontrivial physically relevant consequences [323].

Finally, as mentioned in the updating passages on pp. 22 and 62, machine learning (ML) and artificial intelligence (AI) are already changing the methodology even in this field, likely precipitating another "string revolution." They have finally enabled the first ever computations of *physically normalized* Yukawa couplings in Calabi–Yau compactified $E_8 \times E_8$ heterotic 3-generation models, both with the "standard" [345] and a "non-standard" embedding [363]! As noted already, astounding numbers of candidate models [365] await more detailed study, the traversing and exploring of which, in search of the spot-on stringy Standard Model, most assuredly require ML/AI magic. The key achievement of the recent computation of physically normalized Yukawa couplings [345, 363] is that they are both *parametric*—as functions over the moduli spaces—and adaptable to ML/AI methods to be unleashed throughout the moles upon moles of models to explore.

The experience of the three decades since the first edition of the *Bestiary* [272] and some of its fumbled forecasts suggest paramount prudence in prognostication but exuberant enthusiasm for exploration:

Moles of Models Ho!

[⬦⬦] This automatic inclusion of both a Calabi–Yau 3-fold and its mirror, spanned respectively by subsets of standard coordinates, x^a, and their corresponding "twisted" partners, y_a, reminds us of Landau–Ginzburg orbifolds [227]: there, the distinction between "untwisted" and "twisted" states is a matter of convention, as is that between x^a and y_a, and their swap yields the mirror map.

Lexicon

The following is a collection of definitions and short explanations of various terms used in the text. Some of the entries in the Lexicon duplicate partly the information given in the text, but in a location and form that are more suitable for a quick lookup. Terms which are typeset as for example '*ample* $^{(L)}$' have their own entries elsewhere in this Lexicon. 2024⌈*The Reader will also find Wikipedia's regularly updated online glossary** *very useful.*⌋

— ❦ —

▷ **0-Mode**: Originally, this derives from Fourier expansion $f(x) = \sum_{k=-\infty}^{+\infty} \phi_k e^{ikx}$, where it refers to ϕ_0. More generally, other (integral or otherwise) transforms stand in place of the Fourier transform, and 0-modes refer to 'zeros' in the transform spectrum. As the latter typically includes energy, momentum and/or mass, these are also synonyms for 'massless modes'; see for example § 0.3.

▷ **1-Dot-Decomposable** (1DD): A diagram (see § 2.1.2) which disconnects upon deletion of a dot.

▷ **1-Leg-Decomposable** (1LD): A diagram (see § 2.1.2) which disconnects upon deletion of a leg.

▷ **Adjunction Formulae**: Let $\mathcal{X}$ be a *manifold* $^{(L)}$ with a holomorphic line *bundle* $^{(L)}$ $\mathcal{E}$ over $\mathcal{X}$ and let ϕ be a *section* $^{(L)}$ of $\mathcal{E}$. We can define the corresponding submanifold, called *hypersurface*, $\mathcal{M}_\phi \subset \mathcal{X}$ as the subspace of solutions to $\phi(x) = 0$, $x \in \mathcal{X}$. If $\mathcal{M}$ is smooth, then

$$\text{Adjunction Formula 1:} \quad \mathcal{N}_{\mathcal{X}/\mathcal{M}} = \mathcal{E}|_{\mathcal{M}}, \qquad (1.3.9')$$

---
* `https://en.wikipedia.org/wiki/Glossary_of_algebraic_geometry`

equates the *normal bundle*$^{(L)}$ of $\mathcal{M} \hookrightarrow \mathcal{X}$ with the restriction to $\mathcal{M}$ of $\mathcal{E}$ (see § 1.3.1 or Ref. [23], p. 146). Furthermore,

$$\text{Adjunction Formula 2:} \quad \mathcal{K}_{\mathcal{M}} = (\mathcal{K}_{\mathcal{X}} \otimes \mathcal{E})|_{\mathcal{M}}, \tag{1.3.11$'$}$$

equates the *canonical bundle*$^{(L)}$ on $\mathcal{M}$ with the restriction to $\mathcal{M}$ of the tensor product of the canonical bundle of $\mathcal{X}$ and the determinant of $\mathcal{E}$ (see § 1.3.1 or Ref. [23], p. 146). If $\mathcal{M} \hookrightarrow \mathcal{X}$ is a *complete intersection*$^{(L)}$, defined as the common zero-set of a system of sections $\{\phi^a\}$, where each $\phi^a(x)$ is the section of a line bundle $\mathcal{E}_a$ over $\mathcal{X}$, the above formulae become

$$\mathcal{N}_{\mathcal{X}/\mathcal{M}} = \Big(\bigoplus\nolimits_a \mathcal{E}_a\Big)\big|_{\mathcal{M}}\,, \qquad \mathcal{K}_{\mathcal{M}} = \Big(\mathcal{K}_{\mathcal{X}} \otimes \bigotimes\nolimits_a \mathcal{E}_a\Big)\big|_{\mathcal{M}}\,. \tag{L.1}$$

▷ **Algebraic Variety**: Originally and for our purposes, a (compact, complex) n-dimensional space which can be embedded in some $\mathbb{P}^N$, for large enough N, as the common zero-locus of a system of algebraic equations over an algebraically closed ground field—herein, $\mathbb{C}$. It is called a *curve* for n=1, a *surface* for n=2, and (less often) a *solid* for n=3; a *Riemann surface*$^{(L)}$ is in fact a (complex) *curve.*

▷ **Almost Ample**: A holomorphic line *bundle*$^{(L)}$ $\mathcal{L}$ over a complex *manifold*$^{(L)}$ $\mathcal{M}$ is called almost ample if it is non-trivial and, at each $x \in \mathcal{M}$, admits a global holomorphic *section*$^{(L)}$ which does not vanish there. $\mathcal{L}$ is thereby basepoint-free; see *base locus*$^{(L)}$. In particular, $\mathcal{L}$ must have enough global holomorphic sections so that at least one of them is non-zero at any given point $x \in \mathcal{M}$. The absolute minimum is two global holomorphic sections: their common zero-locus is in *codimension*$^{(L)}$ 2 and may be blown-up; see § 3.1.1. 2024⌈*Elsewhere in the literature (incl. Wikipedia),* almost ample *is defined to be nef*$^{(L)}$ *and big*$^{(L)}$ *but not ample*$^{(L)}$.⌋

▷ **Almost Complex Structure**: If $\mathcal{M}$ is an even dimensional *manifold*$^{(L)}$, an almost complex structure on $\mathcal{M}$ is an *endomorphism*$^{(L)}$ J of the *tangent bundle*$^{(L)}$ of $\mathcal{M}$ satisfying $J^2 = -\mathbb{1}$. It is important to realize that, unlike for a *complex structure*$^{(L)}$, an almost complex structure J need not be integrable. Every complex manifold admits an almost complex structure, given in terms of local complex coordinates $\{z_k = x_k + iy_k\}$ by

$$J\Big(\frac{\partial}{\partial x_k}\Big) = \frac{\partial}{\partial y_k}\,; \qquad J\Big(\frac{\partial}{\partial y_k}\Big) = -\frac{\partial}{\partial x_k}\,. \tag{L.2}$$

There may, however, exist almost complex structures which do not arise in this way.

▷ **Almost del Pezzo Surface**: A non-singular algebraic surface (compact, complex 2-dimensional *manifold*$^{(L)}$) $\mathcal{S}$ is called almost del Pezzo if its anti-canonical bundle $\mathcal{K}^*_{\mathcal{S}}$ is *almost ample*$^{(L)}$ (see there!). Synonyms: almost ample surface (2-fold).

▷ **Almost Fano 3-Fold**: A non-singular algebraic 3-fold (compact, complex 3-dimensional *manifold*$^{(L)}$) $\mathcal{M}$ is called almost Fano if its anti-canonical bundle $\mathcal{K}^*_{\mathcal{M}}$ is *almost ample*$^{(L)}$. Synonyms: almost ample solid, almost ample 3-fold.

▷ **Almost Positive**: A bundle over $\mathcal{X}$ is almost positive if it is *positive*$^{(L)}$ over $\mathcal{X}$ except over a special subset $S \subset \mathcal{X}$ ($\dim S < \dim \mathcal{X}$), where it is flat.

▷ **Ample**: A *holomorphic*$^{(L)}$ line *bundle*$^{(L)}$ $\mathcal{L}$ over a complex manifold $\mathcal{M}$ is called ample if at every point of $\mathcal{M}$ there exist at least two *sections*$^{(L)}$ of $\mathcal{L}$ such that one vanishes, but its gradients and the other section are non-zero. More precisely, for every point $x \in \mathcal{M}$ and for every tangent vector $X = \xi^i \partial_i$ to $\mathcal{M}$ at x, there must exist global *holomorphic*$^{(L)}$ sections σ and τ of $\mathcal{L}$ such that

$$\sigma(x) \neq 0\,, \quad \text{(almost ample)}; \tag{L.3}$$

$$\tau(x) = 0\,; \quad (X\tau)(x) = (\xi^i \partial_i \tau)(x) \neq 0\,. \tag{L.4}$$

Stated differently, $\mathcal{L}$ is ample if the bundle of one-*jets*$^{(L)}$ of local holomorphic sections of $\mathcal{L}$, i.e., the local Taylor series up to second order is generated at every point by global holomorphic sections. See also *Almost ample*$^{(L)}$, *semi-ample*$^{(L)}$, *very ample*$^{(L)}$.

▷ **Analytic**: A function $f(z)$ is analytic in a domain $\mathcal{D}$ if it has a local Taylor expansion $f(z) = \sum_{n=0}^{\infty} a_n (z - z_0)^n$ for all $z_0 \in \mathcal{D}$. This is true precisely if $f(z)$ is holomorphic in $\mathcal{D}$.

▷ **Automorphism**: An *endomorphism*$^{(L)}$ which is also an *isomorphism*$^{(L)}$, that is, a 1–1 onto mapping of a group G to itself which preserves the group structure.

▷ **Base Locus** (of a line bundle, $\mathcal{L}$): the intersection of the zero-sets of all global sections of $\mathcal{L}$. If this intersection is empty (not in the base-space of $\mathcal{L}$), $\mathcal{L}$ is *basepoint-free*; see §§ 2.2.2 and 5.2 for examples.

▷ **Bertini's Theorem** (in its most general form): *A generic element of a linear system is smooth away from the base locus of the system.*

Essentially, it means that a system of sufficiently generally chosen algebraic equations over a space $\mathcal{X}$ has a subspace $\mathcal{M}$ of simultaneous solutions which is a smooth subspace of $\mathcal{X}$ except perhaps at the *base locus* of the system. The latter is the simultaneous zero-set in $\mathcal{X}$ of all polynomials the linear combinations of which form the given system. See §§ 1.1, 1.4 and especially 2.2.2 for application and clarifications.

▷ **Betti Numbers**: Given any topological space $\mathcal{X}$ of dimension n, the Betti numbers $b_0, \dots, b_n$ are defined to be the dimensions of its integral, real or complex *homology*$^{(L)}$ or *cohomology*$^{(L)}$ groups. So,

$$b_q \stackrel{\text{def}}{=} \dim_{\Bbbk} H^q(\mathcal{X}, \Bbbk) = \dim_{\Bbbk} H_q(\mathcal{X}, \Bbbk) , \qquad \Bbbk = \mathbb{Z}, \mathbb{Q}, \mathbb{R}, \mathbb{C} \dots \quad \text{(L.5)}$$

▷ **Big** (line bundle): A *line bundle*$^{(L)}$ $\mathcal{L}$ over $\mathcal{X}$ is big if $\dim H^0(\mathcal{X}, \mathcal{L}^{\otimes k}) \geq r\, k^{\dim \mathcal{X}}$, for some $r \in \mathbb{R}_{>0}$ and $k \geq k_0$ with $k_0, k \in \mathbb{Z}_{>0}$. Also, $\mathcal{L}$ is big if its *Iitaka dimension*$^{(L)}$ is maximal, i.e., equal to $\dim \mathcal{X}$.

▷ **Birational Map**: A pair of *Rational Maps*$^{(L)}$ $\alpha : A \to B$ and $\beta : B \to A$, such that $\alpha \circ \beta = \mathrm{Id}_B$ and $\beta \circ \alpha = \mathrm{Id}_A$—as rational maps.

▷ **Blow-Up**: A process by which a point p in a complex n-dimensional space $\mathcal{X}$ is replaced by a compact complex $(n{-}1)$-dimensional space E in such a way that E is a complex subspace of the resulting blowup, $\widetilde{\mathcal{X}}_p = \mathrm{Bl}_p[\mathcal{X}]$. In other words, the space E must admit a complex 1-dimensional fibration such that the n-dimensional total space of this fibration fits in the 'hole' left in $\mathcal{X}$ when p was removed. Note that p could have been a smooth or a singular point of $\mathcal{X}$; its replacement, E is called the exceptional divisor of $\widetilde{\mathcal{X}}_p$.

Given a complex k-dimensional subspace $C \subset \mathcal{X}$, at every point $p \in C \subset \mathcal{X}$, we can stack along C local $(n{-}k)$-dimensional 'neighborhoods' which are transversal to C and do not intersect. Each point on C can then be blown up in such a way that the $(n{-}k{-}1)$-dimensional exceptional divisors E_p vary holomorphically as we vary $p \in C$ and form the $(n{-}1)$-dimensional exceptional divisor in $\widetilde{\mathcal{X}}_C$. Usually, C is called the *center* of the blowing up process. Most notably, regardless of the dimension of C, the exceptional divisor has *codimension*$^{(L)}$ 1 in $\mathcal{X}$; one says that blowing up is a codimension-1 process.

▷ **Bosons**: Particles and fields ensembles of which obey Bose–Einstein statistics. Physical (not ghost) particles and fields, and have integral spin, that is, transforming as integral representations of the *Lorentz group* [(L)], and its covering *Spin*-group [282, 259, 261]. These are then usual commuting (c-number-valued) functions or sections of appropriate bundles.

▷ **Bott–Borel–Weil Theorem**: *Consider a flag space* [(L)]

$$\mathcal{X} \;=\; \mathbb{F}_{[n_1|\dots|n_F]} \;\approx\; \frac{U(N)}{U(n_1)\times\cdots\times U(n_F)}\;, \qquad N=\sum_{i=1}^{F} n_F\;. \tag{L.6}$$

(1) The homogeneous vector bundles over $\mathcal{X}$ *are in 1–1 correspondence with the* $U(n_1)\times\cdots\times U(n_F)$ *representations. We label irreducible such vector bundles by*

$$\mathcal{V}=(a_1,\dots,a_{n_1}|\cdots|b_1,\dots,b_{n_F})\;. \tag{L.7}$$

(2) The $H^q(\mathcal{X},\mathcal{V})$ *is non-zero for at most one value of* q*, in which case it furnishes an irreducible representation of* $U(N)$*,* $H^q(\mathcal{X},\mathcal{V})\approx(c_1,\dots,c_N)\mathbb{C}^N$*.*
(3) $(a_1,\dots,a_{n_1}|\cdots|b_1,\dots,b_{n_F})$ *determines* $(c_1,\dots,c_N)$ *according to the following algorithm:*

1. *Add the sequence* $1,\dots,N$ *to the entries in* $(a_1,\dots,a_{n_1}|\cdots|b_1\dots b_{n_F})$*.*
2. *If any two entries in the result of Step 1 are equal, all cohomology vanishes; otherwise proceed.*
3. *Swap the minimum number* $(=q)$ *of neighboring entries required to produce a strictly increasing sequence.*
4. *Subtract the sequence* $1,\dots,N$ *from the result of 3, to obtain* $(c_1,c_2,\dots,c_N)$*.*

For more details, see Ref. [18] and for generalizations to other (generalized) flag spaces, see Ref. [6].

▷ **Bott Vanishing Theorem**: Several vanishing theorems are generally understood under this title and they are precursors of the more general Bott–Borel–Weil Theorem, so they may be viewed as special cases thereof. For example:

Theorem. *Consider* $\mathcal{X}=\mathbb{F}_{[1|n]}=\mathbb{P}^n$*, and* $\mathcal{V}=\mathcal{O}(k)$*. Then*

$$H^q(\mathbb{P}^n,\mathcal{O}(k)) \;=\; 0\;, \qquad \begin{cases} q\neq 0, & \text{when } 0\leqslant k;\\ \forall q, & \text{when } -n\leqslant k<0;\\ q\neq n, & \text{when } k\leqslant -(n{+}1). \end{cases} \tag{L.8}$$

Since $\mathcal{O}_{\mathbb{P}^n}(k) = (-k|0, \ldots, 0)$, this easily follows from the Bott–Borel–Weil Theorem.

▷ **Branched Covering**: Consider a complex n-dimensional space $\mathcal{X}$ and $B \subset \mathcal{X}$, a complex $(n-1)$-dimensional subspace thereof. Then $\mathcal{Y}$ is a p-fold ($p \in \mathbb{Z}_+$) cover of $\mathcal{X}$ if there exists a mapping $\pi : \mathcal{Y} \to \mathcal{X}$ such that to every point $x \in \mathcal{X}$, it associates p points $y_1(x), \ldots, y_p(x) \in \mathcal{Y}$. If however, this is true except along $B \subset \mathcal{X}$, where to every point $x_{(B)} \in B$ only a single point $y(x_{(B)}) \in \mathcal{Y}$ is associated, $\mathcal{Y}$ is a p-fold cover of $\mathcal{X}$, branched over (along) $B \subset \mathcal{X}$. "Ramified cover" is a synonym.

▷ **BRST**: ☞ *Superspace*$^{(L)}$ and *Supersymmetry*$^{(L)}$.

▷ **Bundle** (fiber∼, line∼, vector∼): A fiber bundle is a triple $(\mathcal{B}, \boldsymbol{E}, \varpi)$, where $\mathcal{B}$ is called the base-space, $\boldsymbol{E}$ the total space and $\varpi : \boldsymbol{E} \to \mathcal{B}$ is a projection.

At any point $x \in \mathcal{B}$, the inverse image of ϖ at x, i.e., the set of points $\{\xi \in \boldsymbol{E} : \varpi(\xi) = x\}$ form the *fiber* at x. Usually, all fibers are in a certain uniform (e.g., homeomorphic for $\mathbb{R}$,- holomorphic for $\mathbb{C}$-bundles) sense isomorphic to some vector space; we then have a (real or holomorphic) *vector bundle*. The simplest vector bundles over a compact complex manifold $\mathcal{B}$ have unit rank, are called holomorphic *line bundles* and their fiber is isomorphic to $\mathbb{C}^1$. Under the (fiber-wise) tensor product, these bundles form a group known as the *Picard group*$^{(L)}$ of $\mathcal{B}$.

▷ **Canonical Bundle**: Given $\mathcal{T}^*_{\mathcal{M}}$, the *holomorphic*$^{(L)}$ *cotangent bundle*$^{(L)}$ of a manifold $\mathcal{M}$, we construct the determinantal holomorphic line *bundle*$^{(L)}$ $\det \mathcal{T}^*_{\mathcal{M}}$, which is called the canonical bundle and is denoted

$$\mathcal{K}_{\mathcal{M}} \stackrel{\text{def}}{=} \det \mathcal{T}^*_{\mathcal{M}} = \wedge^{\dim \mathcal{M}} \mathcal{T}^*_{\mathcal{M}} . \tag{L.9}$$

▷ **Chern Character**: Given a vector bundle $\mathcal{V}$ over a complex manifold $\mathcal{X}$ and its total *Chern class*,$^{(L)}$ $c(\mathcal{V})$, we use the splitting principle,

$$c(\mathcal{V}) = c\Big(\bigoplus_{i=1}^{k} \mathcal{L}_i\Big) = \prod_{i=1}^{k} c(\mathcal{L}_i) = \prod_{i=1}^{k} (1 + \ell_i) , \tag{1.3.19$'$}$$

to introduce $k = \operatorname{rank} \mathcal{V}$ auxiliary line bundles and $\ell_i \stackrel{\text{def}}{=} c(\mathcal{L}_i)$. Then,

$$\operatorname{ch}(\mathcal{V}) \stackrel{\text{def}}{=} \sum_{i=1}^{k} e^{\ell_i} = k + \sum_{r=1}^{\infty} \operatorname{ch}_r(\mathcal{V}') , \tag{L.10}$$

is the *total Chern character*, given as a formal sum of the rth Chern characters. Clearly, $\mathrm{ch}_r(\mathcal{V})$ is a closed (r,r)-form.

▷ **Chern Class**: Given a vector bundle $\mathcal{V}$ over a complex manifold $\mathcal{X}$ and a closed curvature (1,1)-form $\mathbf{\Theta}$ on $\mathcal{X}$ we define

$$c[\mathbf{\Theta}] \stackrel{\text{def}}{=} \det\Big[\mathbb{1} + \frac{i}{2\pi}\mathbf{\Theta}\Big] = \mathbb{1} + c_1[\mathbf{\Theta}] + c_2[\mathbf{\Theta}] + \ldots \qquad (0.2.9')$$

Clearly, each $c_q[\mathbf{\Theta}]$ is a closed (q,q)-form and represents the qth Chern class; $c[\mathbf{\Theta}]$ represents the *total Chern class.*

▷ **Chern Numbers**: Given the Chern classes on a compact complex n-dimensional manifold $\mathcal{X}$, the integrals

$$C_1{}^n \stackrel{\text{def}}{=} \int_X \wedge^n c_1\ , \qquad C_1{}^{n-2}\cdot C_2 \stackrel{\text{def}}{=} \int_X \wedge^{n-2} c_1 \wedge c_2\ , \qquad (\text{L.11})$$

and so on, are topological (deformation) invariants of $\mathcal{X}$ since $c_i = c_i(\mathcal{T}_X)$ are closed. For 3-folds, we write just C_1, C_2 and C_3, understanding the integrals of $(\wedge^3 c_1)$, $(c_1 \wedge c_2)$ and (c_3), respectively.

▷ **Chiral**: Being the positive eigen-spinor of the product of all Dirac matrices. For strictly massless particles, this equals the sign of helicity (the projection of the spin along the linear momentum): helicity is then Lorentz-invariant, as massless particles move at the speed of light, and no observer can pass them to 'flip' the helicity. Positive chirality/helicity is associated with 'left,' while negative chirality is also referred to as 'anti-chirality' and is associated with 'right.' Related to this, a 'chiral superfield' denotes a *superfield*$^{(L)}$ with two component fields: a complex scalar (spin-0) and a left-handed Weil spinor (spin-1/2, positive helicity). In *superspace*$^{(L)}$ formulation, this implies that chiral superfields are independent of right-handed spinorial coordinates, which is in many ways analogous to *holomorphicity*$^{(L)}$.

▷ **CICY**: Complete intersection Calabi–Yau. Unless explicitly otherwise indicated, implies a complete intersection of hypersurfaces in a product of complex projective spaces. Grammatically a modifier, "Calabi–Yau" is often used as a noun, denoting (depending on context) a CICY manifold, n-fold (algebraic variety), space...

▷ **Closed Subset**: In a variety $\mathcal{X}$, a subset $\mathcal{Y} \subset \mathcal{X}$ consisting of all common zeros of finitely many (local, holomorphic) polynomials. In all cases we will consider, these polynomials are (local) realizations of certain global (holomorphic) sections.

▷ **Codimension**: Given a space $\mathcal{X}$ of dimension $\dim \mathcal{X} = n$ and a subspace $\mathcal{Y} \subset \mathcal{X}$ of dimension $\dim \mathcal{Y} = m$, the codimension of $\mathcal{Y}$ in $\mathcal{X}$ is defined simply to be $(n-m)$. One then speaks of how deep (in codimension) $\mathcal{Y}$ lies in $\mathcal{X}$. By convention, the statement $\mathrm{codim}(\mathcal{Y} \subset \mathcal{X}) > \dim \mathcal{X}$ means that $\mathcal{Y}$ is an empty set.

▷ **Cohomology Group**: (for some more details, see § 7.1, but there are classic books written on this subject, such as Ref. [11]; it would be foolish to attempt a self-contained review here). There exist many different ways of defining cohomology groups: de Rham∼, Čech∼, simplicial∼, equivariant∼, Dolbeault∼ (also known as $\bar{\partial}$∼), sheaf∼,... They are all standard and well known in the mathematics literature. On a smooth manifold $\mathcal{M}$, the different definitions lead to isomorphic results. Most of the time we will deal with the latter two mentioned for both of which we simply write $H^q(\mathcal{M}, V)$, with V a vector bundle or a sheaf over $\mathcal{M}$ in which the elements of $H^q(\mathcal{M}, V)$ take values. $H^q(\mathcal{M}, V)$ is isomorphic to the space of harmonic V-valued q-forms, i.e., rank-$(0, q)$ antisymmetric covariant tensors (no dz's, q d$\bar{z}$'s) ω_q which take values in V and satisfy the second order differential equation $\triangle \omega_q = 0$.

If $\mathcal{M}$ is singular, the above mentioned cohomology groups differ and are often (somewhat) ill-behaved. More recently, so-called intersection∼ and L^2-∼ were introduced by Goresky and MacPherson [26] and respectively Cheeger [15], as well as the combined (co)homology of Banagl [250]. On smooth manifolds, these are isomorphic to the previously known ones; on a large class of singular spaces, however, the resulting cohomology groups are better behaved and retain many of the usual properties such as *Poincaré duality*$^{(L)}$, *Hodge decomposition*$^{(L)}$, *Complex conjugation*$^{(L)}$, *Serre duality*$^{(L)}$, *Lefschetz decomposition*$^{(L)}$ (*Lefschetz Hard Theorem*$^{(L)}$), *etc.*

Somewhat confusingly perhaps, q is referred to as the dimension and is not to be confused with the dimension of the vector space $H^q(M, V)$—loosely speaking—the number of linearly independent elements of $H^q(M, V)$. Since $H^q(M, V)$ is an abelian group under addition, the dimension of the *vector space* $H^q(M, V)$ may also be referred to as the rank of the cohomology *group* $H^q(M, V)$.

▷ **Cokernel**: Given two sets A and B and a map $A \xrightarrow{\gamma} B$, the cokernel of γ is defined to be the quotient $\{B/\gamma(A)\}$. In other words, the cokernel of the map $A \xrightarrow{\gamma} B$ may be regarded as spanned by elements of B, taken modulo

$\gamma(a)$, for all $a \in A$. (For *sheaves*$^{(L)}$, this must be corrected; see Ref. [23], p. 36–37.) We will denote the cokernel of γ by $\mathrm{cok}(\gamma)$; the notation $\mathrm{coker}(\gamma)$ is also found in the literature.

▷ **Compact**: A *Hausdorff*$^{(L)}$ space $\mathcal{X}$ is *compact* if for every open covering $\mathfrak{U} = \{U_i\}$, there is a finite sub-collection of open sets $\{U_{i_1}, \ldots, U_{i_n}\}$ which covers $\mathcal{X}$.

▷ **Complete Intersection**: A variety $\mathcal{X}$ of dimension m in $\mathbb{P}^n$ is a *strict complete intersection* if the *Ideal of the variety*$^{(L)}$, $I(\mathcal{X})$ can be generated by $n-m$ elements. $\mathcal{X}$ is a *set-theoretic complete intersection* if it can be realized as the intersection of some $n-m$ *hypersurfaces*$^{(L)}$.

▷ **Complex**: A *complex* $(K^\star, d) = \{K^0 \xrightarrow{d} K^1 \xrightarrow{d} \cdots\}$ is a sequence of Abelian groups with differentials $d : K^r \to K^{r+1}$ satisfying $d \circ d = 0$. The *cohomology* of a complex is $H^\star(K^\star) = \bigoplus_{r\geq 0} H^r(K^\star)$, with

$$H^r(K^\star) \stackrel{\text{def}}{=} \{ [\ker d : K^r \to K^{r+1}] / dK^{r-1} \} . \tag{L.12}$$

▷ **Complex Conjugation** (on the cohomology): The requirement that $H^{p,q} = \overline{H^{q,p}}$; this is usually included in the *Hodge decomposition*$^{(L)}$.

▷ **Complex Projective Space**: The complex projective space, $\mathbb{P}^n \stackrel{\text{def}}{=} \mathbb{P}(\mathbb{C}^{n+1})$ is the complex n-dimensional space of all complex lines (copies of $\mathbb{C}^1$) in $\mathbb{C}^{n+1}$, that is, a point in $\mathbb{P}^n$ is a 1-dimensional linear subspace $L \subset \mathbb{C}^{n+1}$.

Explicitly, let $z^0, \ldots, z^n$ be coordinates on $\mathbb{C}^{n+1}$. Then $(z^0 : \ldots : z^n)$ and $\lambda(z^0 : \ldots : z^n)$ lie on the same complex line $L \subset \mathbb{C}^{n+1}$, parametrized by $\lambda \in \mathbb{C}$ and correspond to the same point of $\mathbb{P}^n$. In the coordinate patch where $z^i \neq 0$, suitable local coordinates are $\zeta^j_{(i)} \stackrel{\text{def}}{=} z^j/z^i$. The (one dimension redundant) coordinates z^i are called homogeneous and the local coordinates $\zeta^j_{(i)}$ are called affine.

Alternatively, $\mathbb{P}^n$ can also be defined as the quotient

$$\frac{U(n+1)}{U(n) \times U(1)} \cong \frac{SU(n+1)}{S(U(n) \times U(1))} , \tag{L.13}$$

where $U(n)$ denotes the Lie group of $n \times n$ unitary matrices while the prefix S denotes unimodularity. This fact is employed to a great extent in Chapters 9–B.

▷ **Complex Structure**: An integrable *almost complex structure*$^{(L)}$. That is, an *endomorphism*$^{(L)}$ J of the tangent bundle of an even-dimensional manifold, which squares to $-\mathbb{1}$ and in addition satisfies the integrability condition, the vanishing of the Nijenhuis tensor

$$N_{ij}{}^k \stackrel{\text{def}}{=} J_{[i}{}^p J_{j]}{}^q (\partial_p J_q{}^k) - \partial_{[i} J_{j]}{}^k \ . \tag{L.14}$$

Note that this is independent of any connection.

▷ **Compound Du Val (cDV) Singularities**: A 3-fold hypersurface singularity $p^\sharp \in \mathcal{X}$, which is locally given by

$$F(x,y,z,t) \stackrel{\text{def}}{=} f(x,y,z) + t\,g(x,y,z,t) \ = \ 0 \ , \tag{L.15}$$

in $\mathbb{C}^4$, where g is an arbitrary polynomial and $f(x,y,z)$ is one of the following

$$\begin{array}{llll} A_k & : & x^{k+1} + y^2 + z^2 & = \ 0 \ , \quad (n \geq 1) \ , \\ D_k & : & x^{k-1} + y^2 z + z^2 & = \ 0 \ , \quad (n \geq 4) \ , \\ E_6 & : & x^4 + y^3 + z^2 & = \ 0 \ , \\ E_7 & : & x^3 y + y^3 + z^2 & = \ 0 \ , \\ E_8 & : & x^5 + y^3 + z^2 & = \ 0 \ . \end{array} \tag{L.16}$$

These are called ‘simple,’ ‘rigid’ or Du Val 2-fold (surface) singularities and a cDV 3-fold singularity is 1-parameter deformation of some isolated A-D-E 2-fold singularity.

▷ **Configuration**: In our context, most generally, a configuration is a pair $[\mathcal{X}\|\mathcal{E}]$, where $\mathcal{X}$ is the *embedding space* and $\mathcal{E}$ is a vector bundle over $\mathcal{X}$ (typically, a sum of line bundles). We say that a variety $\mathcal{M}$ *belongs* to, or is a *member* of, a configuration if it can be realized as the zero-set of a section ϕ of $\mathcal{E}$ and we write $\mathcal{M} = \phi^{-1}(0)$. Naturally, a configuration contains a deformation class of varieties, parametrized by the choice of the section ϕ.

▷ **Conifold**: A space which fails to be smooth at an isolated set of singular points, the neighborhood of each of which is isomorphic to a cone. This includes *hyperquotient* singularities [46] — singular points which, loosely speaking, arise by coalescing orbifold (finite quotient) and isolated hypersurface singularities atop each other.

▷ **Contractible**: Can be diminished in dimension (usually, to a point). In doing so, care must be taken that the category of objects under scrutiny is not left: for example, non-contractibility of a subspace X in some space Y as a potential cycle (homology class representative) in $H_\star(Y)$ implies that X cannot be "shrunk" continuously to a point and remaining throughout the process within Y. So, e.g., a 'small circle' is non-contractible within a (real) 2-torus, although all circles are contractible within, e.g., 2-spheres or $\mathbb{R}^3$.

▷ **Contraction**: Generally, it refers to crushing a subspace $E \subset X$ to a smaller subspace $e \subset \mathcal{X}^\sharp$ in a certain 'uniform' way. Unless explicitly stated otherwise, we think of a collapse of a (real) 2-cycle $E \subset \mathcal{M}$ to an A_k-, D_k- or E_k-singular point (L.16), where the 2-cycle consists of $\mathbb{P}^1$'s configured according to the Dynkin diagram of the respective Lie algebra. Then, the normal bundle of all these $\mathbb{P}^1$'s is $\mathcal{O}(-1)\otimes\mathcal{O}(-1)$ and they cannot be deformed. Since for 3-folds this happens in (complex) codimension 2, the canonical class is not affected and the singular space obtained upon the contraction inherits the holomorphic 3-form. See also the Small Map Theorem 7.4.

▷ **De Rham Duality**: On a real (compact, orientable) differentiable manifold $\mathcal{X}$, let $\alpha^i_{(r)}$ denote closed differential r-forms and $\gamma_i^{(r)}$ elements of $H_r(\mathcal{X})$ (the vector space of r-cycles modulo boundaries of finite $(r+1)$-chains). Then the condition $\int_{\gamma_j^{(r)}} \alpha^i_{(r)} = \delta^i_j$ depends only on the respective (co)homology classes of $\gamma_i^{(r)}$ and $\alpha^o_{(r)}$, associates pairwise a class of r-cycles to each class of r-forms and establishes that $H^r_{DR}(\mathcal{X})$ is dual to $H_r(\mathcal{X})$. Since definitions of cohomology groups other De Rham's result in groups $H^r(\mathcal{X})$ which are similarly dual to $H_r(\mathcal{X})$, it follows that the De Rham cohomology groups are isomorphic to these other cohomology groups.

▷ **Deformation Space**: In general, given a (topological class of) real (compact, orientable) manifold(s) the choice of the complex structure will depend on a number of parameters. That is, we can deform the space so as to become a different complex manifold while remaining in the same topological class. The space spanned by such parameters is called the deformation space and is often rather straightforwardly parametrized by the choice of the embedding, for example.

▷ **Diagram**: We find it sometimes useful to represent a complete intersection in a product of complex projective spaces by means of an organic chemistry-like diagram. For details, see § 2.1.2 and [95, 146, 148]. It is of course straightforward to generalize these to any other type of embedding spaces.

▷ **Discriminant Locus**: The subset $\mathfrak{M}^\sharp$ of the space of *complex structures*$^{(L)}$ $\mathfrak{M}$ which parametrizes singular complex spaces. Generically, $\dim \mathfrak{M}^\sharp = \dim \mathfrak{M} - 1$. That is, $\mathfrak{M}^\sharp$ is a *divisor*$^{(L)}$ in $\mathfrak{M}$; see § D.3.3.

▷ **Divisor**: On a (possibly non-compact) complex manifold $\mathcal{M}$ of dimension n, a locally finite formal linear combination $D \stackrel{\text{def}}{=} \sum_i a_i S_i$ of irreducible analytic hypersurfaces of $\mathcal{M}$.

Here "locally finite" means that every point of $\mathcal{M}$ has some local neighborhood which intersects only a finite number of S_i's appearing in D; if $\mathcal{M}$ is compact, this means that the sum is finite. An "analytic hypersurface" can be given as the zero-set of some single holomorphic function in the local neighborhood of every point of $\mathcal{M}$.

▷ **Dolbeault Theorem**: *$H^q(\mathcal{M}, \wedge^p \mathcal{T}^*_{\mathcal{M}}) \approx H^{p,q}_{\bar\partial}(\mathcal{M})$ for complex manifolds $\mathcal{M}$.*

▷ **Double Solid**: A *branched double cover*$^{(L)}$ of an *Almost Fano 3-fold*$^{(L)}$ 3-fold. We discuss these in § 5.5; see also [14] and the second article in Ref. [149] for generalizations.

▷ **Duality**: A non-degenerate pairing of elements from A and B, such that $(a, b) \mapsto e$, where e is some special (naturally given) quantity. For example, in a real (compact, orientable) n-fold, to every closed class of r-form we can associate a class of $(n-r)$-forms such that their product is cohomologous to the volume-form. We write $A \stackrel{*}{\sim} B$. It is a standard fact that if furthermore some $C \stackrel{*}{\sim} B$, then this double duality induces an isomorphism $A \approx C$. See *De Rham*$^{(L)}\sim$, *Hodge Star*$^{(L)}\sim$, *Poincaré*$^{(L)}\sim$ and *Serre*$^{(L)}\sim$.

▷ **Dualizing Sheaf**: On a smooth complex n-fold $\mathcal{M}$, the dualizing sheaf $\mathfrak{K}_{\mathcal{M}}$ is the *sheaf*$^{(L)}$ corresponding to the canonical (line) bundle $\mathcal{K}_{\mathcal{M}} \stackrel{\text{def}}{=} \bigwedge^n \mathcal{T}^*_{\mathcal{M}}$. For a large class of singular spaces, $\mathcal{K}_{\mathcal{M}}$ is ill-defined at the singular points, but $\mathfrak{K}_{\mathcal{M}}$ is still well-defined and may be thought of as the generalization of $\mathcal{K}_{\mathcal{M}}$.

▷ **Endomorphism**: A mapping of a group G to itself which preserves its group structure. That is, f is an endomorphism if $f : G \to G$ and $f(ab) = f(a)f(b)$, for any $a, b \in G$. For a vector *bundle*$^{(L)}$, this means that the group structure of each fiber is preserved, point by point over the base space.

▷ **Euler Number (Characteristic)**: It may be defined as the index of the exterior derivative operator, d, so that

$$\chi_E(\mathcal{M}) \stackrel{\text{def}}{=} \sum_{r=0}^{\dim\mathcal{M}} (-)^r b_r(\mathcal{M}) = \sum_{p,q=0}^{\dim\mathcal{M}} (-)^{p+q} h^{p,q}(\mathcal{M}) , \qquad \text{(L.17)}$$

where $b_r(\mathcal{M}) \stackrel{\text{def}}{=} \dim H^r_{\text{DR}}(\mathcal{M})$ are the Betti numbers and $h^{p,q}(\mathcal{M}) \stackrel{\text{def}}{=} \dim H^{p,q}(\mathcal{M})$ are the *Hodge numbers*$^{(L)}$. By the well known Gauss–Bonnet formula, we also have

$$\chi_E(\mathcal{M}) = \int_{\mathcal{M}} c_n(\mathcal{M}) , \qquad n = \dim\mathcal{M} . \qquad \text{(L.18)}$$

The closed (n, n)-form c_n is known as the nth Chern class of $\mathcal{M}$. Suitable generalizations of these formulae exist for singular $\mathcal{M}$.

▷ **Euler Sequence**: On a complex projective space $\mathbb{P}^n$, the holomorphic tangent bundle is spanned by tangent vectors $\ell^i(z)\partial_i$, modulo the relation $z^i\partial_i \approx 0$ (overall scaling is trivial in $\mathbb{P}^n$), and where $\ell^i(z)$ are linear polynomials in the homogeneous coordinates. Being linear in z, each $\ell^i(z)$ corresponds to $\mathcal{O}_{\mathbb{P}^n}(1)$ and the above description of the tangent bundle as a quotient is equivalent to the short *exact sequence*$^{(L)}$

$$0 \to \mathbb{C} \to \mathcal{O}_{\mathbb{P}^n}(1)^{\oplus(n+1)} \to \mathcal{T}_{\mathbb{P}^n} \to 0 , \qquad \text{(L.19)}$$

which is called the Euler sequence. See also the *tautological sequence*$^{(L)}$.

▷ **Exact Sequence**: A sequence of spaces (bundles over some base space) and appropriate maps (preserving the structure of the spaces, bundles ...)

$$X_1 \xrightarrow{\alpha_1} X_2 \xrightarrow{\alpha_2} X_3 \xrightarrow{\alpha_3} \cdots \qquad \text{(L.20)}$$

such that $\text{im}(\alpha_k) = \ker(\alpha_{k+1})$, (see *image*$^{(L)}$ and *kernel*$^{(L)}$). Of interest are exact sequences that have a zero at either end, or even more importantly, at both ends. $0 \to A \xrightarrow{\alpha} B$ being exact means that α is 1–1; $B \xrightarrow{\beta} C \to 0$ being exact means that β is onto. The *short exact sequence* (also called a *monad*$^{(L)}$, written without the outer zeros)

$$0 \to A \xrightarrow{\alpha} B \xrightarrow{\beta} C \to 0 \qquad \text{(L.21)}$$

means that $A \subseteq B$ and that $C = \{B/A\}$. B is then called the *extension* of A by C; we will write $B = C + A$ (note the order; this use of '+' is *not commutative*!) and warn that when A, B, C are holomorphic bundles over some space, it is generally impossible to find a metric in which A and C will be orthogonal. So $B \neq C \oplus A$ holomorphically, although differentiably B *is* the direct sum of C and A.

The above short exact sequence is equivalent to $C = \{B/A\}$, and implies that $A = \ker(\beta)$ and that $C = \mathrm{cok}(\alpha)$. Also, $0 \to C^* \xrightarrow{\beta} B^* \xrightarrow{\alpha} A^* \to 0$ is exact, $A^* = \{B^*/C^*\}$, $C^* = \ker(\alpha)$ and $A^* = \mathrm{cok}(\beta)$.

▷ **Fano**: An algebraic variety, $\mathcal{X}$ (primarily for $\dim \mathcal{X} > 2$), is Fano (after Gino Fano) if its anticanonical bundle, $\mathcal{K}^*_{\mathcal{X}}$, is ample. Elsewhere in the literature (incl. Wikipedia), the following definitions are used: $\mathcal{X}$ is *weakly Fano* if $\mathcal{K}^*_{\mathcal{X}}$ is *nef*$^{(L)}$ and *big*$^{(L)}$ (incl. *ample*$^{(L)}$); $\mathcal{X}$ is *almost Fano* if $\mathcal{K}^*_{\mathcal{X}}$ is *nef*$^{(L)}$ and *big*$^{(L)}$ but not *ample*$^{(L)}$—cf. the different and inclusive use of *almost ample*$^{(L)}$ herein; $\mathcal{X}$ is *semi-Fano* if $\mathcal{K}^*_{\mathcal{X}}$ is *nef*$^{(L)}$.

▷ **Favorable Configuration**: Originally [145], a *configuration*$^{(L)}$ $[\mathcal{X} \| \mathcal{E}]$ where the hypersurfaces specified by $\mathcal{E}$ may be ordered so the *Lefschetz hyperplane theorem*$^{(L)}$ is iteratively applicable to insure that $\varkappa : H^2(\mathcal{M},\mathbb{Z}) \hookrightarrow H^2(\mathcal{X},\mathbb{Z})$ is an injection. Subsequent *CICY*$^{(L)}$ literature usage restricts to cases where $\varkappa$ is an isomorphism, so $H^2(\mathcal{M},\mathbb{Z})$ is generated by the Kähler forms of the projective factors in $\mathcal{X}$.

▷ **Fermions**: Particles and fields ensembles of which obey Fermi–Dirac statistics. Physical (not ghost) particles and fields, and have half-integral spin, that is, transforming as half-odd-integral representations of the *Spin*-group [282, 259, 261]. As corresponding representations of the *Lorentz group*$^{(L)}$, spinors are double-valued: they change sign upon a 2π-rotation. These are then anticommuting (a-number-valued) functions or sections of appropriate bundles.

▷ **Fibration**: While formal textbook definitions may be rather abstract and technical, it will suffice for our purposes to regard this as a generalization of the notion of a *bundle*$^{(L)}$, where the fiber may deviate from varying uniformly and may degenerate or otherwise change at certain locations. ☞ *bundle*$^{(L)}$.

▷ **Fiber Bundle**: ☞ *bundle*$^{(L)}$.

▷ **Finite Quotient Singularity**: The finiteness of the quotient in some $\mathcal{X}/G$ refers to the finiteness of the group (action) G. Locally, the singularity is $\mathbb{C}^n/G$. Best known are the easiest examples, where $G = \mathbb{Z}_m$ is a cyclic group; the singularity is then called a *cyclic quotient singularity.* Note that in the process of *passing to the quotient* $\mathcal{X} \to \mathcal{X}/G$, a finite number of points (an orbit of G) is being identified, akin to cutting a sheet of paper from center to edge, then folding it around the center in m layers to make a uniform circular cone. By contrast, if G were continuous, so would be the orbits of G.

▷ **Flag Space**: A sequence of inclusions $\mathcal{X}_1 \subset \mathcal{X}_2 \subset \cdots \subset \mathcal{X}_n = \mathcal{X}$, where $\dim \mathcal{X}_i < \dim \mathcal{X}_{i+1}$, is called a *flag*. A *flag space* then is the space of all flags in $\mathcal{X}$ with the $\dim \mathcal{X}_i$ fixed by the given pattern. An inclusion $\mathcal{X}_i \subset \mathcal{X}_{i+1}$ is called a *stripe* of a flag. Projective spaces are the simplest flag spaces, having a single stripe and of dimension one: $\mathbb{P}^n = \{L \subset \mathbb{C}^n\}$, $L \approx \mathbb{C}^1$.

▷ **Fixed Point**: In a topological space $\mathcal{X}$ equipped with an (symmetry) map $\boldsymbol{g} : \mathcal{X} \to \mathcal{X}$, a point $p \in \mathcal{X}$ left invariant is a *fixed point*. If a collection of maps $\boldsymbol{g}_i$ forms a group G, p is a fixed point of G if it is fixed by at least one element of G other than the identity.

▷ **Free Action** (of a group): Let G be a group acting on a topological space $G : \mathcal{X} \to \mathcal{X}$ in some specified way. This action is called *free* if it has no fixed points in $\mathcal{X}$. Note that this is a property of the action of the group on $\mathcal{X}$, not the group itself. For example, $\mathbb{Z}_2$ acts freely on a circle by 180°-rotations, but the action by reflections about an axis of symmetry has two fixed points.

▷ **Free Sheaf**: A *sheaf*$^{(L)}$ $\mathfrak{S}$ over $\mathcal{M}$, which admits a *free resolution* in terms of a sequence of vector bundles $\{\mathcal{V}_q\}_{q \geq 0}$, over $\mathcal{M}$. That is, the sequence

$$0 \to \ldots \to \mathcal{V}_{q+1} \to \mathcal{V}_q \to \ldots \mathcal{V}_0 \to \mathfrak{S} \to 0 \tag{L.22}$$

is sheaf-exact: it is exact over a sufficiently fine open covering of $\mathcal{M}$ (a collection of sufficiently small open sets which cover $\mathcal{M}$ completely).

▷ **Foliation**: While locally (at most points) still a product of two separate factors, a foliation generalizes the notion of a vector bundle in that it has no globally preferred "base-space" over which a "fiber" varies. The relatively recently explored self-T-dual string theory formalism [379] shows that the non-commutatively and chirally doubled stringy spacetime admits a foliation [380].

▷ **Fubini–Study Metric**: On a complex projective space $\mathbb{P}^n$, we have the homogeneous coordinates $x^0, \ldots, x^n$ and define the Fubini–Study (sometime also called 'round') metric as the Kähler metric

$$\mathrm{g}^{(FS)}_{\mu\bar{\nu}} \stackrel{\text{def}}{=} \partial_\mu \, \partial_{\bar{\nu}} \log \Big(\sum\nolimits_{\mu=0}^{n} |x^\mu|^2 \Big) \,, \tag{L.23}$$

where $\log(\sum_{\mu=0}^{n}|x^{\mu}|^{2})$ is the Fubini–Study Kähler potential. $J = \mathrm{d}x^{\mu}\wedge\mathrm{d}x^{\overline{\nu}}(ig^{(FS)}_{\mu\overline{\nu}})$ is a closed $(1,1)$-form; it however is not exact owing to the branch cut in $\log\|x\|^2$ and its cohomology class in fact generates $H^{1,1}(\mathbb{P}^n)$.

▷ **Generic**: Loosely speaking, means 'almost all'. More precisely, denotes the subspace of available choices such that its *complement* is of strictly smaller dimension. For example, a family of polynomials is parametrized by the coefficients and a 'generic polynomial' refers a choice corresponding to any point which is not in a subset of parameters of a dimension strictly smaller than the whole parameter space. Note that, in the complex (holomorphic) setting, this means that the (real) codimension of the 'bad' subset is at least 2.

▷ **Genus** (arithmetic∼, geometric∼): For a complex n-fold $\mathcal{X}$, $h^{p,q} \stackrel{\text{def}}{=} \dim H^{p,q}(\mathcal{X})$ are the *Hodge numbers*$^{(L)}$. $h^{0,n}$ is called the *geometric genus*, $h^{0,1}$ is the *irregularity* and

$$\chi^{h}(\mathcal{X}) \stackrel{\text{def}}{=} \sum_{q=0}^{n}(-)^{q}\dim H^{0,q}(\mathcal{X}) = 1+(-)^{n}p_{a} \tag{L.24}$$

is the *Todd genus*, also called the *holomorphic Euler characteristic (number)*, while p_a is the arithmetic genus. See also *plurigenera*$^{(L)}$.

▷ **Germ** (of sections of a sheaf): ☞ *Stalk (of a sheaf)*$^{(L)}$.

▷ **Ghost**: Particles and fields which have the "wrong" spin-statistics correlation: *bosons*$^{(L)}$ with half-integral spin and *fermions*$^{(L)}$ with integral spin [282, 259, 261].

▷ **Hausdorff**: A topological space $\mathcal{X}$ is *Hausdorff* if for any two points of $\mathcal{X}$ there are disjoint open sets each containing exactly one of the points.

▷ **Helicity (Sub-)Group**: The maximal subgroup of the proper orthochronous *Lorentz group*$^{(L)}$, $SO(t,s)$ which leaves invariant a light-like vector (for massless particles) or a time-like vector (for massive particles); s and t ($s \geq t$) denote the number of space-like and time-like dimensions, respectively. So, the helicity (sub)group of massless particles is $SO(s{-}t)$, while for massive particles it is $SO(s)$. Clearly, if $s = t$, the helicity (sub)group for massless particles is finite. The concept and use of this subgroup of the Lorentz subgroup is due to E. Wigner, and is also called the 'Wigner little group'.

▷ **Hessian**: The matrix of second derivatives of a function $f(x)$. It is independent of any connection at singular points, since there $f(x) = 0 = \partial f(x)$. The Hessian is therefore used to characterize singular points.

▷ **Hirzebruch Signature**: F. Hirzebruch defines [29] the signature or a manifold $\mathcal{M}$ as

$$\tau_{_H}(\mathcal{M}) \stackrel{\text{def}}{=} \sum_{q=0}^{\dim \mathcal{M}} (-)^q h^{p,q}(\mathcal{M}) \ , \tag{L.25}$$

where $h^{p,q}(\mathcal{M}) \stackrel{\text{def}}{=} \dim H^{p,q}_{\bar\partial}(\mathcal{M})$ are the Hodge numbers. He then proves

$$\tau_{_H}(\mathcal{M}) = \int_{\mathcal{M}} L(\mathcal{M}) \ , \tag{L.26}$$

where $L(\mathcal{M})$ has is known as the Hirzebruch L polynomial as defined in § 1.5.

▷ **Hodge Decomposition** (1, of forms): Every (p,q)-form ω may be decomposed as

$$\omega = \mathscr{H}(\omega) + \bar\partial(\bar\partial^\dagger G(\omega)) + \bar\partial^\dagger(\bar\partial G(\omega)) \ , \tag{L.27}$$

where $\mathscr{H}(\ )$ is the projection on harmonic forms, $G(\ )$ is the Green's operator and $\bar\partial^\dagger$ is adjoint to $\bar\partial$, that is $\langle\alpha, \bar\partial\beta\rangle = \langle\bar\partial^\dagger\alpha, \beta\rangle$. This is equivalent to saying that, given a (p,q)-form η, the equation $\triangle\omega = \eta$ has a solution precisely if $\mathscr{H}(\eta) = 0$ and then $\omega = G(\eta)$.

▷ **Hodge Decomposition** (2, of cohomology): For a compact Kähler manifold $\mathcal{M}$, the complex cohomology satisfies

$$H^r_{\text{DR}}(\mathcal{M}, \mathbb{C}) \approx \bigoplus_{p+q=r} H^{p,q}_{\bar\partial}(\mathcal{M}) \ , \qquad H^{p,q}_{\bar\partial}(\mathcal{M}) = \overline{H^{q,p}_{\bar\partial}(\mathcal{M})} \ . \tag{L.28}$$

▷ **Hodge Number**: $h^{p,q} \stackrel{\text{def}}{=} \dim H^{p,q}(\mathcal{M})$, also denoted $h^{p,q}$, are the *Hodge numbers* of a complex manifold $\mathcal{M}$.

▷ **Hodge Star Duality**: On a real n-dimensional (compact, orientable) differentiable manifold $\mathcal{X}$, we define the *Hodge star* operator by its action on a differential r-form $\alpha = a_{i_1\cdots i_r} \mathrm{d}x^{i_1} \ldots \mathrm{d}x^{i_r}$:

$$\begin{aligned} *\alpha &= a^*_{j_1\cdots j_{n-r}} \mathrm{d}x^{j_1} \wedge \ldots \wedge \mathrm{d}x^{j_{n-r}} \ , \\ a^*_{j_1\cdots j_{n-r}} &= \epsilon^{i_1\cdots i_r}{}_{j_1\cdots j_{n-r}} a_{i_1\cdots i_r} \ . \end{aligned} \tag{L.29}$$

This clearly takes (closed or exact) r-forms into (closed or exact) $(n-r)$-forms.

Since the scalar product $\int \alpha\wedge*\alpha \geq$ and is zero only if α is exact, it establishes that $H^r(\mathcal{X})$ and $H^{n-r}(\mathcal{X})$ are dual. Combining with the De Rham duality, we have the isomorphism $H^{n-r}(\mathcal{X}) \approx H_r(\mathcal{X})$. Moreover, if $\mathcal{X}$ is complex n-dimensional, the Hodge star operator takes (p,q)-forms into $(n-q, n-p)$-forms and we have that $H^{p,q}(\mathcal{X})$ and $H^{n-q,n-p}(\mathcal{X})$ are dual.

▷ **Holomorphic**: A function $f(z)$ is holomorphic in a domain $z \in \mathcal{D}$ if $\frac{\partial f}{\partial \bar{z}_i} = 0$ everywhere in $\mathcal{D}$. Also, this is true precisely if $f(z)$ is analytic in $\mathcal{D}$.

▷ **Holomorphic Euler Number (Characteristic)**: ☞ *genus* $^{(L)}$.

▷ **Holonomy**: On a manifold $\mathcal{X}$, consider parallel-transporting a vector $\vec{v}_p$ tangent to $\mathcal{X}$ at $p \in \mathcal{X}$, along a closed path back to p. The transformation required to bring the parallel-transported vector $\vec{v}_p{}'$ back to the original $\vec{v}_p$ is called the *holonomy.* The collection of such transformations, parametrized by all closed paths in $\mathcal{X}$ containing p, manifestly forms a group and is called the *holonomy group*; it is independent of the choice of $p \in \mathcal{X}$. If the paths are restricted to be contractible, the resulting group is called the *restricted holonomy group.*

▷ **Hom($\mathcal{A}, \mathcal{B}$)**: For two bundles $\mathcal{A}$ and $\mathcal{B}$, the bundle of *homomorphism* $\mathcal{A} \to \mathcal{B}$, that is, maps that preserve the bundle structure. Clearly, $\mathrm{Hom}(\mathcal{A}, \mathcal{B}) = \mathcal{A}^* \otimes \mathcal{B}$: when contracting $\boldsymbol{a} \in \mathcal{A}$ with $\boldsymbol{a}^* \otimes \boldsymbol{b}$, $\boldsymbol{a}$ is replaced with $\boldsymbol{b}$.

▷ **Homology**: Two subspaces $\mathcal{Y}$ and $\mathcal{Y}'$ of a manifold $\mathcal{X}$ are said to be homologous if $\mathcal{Y} \cup \mathcal{Y}'$ is the boundary of a subspace of $\mathcal{X}$. For example, two ‘little’ circles on a torus are homologous because they are the boundaries of a cylindrical portion of the torus.

▷ **Homotopy**: A one (real) parameter family of continuous mappings $f_t : \mathcal{Y} \xrightarrow{\text{into}} \mathcal{X}$, $0 \leq t \leq 1$, which is also continuous with respect to t. So, for two subspaces $\mathcal{Y}$ and $\mathcal{Y}'$ in a manifold $\mathcal{X}$ to be homologous, we need that (1) $\mathcal{Y} = f_0(Y)$ and $\mathcal{Y}' = f_1(Y)$ are images in $\mathcal{X}$ of the same space Y and that there is a family of images $\mathcal{Y}_t = f_t(Y)$ ‘interpolating’ continuously between $\mathcal{Y}$ and $\mathcal{Y}'$. Loosely speaking, it must be possible to deform $\mathcal{Y}$ into $\mathcal{Y}'$ continuously and without breaking or tearing.

▷ **Hopf Index Theorem**: Let $\mathcal{M}$ be a compact oriented n-fold and $\boldsymbol{v}$ a global C^∞ vector field on $\mathcal{M}$. A zero, p, of $\boldsymbol{v}$ is *non-degenerate* if it is isolated and

$$\boldsymbol{v}(x) \quad = \quad x^i \lambda_i{}^j \partial_j \ + \ldots \tag{L.30}$$

in local coordinates around the zero of $\boldsymbol{v}$, where $\lambda_i{}^j$ is non-singular and the ellipses denote terms of at least quadratic in x. Integrating the vector field $\boldsymbol{v}$ to time t produces a flow $f_t : \mathcal{M} \to \mathcal{M}$, such that for $t > 0$ and sufficiently small, $\iota_{f_t}(p) = \iota_v(p)$ at every zero of $\boldsymbol{v}$. However, since f_t is manifestly homotopic to the identity, $f^\star$ acts as the identity on the cohomology and we have

Theorem.

$$\sum_{\substack{p\in\mathcal{M}\\ v(p)=0}} \iota_v \;=\; L(f_t) \;=\; \chi_{_E}(\mathcal{M}) \;\stackrel{\text{def}}{=}\; \int_{\mathcal{M}} c_n(\mathcal{M})\,. \tag{L.31}$$

See Eq. (L.38) and the *Lefschetz Fixed-Point Formulae*$^{(L)}$.

▷ **Hyperplane**: A linear subspace of codimension 1. That is, given an n-fold $\mathcal{X}$, a hyperplane $H \subset \mathcal{X}$ is defined locally by $\ell(z) = 0$, where $\ell(z)$ is a single linear constraint.

▷ **Hyperplane Bundle**: Over an n-fold $\mathcal{X}$, the line *bundle*$^{(L)}$ the fiber of which, at $p \in \mathcal{X}$, is the space of hyperplanes through that point. As hyperplanes are defined by linear constraints, the hyperplane bundle written as $\mathcal{O}(1)$ is also the (line) bundle the (global, holomorphic) sections of which are linear polynomials over $\mathcal{X}$.

▷ **Hypersurface**: Given a complex manifold $\mathcal{X}$ and a function (section of some line bundle) ϕ over $\mathcal{X}$, we define the hypersurface $F \subset \mathcal{X}$ is the subspace of $\mathcal{X}$ as $F \stackrel{\text{def}}{=} \{x \in \mathcal{X} : \phi(x) = 0\}$. We also write, somewhat formally, $F = \phi^{-1}(0) \hookrightarrow \mathcal{X}$. The (complex) *codimension*$^{(L)}$ of F in $\mathcal{X}$ is 1, that is, $\dim \mathcal{X} = \dim F + 1$. It is a basic fact that $F \subset \mathcal{X}$ carries an integral homology class, $[F] \in H_{\dim F}(\mathcal{X}, \mathbb{Z})$, which is isomorphic to some $[f] \in H^2(\mathcal{X}, \mathbb{Z})$ as both are *dual*$^{(L)}$ to some $[*F] \in H^{\dim F}(\mathcal{X}, \mathbb{Z})$. This establishes a correspondence between the *Picard group*$^{(L)}$ $\mathrm{Pic}(\mathcal{X})$, $H^2(\mathcal{X}, \mathbb{Z})$ and $H_{\dim F}(\mathcal{X}, \mathbb{Z})$.

▷ **Ideal**: A right (left) ideal $\Im$ of a *ring*$^{(L)}$ R is an additive subgroup of R such that, for any $a \in R$, $a \cdot e \in \Im$ ($e \cdot a \in \Im$) if $e \in \Im$. The *quotient*$^{(L)}$ $R/\Im$ is called the *residue class ring* or the *factor ring* (*modulo* $\Im$). In the special case when $\Im$ is generated by gradients of a single function, W, the ideal $\Im[\partial W]$ is called the *Jacobian ideal.*

▷ **Ideal of a Variety**: For a (projective) subvariety $\mathcal{Y}$ of dimension m in a (projective) variety $\mathcal{X}$ of dimension n, the collection of all (homogeneous) functions which vanish precisely on $\mathcal{Y}$ forms an *ideal*$^{(L)}$ of the *ring*$^{(L)}$ of all (homogeneous) functions on $\mathcal{X}$ and is denoted $I(\mathcal{Y} \subset \mathcal{X})$. If $\mathcal{X} = \mathbb{P}^n$, it is omitted in conventional notation.

▷ **Iitaka Dimension**: of a line bundle $\mathcal{L}$ over $\mathcal{X}$ equals one less than the dimension of its *section ring*$^{(L)}$, $\mathcal{R}(\mathcal{X}, \mathcal{L}) \stackrel{\text{def}}{=} \bigoplus_{k\geq 0} H^0(\mathcal{X}, \mathcal{L}^{\otimes k})$.

▷ **Image**: The image of a map $\pi\colon A \to B$ is defined as $\mathrm{im}(\pi) \stackrel{\text{def}}{=} \bigcup_{a\in A}\{\pi(a) = b \in B\}$. Naturally, $\mathrm{im}(\pi) \subseteq B$.

▷ **Implicit Function Theorem**: *Let f_a be K local functions of $n > K$ local variables $x_i \in \mathcal{X}$ which can be chose so that $\det\left(\frac{\partial f_a}{\partial x_i}\right) \neq 0$ for $a, i = 1, \dots, K$. Then there exist functions q_a, $a = 1, \dots, K$, such that*

$$f_a(x) = 0\ , \quad \Longleftrightarrow \quad x_a = q_a(x_{K+1}, \dots, x_n)\ , \quad a = 1, \dots, K\ . \tag{L.32}$$

Clearly, the $(n{-}K)$ local coordinates $x_{K+1}, \dots, x_n$ remain for the subspace of $\mathcal{X}$ where $f_a(x) = 0$, $\forall a$.

▷ **Index**: The index of a map $\pi\colon A \to B$ is defined as $\boldsymbol{\iota}(\pi) \stackrel{\text{def}}{=} (\ker(\pi) - \mathrm{cok}(\pi))$. If $\pi^\dagger$ is adjoint to π (with respect to some norm, $\langle b, \pi(a)\rangle = \langle \pi^\dagger(b), a\rangle$, for all $a \in A$, $b \in B$) then $\boldsymbol{\iota}(\pi) = (\ker(\pi) - \ker(\pi^\dagger))$.

▷ **Injective Map**: Synonym for 'a map is 1–1'. That is, a map $A \xrightarrow{f} B$ (also "$\hookrightarrow$") is injective if there exists precisely one $b \in B$ for every $a \in A$ such that $b = f(a)$.

▷ **Invariant Polynomial**: For $\mathbb{M}_n$ denote the space of $n \times n$ complex matrices, the (homogeneous) k-linear form

$$P^k\ :\ \underbrace{\mathbb{M}_n \times \cdots \times \mathbb{M}_n}_{k\ \text{times}} \ \longrightarrow\ \mathbb{C} \tag{L.33}$$

is called *invariant* if for any $\mathbf{A}_1, \dots, \mathbf{A}_k \in \mathbb{M}_n$ and $\mathrm{g} \in GL(n, \mathbb{C})$,

$$P^k(\mathbf{A}_1, \dots, \mathbf{A}_k) \ = \ P^k((\mathrm{g}\mathbf{A}_1\mathrm{g}^{-1}), \dots, (\mathrm{g}\mathbf{A}_k\mathrm{g}^{-1}))\ . \tag{L.34}$$

Then, $S_k(\mathbf{A}) = P^k(\mathbf{A}, \dots, \mathbf{A})$ are the *elementary symmetric polynomials.* It can be shown that $S_k(\mathbf{A})$ are unique; most naturally, the they arise in the characteristic polynomial:

$$\det[\mathbf{A} - \lambda\mathbb{1}] \ = \ \sum\nolimits_{k=0}^{n} \lambda^k\, S_{n-k}(\mathbf{A})\ . \tag{L.35}$$

▷ **Inverse Function Theorem**: *Let U, V be two open sets in $\mathbb{P}^n$ with $0 \in U$ and let $f\colon U \to V$ be a holomorphic map with the Jacobian $\mathfrak{J}(\frac{\partial f_i}{\partial x_j})$ non-singular at 0. Then f is one-to-one in a neighborhood of 0, and f^{-1} is holomorphic at $f(0)$.*

▷ **Jet** (in particular, k-$\sim$): In a local neighborhood around a point p in a manifold $\mathcal{M}$ and for a local function (or section of some bundle) in that neighborhood, the k-jet is its Taylor series around p up to and including the kth order terms. Statements like 'non-zero k-jet' make sense globally provided that the function (section of a bundle) is globally defined and means that at every point p of the manifold $\mathcal{M}$, the local k-jet is non-zero.

▷ **Kähler Class**: The cohomology class defined by the Kähler metric, on a Kähler manifold $\mathcal{M}$, as follows. Given a Kähler metric $g_{\mu\overline{\nu}}$, $J \stackrel{\text{def}}{=} ig_{\mu\overline{\nu}}\mathrm{d}z^{\mu} \wedge \mathrm{d}z^{\overline{\nu}}$ is the associated Kähler (1,1)-form representing the Kähler (cohomology) class. Clearly, J is self-conjugate (real) and *positive* in the sense that $g_{\mu\overline{\nu}}$ is (in any local coordinate chart) a Hermitian and positive matrix. A coordinate independent equivalent statement is that $\int_C J > 0$, $\int_S J \wedge J > 0 \ldots \int J^{\dim \mathcal{M}} > 0$, where $C, S, \ldots$ are submanifolds of $\mathcal{M}$ of complex dimension 1,2,...

▷ **Kähler Cone**: The space of choices of the *Kähler class*$^{(L)}$. It has the structure of a cone because if J and J' are two Kähler classes, so is $J+J'$ and so is αJ for $\alpha \in \mathbb{R}_+$. Guided by string theory, we often extend the Kähler class by considering $(B+iJ) = w^A e_A$ instead of the Kähler form J, where B is a harmonic (1,1)-class, $\{e_A\}$ form a basis of $H^{1,1}(\mathcal{M})$ and w^A are $h^{1,1}$ complex parameters. The space spanned by w^A is referred to as the complexified Kähler cone.

▷ **Kähler Manifold**: A complex manifold which *admits* a *Kähler metric*$^{(L)}$.

▷ **Kähler Metric**: A metric $\mathrm{g}_{\mu\overline{\nu}}$ which locally Euclidean up to second order terms. Associated to a Hermitian metric is the (1,1)-form $J \stackrel{\text{def}}{=} (i\,\mathrm{g}_{\mu\overline{\nu}}\,\mathrm{d}z^{\mu} \wedge \mathrm{d}z^{\overline{\nu}})$; a Hermitian metric is Kähler precisely if the associated (1,1)-form is closed. The cohomology class of J is called the Kähler class. Locally, $J = \partial\bar{\partial}\Phi$, where Φ is the Kähler potential.

▷ **Kernel**: The kernel of a map $\pi\colon A \to B$ is defined as $\ker(\pi) \stackrel{\text{def}}{=} \{a \in A:\ \pi(a) = 0 \in B\}$, the subset of A which is annihilated by π.

▷ **Kodaira Embedding Theorem**: *A compact complex manifold $\mathcal{M}$ is an algebraic variety—is embeddable in a projective space—precisely if it has a closed, positive $(1,1)$-form ω the cohomology class of which is rational.*

▷ **Kodaira–Nakano Vanishing Theorem**: *If $\mathcal{L}$ is a positive holomorphic line bundle over the compact complex algebraic manifold $\mathcal{X}$ of dimension n, then*

$$H^q(\mathcal{X}, \mathcal{L} \otimes \wedge^p \mathcal{T}_{\mathcal{X}}) \;=\; 0 \qquad \text{for } p{+}q > n\ . \tag{L.36}$$

A useful fact to recall when using this theorem is that $\wedge^n \mathcal{T}_{\mathcal{X}} \stackrel{\text{def}}{=} \mathcal{K}_{\mathcal{X}}$ is the canonical line bundle, so $\wedge^p \mathcal{T}_{\mathcal{X}} = \mathcal{K}_{\mathcal{X}} \otimes \wedge^{p-n} \mathcal{T}_{\mathcal{X}}$.

▷ **Kodaira–Serre Duality**: ☞ *Serre Duality*$^{(L)}$.

▷ **Koszul Complex**: In complete generality, this is defined in Ref. [38], p. 127–131. For the purposes of this volume, the following description will suffice. Let $\mathcal{E} = \bigoplus_a \mathcal{L}_{a=1}^K$, where $\mathcal{L}_a$ are line bundles over a complex manifold $\mathcal{X}$ and have global holomorphic sections, which we denote by ξ_a. Let $\xi = \oplus_a \xi_a$ and let $\mathcal{M}$ be the zero-set of ξ. Then, over $(\mathcal{X}{-}\mathcal{M})$, the sequence of bundles and mappings

$$0 \to \wedge^K \mathcal{E}^* \xrightarrow{\wedge\xi} \cdots \xrightarrow{\wedge\xi} \wedge^2 \mathcal{E}^* \xrightarrow{\wedge\xi} \mathcal{E}^* \xrightarrow{\wedge\xi} \mathcal{O}_{\mathcal{X}} \to 0 \tag{9.3.7'}$$

is exact; $\mathcal{O}_{\mathcal{X}}$ denotes the trivial (constant) bundle over $\mathcal{X}$. To show exactness, note that ξ is non-zero over $(\mathcal{X}{-}\mathcal{M})$ and that $\wedge\xi$ results in contraction with one factor in $\wedge^k\mathcal{E}$. Owing to the skew symmetry of this product, $(\wedge\xi)^2 = 0$ and the above sequence is the *Koszul complex.* See Chapter 9 for more information.

▷ **Künneth Formula**: Given two complex compact spaces, $\mathcal{X}$ and $\mathcal{Y}$, and a vector *bundle*$^{(L)}$ (*sheaf*$^{(L)}$) $\mathcal{V}$ over the product space $\mathcal{X} \times \mathcal{Y}$, we have that

$$H^{p,q}(\mathcal{X} \times \mathcal{Y}) \approx \bigoplus_{\substack{r+u=p \\ s+v=q}} \Big(H^{r,s}(\mathcal{X}, \mathcal{V}|_{\mathcal{X}}) \otimes H^{r,s}(\mathcal{Y}, \mathcal{V}|_{\mathcal{Y}})\Big). \tag{L.37}$$

Note that the *holonomy group*$^{(L)}$ of the product $\mathcal{X} \times \mathcal{Y}$ is just the product of the holonomy groups of the factor spaces. Given an action of the factor holonomy groups on $\mathcal{V}$ and the naturally induced action on the $\mathcal{V}$ valued cohomology, the Künneth formula can be refined further, into a direct sum of (products of) representations of the factor holonomy groups. Clearly, a further decomposition may be obtained if there is a symmetry acting on $\mathcal{X} \times \mathcal{Y}$ and $\mathcal{V}$.

▷ **Lefschetz Fixed-Point Formulae**: Let $\mathcal{M}$ and $G : \mathcal{M} \to \mathcal{M}$ a symmetry with fixed points p_i; write $G^\star$ stand for the action on the cohomology on $\mathcal{M}$. Locally at the ith fixed point, p_i, let the action be given as $G : z_{(i)}{}^\mu \to \mathrm{g}_{(i)}{}^\mu{}_\nu\, z_{(i)}{}^\nu$ and write $\boldsymbol{\iota}_{\mathbf{g}}(p_i) \stackrel{\text{def}}{=} \operatorname{sign}\det(\mathbf{g}_{(i)} - \mathbb{1})$, where $\mathbf{g}_{(i)} \stackrel{\text{def}}{=} [\mathrm{g}_{(i)}{}^\mu{}_\nu]$. Then

$$L(G) \stackrel{\text{def}}{=} \sum_r (-)^r \mathrm{Tr}\big[G^\star|_{H^r_{\mathrm{DR}}(\mathcal{M})}\big] = \sum\nolimits_{\mathbf{g}(p)=p} \boldsymbol{\iota}_{\mathbf{g}}(p) \tag{L.38}$$

is the *Lefschetz fixed point formula.* Using the *Hopf Index Theorem*$^{(L)}$, it may be rewritten as

$$L(f) = \sum\nolimits_{\mathbf{g}(\mathcal{N})=\mathcal{N}} \chi_{E}(\mathcal{N}) \,. \tag{L.39}$$

On a complex manifold $\mathcal{M}$,

$$L^h(G) \stackrel{\text{def}}{=} \sum_q (-)^q \text{Tr}\big[G^\star\big|_{H^{0,q}(\mathcal{M})}\big] = \sum_{\mathbf{g}(p)=p} \big[\mathbb{1} - \mathbf{g}(p)\big]^{-1} \qquad \text{(L.40)}$$

is called the *holomorphic Lefschetz fixed point formula.*

▷ **Lefschetz Hard Theorem**: *Given the Kähler class J on a compact Kähler n-fold $\mathcal{X}$, the map $(\wedge J)^k : H^{p,q} \to H^{p+k,q+k}$ is an isomorphism. Writing $J^\dagger$ for the adjoint of the Kähler form and $J^\dagger\cdot$ for the adjoint of $\wedge J$,*

$$H^{p,q}_\perp \stackrel{\text{def}}{=} \ker\big[(J^\dagger\cdot)^k : H^{p-k,q-k} \to H^{p,q}\big] \qquad \text{(L.41)}$$

is the primitive cohomology, that is, the J-traceless cohomology. Then

$$H^{p,q} = \bigoplus_{k=0}^{\min(p,q)} \big[J^k \wedge H^{p-k,q-k}_\perp\big] \qquad \text{(L.42)}$$

is called the Lefschetz decomposition.

▷ **Lefschetz Hyperplane Theorem** (as proven by Bott): *Let $\mathcal{L}$ be a positive holomorphic line bundle over an $n{+}1$-dimensional complex compact manifold $\mathcal{X}$ and let λ be a holomorphic section of $\mathcal{L}$. Then:*

$$\begin{aligned} &\pi_q\big(\lambda^{-1}(0)\big) \approx \pi_q(\mathcal{X}) && 0 \le q < n = \dim \lambda^{-1}(0), \\ &\pi_n\big(\lambda^{-1}(0)\big) \xrightarrow{j} \pi_n(\mathcal{X}) && \textit{is onto/surjective.} \end{aligned} \qquad (1.6.1')$$

▷ **Lorentz Group**: The (connected component of the) group of linear transformations of the tangent space of spacetime. In usual 4-dimensional spacetime, this is $SO(1,3)$; for the 10-dimensional spacetime, this is $SO(1,9)$. In both cases, a pseudo-Riemannian metric is assumed to exist in spacetime, with signature $(1,3)$ and $(1,9)$ respectively. Of interest are also the covering, so-called *Spin*-groups, denoted $Spin(1,3) \approx SL(2,\mathbb{C})$ and $Spin(1,9)$.

▷ **Manifold**: An everywhere smooth space. More precisely, a real (complex) n-dimensional manifold $\mathcal{M}$ is defined to have a finite set of subsets $\{\mathcal{U}_i \subset M\}$ so that the following is true:

1. Each $p \in M$ belongs to at least one U_i.
2. For each i, there is a 1–1, onto, map $\psi_i\colon U_i \to O_i$, where O_i is an open neighborhood $\mathbb{R}^n$ ($\mathbb{C}^n$).
3. For any $U_i \cap U_j \neq \varnothing$, the composition of maps $\psi_i \circ \psi_j^{-1}$ is C^∞.

See *compact*$^{(L)}$, *Hausdorff*$^{(L)}$ and *separable*$^{(L)}$.

▷ **Minimal Configuration**: A *configuration* $^{(L)}$ the members of which do not belong to a smaller configuration (see Chapter 2).

▷ **Mirror Map**: Given a Calabi–Yau 3-fold $\mathcal{M}$, the *mirror map* associates to it another Calabi–Yau 3-fold, $\mathcal{W}$, such that

$$\bigoplus_{p+q=3} H^{p,q}(\mathcal{M}) \cong_Q \bigoplus_{p=q} H^{p,q}(\mathcal{W}) \ , \tag{L.43}$$

and '$\cong_Q$' denotes an equivalence of the *ring structures* as determined by the *exact quantum* Yukawa couplings. In other words, the exact value (with all quantum corrections) of all possible n-tuple product integrals computed on the right hand side equal those computed likewise on the left hand side. Straightforwardly, $H^{p,q}(\mathcal{M}) = H^{3-p,q}(\mathcal{W})$ and so $\chi_{_E}(\mathcal{M}) = -\chi_{_E}(\mathcal{W})$. More generally for n-folds, $H^{p,q}(\mathcal{M}) = H^{n-p,q}(\mathcal{W})$ and so $\chi_{_E}(\mathcal{M}) = (-)^n \chi_{_E}(\mathcal{W})$.

▷ **Mirror Particle**: The "parity"-reflection of any Standard Model particle X, thus having the same mass and (gauge interaction) charges but opposite helicity; it is not (the Hermitian/CPT conjugate of) the antiparticle, and where X has left-handed (such as weak) interactions, its mirror has right-handed analogous interactions. Equivalently (by Hermitian/CPT conjugation), the mirror particle has the same mass and helicity, but all its (gauge interaction) charges are opposite. In the standard Calabi–Yau compactification framework of Chapter 0, (generations of) Standard Model fermions are associated with the **27** representation of the E_6 gauge group, while $\mathbf{27}^*$ correspond to (the mirror generations) of their mirror particles, and **1** to "gauge-singlets"—particles with no gauge interaction; see Table 0.2.

▷ **Moduli Space**: The effective parameter space. Concretely, for the complex n-folds possibly endowed with a Kähler class, the moduli space consists of the (effective) space of *complex structures* $^{(L)}$, at every point of which there is the a *Kähler cone* $^{(L)}$, a space of choices of the *Kähler class* $^{(L)}$. Clearly, as the complex structures are being varied, the associated Kähler cones vary into each other in a uniform way, so that the total space has the structure of a fibration. See also *Kähler cone* $^{(L)}$.

▷ **Monad**: in homological algebra, a short *exact sequence* $^{(L)}$ (L.21) of abelian groups (or group-like structures), $A \xrightarrow{\alpha} B \xrightarrow{\beta} C$, where α is injective (into, 1–1, also denoted "$\hookrightarrow$") and β is surjective (onto, also denoted "$\twoheadrightarrow$").

▷ n-**Fold**: Unless otherwise noted, throughout this volume, a compact complex n-dimensional space, possibly singular at some relatively nicely defined subset (isolated points, curves which themselves are not too badly singular and so on). Algebraic geometry restrict to n-folds that may be defined as solutions of systems of algebraic equations.

▷ **Nef** (line bundle): A line bundle $\mathcal{L}$ over X is *nef* if it has nonnegative degree on every (closed irreducible) curve $C \subset X$. *Semi-Ample*$^{(L)}$ (and basepoint-free, ☞ *base locus*$^{(L)}$) implies nef, but not the other way around.

▷ **Node**: Also called a "double point," is the mildest hypersurface singularity. That is, the hypersurface $\phi(x) = 0$ ($x \in \mathbb{C}^n$) has a node at $x = 0$ if all gradients vanish but the Hessian is non-singular. Up to a change of coordinates then, $\phi(x) = \sum_i x_i{}^2$; a node is therefore an A_1 singularity in the classification of Ref. [1].

▷ **Noether Formula**: For surfaces, $C_1{}^2 + C_2 = 12\chi_h$, where $C_1{}^2 = \int c_1{}^2$ and $C_2 = \int c_2$ are the *Chern numbers*$^{(L)}$ and χ_h is the *holomorphic Euler characteristic*$^{(L)}$.

▷ **Normal Bundle**: Given a submanifold $\mathcal{M}$ of a manifold X, at every point $p \in \mathcal{M}$, we define the normal space N_p to be spanned by the tangent vectors of X, considering however two vectors equivalent if they differ by a vector tangent to $\mathcal{M}$. In other words,

$$N_p \stackrel{\text{def}}{=} \{T_p(X)/T_p(\mathcal{M})\} , \qquad \mathcal{N}_{X/\mathcal{M}} \stackrel{\text{def}}{=} \bigcup_{p\in\mathcal{M}} N_p = \{\mathcal{T}_X/\mathcal{T}_\mathcal{M}\} \quad \text{(L.44)}$$

are, respectively, the normal space at $p \in \mathcal{M}$ and the normal bundle of the embedding $\mathcal{M} \hookrightarrow X$. Taking the union with p ranging all over $\mathcal{M}$, we also obtain $\mathcal{T}_\mathcal{M} = \bigcup_{p\in\mathcal{M}} T_p(\mathcal{M})$, $\mathcal{T}_X = \bigcup_{p\in\mathcal{M}} T_p(X)|_\mathcal{M}$. See also *exact sequence*$^{(L)}$.

▷ **Orbifold**: In general, a space which is smooth except at a finite number of isolated points, at each of which the space is locally of the form $\mathbb{C}^n/D$, where D is a finite group.

▷ **Period Matrix**: Given a basis of exterior r-forms $\{\alpha\}$ and a basis of r-cycles $\{\gamma_i\}$, $\varpi_{(r)}{}^i{}_j \stackrel{\text{def}}{=} \int_{\gamma_j} \alpha^i$ is the period matrix. Of special interest, in a complex n-dimensional manifold, is the period matrix $\boldsymbol{\varpi}_{(n)}$.

▷ **Periods**: Given an exterior r-form α and let $\{\gamma_i\}$ be a basis of r-cycles, $\varpi_i(\alpha) \stackrel{\text{def}}{=} \int_{\gamma_i} \alpha$ are the periods of α. If all periods of an exterior r-form vanish (sweeping γ_i through all of H_r), the form is exact.

▷ **Picard Group**: Holomorphic line bundles over a compact complex manifold $\mathcal{X}$ form, under the (fiber-wise) tensor product, a multiplicative group called the *Picard group*, denoted $\mathrm{Pic}(\mathcal{X})$.

▷ **Picard Number**: The dimension of the *Picard group*$^{(L)}$ $\mathrm{Pic}(\mathcal{X})$. Also equal to $\dim\big(H^{1,1}(\mathcal{X})\cap H^2(\mathcal{X},\mathbb{Z})\big)$; see Ref. [23], p. 456-7.

▷ **Plurigenera**: Let $\mathcal{K}_{\mathcal{X}} = \det\mathcal{T}^*_{\mathcal{X}}$ be the canonical bundle over $\mathcal{X}$ and let $\mathcal{K}^{\otimes n}_{\mathcal{X}}$ denote the nth tensor power line bundle over $\mathcal{X}$. Then $\mathfrak{P}_n \stackrel{\text{def}}{=} \dim H^0(\mathcal{X},\mathcal{K}^{\otimes n}_{\mathcal{X}})$, the number of holomorphic sections of $\mathcal{K}^{\otimes n}_{\mathcal{X}}$ is the nth plurigenus of $\mathcal{X}$. By analogy (and contravariantly): $\mathfrak{P}^m \stackrel{\text{def}}{=} \dim H^0(\mathcal{X},\mathcal{K}^{*\otimes n}_{\mathcal{X}})$.

▷ **Poincaré Duality**: On a compact oriented real n-dimensional manifold $\mathcal{X}$, the intersection pairing

$$H_k(\mathcal{X},\mathbb{Z})\times H_{n-k}(\mathcal{X},\mathbb{Z}) \longrightarrow \mathbb{Z} \tag{L.45}$$

is unimodular; i.e., any linear functional $H_{n-k}(\mathcal{X},\mathbb{Z})\to\mathbb{Z}$ is expressible as the intersection with some class $\alpha\in H_k(\mathcal{X},\mathbb{Z})$ and any class $\alpha\in H_k(\mathcal{X},\mathbb{Z})$ having intersection number 0 with all classes in $H_{n-k}(\mathcal{X},\mathbb{Z})$ is a torsion class.

It is easy to see that this is (De Rham-) dual to the Hodge star duality.

▷ **Positive Bundle**: A *bundle*$^{(L)}$ over $\mathcal{X}$ which admits a curvature 2-form $\boldsymbol{\Theta}$ which is everywhere on $\mathcal{X}$ locally represented by a positive (Hermitian) matrix. A positive first *Chern class*$^{(L)}$ implies positivity of the bundle, but is not necessary; see Lemma 3.1.

▷ **Preimage**: The preimage (inverse-image) of an element $b\in B$ under a map $\pi\colon A\to B$ is defined as $\pi^{-1}(b) \stackrel{\text{def}}{=} \{a\in A : \pi(a)=b\}$. Naturally, $\pi^{-1}(b)\subseteq A$.

▷ **Ramified Covering**: ☞ *Branched Covering*$^{(L)}$.

▷ **Rank of a Mapping**: Given a mapping $\pi\colon A\to B$, the rank of π is defined to be the dimension of the image, $\dim[\pi(A)\in B]$.

▷ **Rank of a Vector Bundle**: The dimension of the fiber. Recall that a vector bundle $\mathcal{V}$ over the manifold $\mathcal{X}$ may be constructed by attaching a copy of the vector space V_x to each point $x\in\mathcal{X}$ in such a way that this vector space varies smoothly (differentiably, holomorphically ...) as we sweep the point x over the base space $\mathcal{X}$. V_x is called the *fiber* at $x\in\mathcal{X}$.

▷ **Rational Curve**: An algebraic (embeddable in some $\mathbb{P}^n$) curve which is birationally isomorphic to $\mathbb{P}^1$ (see *Birational Map*$^{(L)}$).

▷ **Rational Function**: A rational (or meromorphic) function on an open set $U \subset \mathcal{M}$ is given locally as the quotient of two holomorphic functions. So, for some open covering $\mathfrak{U} = \{U_i\}$, $f|_{U_i} = g_i/h_i$, where g_i and h_i have no common (holomorphic) factor and $g_ih_j = g_jh_i$ over $(U_i \cap U_j$. A rational function is not a function in the usual sense, even if '∞' is included as a value; f is simply not defined where both g_i and h_i vanish.

▷ **Rational Map**: A rational (or meromorphic) map of a complex manifold $\mathcal{M}$ to a projective space $\mathbb{P}^n$ is a map

$$f : z \to [1, f_1(z), \ldots, f_n(z)] \;, \qquad z \in \mathcal{M} \;, \tag{L.46}$$

given by n global meromorphic functions on $\mathcal{M}$. A *rational* map $f : \mathcal{M} \to \mathcal{N}$ to an algebraic variety $\mathcal{N} \subset \mathbb{P}^n$ is a rational map to $\mathbb{P}^n$, such that its *image*$^{(L)}$ lies in $\mathcal{N} \subset \mathbb{P}^n$.

Caution: a rational map $f : \mathcal{M} \to \mathcal{N}$ need not be defined everywhere but is instead given by a *holomorphic* map

$$f_h : (\mathcal{M} \smallsetminus \mathfrak{E}) \to \mathcal{N} \;, \tag{L.47}$$

where $\mathfrak{E}$ is the exceptional set (where the rational map fails to be defined) and which has *codimension*$^{(L)}$ 2 in $\mathcal{M}$.

▷ **Rational Surface**: An algebraic (embeddable in some $\mathbb{P}^n$) surface which is birationally isomorphic to $\mathbb{P}^2$ (see *Birational Map*$^{(L)}$).

▷ **Residue Class Ring**: ☞ *Ideal*$^{(L)}$.

▷ **Resolution** (1, of a singularity): Given a singular model X (a singular point with its local complex neighborhood or a complex n-fold with singularities), its resolution is a smooth model Y together with a map $\sigma : Y \to X$, such that $(X - \mathrm{Sing}X) = (Y - \mathfrak{E})$ as (non-compact) complex manifolds. $\mathfrak{E}$ here denotes the (collection of) exceptional set(s) into which the singularities of X have been resolved. When Y is a smooth complex 3-fold (with no boundaries), $\mathfrak{E}$ consists of complex 1- and/or 2-dimensional subspaces of Y.

▷ **Resolution** (2, of a sheaf): A *sheaf*$^{(L)}$ $\mathfrak{S}$ over a manifold $\mathcal{X}$ is said to have a resolution if it fits into a sequence

$$0 \to \mathcal{V}_K \to \mathcal{V}_{K-1} \to \cdots \to \mathcal{V}_1 \to \mathcal{V}_0 \xrightarrow{\varrho} \mathfrak{S} \to 0 \;, \tag{L.48}$$

where $\mathcal{V}_k$ are fiber bundles over $\mathcal{X}$ and the sequence is *sheaf-exact*, i.e., is an *exact sequence* $^{(L)}$ for a sufficiently fine open cover of $\mathcal{X}$. Equivalently, this sequence is exact with $\mathfrak{S}$ set to 0 over the generic part of $\mathcal{X}$, and being non-zero only over a special subset of $\mathcal{X}$.

▷ **Ricci Tensor and Form**: On a Hermitian manifold (complex manifold with a Hermitian metric), the Ricci tensor, $R_{mnp}{}^n$, becomes

$$R_{\mu\overline{\nu}} \stackrel{\text{def}}{=} \mathrm{g}^{\pi\overline{\varrho}}\, R_{\mu\overline{\varrho}\pi\overline{\nu}} \;=\; -\,\mathrm{g}^{\pi\overline{\varrho}}\, R_{\mu\overline{\nu}\pi\overline{\varrho}} \;=\; -\,R_{\mu\overline{\nu}\rho}{}^{\rho} \tag{L.49}$$

and defines the Ricci curvature two-form

$$i\, R_{\mu\overline{\nu}\rho}{}^{\rho}\, dz^{\mu} \wedge dz^{\overline{\nu}} \;=\; i\,\partial\bar{\partial}\log\sqrt{\det \mathrm{g}}\;, \tag{L.50}$$

which is both ∂- and $\bar{\partial}$-closed but not exact because of the logarithm. It is also the trace of the Riemann curvature two-form $\Theta_{\rho}{}^{\sigma} \stackrel{\text{def}}{=} iR_{\mu\overline{\nu}\rho}{}^{\sigma}\, dz^{\mu} \wedge dz^{\overline{\nu}}$.

▷ **Riemann–Roch–Hirzebruch Theorem**: *Let $\mathcal{X}$ be an algebraic manifold of dimension n and let $\mathcal{L}$ be a holomorphic line bundle over $\mathcal{X}$, with $c_1[\mathcal{L}] = \lambda$. The cohomology groups $H^q(\mathcal{X}, \mathcal{L})$ are finite dimensional and the (generalized) Euler–Poincaré characteristic is*

$$\sum_{q=0}^{n} H^q(\mathcal{X},\mathcal{L}) \;\stackrel{\text{def}}{=}\; \chi_{E}(\mathcal{X},\mathcal{L}) \;=\; \int_{\mathcal{X}} \Big[e^{\lambda} \prod_{i=1}^{n} \frac{\ell_i}{1-e^{-\ell_i}}\Big]\,, \tag{3.2.6}$$

where

$$c(\mathcal{X}) \;=\; 1 + c_1 + c_2 + \ldots + c_n \;=\; \prod_{i=1}^{n} (1+\ell_i)\;. \tag{1.3.19$'$}$$

Note that integration over $\mathcal{X}$ picks out terms which are of degree n in λ, ℓ_i.

▷ **Riemann Tensor and Form**: Writing

$$\nabla_m v^k \;\stackrel{\text{def}}{=}\; \partial_m v^k \;+\; \Gamma_{mn}{}^{k} v^n \tag{L.51}$$

for a covariant derivative (connection), the Riemann curvature tensor is defined by

$$\big[\nabla_m\,,\,\nabla_n\big] v^l \;\stackrel{\text{def}}{=}\; T_{mn}{}^{k} \nabla_k v^l \;+\; R_{mnk}{}^{l}\, v^k\;, \tag{L.52}$$

where $T_{mn}{}^{k}$ is the *torsion* tensor. The

$$\Psi_k{}^l \stackrel{\text{def}}{=} (\mathrm{d}x^m\, \Gamma_{mk}{}^{l}) \qquad \text{and} \qquad \Theta_k{}^l \stackrel{\text{def}}{=} (\mathrm{d}x^m \wedge \mathrm{d}x^n\, R_{mnk}{}^{l}) \tag{L.53}$$

are the (matrix-valued) *torsion 1-form* and *curvature 2-form*, respectively. For a *Riemannian* ∇_m, $T_{mn}{}^{k} = 0$ and

$$R_{mnp}{}^{q}\mathrm{g}_{rq} = R_{prm}{}^{q}\mathrm{g}_{nq}\;, \qquad \text{and} \qquad R_{m[np}{}^{q}\,\mathrm{g}_{r]q} = 0\;. \tag{L.54}$$

A typical choice for a Riemannian connection involves the Christoffel connection

$$\Gamma_{mn}{}^{k} \stackrel{\text{def}}{=} \tfrac{1}{2}g^{kl}(\partial_m g_{nl} + \partial_n g_{ml} - \partial_l g_{mn}) \ . \qquad \text{(L.55)}$$

On a Hermitian manifold, $g_{\mu\nu} = 0 = g_{\overline{\mu}\,\overline{\nu}}$ and so $\Gamma_{\mu\nu}{}^{\sigma}$, $R_{\mu\overline{\nu}\rho}{}^{\sigma}$, (and their complex conjugates) are the only non-zero components of the connection and Riemann tensor.

▷ **Ring**: An non-empty set R endowed with two binary operations, '+' and '·' such that (1) for any $a, b \in R$, $(a{+}b), (a{\cdot}b) \in R$; (2) $(a{+}b) = (b{+}a)$; (3) $(a{+}b){+}c = a{+}(b{+}c)$ and $(a{\cdot}b){\cdot}c = a{\cdot}(b{\cdot}c)$; (4) $a{\cdot}(b{+}c) = (a{\cdot}b){+}(a{\cdot}c)$ and $(a{+}b){\cdot}c = (a{\cdot}c){+}(b{\cdot}c)$; (5) for every $a, b \in R$, there is a unique $c \in R$ such that $a{+}c = b$. Note, in particular, that 'multiplicative unity' ($1 \in R$) or 'division' need not be defined. If there exists $1 \in R$, R is a *unitary ring*. '1' must be distinct from '0' unless $R = \{0\}$.

▷ **Section**: A (holomorphic) *section* of a vector *bundle*$^{(L)}$ $(\mathcal{B}, \boldsymbol{E}, \varpi)$ over $U \subseteq \mathcal{B}$ is a C^∞ (holomorphic) map $\sigma : U \to \boldsymbol{E}$ such that $\sigma(x) \in E_x$ for all $x \in U$. That is, while the map ϖ projects the total space of the bundle $\boldsymbol{E}$ to the base manifold $\mathcal{B}$, a section takes a point $x \in \mathcal{B}$ as the argument and produces a point in the fiber E_x over that point. Clearly, a section is *global* if it can be extended over all of $\mathcal{B}$. A complete set of sections $\{\sigma_i(x)\}$ which provide a basis for E_x for all $x \in U \subseteq \mathcal{B}$ is called at *frame*.

▷ **Segré Embedding**: The embedding of the product of two *complex projective spaces*$^{(L)}$ $\mathbb{P}^m \times \mathbb{P}^n \hookrightarrow \mathbb{P}^{(m+1)(n+1)-1}$ given by sending the coordinates

$$(x_a; y_i) \ \mapsto \ z_{ai} \ , \qquad a = 0, \ldots, m \ , \quad i = 0, \ldots, m \qquad \text{(L.56)}$$

and imposing the obvious constraints

$$z_{ai}z_{bj} - z_{aj}z_{bi} \ = \ 0 \ , \qquad (\text{that is,} \ \ x_a y_i x_b y_j = x_a y_j x_b y_i) \ . \qquad \text{(L.57)}$$

Note that $\mathbb{P}^m \times \mathbb{P}^n$ has *codimension*$^{(L)}$ mn inside $\mathbb{P}^{(m+1)(n+1)-1}$, while there are $\binom{m+1}{2}\binom{n+1}{2}$ constraints. Clearly, unless $m = n = 1$, there are too many constraints for them to be independent and the Segré embeddings are not *complete intersections*$^{(L)}$.

▷ **Semi-Ample**: A *line bundle*$^{(L)}$ $\mathcal{L}$ over $\mathcal{X}$ is semi-ample if some positive power of it is basepoint-free. ☞ *ample*$^{(L)}$

▷ **Separable**: A topological space is *separable* if it admits a countable basis for topology.

▷ **Serre Duality**: *For a compact complex n-fold* $\mathcal{M}$ *and a holomorphic vector bundle* $\mathcal{V}$ *over it,*

1. $H^n(\mathcal{M}, \mathcal{K}_\mathcal{M}) = H^{n,n}(\mathcal{M}) \xrightarrow{\int_\mathcal{M}} \mathbb{C}$ *is an isomorphism, and*
2. $H^q(\mathcal{M}, \mathcal{V})^* \approx H^{n-q}(\mathcal{M}, \mathcal{V}^* \otimes \mathcal{K}_\mathcal{M})$.

▷ **Sheaf**: Over a topological space X, a *sheaf* $\mathfrak{S}$ assigns (a) for every open neighborhood $U \subset X$ an abelian group (in most examples, under addition) $\mathfrak{S}_U$ and (b) for every inclusion $V \subseteq U$ of open neighborhoods provides a *restriction* map $\mathbf{r}_V{}^U : \mathfrak{S}_U \to \mathfrak{S}_V$, subject to the conditions

1. $\mathfrak{S}_\varnothing = 0$;
2. $\mathbf{r}_U{}^U$ is the identity;
3. if $U \subseteq V \subseteq W$ are open sets, $\mathbf{r}_U{}^W = \mathbf{r}_U{}^V \circ \mathbf{r}_V{}^W$;
4. if U is an open set, $\{V_i\}$ an open cover of U and $\mathfrak{s} \in \mathfrak{S}_U$—then
$$\mathbf{r}_{V_i}{}^U(\mathfrak{s}) = 0 \;, \quad \forall i \;, \qquad \iff \qquad \mathfrak{s} = 0 \;; \tag{L.58}$$
5. if U is an open set, $\{V_i\}$ an open cover of U and there are elements $\mathfrak{s}_i \in \mathfrak{S}_{V_i}$ such that
$$\mathbf{r}_{(V_i \cap V_j)}{}^{V_i}(\mathfrak{s}_i) = \mathbf{r}_{(V_j \cap V_i)}{}^{V_j}(\mathfrak{s}_j) \;, \qquad \forall i, j \;, \tag{L.59}$$
holds true—then there is a unique $\mathfrak{s} \in \mathfrak{S}_U$ such that $\mathfrak{s}_i = \mathbf{r}_{V_i}{}^U(\mathfrak{s})$.

Note: if the last two conditions are not necessarily satisfied, $\mathfrak{S}$ is called a *presheaf*. We may define a *(pre)sheaf with values in a category* $\mathfrak{C}$ by replacing 'abelian group' with 'object of $\mathfrak{C}$' and 'map' with 'morphism of $\mathfrak{C}$'. For example, in addition to the above structure we often use a multiplication between the elements of $\mathfrak{S}_U$, hence the corresponding sheaf of rings.

▷ **Spectral Sequence**: A sequence $\{E_r, d_r\}_{r\geq 0}$ of bi-graded groups $E_r \overset{\text{def}}{=} \bigoplus_{i,j\geq 0} E_r^{i,j}$ together with the *differentials*
$$d_r \;:\; E_r^{i,j} \;\to\; E_r^{i-r,j-r-1} \;, \qquad d_r{}^2 = 0 \;, \tag{L.60}$$
such that $E_{r+1}^{i,j}$ consists of the d_r-closed elements of $E_r^{i,j}$, taken modulo d_r-exact terms. A spectral sequence is *finite* if $E_r = E_{r+1} = \cdots$ for $r \geq r_0$; this limit group is called E_∞ and one says that the spectral sequence *converges* to E_∞. Clearly, the steps E_r, $r \leq r_0$, may be considered as successive approximations for E_∞.

Note that, while E_{r+1} is completely determined in terms of E_r and d_r, the 'next level' differential is not. Thus, the main utility of spectral sequences is in situation when the arrays $E_r^{i,j}$, for each r, are sparse enough so that a detailed knowledge of the differentials is not necessary to pass to the next level, $r+1$. Alternatively, as in Chapters 9, A and B, it is sometimes possible to discern the action of the differentials from some general principles and reach E_∞ with complete certainty.

▷ **Splitting** (1, for exact sequences): Given a short *exact sequence*$^{(L)}$

$$0 \to A \xrightarrow{\alpha} B \xrightarrow{\beta} C \to 0$$

and in addition a map $\gamma : C \to B$ such that $\beta \circ \gamma = \mathrm{Id}_B$ and $\gamma \circ \beta = \mathrm{Id}_C$, then $B = A \oplus C$ and the sequence is said to split.

▷ **Splitting** (2, for configurations): Given a *configuration*$^{(L)}$ $[\mathcal{X}\|\mathcal{E}]$, a *split* is another configuration of the type $\left[\begin{smallmatrix}\mathcal{X}\\ \mathbb{P}^n\end{smallmatrix}\middle\|\begin{smallmatrix}\mathcal{E}_1 & \cdots & \mathcal{E}_{n+1}\\ \mathcal{O}(1) & \cdots & \mathcal{O}(1)\end{smallmatrix}\right]$, such that $\mathcal{E} = \bigotimes_{r=1}^{n+1} \mathcal{E}_r$. The former configuration is called the *contraction* of the latter one. Their members are related as follows. Given a variety belonging to the latter configuration, it is defined by a system of constraints which may be regarded as a system of linear equations in the $n+1$ homogeneous variables of $\mathbb{P}^1$. Passing to the determinant of this system, a (singular if $\dim > 2$) member of the former configuration is obtained.

▷ **Stalk** (of a sheaf): In a *sheaf*$^{(L)}$ $\mathfrak{S}$ over $\mathcal{X}$, the stalk $\mathfrak{S}_p$ at the point $p \in \mathcal{X}$ is defined as the direct limit of $\mathfrak{S}_{U_i}$, for all open sets $U_i \supset U_{i+1} \ni p$, via the restriction map. Elements $\mathfrak{s}_p \in \mathfrak{S}_p$ of a stalk are called *germs of sections of* $\mathfrak{S}$ at the point $p \in \mathcal{X}$.

▷ **Standard Model**: This by now iconic phrase in elementary particle physics denotes the particular model involving three generations (families) of fundamental fermions, their $SU(3)_c \times SU(2)_w \times U(1)_y$ gauge interactions, together with the Higgs field that implements the electroweak phase transition, $SU(2)_w \times U(1)_y \to U(1)_{EM}$ and makes fermions massive via *Yukawa couplings*$^{(L)}$. By now detailed in many textbooks and its own Wikipedia page† with plenty of good references, it notably omits gravity and related phenomena of the other, cosmic proportions; for a unifying text, see [273].

▷ **Stratified** (of a space): Decomposed into collections of subspaces (called strata) that are distinguished by certain common characteristics such as their dimension; see this used in Definition 7.1. For example, the 1-dimensional X-shaped space has four 1-dimensional line-segments (forming the 1-dimensional stratum) and the point where they meet (forming the

† See `https://en.wikipedia.org/wiki/Standard_Model`

0-dimensional stratum). A leaf has a 2-dimensional body, an approximately 1-dimensional stem, and a 0-dimensional point where the two connect.

▷ **Structure Sheaf**: The *sheaf*$^{(L)}$ of germs of holomorphic functions over a complex manifold $\mathcal{X}$, that is, the sheaf the *stalks*$^{(L)}$ of which are just complex numbers. In other words, for every open neighborhood $U \subset \mathcal{X}$, it associates to U local holomorphic functions on U.

▷ **Superfield**: A field Φ over superspace spanned by $(x;\theta)$, where x are commuting and θ are anticommuting coordinates, is defined by the formal expansion

$$\Phi(x;\theta) \stackrel{\text{def}}{=} \phi(x) + \theta\cdot\psi(x) + \theta\cdot\theta\cdot F(x) + \dots \tag{L.61}$$

where we have suppressed the indices labeling the θ's. Owing to the relation $\theta^i\theta^j + \theta^j\theta^i = 0$, the above products are skew-symmetric and the series terminates. The fields $\phi(x), \psi(x), F(x)\dots$ are referred to as the 'component fields' of $\Phi(x;\theta)$. If there are several different coordinates θ^i, a superfield which is independent of some of them is generally referred as chiral. Since all $\Phi(x;\theta)$ are polynomials in θ, the notion of chiral is practically synonymous to holomorphic.

▷ **Superspace**: For our purposes, superspace may be thought of as a supersymmetric extension of ordinary space in that it is spanned by both commuting and anticommuting coordinates.

(**A**) In applications for supersymmetry, the commuting spacetime coordinates are transforming as a vector under the *Lorentz group*$^{(L)}$, while the anticommuting coordinates furnish N copies of the spinor representation; N is referred to as the extendedness of the superspace. 2-dimensional spacetime is exceptional in that the helicity subgroup of the Lorentz group finite. As a consequence, the right-extendedness may differ from the left-extendedness, so one speaks of (p,q)-superspace which is spanned by a commuting $SO(1,1)$-vector coordinate, p anticommuting left-handed spinor coordinates and q anticommuting right-handed spinor coordinates. Moreover, the spinors can be chosen to be of the Weyl-Majorana type.

(**B**) In applications for the Becchi–Rouet–Stora–Tyupin (BRST) symmetry, the commuting spacetime coordinates are again spanning a vector representation of the Lorentz group while the anticommuting coordinates are taken to furnish N copies of the trivial (scalar) representation. One may combine pairs of the anticommuting coordinates into complex pairs, and the associated holomorphic and antiholomorphic translations give rise to BRST and anti-BRST symmetries.

▷ **Supersymmetry**: A continuous symmetry of a system such that: (1) *fermions*$^{(L)}$ and *bosons*$^{(L)}$ are mapped into each other, and (2) the composition of two super-transformations generates (ordinary) spacetime translation. Generators of supersymmetry: Q and $Q^\dagger$ (the adjoint), and also the super-shift parameters ϵ and $\bar{\epsilon}$, must then be fermionic: spin-1/2 anticommuting quantities. Then

$$\big[\,\epsilon.Q\,,\,\bar{\epsilon}.Q^\dagger\,\big] = (\epsilon{\cdot}\bar{\epsilon}).P\ . \tag{L.62}$$

Since ϵ and $\bar{\epsilon}$ are anticommuting, this is equivalent to

$$\{\,Q_\alpha\,,\,Q^\dagger_{\dot{\beta}}\,\} = \gamma^\mu_{\alpha\dot{\beta}} P_\mu\ , \tag{L.63}$$

where $\gamma^\mu_{\alpha\dot{\beta}}$, $\mu = 0,1,2,3$ are the Dirac matrices and P_μ the 4-vector components of energy-momentum. Notice the similarity with the algebra of the exterior differentials:

$$\{\,\mathrm{d}\,,\,\mathrm{d}^\dagger\,\} = \triangle\ , \tag{L.64}$$

with all other anticommutators vanishing. Supersymmetry may be regarded as a translation in *superspace*$^{(L)}$, in the fermionic direction. However, the non-trivial relation between such 'fermionic' translations and ordinary 'bosonic' translations represented by Eq. (L.63) is essential. Without it, the transformation (and symmetry) is merely a fermionic symmetry—the BRST and anti-BRST symmetries are examples of such (see *superspace*$^{(L)}$).

▷ **Surjective Map**: Synonym for 'a map is onto'. That is, a map $A \xrightarrow{f} B$ (also "$\overset{f}{\twoheadrightarrow}$") is surjective (onto) if there exists an $a \in A$ for every $b \in B$ such that $b = f(a)$.

▷ **Tangent Bundle**: Given a manifold $\mathcal{M}$, consider all the curves passing through a point $p \in \mathcal{M}$. The tangent space at p, $T_p(\mathcal{M})$, is then defined as the vector space spanned by the tangents to the curves through p, that is, by $\ell^i(x)\frac{\partial}{\partial x^i}$, where $\ell^i(x)$ are linear combinations of the local coordinates x^i. The tangent bundle is then defined as the union thereof, $\mathcal{T}_\mathcal{M} = \bigcup_{p\in\mathcal{M}} T_p(\mathcal{M})$. When $\mathcal{M}$ is a complex manifold, $\mathcal{T}_\mathcal{M}$ will denote the holomorphic tangent bundle, spanned locally by $\ell^i(z)\partial_i$, where $\ell^i(z)$ are holomorphic and linear in z^i and $\partial_i = \frac{\partial}{\partial z^i}$.

▷ **Tautological Sequence**: Consider the Grassmannian $G_{(k,n)}$, that is the space of k-planes H through the origin of $\mathbb{C}^n$. Let $\mathbb{C}^n$ denote the trivial bundle $G_{(k,n)} \times \mathbb{C}^n$ over $G_{(k,n)}$ and define its *universal sub-bundle*

$$\mathcal{S} \stackrel{\text{def}}{=} \{(H, x) \in G_{(k,n)} \times \mathbb{C}^n : \; x \in H\}. \tag{L.65}$$

Defining the universal quotient bundle $\mathcal{Q} = \{\mathbb{C}^n/\mathcal{S}\}$, we have a short *exact sequence* [(L)]

$$0 \to \mathcal{S} \to \mathbb{C}^n \to \mathcal{Q} \to 0 \,, \tag{L.66}$$

which is called the *tautological sequence.* Know also that $\mathcal{T}_{G_{(k,n)}} = \mathcal{S}^* \otimes \mathcal{Q}$; see *Euler sequence* [(L)] too. On $\mathbb{P}^n = G_{(1,n+1)}$, $\mathcal{S} = \mathcal{O}_{\mathbb{P}^n}(-1)$.

▷ **Technique of Exact and Spectral Sequences** (TESS): General reference for a computation of cohomology groups $H^\star(\mathcal{M}, \mathcal{V})$ in the following situation: (1) $\mathcal{M}$ is a submanifold of $\mathcal{X}$ and the resolution of the structure sheaf $\mathcal{O}_\mathcal{M}$ in terms of a sheaf-*exact sequence* [(L)] of bundles over $\mathcal{X}$ is known. (2) The bundle $\mathcal{V}$ over $\mathcal{M}$ is related to (the restriction to $\mathcal{M}$ of) certain bundles (sheaves) over $\mathcal{X}$ in terms of exact sequences over $\mathcal{M}$. See Chapters 9, A and B for examples.

▷ **Threefold**: ☞ n-fold.

▷ **Transverse**: Two subspaces $\mathcal{Y}$ and $\mathcal{Y}'$ of $\mathcal{X}$, with dimensions k, k' and n, respectively, intersect transversely if $\dim(N_p(\mathcal{Y}) + N_p(\mathcal{Y}')) = (n-k)+(n-k')$ at every point of the intersection, where $N_p(\mathcal{Y})$ is the normal space of $\mathcal{Y} \subset X$. If $\mathcal{Y}$ and $\mathcal{Y}'$ are defined as by $\phi(x) = 0$ and $\phi'(x) = 0$, respectively, then $N_p(\mathcal{Y})$ is spanned by the local gradients $\partial_i\phi(p)$, $p \in \mathcal{Y}$ and similarly for $\mathcal{Y}'$. The gradients are independent of any connection since $\nabla\phi = \partial\phi + \Gamma\cdot\phi = \partial\phi$ at $\mathcal{Y}$, where $\phi = \Gamma \cdot \phi = 0$.

▷ **Twofold**: ☞ n-fold.

▷ **Universal Sub-Bundle**: ☞ tautological sequence.

▷ **Variety**: A affine (projective) variety is an irreducible zero-set of a finite collection of (homogeneous) polynomials on $\mathbb{C}^n$ ($\mathbb{P}^n$). A nonempty subset $\mathcal{Y}$ of a topological space $\mathcal{X}$ is *irreducible* if it cannot be expressed as the union $(\mathcal{Y}_1 \cup \mathcal{Y}_2)$ of two proper subsets each of which closed in $\mathcal{Y}$. (The empty set is not considered to be irreducible.) For a more detailed definition of a variety and related analysis, see Ref. [41, 27, 31, 38].

▷ **Vector Bundle**: ☞ *bundle* [(L)].

▷ **Veronese Map**: The embedding $\mathbb{P}^n \hookrightarrow \mathbb{P}^{\binom{n+d}{d}-1}$ by means of sections of $\mathcal{O}_{\mathbb{P}^n}(d)$. For example,

$$\mathbb{P}^1 \hookrightarrow \mathbb{P}^n \quad : \qquad (x:y) \to (x^n : x^{(n-1)}y : \cdots : y^n)\ , \tag{L.67}$$

$$\mathbb{P}^2 \hookrightarrow \mathbb{P}^5 \quad : \qquad (x:y:z) \to (x^2 : xy : xz : y^2 : yz : z^2)\ . \tag{L.68}$$

▷ **Very Ample** (line bundle): A *line bundle*$^{(L)}$ $\mathcal{L}$ over $\mathcal{X}$ is very ample if it has enough sections to provide a closed immersion (or "embedding") $\mathcal{X} \hookrightarrow \mathbb{P}^N$ for some N. ☞ *ample*$^{(L)}$.

▷ **Weakly Ample** (line bundle): A *line bundle*$^{(L)}$ $\mathcal{L}$ over $\mathcal{X}$ is weakly ample if it is *nef*$^{(L)}$ and *big*$^{(L)}$; weakly ample includes ample.

▷ **Whitney Product Formula**: For two vector bundles $\mathcal{V}$ and $\mathcal{V}'$ over a compact complex space $\mathcal{X}$, total Chern classes are multiplicative

$$c[\mathcal{V} \oplus \mathcal{V}'] = c[\mathcal{V}] \wedge c[\mathcal{V}']. \tag{L.69}$$

(It suffices if '$\oplus$' denotes a direct sum differentiably.)

▷ **Wick-Rotate**: Formally associate a relativistic (quantum) field theory in a spacetime with Lorentzian signature to an Euclidean (statistical) field theory in a Riemannian space by way of analytical continuation, $t \mapsto x^0 := ict$.

▷ **Yukawa Couplings**: In general, any fermion-scalar interaction coupling; in the *Standard Model*$^{(L)}$, these encode the interactions between the Higgs scalar field and the fundamental fermions. The non-zero vacuum expectation value of the Higgs field thus also provides the so-interacting fermions their mass. In Calabi–Yau compactification, Yukawa couplings are structure constants of the (quantum) rings spanned by certain cohomology groups on the compactification Calabi–Yau space; see § 0.5.

Afterword

by Yang-Hui He

When Professor Tristan Hübsch and World Scientific asked me to write a few words for the publication of the Second Edition of *Calabi–Yau Manifolds: A Bestiary For Physicists*, I felt tremendously honored. It was as if an Aquinian monk had been summoned to scribe a commentative addendum to the *Summa*, or a neo-Confucian scholar, to the *Analects*. In my generation of mathematical physicists—pursuing PhDs at the turn of the millennium—the *Bestiary* was a unique and valued book. If anything, it was perhaps the only textbook on our shelves—where the typical title would be "Introduction to x", where x is some technical word in theoretical physics or pure mathematics—to have so exotic a word as "bestiary". Even the font (Gothic) and the cover picture (a photo of origami and a notebook) added an eccentric but inviting charm.

The book, as it turned out, was exactly that; it was *inviting*. In the early decades of the twentieth century, the ground-breaking work of Einstein and Hilbert made it clear that physicists should learn differential geometry. By the mid-decades, that of Dirac, Wigner, Yang, *et al.* did likewise for abstract algebra and representation theory. In the last decades, that of Atiyah, Chern, Witten, *et al.*—highlighted perhaps by Witten's Fields Medal in 1990—inspired theorists that algebraic geometry, too, should be an essential tool in a physicist's machinery. But, there's the rub.

To quote a perspicacious comment from Terence Tao in a recent interview [466], "... in the sixties and seventies ... there was an era of mathematics where abstraction was making huge strides in simplifying and unifying a lot of mathematics that was previously very empirical ... this is sometimes what we call the Bourbaki era in mathematics. And it did veer a little bit too far from being grounded." This Bourbaki style resulted in tomes of the revered GTM yellow [246], definitive but intimidating, rigorous but impenetrable, serving far more as decorative references on the shelves to impress a visitor than to initiate a neophyte into the subject. In algebraic geometry, at the time of the 1980s and 1990s when physicists realized the necessity of studying this amazing field, there was Hartshorne's canon [27], which was considered even amongst graduate students of pure mathematics as an Olympian challenge. For physicists, it might as well have been written in Sumerian. The friendlier alternatives would have been Reid [285], Harris [265], or [247] (which did not have the advanced material requisite to gauge theory), or Griffith-Harris [23] (832 pages!). Thus, as wonderful as these books are, they were not what was desperately needed by the theoretical physics community.

Hübsch's *Bestiary* was the saving grace. Though the title suggested a rather technical focus on Calabi–Yau manifolds, it really was an invitation to complex algebraic geometry for the non-specialist. As esoteric as the word "Calabi–Yau" might sound, there is important historical context. In the nascence of topology, Euler classified (smooth, compact, orientable, and boundary-less) surfaces by a single integer, the *genus* g, which characterizes the number of "holes". Thereafter, Gauss, in his Theorema Egregium, related the Euler number $\chi = 2 - 2g$ to the integral of curvature, whereby linking topology to analysis. Here, we see a *trichotomy*: (i) $g = 0$ for $\chi > 0$ (positive curvature); (ii) $g = 1$ for $\chi = 0$ (zero curvature); and (iii) $g > 1$ for $\chi < 0$ (negative curvature). For surfaces, there are respectively the sphere, the torus, and those of higher genera (including the famous anti-de-Sitter space in real dimension two, as well as modular curves as quotients of the upper-half complex plane). The intermediate case of zero-curvature obviously distinguishes itself. Indeed, the torus is precisely a (the) Calabi–Yau manifold of complex dimension 1.

Generalizing this story to beyond surfaces had been a central problem in mathematics since Euler–Gauss–Riemann. At least in complex geometry (even-dimensional manifolds that admit a complex structure), and in particular those with an additional Kähler structure, the above trichotomy persists. Through the index theorem, the Euler number χ is given by the integral of the Chern class (of the tangent bundle), giving us (i) Fano varieties of positive curvature; (ii) Calabi–Yau varieties of zero curvature; and (iii) varieties of general type. The uniqueness and conjectural existence of the Kähler metric that determines the Chern class, by Calabi [253] in the 1950s, and the subsequent Fields-winning demonstration of its existence by Yau [54] in the late 1970s, were great triumphs in modern mathematics.

All beautiful mathematics must find their place in physics. In the 1980s, when string theorists came across the conditions of "Ricci-flat" and "Kähler" in their search for a consistent, supersymmetric, low-energy effective theory from the 10-dimensional heterotic string [93, 409] in order to attain the Standard Model of particle physics, Strominger's office was next to Yau's at the IAS. Geometry, then unknown to physicists, was at once available from the expert who just obtained the Fields Medal. The two worlds coincided, and sparkled ideas and collaborations that would persist till the very present. Indeed, the name "Calabi–Yau" was coined by the physicists in honor of the two great mathematicians.

While Yau gave some immediate examples of Calabi–Yau manifolds on which one might compactify string theory, even meeting some of the rudimentary constraints, such as getting the right number of generations of fermions, turned out to be a non-trivial challenge. How to find (smooth, compact) Calabi–Yau 3-folds (complex dimension 3, real dimension 6) that satisfied physical constraints which are translated into exact conditions in algebraic and differential geometry? The hunt was on.

From the earliest cases of complete intersections in a single projective space, to a single hypersurface [96] in weighted projective $\mathbb{CP}^4$, to complete intersections in products in projective spaces—the so-called CICYs [85, 352, 385, 148, 425], the number of explicitly constructed Calabi–Yau manifolds rapidly rose to the thousands. The fact that it was the physicists who had to pursue such constructions proved very much the aforementioned point about the predominant air of abstraction in the pure mathematics community left from the Bourbaki era. Somehow it was considered—a sentiment which is luckily no longer the mainstream today—uncivilized to stoop from abstraction to concrete examples in mathematics. In a way, the *Bestiary* documented this flurry of activity in the last decade of the twentieth century and threw down the gauntlet back at the Bourbakians.

Some three decades have passed since the publication of the first edition of the *Bestiary* [272]. Enormous strides, of course, have since been made in physics and in mathematics. The *First String Revolution*, wherein the different formulations of string theory were shown to be dual to each other, with the compactifications of which leading to Calabi–Yau manifolds, has been roughly marked by historians to be circa 1984–1994. This was the inspiration of the *Bestiary*. Its last chapter alluded to *mirror symmetry*, where the complex and Kähler structures of a given Calabi–Yau manifold

M should be exchanged. While in physics, this followed from a duality between super-conformal field theories [89], the implication to geometry, where complex and symplectic structures could be mixed, was astounding. Mirror symmetry is now a vast field in itself, and the Reader is referred to Refs. [255, 407, 276, 292] for more recent textbooks.

The search for mirror pairs was a parallel skein in mathematics to the search for the Standard Model from Calabi–Yau compactifications in physics. The search led to the remarkable works of Batyrev–Borisov on complete intersections in toric varieties coming from reflexive polytopes [309, 68], the tour-de-force classification thereof [429], and the more recent works on generalized complete intersections [296, 319]. All at once, there is a simple combinatorial criterion for mirror symmetry for large classes of manifolds, and the number of explicit smooth Calabi–Yau 3-folds grew to billions.

The *Second String Revolution* came in the period of 1994–2003, when M-theory, D-branes and the AdS/CFT holographic correspondence enriched strings to extended objects. For Calabi–Yau manifolds, it opened up vast interests in non-compact and singular ones (of course, M-theory also ignited interests in G_2 manifolds, many of which can be constructed from fibrations over Calabi–Yau 3-folds). These "local" Calabi–Yau manifolds are key to AdS/CFT [438] because the D-branes can "probe" them [426, 394]. In this geometric context, the profound holographic principle relating quantum gravity in dimension d to gauge theory on its $(d-1)$-dimensional boundary exhibits as the correspondence between the algebraic geometry of a moduli space and the representation theory of quivers with potential. The Reader is referred to pedagogical books in [280, 248, 281] for the physics, and to [397] for some of the mathematics, especially to the McKay Correspondence [440] and to bi-partite structures [480, 377].

The *Third String Revolution* started in 2003, when the *string landscape* and the statistics of vacuum configurations were the concepts to seep into the minds of scientists and the public alike. Prompted by the plethora of the Kreuzer–Skarke Calabi–Yau manifolds, as well as the increasing number of further structures one could place on these manifolds in compactification—such as stable vector bundles, D-branes wrapping algebraic cycles, etc.—Doulgas [374] and KKLT [418] contemplated, on a concrete footing, the vastness of possible four-dimensional universes obtainable from string theory.

Whilst astronomical numbers such as 10^{500} began to dishearten public confidence in string theory, it must be pointed out that such numbers came from back-of-the-envelope estimates (typical Betti numbers of Calabi–Yau manifolds in a generic calculation). More importantly, the whole point of the landscape (and related swampland [447]) is not to plague us with a plethora but to rule out the overwhelming majority of scenarios. Indeed, even without imposing cosmological constraints and the correct Yukawa couplings, reduction via the constraint of getting the right particle content alone should be by some eight orders of magnitude [373, 389, 349, 347]. Thus, though there seems to be an unfathomably large landscape of possible four-dimensional universes obtainable from string theory, so far, a single one that is consistent with our observations—the Standard Model of particles, general relativity dictating large-scale universes with a small positive cosmological constant (de Sitter space), and quantum gravity governing black holes—is yet to be found.

In mathematics, this situation is curiously paralleled. Recall the trichotomy discussed above of positive (Fano), zero (Calabi–Yau) and negative curvature (general type) complex manifolds. The Fano case is known to have a finite number of topological types in every dimension. The general type is known to be wildly unclassifiable and infinite in number. For the boundary case of Calabi–Yau, it is a conjecture of Yau that in every dimension, the number of different topological types (Betti numbers, Hodge numbers, intersection numbers, etc.) is finite. In complex dimension one, we see this is true: there is only the torus T^2. In complex dimension two, there are only the 4-torus T^4 and the K3-surface. In complex dimension three (and above), despite the vast numbers, one still suspects finite possibilities in topology. The statement is even stronger [203]: the moduli space of (smooth, compact, simply connected) Calabi–Yau 3-folds is fantasized to be connected, so that all of them are related by bi-rational transformations such as flops.

We see, therefore, that Calabi–Yau manifolds still very much occupy a central place in contemporary mathematics and theoretical physics (even completely outside of string theory, Calabi–Yau structures are unexpectedly manifesting in the Standard Model [401] and in ordinary Feynman amplitudes [339]). Irrespective of one's interests in Calabi–Yau spaces in any context, it should be noted that in many ways, the *Bestiary* was not merely a pioneer in inviting researchers to algebraic geometry but to *computational* algebraic geometry. This brings us to the final, and by far the most important, aspect of our discussion.

The second half of the twentieth century has convinced us of the indispensable *role of computers* in mathematics (broadly defined to include any theoretical research) [367]. The speeding up of computations and simulations certainly goes without saying, but more pertinent are the ever-increasing cases where major fundamental results *could not* be attained without the computer. The four-color theorem, the Kepler Conjecture, the classification theorem of finite simple groups, to name but a few, could not have been possible to humans alone, simply because the proofs or case-checking would take longer than a life-span to go through.

Indeed, the data of Calabi–Yau manifolds have been compiled since the 1990s by impressive computer work. For instance, the CICYs, arguably the first database ever created for geometry, were found on the CERN supercomputer. With the astounding growth and availability of the personal computer, already on the Kreuzer–Skarke list in 2000, half a billion manifolds were found on a Pentium desktop. Since then, there have been a host of databases and freely available software dedicated to studying the Calabi–Yau (and related) landscape, such as:

The Graded Ring Database: http://www.grdb.co.uk/
Kreuzer–Skarke: http://hep.itp.tuwien.ac.at/~kreuzer/CY/
KS (some triangulations/orientifolds): http://www.rossealtman.com/toriccy/
KS Explorer: https://cy.tools/
Picard-Fuchs operators: https://cydb.mathematik.uni-mainz.de/
CICY+ (at bottom): https://www.physics.ox.ac.uk/our-people/lukas
Various resources, e.g., numerical metrics, etc.: https://github.com/topics/calabi-yau-manifolds

Meanwhile, the paucity of friendly introductions to algebraic geometry at the time of Ref. [272] is well supplanted by a number of excellent textbooks that emphasize explicit calculations rather than abstruse sophistry [256, 287, 257], as well as some dedicated to physicists and engineers [268, 245, 247].

Something was in the air in 2017, and four independent groups thought of using machine learning to study problems in string compactifications [398, 428, 456, 354]. In particular, Ref. [398] proposed that algebraic geometry, if not any mathematical problem, could be phrased as an image-processing query for a neural network and that the structure of mathematics ought to be studied by AI [399]. A flurry of activity on AI-guided theoretical discovery in related and unrelated areas in the past lustrum is an encouraging testimony that Calabi–Yau manifolds reside at the confluence of pure mathematics, theoretical physics, and AI research. One illustrative time-line might give the Reader an informative glimpse. There have been a number of branch-offs of the annual string theory conference, which since 1986 has been bringing together various sub-communities in high-energy theory and in mathematics. The following are the beginning dates of these international conference series:

1986– *"Strings"*
2002– *"StringPheno"*: string theorists interested in obtaining phenomenologically viable physics
2006–2010 *String Vacuum Project*: an NSF funded multi-institution collaboration for the string landscape
2008– *"ISGT"* Integrability in String/Gauge Theory
2011– *"String-Math"*: interactions between string theory and mathematics
2012– *"Amplitudes"*: computing QFT amplitudes using string-inspired techniques
2014– *SIAM annual meeting* with a String/Theoretical Physics session
2017– *"String-Data"*: machine learning (ML) for the string landscape
2020– *"DANGER"*: (Data in ANalysis, GEometry & Representation Theory)

The trend toward data- and AI-driven research is self-evident. We refer the Reader to the first textbook [266] and editorial [267] on machine learning in theoretical physics and pure mathematics, as well as a wonderful pedagogical monograph on modern data scientific methods [457] for physicists (cf. [400, 316, 403]).

In many ways [266], focusing on machine learning for the Calabi–Yau landscape, as an attempt to invite three communities (mathematics, physics and computer science/AI) to a cocktail party was my homage to the *Bestiary*, which had been so influential to my intellectual youth. I am thrilled to see the appearance of a new edition, with updated additions and revisions by Prof. Hübsch, and am flattered to be given the opportunity to write these words. I certainly hope the current generation of theoretical physicists and pure mathematicians would embrace the *Bestiary* with the same excitement and curiosity that I did as a student.

Bibliography

The list of books, reviews and research articles compiled here is meant to be a helpful guide through the literature rather then a complete catalogue. I apologize to the many researchers whose work is underrepresented here; omissions were made only to bring the list to a tractable length. Collective bibliographic entries are listed chronologically and alphabetically. Many of the items contain several related references; these are ordered according to date of publication. "American Mathematical Society" is commonly abbreviated "AMS" and "Theoretical Advanced Study Institute in Elementary Particle Physics" is abbreviated "TASI."

Books and Key Reference Articles:

[1] V.I. Arnold: in *Proceedings of the International Congress of Mathematicians, Vancouver, 1974*, p. 19; *Singularity Theory, London Math. Soc. Lect. Note Ser.* no. 53 (Cambridge University Press, Cambridge, 1981).
V.I. Arnold, S.M. Gusein-Zade and A.N. Varchenko: *Singularities of Differentiable Maps* (Birkhäuser, Boston, 1985).

[2] M.F. Atiyah: "On Analytic Surfaces With Double Points," *Proc. Royal Soc.* **A 247** (1958) 237.

[3] M.F. Atiyah and I.G. Macdonald: *Introduction to Commutative Algebra* (Addison–Wesley, Reading, MA, 1969).

[4] W. Barth, C. Peters and A. van de Ven: *Compact Complex Surfaces* (Springer, Berlin, 1984).

[5] A.O. Barut and R. Rączka: *Theory of Group Representations and Applications* (World Scientific, Singapore, 1986).

[6] R.J. Baston and M.G. Eastwood: *The Penrose Transform—Its Interaction With Representation Theory* (Claredon Press, Oxford, 1989).

[7] A. Beauville: "Variétés Kähleriennes dont la premiére classe de Chern est nulle," *J. Diff. Geom.* **18** (1983) 755–782.

[8] M. BERGER: "Sur les groupes d'holonomie homogène des variétés a connexion affine et des variétés riemanniennes," *Soc. Math. Franc. Bull.* **83** (1953) 79.

[9] R. BOTT: "Homogeneous Vector Bundles," *Ann. Math.* **66** (1957) 203.

[10] R. BOTT: "On a Theorem of Lefschetz," *Mich. Math. J.* **6** (1959) 211.

[11] R. BOTT and L. TU: *Differential Forms in Algebraic Topology* (Springer, New York, 1982).

[12] E. CARTAN: "Sur les domaines bornés homogènes de l'espace de n variables complexes," in *Actualités Scientifiques et Industrielles* no. 358 (Hermann, Paris, 1936).

[13] F.M.E. CATANESE: Ch. VIII in *Topics in Transcendental Algebraic Geometry*, p. 143, P. Griffiths, ed. (Princeton University Press, Princeton, 1984).

[14] H. CLEMENS: "Double Solids," *Publ. Math. IHES* **58** (1983) 19.

[15] J. CHEEGER: "On the Hodge Theory of Riemannian Pseudomanifolds." *Proc. Symp. Pure Math.* **36** (1980) 91–146.
J. CHEEGER, M. GORESKY and R. MACPHERSON: "L^2-Cohomology and Intersection Homology of Singular Algebraic Varieties," in *Seminar on Differential Geometry*, S.-T. Yau, ed. *Ann. Math. Studies* **102** (1982).

[16] I. DOLGACHEV: "Weighted Projective Varieties," *Group Actions and Vector Fields*, pp. 34–71, *Lecture Notes in Math.* **956** (Springer, New York, 1982).

[17] B.A. DUBROVIN, A.T. FOMENKO and S.P. NOVIKOV: *Modern Geometry—Methods and Applications, Parts I, II and III* (Springer, New York, 1984).

[18] M.G. EASTWOOD: "The Generalized Penrose–Ward Transform." *Math. Proc. Camb. Phil. Soc.* **97** (1985) 165.

[19] T. EGUCHI, P.B. GILKEY and A.J. HANSON: "Gravitation, Gauge Theories and Differential Geometry." *Phys. Rep.* **C66** (1980) 213–393.

[20] W. FULTON: *Introduction to Toric Varieties* (Princeton University Press, Princeton, 1993).

[21] S.J. GATES, JR., M.T. GRISARU, M. ROČEK and W. SIEGEL: *Superspace* (The Benjamin/Cummings Pub. Co., Reading, MA, 1983).

[22] M.B. GREEN, J.H. SCHWARZ and E. WITTEN: *Superstring Theory II* (Cambridge University Press, Cambridge, 1987).

[23] P. GRIFFITHS and J. HARRIS: *Principles of Algebraic Geometry* (John Wiley, New York, 1978).

[24] R. GODEMENT: *Théorie des faisceaux* (Hermann, Paris, 1958).

[25] S.I. GOLDBERG: *Curvature and Homology* (Dover, New York, 1982).

[26] M. GORESKY: "Triangulation of Stratified Objects." *Proc. Am. Math. Soc.* **72** (1978) 193–200.
M. GORESKY and R. MACPHERSON: "Intersection Homology Theory." *Topology* **19** (1980) 135–162; "Intersection Homology II." *Invent. Math.* **71** (1983) 77–129.

[27] R. HARTSHORNE: *Algebraic Geometry* (Springer, New York, 1977).

[28] N. HITCHIN: "Harmonic Spinors." *Adv. Math.* **14** (1974) 1–55.

[29] F. HIRZEBRUCH: *Topological Methods in Algebraic Geometry* (Springer, New York, 1978).

[30] T. Hübsch: "All the String's Vacua," in *Proc. of the XIII International Particle Theory School*, Sep. 1989, Szczyrk, Poland.
[31] S. Iitaka: *Algebraic Geometry* (Springer, New York, 1982).
[32] V.A. Iskovskih: "Fano 3-Folds I." *Izv. Akad. Nauk SSSR Ser. Mat.* **41** (1977) 516–562, English transl. in *Math. USSR Izv.* **11** (1977); "Fano 3-Folds II." *Izv. Akad. Nauk SSSR Ser. Mat.* **42** (1978) 469–506, English transl. in *Math. USSR Izv.* **11** (1977).
[33] M. Kaku: *Introduction to Superstrings* (Springer, New York, 1988).
[34] F. Kirwan: *An Introduction to Intersection Homology Theory* (Longman Sci. & Technical, Essex, 1988).
[35] K. Kodaira: *Complex Manifolds and Deformations of Complex Structures* (Springer, New York, 1986).
[36] K. Kodaira and D.C. Spencer: "On Deformations of Complex Analytic Structures." *Annals of Math.* **67** (1958) 328–466; *ibid.* **71** (1960) 43–76.
[37] J. Kollar: "The Structure of Algebraic Threefolds: An Introduction to Mori's Program." *Bull. Am. Math. Soc.* **17** (1987) 211–273.
[38] H. Matsumura: *Commutative Ring Theory* (Cambridge University Press, Cambridge, 1986).
[39] J. McCleary: *User's Guide to Spectral Sequences* (Publish or Perish, Inc., Willmington, 1985).
[40] S. Mori and S. Mukai: "On Fano 3-Folds With $b_2 \geq 2$," *Algebraic Varieties and Analytic Varieties*, S. Iitaka, ed. *Adv. Stud. Pure Math.* **1** (1983) 101–129.
[41] D. Mumford: *Algebraic Geometry I—Complex Projective Varieties* (Springer, New York, 1976).
[42] T. Oda: *Convex Bodies and Algebraic Geometry* (Springer, New York, 1985).
[43] C. Okonek, M. Schneider and H. Spindler: *Vector Bundles on Complex Projective Spaces* (Birkhäuser, Boston, 1980).
[44] L. O'Reifeartaigh: *Group Structure of Gauge Theories* (Cambridge University Press, Cambridge, 1986) esp. chapters 8 and 12 and references therein.
[45] R.M. Range: *Holomorphic Functions and Integral Representations in Several Complex Variables* (Springer, New York, 1986).
[46] M. Reid: "Young Persons Guide to Canonical Singularities." *Proc. Sump. Pure Math.* **46** (1987) 345.
[47] I.R. Shafarevich: *Basic Algebraic Geometry* (Springer, New York, 1974).
[48] R. Slanski: "Group Theory for Unified Model Building." *Phys. Rep.* **C79** (1981) 1.
[49] R.P. Stanley: *Combinatorics and Commutative Algebra* (Birkhäuser, Boston, 1983).
[50] C.T.C. Wall: "Classification Problems in Differential Topology, V: On Certain 6-Manifolds." *Invent. Math.* **1** (1966) 355.
[51] J. Wess and J. Bagger: *Supersymmetry and Supergravity* (Princeton University Press, Princeton, 1984).

[52] C. von Westenholz: *Differential Forms in Mathematical Physics* (North Holland Pub. Co., Amsterdam, 1978).

[53] B.G. Wybourne: *Classical Groups for Physicists* (Wiley, New York, 1974).

[54] S.-T. Yau: "Calabi's Conjecture and Some New Results in Algebraic Geometry," *Proc. Natl. Acad. Sci. USA* **74** (1977) 1798; "On the Ricci Curvature of a Compact Kähler Manifold and the Complex Monge–Ampère Equation I," *Comm. Pure and App. Math.* **31** (3) (1978) 339–411.

Research Articles:

[55] L. Alvarez-Gaumé, D.Z. Freedman and S. Mukhi: "The Background Field Method and the Ultraviolet Structure of the Supersymmetric Non-Linear σ-Model," *Ann. Phys.* (NY) **134** (1981) 85.
S. Mukhi: "Non-Linear σ-Models, Scale Invariance and String Theories: A Pedagogical Review," in the Proc. of the TIFR Winter School, Panchgani 1986, V. Singh and S. Wadia, eds. (World Scientific, Singapore, 1986); "The Geometric Background Field Method, Renormalization and the Wess–Zumino Term in Non-Linear σ-Models," *Nucl. Phys.* **B264** (1986) 640.

[56] L. Alvarez-Gaumé and E. Witten: "Gravitational Anomalies," *Nucl. Phys.* **B234** (1983) 269.

[57] P.S. Aspinwall: "Superconformal Field Theories Near Orbifold Points," *Comm. Math. Phys.* **128** (1990) 593–611; "Resolution of Orbifold Singularities in String Theory," pp. 355–379, *Mirror Symmetry II*, B. Greene and S.-T. Yau, eds. *Stud. Adv. Math.* **1** (AMS, Providence, RI and Intl. Press, Cambridge, MA, 1996).

[58] P.S. Aspinwall, B.R. Greene, K.H. Kirklin and P.J. Miron: "Searching for Three-Generation Calabi–Yau Manifolds," *Nucl. Phys.* **B294** (1987) 193.

[59] P.S. Aspinwall, B.R. Greene and D.R. Morrison: "Calabi–Yau Moduli Space, Mirror Manifolds and Spacetime Topology Change in String Theory," *Nucl. Phys.* **B416** (1994) 414; "The Monomial-Divisor Mirror Map," `arXiv:alg-geom/9309007`; "Spacetime Topology Change: the Physics of Calabi–Yau Moduli Space," in *Berkeley 1993, Proceedings, Strings '93,* pp. 241–262, M.B. Halpern, G. Rivlis, and A. Sevrin, eds. (World Scientific, Singapore, 1995).

[60] P.S. Aspinwall, B.R. Greene and D.R. Morrison: "Measuring Small Distances in $N=2$ Sigma Models," *Nucl. Phys.* **B420** (1994) 184–242.
P.S. Aspinwall: "Minimum Distances in Non-Trivial String Target Spaces," *Nucl. Phys.* **B431** (1994) 78–96.

[61] P.S. Aspinwall and C.A. Lütken: "Geometry of Mirror Manifolds," *Nucl. Phys.* **B353** (1991) 427–461.

[62] P.S. Aspinwall and C.A. Lütken: "Quantum Algebraic Geometry of Superstring Compactifications," *Nucl. Phys.* **B355** (1991) 482–510; "A New Geometry From Superstring Theory," in *Essays on Mirror Manifolds,* p. 316–341, S.-T. Yau, ed. (Intl. Press, Hong Kong, 1992).

[63] P.S. Aspinwall and D.R. Morrison: "String Theory on K3 Surfaces," in *Mirror Symmetry II*, B. Greene and S.-T. Yau, eds. *AMS/IP Stud. Adv. Math.* **1** (1996) 703–716.

[64] M. Atiyah, R. Bott and L. Gårding: Acta Math. **131** (1973) 145.

[65] M. Atiyah and G. Segal: "On Equivariant Euler Characteristics," *J. Geom. Phys.* **6** (1989) 671–677.

[66] M. Bando, T. Kugo and K. Yamawaki: "Nonlinear Realization and Hidden Local Symmetries," *Phys. Lett.* **C164** (1988) 218.

[67] V. Batyrev: "Variations of the Mixed Hodge Structure of Affine Hypersurfaces in Algebraic Tori," *Duke Math. J.* **69** (1993) 349.

[68] V. Batyrev: "Dual Polyhedra and Mirror Symmetry of Calabi–Yau Hypersurfaces in Toric Varieties," *J. Alg. Geom.* **3** (3) (1994) 493–535.
L.A. Borisov: "Towards the Mirror Symmetry for Calabi–Yau Complete Intersections in Gorenstein Toric Fano Varieties," `arXiv:alg-geom/9310001`.
V. Batyrev and L.A. Borisov: "Dual Cones and Mirror Symmetry for Generalized Calabi–Yau Manifolds," in *Mirror Symmetry II*, B. Greene and S.-T. Yau, eds. *AMS/IP Stud. Adv. Math.* **1** (1996) 71–86.

[69] V. Batyrev and D. van Straten: "Generalized Hypergeometric Functions and Rational Curves on Calabi–Yau Complete Intersections in Toric Varieties," *Comm. Math. Phys.* **168** (3) (1995) 493-–533.

[70] A.A. Belavin and V.G. Knizhnik: "Algebraic Geometry of Quantum Strings," *Phys. Lett.* **168B** (1986) 201.

[71] P. Berglund, P. Candelas, X.C. de la Ossa, A. Font, T. Hübsch, D. Jančić and F. Quevedo: "Periods for Landau–Ginzburg and Calabi–Yau Models," *Nucl. Phys.* **B419** (1994) 352–403.
P. Berglund, E. Derrick, T. Hübsch and D. Jančić: "Periods for String Compactifications," *Nucl. Phys.* **B420** (1994) 268–288.

[72] P. Berglund, B. Greene and T. Hübsch: "Classical *vs.* Landau–Ginzburg Orbifold Geometry of Compactification," *Mod. Phys. Lett.* **A7** (1992) 1885–1896.

[73] P. Berglund and T. Hübsch: "Twisted Three-Generation Compactification," *Phys. Lett.* **B260** (1991) 32–38.

[74] P. Berglund and T. Hübsch: "A Generalized Construction of Mirror Manifolds," *Nucl. Phys.* **B393** (1993) 377–391.

[75] P. Berglund and T. Hübsch: "Couplings for Compactification," *Nucl. Phys.* **B411** (1994) 223–254.

[76] P. Berglund, T. Hübsch and L. Parkes: "Gauge-Neutral Matter in a Three-Generation Superstring Compactification," *Mod. Phys. Lett.* **A5** (1990) 1485–1488.

[77] P. Berglund, T. Hübsch and L. Parkes: "The Complete Matter Sector in a Three-Generation Compactification," *Comm. Math. Phys.* **148** (1992) 57.

[78] P. Berglund and S.H. Katz: "Mirror Symmetry for Hypersurfaces in Weighted Projective Space and Topological Couplings," *Nucl. Phys.* **B420** (1994) 289–314.

[79] F. Bogomolov: "Kähler Manifolds With Trivial Canonical Class," *Izv. Akad. Nauk SSSR Ser. Mat.* **38** (1974) 11–21.

[80] W. Boucher, D. Friedan and A. Kent: "Determinant Formulae and Unitarity for the $N=2$ Superconformal Algebras in Two Dimensions or Exact Results on String Compactification," *Phys. Lett.* **172B** (1986) 316.

A.B. Zamolodchikov and V.A. Fateev: "Disorder Fields in Two-Dimensional Conformal Quantum Field Theories," *Zh. Eksp. Theor. Fiz.* **90** (1986) 1553–1566.

S. Nam: "The Kac Formula for the $N=1$ and $N=2$ Superconformal Algebras," *Phys. Lett.* **172B** (1986) 323.

P. Di Vecchia, J.L. Petersen, M. Yu and H.B. Zheng: "Explicit Construction of Unitary Representations of the $N=2$ Superconformal Algebra," *Phys. Lett.* **174B** (1986) 280.

Z. Qiu: "Non-Local Current Algebra and $N=2$ Superconformal Field Theory in Two Dimensions," *Phys. Lett.* **188B** (1987) 207; "Modular Invariant Partition Functions for $N=2$ Superconformal Field Theories," *Phys. Lett.* **198B** (1987) 497.

[81] R. Brooks, F. Muhammad and S.J. Gates, Jr.: "Unidexterous $d=2$ Supersymmetry in Superspace," *Nucl. Phys.* **B268** (1986) 599–620.

[82] R. Bryant and P. Griffiths: *Progress in Mathematics* **36**, pp. 77–102 (Birkhäuser, Boston, 1983).

[83] E. Calabi: "On Kähler Manifolds With Vanishing Canonical Class, Algebraic Geometry and Topology," *Proc. Intl. Congr. Math. Amsterdam* **2** (1954) 206–207.

[84] P. Candelas: "Yukawa Couplings Between (2, 1)-Forms," *Nucl. Phys.* **B298** (1988) 458–492.

[85] P. Candelas, A.M. Dale, C.A. Lütken and R. Schimmrigk: "Complete Intersection Calabi–Yau Manifolds," *Nucl. Phys.* **B298** (1988) 493–525.

[86] P. Candelas and X.C. de la Ossa: "Comments on Conifolds," *Nucl. Phys.* **B342** (1990) 246–268.

[87] P. Candelas and X.C. de la Ossa: "Moduli Space of Calabi–Yau Manifolds," *Nucl. Phys.* **B355** (1991) 455–481.

[88] P. Candelas, X.C. de la Ossa, A. Font, S.H. Katz and D.R. Morrison: "Mirror Symmetry for Two-Parameter Models, 1," *Nucl. Phys.* **B416** (1994) 481–538.

P. Candelas, A. Font, S.H. Katz and D.R. Morrison: "Mirror Symmetry for Two-Parameter Models, 2," *Nucl. Phys.* **B429** (1994) 626–674.

[89] P. Candelas, X.C. de la Ossa, P.S. Green and L. Parkes: "An Exactly Soluble Superconformal Theory From a Mirror Pair of Calabi–Yau Manifolds," *Phys. Lett.* **258B** (1991) 118–126; A Pair of Calabi–Yau "Manifolds as an Exactly Soluble Superconformal Theory," *Nucl. Phys.* **B359** (1991) 21–74.

[90] P. Candelas, P. Green and T. Hübsch: "Finite Distance Between Distinct Calabi–Yau Manifolds," *Phys. Rev. Lett.* **62** (1989) 1956–1959.

[91] P. CANDELAS, P. GREEN and T. HÜBSCH: "Connected Calabi–Yau Compactifications," in "Strings '88," p. 155, S.J. Gates Jr., C.R. Preitschopf and W. Siegel, eds. (World Scientific, Singapore, 1989).

[92] P. CANDELAS, P.S. GREEN and T. HÜBSCH: "Rolling Among Calabi–Yau Vacua," *Nucl. Phys.* **B330** (1990) 49–102.

[93] P. CANDELAS, G. HOROWITZ, A. STROMINGER and E. WITTEN: "Vacuum Configurations for Superstrings," *Nucl. Phys.* **B258** (1985) 46.

[94] P. CANDELAS, T. HÜBSCH and R. SCHIMMRIGK: "Relation Between the Weil–Petersson and Zamolodchikov Metrics," *Nucl. Phys.* **B329** (1990) 583–590.

[95] P. CANDELAS, C.A. LÜTKEN and R. SCHIMMRIGK: "Complete Intersection Calabi–Yau Manifolds. 2 Three Generation Manifolds," *Nucl. Phys.* **B306** (1988) 113.

[96] P. CANDELAS, M. LYNKER and R. SCHIMMRIGK: "Calabi–Yau Manifolds in Weighted $\mathbb{P}^4$," *Nucl. Phys.* **B341** (1990) 383–402.

[97] S. CECOTTI, S. FERRARA and L. GIRARDELLO: "A Topological Formula for the Kähler Potential of 4-d $N=1$, $N=2$ Strings and its Implications for the Moduli Problem," *Phys. Lett.* **213B** (1989) 443.

[98] S. CECOTTI and L. GIRARDELLO: "Functional Measure, Topology and Dynamical Supersymmetry Breaking," *Phys. Lett.* **B110** (1982) 39–43.
L. ALVAREZ-GAUMÉ: "Supersymmetry and the Atiyah–Singer Index Theorem," *Comm. Math. Phys.* **90** (1983) 161–173.

[99] S. CECOTTI, L. GIRARDELLO and A. PASQUINUCCI: "Non-Perturbative and Exact Results for the $N=2$ Landau–Ginzburg Models," *Nucl. Phys.* **B328** (1989) 701–722; "Singularity Theory and $N=2$ Supersymmetry," *Int. J. Mod. Phys.* **A6** (1991) 2427–2496.
S. CECOTTI: "$N=2$ Landau–Ginzburg *vs.* Calabi–Yau σ-Models: Non-Perturbative Aspects," *Int. J. Mod. Phys.* **A6** (1991) 1749–1814; "Geometry of $N=2$ Landau–Ginzburg Families," *Nucl. Phys.* **B355** (1991) 755–775.

[100] S. CECOTTI and C. VAFA: "Topological Anti-Topological Fusion," *Nucl. Phys.* **B367** (1991) 359–461.

[101] S. CECOTTI and C. VAFA: "Exact Results for Supersymmetric Sigma Models," *Phys. Rev. Lett.* **68** (1992) 903–906.

[102] S. CECOTTI and C. VAFA: "A New Supersymmetric Index," *Nucl. Phys.* **B386** (1992) 405–452.

[103] B. CRAUDER and S.H. KATZ: "Cremona Transformations and Hartshorne's Conjecture," *Am. J. Math.* **113** (2) (1991) 269–285.
L. EIN and N. SHEPHERD-BARON: "Some Special Cremona Transformations," *Am. J. Math.* **111** (1989) 783–800.

[104] M. CVETIČ: "Phenomenological Implications of the Blown-Up Orbifolds," in *Proceedings of the Eight Workshop on Grand Unification*, Syracuse, New York, April 1987, in *Superstrings, Unified Theories and Cosmology 1987*, p. 138, (World Scientific, Singapore, 1988); "Suppression of Non-Renormalizable Terms in the Effective Superpotential for (Blown-Up)

Orbifold Compactification," *Phys. Rev. Lett.* **59** (1987) 1795; "Exact Construction of (2,0) Calabi–Yau Manifolds," *ibid.* **59** (1987) 2989.
L.E. Ibáñez, H.P. Nilles and F. Quevedo: "Orbifolds and Wilson Lines," *Phys. Lett.* **187B** (1987) 25; "Reducing the Rank of the Gauge Group in Orbifold Compactifications of the Heterotic Superstring," *Phys. Lett.* **192B** (1987) 332.
see also Ref. [188].

[105] M. Cvetič, J. Louis and B. Ovrut: "A String Calculation of the Kähler Potentials for Moduli of $\mathbb{Z}_n$ Orbifolds," *Phys. Lett.* **206B** (1988) 227.

[106] M. Cvetič, J. Louis and B. Ovrut: "The Zamolodchikov Metric and Effective Lagrangians in String Theory," *Phys. Rev.* **D40** (1989) 684.

[107] A. d'Adda, A.C. Davis, P. Di Vecchia and P. Salomonson: "An Effective Action for the Supersymmetric $\mathbb{P}^{n-1}$ Model," *Nucl. Phys.* **B222** (1983) 45.

[108] D.I. Dais: *Enumerative Combinatorics of Invariants of Certain Complex Threefolds With Trivial Canonical Bundle,* vol. 279 of *Bonner Math. Schriften,* 1995. PhD dissertation, Rheinische Friedrich-Wilhelms-Universität Bonn, 1994.

[109] D.A. Depireux, S.J. Gates, Jr. and Q-H. Park: "Lefton-Righton Formulation of Massless Thirring Models," *Phys. Lett.* **224** (1989) 364.
S. Bellucci, D.A. Depireux, and S.J. Gates, Jr.: "(1,0) Thirring Models and the Coupling of Spin-0 Fields to the Heterotic String," *Phys. Lett.* **232B** (1989) 67–74.

[110] A. Dimca: "Singularities and Coverings of Weighted Complete Intersections," *Reine Ang. J. Math.* **366** (1986) 184–193.

[111] M. Dine, N. Seiberg, X.G. Wen and E. Witten: "Non-Perturbative Effects on the String World Sheet," *Nucl. Phys.* **B278** (1986) 769; "*ditto*, 2.," *ibid.* **B289** (1987) 319.

[112] J. Distler: "Resurrecting (2,0) Compactifications." *Phys. Lett.* **188B** (1987) 431.
J. Distler and B.R. Greene: "Aspects of (2,0) String Compactification," *Nucl. Phys.* **B304** (1988) 1.

[113] J. Distler and B.R. Greene: "Some Exact Results on the Superpotential From Calabi–Yau Compactifications," *Nucl. Phys.* **B309** (1988) 295.

[114] J. Distler, B. Greene, K. Kirklin and P. Miron: "Evaluation of $\overline{27}^3$ Yukawa Couplings in a Three Generation Superstring Model," *Phys. Lett.* **195B** (1987) 41.
P. Green and T. Hübsch: "$(1,1)^3$ Couplings in Calabi–Yau Threefolds," *Class. Quant. Grav.* **6** (1989) 311.

[115] J. Distler, B. Greene, K. Kirklin and P. Miron: "Calculating Endomorphism Valued Cohomology: Singlet Spectrum in Superstring Models," *Comm. Math. Phys.* **112** (1989) 117.

[116] I. Di Vecchia, V. Knizhnik, J. Patersen and P. Rossi: "A Supersymmetric Wess–Zumino Lagrangian in Two Dimensions." *Nucl. Phys.* **B253** (1985) 701.

I. ANTONIADIS, C. BACHAS, C. KOUNNAS and P. WINDEY: "Supersymmetry Among Free Fermions and Superstrings." *Phys. Lett.* **171B** (1986) 51.

H. KAWAI, D. LEWELLEN and S.-H. TYE: "Construction of Four-Dimensional Fermionic Strings," *Phys. Rev. Lett.* **57** (1986) 1832, *ibid.* **58(E)** (1987) 429.

[117] L. DIXON: "Some World Sheet Properties of Superstring Compactifications, on Orbifolds and Otherwise," in *Superstrings, Unified Theories and Cosmology 1987*, p. 67–127, G. Furlan *et al.*, eds. (World Scientific, Singapore, 1988).

[118] L. DIXON and J.A. HARVEY: "String Theories in Ten Dimensions Without Spacetime Supersymmetry," *Nucl. Phys.* **B274** (1986) 93.

L. ALVAREZ-GAUMÉ, P. GINSPARG, G. MOORE and C. VAFA: "An $SO(16) \times SO(16)$ Heterotic String," *Phys. Lett.* **171B** (1986) 155.

[119] L. DIXON, J.A. HARVEY, C. VAFA and E. WITTEN: "Strings on Orbifolds," *Nucl. Phys.* **B261** (1985) 678, *ibid.* **B274** (1986) 285.

M. MUELLER and E. WITTEN: "Twisting Toroidally Compactified Heterotic Strings With Enlarged Symmetry Groups." *Phys. Lett.* **182B** (1986) 28.

K.S. NARAIN, M.H. SARMADI and C. VAFA: "Asymmetric Orbifolds." *Nucl. Phys.* **279** (1987) 369.

L.J. DIXON, E. MARTINEC D. FRIEDAN and S. SHENKER: "The Conformal Field Theory of Orbifolds," *Nucl. Phys.* **B282** (1987) 13–73.

L.E. IBÁÑEZ, J. MAS, H.P. NILLES and F. QUEVEDO: "Heterotic Strings in Symmetric and Asymmetric Orbifold Backgrounds." *Nucl. Phys.* **B301** (1988) 157.

[120] L. DIXON, V. KAPLUNOVSKY and J. LOUIS: "On Effective Field Theories Describing (2,2) Vacua of the Heterotic String." *Nucl. Phys.* **B329** (1990) 27.

[121] M.G. EASTWOOD and T. HÜBSCH: "Endomorphism Valued Cohomology and Gauge-Neutral Matter," *Comm. Math. Phys.* **132** (1990) 383.

[122] S. ELIZUR and A. SCHWIMMER: "$N=2$ Two-Dimensional Wess–Zumino Model on the Lattice," *Nucl. Phys.* **B226** (1983) 109–120.

M. CLAUDSON and M.B. HALPERN: "Supersymmetric Ground State Wave Functions," *Nucl. Phys.* **B250** (1985) 689–715.

[123] S. FERRARA and M. PORRATTI: "The Manifold of Scalar Background Fields in $\mathbb{Z}_n$ Orbifolds," *Phys. Lett.* **216B** (1989) 289.

M. CVETIČ, J. MOLERA and B. OVRUT: "Kähler Potentials for Matter Scalars and Moduli of $\mathbb{Z}_n$ Orbifolds." *Phys. Rev.* **D40** (1989) 1140.

[124] S. FERRARA and A. STROMINGER: "$N=2$ Spacetime Supersymmetry and Calabi–Yau Moduli Space," in the Proceedings of the "Strings '89" meeting in College Station, Texas.

[125] M. FISCHLER: "Young Tableaux Methods for Kronecker Products of Representations in the Classical Groups," *J. Math. Phys.* **22** (1981) 637.

[126] A. FONT, L.E. IBÁÑEZ and F. QUEVEDO: "$\mathbb{Z}_n \times \mathbb{Z}_m$ Orbifolds and Discrete Torsion," *Phys. Lett.* **217B** (1989) 272.

[127] I.B. Frenkel, H. Garland and G.J. Zuckerman: "Semi-Infinite Cohomology and String Theory," *Proc. Natl. Acad. Sci. USA* **83** (1986) 8442.
M.J. Bowick and S.G. Rajeev: "String Theory as The Kähler Geometry of Loop Space," *Phys. Rev. Lett.* **58** (1987) 535; *ibid.* **58(E)** (1987) 1158; "The Holomorphic Geometry of Closed Bosonic String Theory and DiffS^1," *Nucl. Phys.* **B293** (1987) 348.
L. Alvarez-Gaumé, C. Gomez and C. Reina: "Loop Groups, Grassmannians and String Theory," *Phys. Lett.* **190B** (1987) 55.
K. Pilch and N.P. Warner: "Holomorphic Structure of Superstring Vacua," *Class. Quant. Grav.* **4** (1987) 1183.
P. Oh and P. Ramond: "Curvature of Super-DiffS^1/S^1," *Phys. Lett.* **195B** (1987) 130.
D. Harari, D.K. Hong, P. Ramond and V.G.J. Rogers: "The Superstring, Diff S^1/S^1 and Holomorphic Geometry," *Nucl. Phys.* **B294** (1987) 556.

[128] D. Friedan: "Nonlinear Sigma Models in $2+\epsilon$ Dimensions," *Phys. Rev. Lett.* **45** (1980) 1057.
C.G. Callan, D. Friedan, E.J. Martinec and M.J. Perry: "Strings in Background Fields," *Nucl. Phys.* **B262** (1985) 593.
T. Curtright and C. Zachos: Geometry, Topology and "Supersymmetry in Nonlinear σ-Models," *Phys. Rev. Lett.* **53** (1984) 1799.

[129] D. Friedan and S. Shenker: "The Analytic Geometry of Two-Dimensional Conformal Field Theory," *Nucl. Phys.* **B281** (1987) 509.

[130] K. Fujikawa: "Path Integral Measure for Gauge Invariant Fermion Theories," *Phys. Rev. Lett.* **42** (1979) 1195.
L. Alvarez-Gaumé and P. Ginsparg: "The Structure of Gauge and Gravitational Anomalies," *Ann. Phys.* (NY) **161** (1985) 423.

[131] M. Gagnon and Q. Ho-Kim: "An Exhaustive List of Complete Intersection Calabi–Yau Manifolds," *Mod. Phys. Lett.* **A9** (1994) 2235–2243.

[132] S.J. Gates, Jr., R. Brooks and F. Muhammad: "Unidexterous Superspace: The Flax of (Super)Strings," *Phys. Lett.* **194B** (1987) 35.
S.J. Gates, Jr., and W. Siegel: "Leftons, Rightons, Non-Linear σ-Models and Superstrings," *Phys. Lett.* **206B** (1988) 631.
L. Mezincescu and R.I. Nepomechie: "Critical Dimensions for Chiral Bosons," *Phys. Rev.* **D37** (1988) 3067.
R.I. Nepomechie: "Chiral Bosonization," in *NATO Advanced Research Workshop on Superstrings,* p. 229 (Boulder, August, 1987).

[133] S.J. Gates, Jr. and F. Gieres: "Unidexterous Supergravity, Beltrami Parametrization and BRST Quantization," *Nucl. Phys.* **B320** (1989) 310.

[134] S.J. Gates and T. Hübsch: "Unidexterous Locally Supersymmetric Actions for Calabi–Yau Compactifications," *Phys. Lett.* **226** (1989) 100.

[135] S.J. Gates and T. Hübsch: "Calabi–Yau Heterotic Strings and Unidexterous σ-Models," *Nucl. Phys.* **B343** (1990) 741–774.

[136] S.J. Gates, Jr., C.M. Hull and M. Roček: "Twisted Multiplets and New Supersymmetric Non-Linear σ-Models," *Nucl. Phys.* **B248** (1984) 157.

[137] J.S. Gates Jr. and H. Nishino: "Manifestly Supersymmetric $O(\alpha')$ Superstring Corrections in New $D=10$, $N=1$ Supergravity Yang–Mills Theory," *Phys. Lett.* **173B** (1986) 52; *ibid.* **189B** (1987) 45; *Nucl. Phys.* **B291** (1987) 205.

S.J. Gates Jr. and S. Vashakidze: "On $D=10$, $N=1$ Supersymmetry, Superspace Geometry and Superstring Effects," *Nucl. Phys.* **B291** (1987) 172.

[138] D. Gepner: "Exactly Solvable String Compactifications on Manifolds of $SU(N)$ Holonomy," *Phys. Lett.* **199B** (1987) 380.

[139] D. Gepner: "Space-Time Supersymmetry in Compactified String Theory and Superconformal Models," *Nucl. Phys.* **B296** (1988) 732.

M. Lynker and R. Schimmrigk: "Heterotic String Compactification on $N=2$ Superconformal Theories With $c=9$," *Phys. Lett.* **208B** (1988) 216; "On the Spectrum of $(2,2)$ Compactification of the Heterotic String on Conformal Field Theories," *Phys. Lett.* **215B** (1988) 681.

[140] D. Gepner: "String Theory on Calabi–Yau Manifolds: the Three Generations Case," *Princeton Univ. report* (Dec. 1987), `arXiv:hep-th/9301089`.

[141] P. Goddard, A. Kent and D. Olive: "Virasoro Algebras and Coset Space Models," *Phys. Lett.* **152B** (1985) 88; "Unitary Representations of the Virasoro and Super-Virasoro Algebras," *Comm. Math. Phys.* **103** (1986) 105.

P. Goddard and A. Schwimmer: *Phys. Lett.* **206B** (1988) 62.

[142] H. Grauert: "Über Modifikationen und exceptionelle analytishe Mengen," *Math. Ann.* **146** (1962) 331.

M. Artin: "Algebraization of Formal Moduli. II. Existence of Modifications," *Ann. Math.* **91** (1970) 88.

[143] M.B. Green and J.H. Schwarz: "Anomaly Cancellations in Supersymmetric $D=10$ Gauge Theory and Superstring Theory," *Phys. Lett.* **149B** (1984) 117.

[144] P.S. Green: "Singular Degenerations of Calabi–Yau Manifolds and the Weil–Petersson Metric," *Proc. Am. Math. Soc.* **111** (1991) 599–605.

[145] P. Green and T. Hübsch: "Calabi–Yau Manifolds as Complete Intersections in Products of Complex Projective Spaces," *Comm. Math. Phys.* **109** (1987) 99.

[146] P. Green and T. Hübsch: "Polynomial Deformations and Cohomology of Calabi–Yau Manifolds," *Comm. Math. Phys.* **113** (1987) 505.

T. Hübsch: "Superstring Phenomenology and Cohomology on Calabi–Yau Manifolds," in *Superstrings, Unifying Theories and Cosmology 1987,* p. 164, G. Furlan *et al.*, eds. (World Scientific, Singapore, 1988).

[147] P. Green and T. Hübsch: "Calabi–Yau Hypersurfaces in Products of Semi-Ample Surfaces," *Comm. Math. Phys.* **115** (1988) 231.

[148] P. Green, T. Hübsch and A. Lütken: "All the Hodge Numbers for all Calabi–Yau Complete Intersections," *Class. Quant. Grav.* **6** (1989) 105.

[149] P.S. Green and T. Hübsch: "Possible Phase Transitions Among Calabi–Yau Compactifications," *Phys. Rev. Lett.* **61** (1988) 1163; "Connecting

Moduli Spaces of Calabi–Yau Threefolds," *Comm. Math. Phys.* **119** (1989) 431.

[150] P. Green and T. Hübsch: "Spacetime Variable Superstring Vacua," *Int. J. Mod. Phys.* **A9** (1994) 3203–3228.

[151] B.R. Greene: "Special Points in Three Generation Moduli Space," *Phys. Rev.* **D40** (1989) 1145.

[152] B.R. Greene: "Superconformal Compactifications in Weighted Projective Space," *Comm. Math. Phys.* **130** (1990) 335–355.

[153] B.R. Greene, K.H. Kirklin, P.J. Miron and G.G. Ross: "A Three-Generation Superstring Model," *Nucl. Phys.* **B278** (1986) 667.

[154] B.R. Greene, C.A. Lütken and G.G. Ross: "Couplings in the Heterotic Superconformal Three Generation Model," *Nucl. Phys.* **B325** (1989) 101.

S.F. Cordes and Y. Kikuchi: "Correlation Functions and Selection Rules in Minimal $N = 2$ String Compactifications," *Univ. of Texas A&M report* CTP-TAMU-92/88 (1988, unpublished); "Non-Renormalizable Terms From a Selection Rule in the Three Generation Model," *Mod. Phys. Lett.* **A4** (1989) 1365.

R.Schimmrigk: "Heterotic RG Flow Fixed Points With Non-Diagonal Affine Invariants," *Phys. Lett.* **229B** (1989) 227–237.

[155] B.R. Greene, D.R. Morrison and M.R. Plesser: "Mirror Manifolds in Higher Dimension," *Comm. Math. Phys.* **173** (1995) 559–598.

[156] B.R. Greene and M.R. Plesser: "Duality in Calabi–Yau Moduli Space," *Nucl. Phys.* **B338** (1990) 15–37, in *Essays on Mirror Symmetry*, p. 1, S.-T. Yau, ed. (Intl. Press, Hong Kong, 1992).

[157] B.R. Greene, S.-S. Roan and S.-T. Yau: "Geometric Singularities and Spectra of Landau–Ginzburg Models," *Comm. Math. Phys.* **142** (1991) 245–259.

[158] B. Greene, A. Shapere, C. Vafa and S.-T. Yau: "Stringy Cosmic Strings and Non-Compact Calabi–Yau Manifolds," *Nucl. Phys.* **B337** (1990) 1–36.

[159] B.R. Greene, C. Vafa and N.P. Warner: "Calabi–Yau Manifolds and Renormalization Group Flows," *Nucl. Phys.* **B324** (1989) 371.

[160] D. Gross, J. Harvey, E. Martinec and R. Rohm: "Heterotic String." *Phys. Rev. Lett.* **54** (1985) 502; "Heterotic String Theory, I," *Nucl. Phys.* **B265** (1985) 253, and "*ditto*, II," *ibid.* **B267** (1986) 75.

[161] D. Gross and E. Witten: "Superstring Modifications of Einstein's Equations," *Nucl. Phys.* **B277** (1986) 1.

M.T. Grisaru, A. van de Ven and D. Zanon: "Four Loop Beta Function for the $N = 1$ and $N = 2$ Supersymmetric Non-Linear σ-Model in Two Dimensions," *Phys. Lett.* **173B** (1986) 423; "Two-Dimensional Supersymmetric Models on Ricci-Flat Kähler Manifolds are not Finite," *Nucl. Phys.* **277** (1986) 388; "Four Loop Divergences for the $N = 1$ Supersymmetric Non-Linear σ-Model in Two Dimensions," *ibid.* **277** (1986) 409.

M.D. Freeman and C.N. Pope: "Beta Functions and Superstring Compactifications," *Phys. Lett.* **174B** (1986) 48.

M.D. FREEMAN, C.N. POPE, M.F. SOHNIUS and K.S. STELLE: "Higher Order σ-Model Counterterms and the Effective Action for Superstrings," *Phys. Lett.* **178B** (1986) 199.
C.N. POPE, M.F. SOHNIUS and K.S. STELLE: "Counterterm Counterexamples," *Nucl. Phys.* **B283** (1987) 192.

[162] J. HARRIS and L. TU: "On Symmetric and Skew-Symmetric Determinantal Varieties," *Topology* **23** (1984) 71–84; "Chern Numbers of Kernel and Cokernel Bundles," *Inv. Math.* **75** (1984) 467–475.

[163] A. HE and P. CANDELAS: "On the Number of Complete Intersection Calabi–Yau Manifolds," *Comm. Math. Phys.* **135** (1990) 193–199.

[164] F. HIRZEBRUCH and J. WERNER: "Some Examples of Threefolds With Trivial Canonical Class," *Max-Planck-Institut report* SFB/MPI-85-58.
J. WERNER: "Small Resolutions of Special Three-Dimensional Varieties," vol. 186 of *Bonner Math. Schriften,* 1997. PhD dissertation, Rheinische Friedrich-Wilhelms-Universität Bonn, 1994. arXiv:2208.01383.

[165] S. HOSONO, A. KLEM, S. THEISEN and S.-T. YAU: "Mirror Symmetry, Mirror Map and Applications to Calabi–Yau Hypersurfaces," *Comm. Math. Phys.* **167** (1995) 301–350; "Mirror Symmetry, Mirror Map and Applications to Complete Intersection Calabi–Yau Spaces," *Nucl. Phys.* **B433** (1995) 501–554, and *AMS/IP Stud. Adv. Math.* **1** (1996) 545.

[166] Y. HOSOTANI: *Phys. Lett.* **129B** (1983) 193.
T. HÜBSCH: "Flux-Lines Through Calabi–Yau Manifolds and Related Couplings," *J. Phys.* **A21** (1988) 3051.
G.G. ROSS: "Hierarchy Generation in Compactified Supersymmetric Models," *Phys. Lett.* **221B** (1988) 315.

[167] C.M. HULL: "Covariant Quantization of Chiral Bosons and Anomaly Cancellation," *Phys. Lett.* **206B** (1988) 234; "Chiral Conformal Field Theory and Asymmetric String Compactification," *ibid.* **212B** (1988) 437.
D.A. DEPIREUX, S.J. GATES, JR. and B. RADAK: "Yes, Leftons for Heterotic Superstrings!" *Phys. Lett.* **236B** (1990) 411.
see also Ref. [109].

[168] T. HÜBSCH: "Calabi–Yau Manifolds—Motivations and Constructions," *Comm. Math. Phys.* **108** (1987) 291; "Constructions of Calabi–Yau Manifolds," in *Superstrings, Anomalies and Unification*, p. 495, M. Martinis and I. Andrić, eds. (World Scientific, Singapore, 1987); "Manifold Compactifications of Superstrings," in *Superstrings, Unified Theories and Cosmology,* p. 274, G.Furlan *et al.*, eds. (World Scientific, Singapore, 1987).

[169] T. HÜBSCH: "Chameleonic σ-Models," *Phys. Lett.* **247B** (1990) 317–321.

[170] T. HÜBSCH: "How Singular a Space Can Superstrings Thread?" *Mod. Phys. Lett.* **A6** (1991) 207–216.

[171] T. HÜBSCH: "Of Marginal Kinetic Terms and Anomalies," *Mod. Phys. Lett.* **A6** (1991) 1553–1559.

[172] T. HÜBSCH: "Elusive Conifold Compactification," *Class. Quant. Grav.* **8** (1991) L31–L35.

[173] T. HÜBSCH and S.-T. YAU: "An $SL(2,\mathbb{C})$ Action on Certain Jacobian Rings and the Mirror Map," in *Essays in Mirror Symmetry,* S.-T. Yau,

ed.(Intl. Press, Hong-Kong, 1992); "An $SL(2,\mathbb{C})$ Action on Chiral Rings and the Mirror Map," *Mod. Phys. Lett.* **A7** (1992) 3277.

[174] S. Kachru and E. Witten: "Computing the Complete Massless Spectrum of a Landau–Ginzburg Orbifold," *Nucl. Phys.* **B407** (1993) 637–666.

[175] S. Kalara and R.N. Mohapatra: "Yukawa Couplings and Phenomenology of a Three Generation Superstring Model," *Phys. Rev.* **D36** (1987) 3474, *ibid.* **D38 (E)** (1988) 411.

[176] S. Kalara and R.N. Mohapatra: "CP Violation and Yukawa Couplings in Superstring Models: A Four-Generation Example," *Phys. Rev.* **D35** (1987) 3143, *Z. Physik* **C37** (1988) 395.

[177] V. Kaplunovsky: "Mass Scales of the String Unification," *Phys. Rev. Lett.* **55** (1985) 1036.

[178] V. Kaplunovsky and J. Louis: "Field Dependent Gauge Couplings in Locally Supersymmetric Effective Quantum Field Theories," *Nucl. Phys.* **B422** (1994) 57–124.

[179] D. Karabali, Q-H. Park and H.J. Schnitzer: "A GKO Construction Based on a Path Integral Formulation of Gauged Wess–Zumino–Witten Actions," *Phys. Lett.* **216B** (1989) 307.
H.J. Schnitzer: "A Path-Integral Construction of Superconformal Field Theories From a Gauged Supersymmetric Wess–Zumino–Witten Action," *Nucl. Phys.* **B324** (1989) 412.
K. Gawędzki and A. Kupiainen: "Coset Construction From Functional Integrals," *Nucl. Phys.* **B320** (1989) 625.

[180] A. Kasparian: "Calabi–Yau Manifolds of Some Special Forms," *Lett. Math. Phys.* **15** (1988) 171–174.

[181] J.K. Kim, I.G. Koh and Y. Yoon: "Calabi–Yau Manifolds From Arbitrary Weighted Projective Spaces," *Phys. Rev.* **D33** (1986) 2893.

[182] A. Klemm, W. Lerche and S. Theisen: "Nonperturbative Effective Actions of $N=2$ Supersymmetric Gauge Theories," *Int. J. Mod. Phys.* **A11** (1996) 1929–1974.

[183] M. Kreuzer: "The Mirror Map for Invertible LG Models," *Phys. Lett.* **328B** (1994) 312–318.

[184] M. Kreuzer and H. Skarke: "On the Classification of Quasihomogeneous Functions," *Comm. Math. Phys.* **150** (1992) 137; "No Mirror Symmetry in Landau–Ginzburg Spectra!" *Nucl. Phys.* **B388** (1992) 113–130.
A. Klemm and R. Schimmrigk: "Landau–Ginzburg String Vacua," *Nucl. Phys.* **B411** (1994) 559–583.

[185] J.I. Latorre and C.A. Lütken: "Constrained CP^N Models," *Phys. Lett.* **222B** (1989) 55.

[186] G. Lazarides and Q. Shafi: "Three-Generation Superstring Models With Maximal Discrete Symmetries," *J. Math. Phys.* **36** (1989) 711.

[187] W. Lerche, D.-J. Smit and N. Warner: "Differential Equations for Periods and Flat Coordinates in Two Dimensional Topological Matter Theories," *Nucl. Phys.* **B372** (1992) 87–112.

[188] C.A. Lütken: "Geometry of the Z-Fold," *J. Phys.* **A21** (1988) 1889.

[189] D. Markushevich, M. Olshanetsky and A. Perelomov: "Description of a Class of Superstring Compactifications Related to Semi-Simple Lie Algebras," *Comm. Math. Phys.* **111** (1987) 247.

[190] E. Martinec: "Non-Renormalization Theorems and Fermionic String Finiteness," *Phys. Lett.* **171B** (1986) 189.

[191] E. Martinec: "Algebraic Geometry and Effective Lagrangians," *Phys. Lett.* **217** (1989) 431.

[192] E. Martinec: "Criticality, Catastrophes and Compactifications," in *Physics and Mathematics of Strings*, p. 389–433, L. Brink *et al.*, eds. (World Scientific, Singapore, 1990).

[193] B. McInnes: "Superstrings and Holonomy Groups of Kähler Manifolds," *Class. Quant. Grav.* **5** (1988) 561.

[194] D.R. Morrison: "Picard–Fuchs Equations and Mirror Maps for Hypersurfaces," in *Essays on Mirror Manifolds* p. 241–264, S.-T. Yau, ed. (International Press, Hong Kong, 1992).

[195] K.S. Narain: "New Heterotic String Theories in Uncompactified Dimensions <10," *Phys. Lett.* **169B** (1986) 41.
A.N. Schellekens and N.P. Warner: "Weyl Groups, Supercurrents and Covariant Lattices," *Nucl. Phys.* **B308** (1988) 397.
W. Lerche, D. Lüst and A.N. Schellekens: "Chiral Four-Dimensional Heterotic String From Self-Dual Lattices," *Nucl. Phys.* **B287** (1987) 477.
W. Lerche and A.N. Schellekens: "The Covariant Lattice Construction of Four-Dimensional Strings," *CERN report* CERN-TH.4925 (1987), also in *Proceedings of the Summer School on Strings and Superstrings*, Poiana Brasov, Romania, Sep. 1–12, 1987.
(See Ref. [119]).

[196] D. Nemeschansky and A. Sen: "Conformal Invariance of Supersymmetric σ-Models on Calabi–Yau Manifolds," *Phys. Lett.* **178B** (1986) 365.
P. Candelas, M.D. Freeman, C. Pope, M.F. Sohnius and K.S. Stelle: "Higher Order Corrections to Supersymmetry and Compactifications of the Heterotic String," *Phys. Lett.* **177B** (1986) 341.

[197] H. Osborn: "String Theory Effective Actions From Bosonic σ-Models," *Nucl. Phys.* **308B** (1988) 629.

[198] N.K. Pak and R. Percacci: "Hamiltonian Methods for Non-Linear σ-Models," *J. Math. Phys.* **30** (1989) 2951.

[199] L. Pilch and A.N. Schellekens: "Fermion Spectra From Superstrings," *Nucl. Phys.* **B259** (1985) 637.

[200] H. Pinkham: "Factorization of Birational Maps in Dimension 3," *Proc. Sym. Pure Math.* **40(2)** (1983) 343.

[201] A.M. Polyakov: "Quantum Geometry of Bosonic Strings," *Phys. Lett.* **103B** (1981) 207: "Quantum Geometry of Fermionic Strings." *ibid.* **103B** (1981) 211.

[202] M. Reid: "Tendencious Survey of 3-Folds," *Proc. Symp. Pure Math.* **46** (1987) 333–344.

[203] M. Reid: "The Moduli Space of 3-Folds With $K\equiv 0$ May Nevertheless be Irreducible," *Math. Ann.* **278** (1987) 329.

[204] S.-S. Roan: "On Calabi–Yau Orbifolds in Weighted Projective Spaces," *Int. J. Math.* **1** (1990) 211–232.

[205] S.-S. Roan: "The Mirror of Calabi–Yau Orbifold," *Int. J. Math.* **2** (1991) 439–455.

[206] S.-S. Roan and S.-T. Yau: "On Ricci-Flat 3-Folds," *Acta Math. Sin.* **3** (1987) 256.

[207] M. Roček: "Supersymmetry and Nonlinear σ-Models," *Physica* **15D** (1985) 75–82; "Modified Calabi–Yau Manifolds With Torsion," in *Essays on Mirror Manifolds* p. 480–488, S.-T. Yau, ed. (International Press, Hong Kong, 1992).

[208] A. Salam: "Lagrangian Theory of Composite Particles," *il N. Cim.* **25** (1962) 224.

S. Weinberg: "Elementary Particle Theory of Composite Particles," *Phys. Rev.* **130** (1963) 776; "Evidence That the Deuteron is Not an Elementary Particle," *ibid.* **137** (1965) B672.

[209] A.N. Schellekens and N.P. Warner: "Anomalies, Characters and Strings," *Nucl. Phys.* **B287** (1987) 317.

[210] R. Schimmrigk: "A New Construction of a Three Generation Calabi–Yau Manifold," *Phys. Lett.* **193B** (1987) 175.

[211] C. Schoen: "On Fiber Products of Rational Elliptic Surfaces With Section," *J. für Math.* **364** (1986) 85–111.

[212] C. Schoen: "On Fiber Products of Rational Elliptic Surfaces With Section," *Math.Z.* **197** (1988) 177.

[213] N. Seiberg: "Observations on the Moduli Space of Superconformal Field Theories," *Nucl. Phys.* **B303** (1988) 286.

[214] N. Seiberg and E. Witten: "Spin Structures in String Theory," *Nucl. Phys.* **B276** (1986) 272.

[215] A. Sen: "Heterotic String Theory on Calabi–Yau Manifolds in the Green–Schwarz Formalism," *Nucl. Phys.* **284B** (1987) 423.

[216] I.R. Shafarevich: (English translation) in *Math. Int.* **3** (1980/81), no. 4, p. 182–184.

[217] W. Siegel: "Manifest Lorentz Invariance Sometimes Requires Non-Linearity," *Nucl. Phys.* **B238** (1984) 307.

[218] A. Strominger: "Yukawa Couplings in Superstring Compactification," *Phys. Rev. Lett.* **55** (1985) 2547.

E. Cremmer, C. Kounnas, A. van Proeyen, J.P. Derendinger and S. Ferrara: "Vector Multiplets Coupled to $N=2$ Supergravity: Super-Higgs Effect, Flat Potentials and Geometric Structure," *Nucl. Phys.* **B250** (1985) 385.

[219] A. Strominger: "Special Geometry," *Comm. Math. Phys.* **133** (1990) 163–180.

[220] A. Strominger and E. Witten: "New Manifolds for Superstring Compactification," *Comm. Math. Phys.* **101** (1985) 341.

[221] W.-W. Sung: "On Calabi–Yau 3-Folds Fibered Over Compact Complex Surfaces," *Proc. Symp. Pure Math.* **54**, Part 2, (1993) 577–594.

[222] G. TIAN: "Smoothness of the Universal Deformation Space of Compact Calabi–Yau Manifolds and its Petersson Weil Metric," in *Mathematical Aspects of String Theory*, p. 629–646, S.-T. Yau, ed. (World Scientific, Singapore, 1987).

[223] G. TIAN and S.-T. YAU: "Three-Dimensional Algebraic Manifolds With $c_1 = 0$ and $\chi_E = -6$," in *Mathematical Aspects of String Theory*, p. 543–559, S.-T. Yau, ed. (World Sci., Singapore, 1987).

[224] G. TIAN and S.-T. YAU: "Existence of Kähler–Einstein Metrics on Complete Kähler Manifolds and Their Applications to Algebraic Geometry," in *Mathematical Aspects of String Theory*, p. 574–628, S.-T. Yau, ed. (World Scientific, Singapore, 1987).

[225] A.N. TODOROV: "The Weil–Petersson Geometry of the Moduli Space of $SU(n \geq 3)$ (Calabi–Yau) Manifolds I," *Comm. Math. Phys.* **126** (1989) 325–246.

[226] K. UHLENBECK and S.-T. YAU: "On the Existence of Hermitian Yang–Mills Connections in Stable Vector Bundles," *Comm. Pure and Appl. Math.* **39** (1986) S257–S293.

[227] C. VAFA: "String Vacua and Orbifoldized Landau–Ginzburg Models," *Mod. Phys. Lett.* **A4** (1989) 1169.
K. INTRILLIGATOR and C. VAFA: "Landau–Ginzburg Orbifolds," *Nucl. Phys.* **B339** (1990) 95.

[228] C. VAFA: "Quantum Symmetries of String Vacua," *Mod. Phys. Lett.* **A4** (1989) 1615.

[229] C. VAFA and N. WARNER: "Catastrophes and the Classification of Conformal Theories," *Phys. Lett.* **218B** (1989) 51.
W. LERCHE, C. VAFA and N. WARNER: "Chiral Rings in $N=2$ Superconformal Theories," *Nucl. Phys.* **B324** (1989) 427.

[230] C.T.C. WALL: "Is Every Quartic a Conic of Conics?" *Math. Proc. Cambridge Phil. Soc.* **109** (3) (1991) 419–424.

[231] H.-C. WANG: "Closed Manifolds With Homogeneous Complex Structure," *Am. J. Math.* **76** (1956) 1–32.

[232] X.G. WEN and E. WITTEN: "World-Sheet Instantons and the Peccei–Quinn Symmetry," *Phys. Lett.* **166B** (1986) 397.

[233] C. WETTERICH: "The Cosmological Constant and Non-Compact Internal Spaces in Kałuża–Klein Theories," *Nucl. Phys.* **B255** (1985) 480; "Fermion Chirality From Higher Dimensions and Kałuża–Klein Theories With Non-Compact Internal Space," in the *Proceedings of the Jerusalem Winter School 1984*, p. 204.
H. NICOLAI and C. WETTERICH: "The Spectrum of Kałuża–Klein Theories With Non-Compact Internal Spaces," *Phys. Lett.* **150B** (1985) 347.
S. RANDJBAR-DAEMI and C. WETTERICH: "Kałuża–Klein Solutions With Non-Compact Internal Spaces," *Phys. Lett.* **166B** (1985) 65.

[234] P.M.H. WILSON: "Calabi–Yau Manifolds With Large Picard Number," *Invent. Math.* **98** (1989) 139–155.

[235] P.M.H. WILSON: "The Kähler Cone on Calabi–Yau Threefolds," *Inv. Math.* **107** (1) (1992) 561–583.

[236] E. WITTEN: "Supersymmetry and Morse Theory," *J. Diff. Geom.* **17** (1982) 661–692.
[237] E. WITTEN: "Constraints on Supersymmetry Breaking," *Nucl. Phys.* **B202** (1982) 253–316.
[238] E. WITTEN: "Dimensional Reduction of Superstring Models," *Phys. Lett.* **155B** (1985) 151.
[239] E. WITTEN: "Symmetry Breaking Patterns in Superstring Models," *Nucl. Phys.* **B258** (1985) 75.
[240] E. WITTEN: "New Issues in Manifolds of $SU(3)$ Holonomy," *Nucl. Phys.* **B268** (1986) 79.
M. DINE and N. SEIBERG: "Non-Renormalization Theorems in Superstring Theory," *Phys. Rev. Lett.* **57** (1986) 2625.
M. DINE, N. SEIBERG and E. WITTEN: "Fayet–Iliopoulos Terms in String Theory," *Nucl. Phys.* **B289** (1987) 589.
see also Ref. [190].
[241] E. WITTEN: "Phases of $N=2$ Theories in Two Dimensions," *Nucl. Phys.* **B403** (1993) 159.
[242] S.-T. YAU: "Compact Three-Dimensional Kähler Manifolds With Zero Ricci Curvature," in Proc. of *Symposium on Anomalies, Geometry, Topology*, p. 395, esp. the Appendix, by G. TIAN and S.-T. YAU, pp. 402–405, W.A. Bardeen and A.R. White, eds. (World Scientific, Singapore, 1985).
[243] A.B. ZAMOLODCHIKOV: "'Irreversibility' of the Flux of the Renormalization Group in a 2-Dimensional Field Theory," *JEPT Lett.* **43** (1986); "Renormalization Group and Perturbation Theory Near Fixed Points in Two-Dimensional Field Theory," *Yad. Phys.* **46** (1987) 1819–1831.
[244] B. ZUMINO: "Supersymmetry and Kähler Manifolds," *Phys. Lett.* **87B** (1979) 203–211.

Books and Key References Added in the 2024 Edition:

(the leading symbol ❦ indicates books I would recommend for further reading)

[245] *Applicable Algebra in Engineering, Communication and Computing* (Springer, New York, since 1990).
[246] *Graduate Texts in Mathematics* (Springer, Berlin and New York, since 1982) https://en.wikipedia.org/wiki/Graduate_Texts_in_Mathematics.
[247] S.S. ABHYANKAR: *Algebraic Geometry for Scientists and Engineers* (AMS, Providence, RI 35, 1990).
[248] M. AMMON and J. ERDMENGER: *Gauge/Gravity Duality: Foundations and Applications* (Cambridge University Press, Cambridge, 2015).
[249] ❦P.S. ASPINWALL, T. BRIDGELAND, A. CRAW, M.R. DOUGLAS, A. KAPUSTIN, G.W. MOORE, M. GROSS, G. SEGAL, B. SZENDRŐI and P.M.H. WILSON: *Dirichlet Branes and Mirror Symmetry*, vol. 4 of *Clay Mathematics Monographs* (AMS, Providence, RI, 2009).
[250] M. BANAGL: *Intersection Spaces, Spatial Homology Truncation, and String Theory*, *Lecture Notes in Math.* **1997** (Springer, New York, 2010).
[251] K. BECKER, M. BECKER and J.H. SCHWARZ: *String Theory and M-Theory: A Modern Introduction*, (Cambridge University Press, Cambridge, 2007).

[252] R.L. Bryant: "Remarks on the Geometry of Almost Complex 6-Manifolds," *Asian J. of Math.* **10** (3) (2006) 561–606.

[253] E. Calabi: "The Space of Kähler Metrics," *Proc. Internat. Congress Math. Amsterdam,* **2**, pp. 206–207 (1954); "On Kähler Manifolds With Vanishing Canonical Class," in *Algebraic Geometry and Topology. A Symposium in Honor of S. Lefschetz,* pp. 78–89, R.H. Fox, D.C. Spencer, A.W. Tucker, eds. *Princeton Legacy Lib.* **1873**, (Princeton University Press, Princeton, 1957; repub. 2015).

[254] E. Cattani, F. El Zein, P.A. Griffiths and D.T. Lê, eds.: *Hodge Theory,* vol. 49 of *Mathematical Notes* (Princeton University Press, Princeton, NJ, 2014).

[255] ❦D.A. Cox and S.H. Katz: *Mirror Symmetry and Algebraic Geometry,* vol. 68 of *Mathematical Surveys and Monographs,* (AMS, Providence, RI, 1999).

[256] D.A. Cox, J.B. Little and D. O'Shea: *Ideals, Varieties, and Algorithms: An Introduction to Computational Algebraic Geometry and Commutative Algebra* (Springer, New York, 2007); *Using Algebraic Geometry, Grad. Texts in Math.* no. 185 (Springer, New York, 2005).

[257] D.A. Cox, J.B. Little and H.K. Schenck: *Toric Varieties,* vol. 124 of *Grad. Studies in Math.* (AMS, Providence, RI, 2011).

[258] V.I. Danilov: "The Geometry of Toric Varieties," *Russian Math. Surv.* **33** (2) (1978) 97–154.

[259] ❦P. Deligne, P. Etingof, D.S. Freed, L.C. Jeffrey, D.K. Kazhdan, J.W. Morgan, D.R. Morrison and E. Witten, eds.: *Quantum Fields and Strings: a Course for Mathematicians, Vol. 1 & 2,* (AMS, Providence, RI, 1999).

[260] G. Ewald: *Combinatorial Convexity and Algebraic Geometry* (Springer, New York, 1996).

[261] E. Fradkin: *Quantum Field Theory: An Integrated Approach* (Princeton University Press, Princeton, 2021).

[262] D.S. Freed: *Five Lectures on Supersymmetry* (AMS, Providence, RI, 1999).

[263] K. Fukaya, Y.G. Oh, H. Ohta and K. Ono: *Lagrangian Intersection Floer Theory: Anomaly and Obstructions, Part I & II, Stud. Adv. Math.* **46** (AMS, Providence, RI and Intl. Press, Cambridge, MA, 2009); "Lagrangian Floer Theory on Compact Toric Manifolds, I," *Duke Math. J.* **151** (1) (2010) 23–174; "Lagrangian Floer Theory on Compact Toric Manifolds, II: Bulk Deformations," *Sel. Math. (N.S.)* **17** (3) (2011) 609–711; *Lagrangian Floer Theory and Mirror Symmetry on Compact Toric Manifolds, Astérisque* **376** (Société Mathématique de France, 2016); "Lagrangian Floer Theory on Compact Toric Manifolds: A Survey," *Surv. Diff. Geom.* **17** (2012) 229–298.

[264] I.M. Gelfand, M.M. Kapranov and A.V. Zelevinsky: *Discriminants, Resultants and Multidimensional Determinants* (Birkhäuser, Boston, MA, 1994).

[265] J. Harris: *Algebraic Geometry: a First Course*, *Grad. Texts in Math.* no. 133 (Springer, New York, 1992).

[266] Y.-H. He: *The Calabi–Yau Landscape: From Geometry, to Physics, to Machine Learning*, *Lecture Notes in Math.* **2293** (Springer, New York, 2021).

[267] Y.-H. He, ed.: *Machine Learning In Pure Mathematics and Theoretical Physics* (World Scientific, Singapore, 2023).

[268] Y.-H. He, P. Candelas, A. Hanany, A. Lukas and B. Ovrut, eds.: *Computational Algebraic Geometry in String and Gauge Theory*, *Adv. High E. Phys.* Special Issue 431898 (Hindawi, London, 2012).

[269] F. Hirzebruch: "Über eine Klasse von einfach-zusammenhängenden komplexen Mannigfaltigkeiten," *Math. Ann.* **124** (1951) 77–86.

[270] ❦K. Hori, S.H. Katz, A. Klemm, R. Pandharipande, R. Thomas, C. Vafa, R. Vakil and E. Zaslow: *Mirror Symmetry*, vol. 1 of *Clay Mathematics Monographs* (AMS, Providence, RI, 2003).

[271] A. Klemm: "The B-Model Approach to Topological String Theory on Calabi–Yau n-Folds," pp. 80–398 in *B-Model Gomov-Witten Theory*, E. Clader and Y. Ruan, eds., Trends in Mathematics (Birkhäuser, Cham, Switzerland, 2018).

[272] T. Hübsch: *Calabi–Yau Manifolds: A Bestiary for Physicists* (World Scientific, Singapore, 1992).

[273] T. Hübsch: *Advanced Concepts in Particle and Field Theory* (Cambridge University Press, 2015); since 2022 freely available via Open Access Cambridge Core and mirrored at inSPIRE-hep.

[274] L. Ji, B. Wu and S.-T. Yau, eds.: *Handbook for Mirror Symmetry of Calabi–Yau and Fano Manifolds*, vol. 47 of *Adv. Lect. Math. (ALM)*, (Intl. Press, Somerville, MA; Higher Ed. Press, Beijing, 2020).

[275] C. Johnson: *D-Branes* (Cambridge University Press, Cambridge, 2006).

[276] S. Katz: *Enumerative Geometry and String Theory*, *Student Math. Lib.* vol. 32 (AMS, Providence, RI, 2006).

[277] E. Kiritsis: *String Theory in a Nutshell* (Princeton University Press, Princeton, 2007).

[278] R. Lazarsfeld: *Positivity in Algebraic Geometry, vol. I & II* (Springer, New York, 2004).

[279] M. Kontsevich: "Homological Algebra of Mirror Symmetry," in *Proceedings of the International Congress of Mathematicians*, S.D. Chatterji, ed., pp. 120–139 (Birkhäuser Basel, Basel, 1995); "Mirror symmetry in dimension 3," *Séminaire Bourbaki (1994–95) exp. n° 801* in *Astérisque* **237** (1996) 275–293.
M. Kontsevich and Y. Soibelman: "Homological Mirror Symmetry and Torus Fibrations," in *Symplectic Geometry and Mirror Symmetry*, Proceedings of the KIAS Annual International Conference on Symplectic Geometry and Mirror Symmetry, pp. 203–263. (World Scientific, Singapore, 2000).

A. Kapustin, M. Kreuzer and K. G. Schlesinger, eds.: *Homological Mirror Symmetry: New Developments and Perspectives*, *Lecture Notes in Physics* **757** (Springer Berlin Heidelberg, 2009).

M. Gross and B. Siebert: "Intrinsic Mirror Symmetry," `arXiv:1909.07649`.

[280] H. Nastase: *Introduction to the ADS/CFT Correspondence* (Cambridge University Press, Cambridge, 2015).

[281] M. Natsuume: *AdS/CFT Duality User Guide*, *Lecture Note in Phys.* vol. 903 (Springer, New York, 2015).

[282] M.E. Peskin and D.V. Schroeder: *An Introduction to Quantum Field Theory* (Addison-Wesley, Reading, MA, 1995).

[283] ❦J. Polchinski: "TASI Lectures on D-Branes," pp. 293–356 in *Proceedings,* **TASI** *(1996): Fields, Strings, and Duality*, `arXiv:hep-th/9611050`.

[284] ❦J. Polchinski: *String Theory, Vol. 1 & 2*, *Cambridge Monographs on Mathematical Physics* (Cambridge University Press, Cambridge, 1998).

[285] M. Reid: *Undergraduate Algebraic Geometry* (Cambridge University Press, Cambridge, 1988).

[286] M. Reid: "Update on 3-Folds," in *Proceedings of the ICM*, vol. 2, pp. 513–524, (Higher Ed. Press, Beijing, 2002).

[287] H. Schenck: *Computational Algebraic Geometry* (Cambridge University Press, Cambridge, 2003).

[288] ❦R.J. Szabó: *An Introduction to String Theory and D-Brane Dynamics: With Problems and Solutions*, 2nd ed., (Imperial College Press, London, 2011).

[289] L. Susskind: *The Cosmic Landscape: String Theory and the Illusion of Intelligent Design*, (Back Bay Books, 2006).

[290] V.S. Varadarajan: *Supersymmetry for Mathematicians: An Introduction* (AMS, Providence, RI, 2004).

[291] S.-T. Yau, ed.: *Essays on Mirror Manifolds*, (Intl. Press, Cambridge, MA, 1992).

❦S.-T. Yau, ed.: *Mirror Symmetry I*, vol. 9 of *Studies in Advanced Mathematics*, corrected reprint of the 1992 "Essays on Mirror Manifolds" (here listed just above) (AMS, Providence, RI and Intl. Press, Cambridge, MA, 1998).

❦B. Greene and S.-T. Yau, eds.: *Mirror Symmetry II*, vol. 1 of *Studies in Advanced Mathematics*, (AMS, Providence, RI and Intl. Press, Cambridge, MA, 1997).

❦D.H. Phong, L. Vinet and S.-T. Yau, eds.: *Mirror Symmetry III*, vol. 10 of *Studies in Advanced Mathematics*, (AMS, Providence, RI and Intl. Press, Cambridge, MA, 1999).

❦E. D'Hoker, D. Phong and S.-T. Yau, eds.: *Mirror Symmetry IV*, vol. 33 of *Studies in Advanced Mathematics*, (AMS, Providence, RI and Intl. Press, Cambridge, MA, 2002).

❦N. Yui, S.-T. Yau and J.D. Lewis, eds.: *Mirror Symmetry V*, vol. 38 of *Studies in Advanced Mathematics*, (AMS, Providence, RI and Intl. Press, Cambridge, MA, 2006).

[292] E. Zaslow: "Mirror Symmetry," in *The Princeton Companion to Mathematics*, pp. 523–539, T. Gowers, J. Barrow-Green and I. Leader, eds. (Princeton University Press, July, 2010).

[293] ❦B. Zwiebach: *A First Course in String Theory* (Cambridge University Press, Cambridge, 2004).

Research Articles Added in the 2024 Edition:

(a woefully incomplete collection of literature search pointers)

[294] M. Alexandrov, M. Kontsevich, A. Schwarz and O. Zaboronsky: "The Geometry of the Master Equation and Topological Quantum Field Theory," *Int. J. Mod. Phys.* **A12** (1997) 1405–1430.
A.S. Cattaneo and G. Felder: "On the AKSZ Formulation of the Poisson Sigma Model," *Lett. Math. Phys.* **56** (2001) 163–179.
Z. Kökényesi, A. Sinkovics and R.J. Szabo: "Double Field Theory for the A/B-Models and Topological S-Duality in Generalized Geometry," *Fortsch. Phys.* **66** (11–12) (2018) 1800069.

[295] R. Altman, J. Gray, Y.-H. He, V. Jejjala and B.D. Nelson: "A Calabi–Yau Database: Threefolds Constructed From the Kreuzer-Skarke List," *J. High E. Phys.* **02** (2015) 158.

[296] L.B. Anderson, F. Apruzzi, X. Gao, J. Gray and S.-J. Lee: "A New Construction of Calabi–Yau Manifolds: Generalized CICYs," *Nucl. Phys.* **B906** (2016) 441–496.

[297] L.B. Anderson, C.R. Brodie and J. Gray: "Branes and Bundles Through Conifold Transitions and Dualities in Heterotic String Theory," *Phys. Rev.* **D108** (10) (2023) 106019.

[298] L.B. Anderson, X. Gao, J. Gray, S.-J. Lee: "Fibrations in CICY Threefolds," *J. High E. Phys.* **10** (2017) 077.

[299] L.B. Anderson, M. Gerdes, J. Gray, S. Krippendorf, N. Raghuram and F. Ruehle: "Moduli-Dependent Calabi–Yau and $SU(3)$-Structure Metrics From Machine Learning," *J. High E. Phys.* **05** (2021) 013.

[300] B. Andreas and M. Garcia-Fernandez: "Solutions of the Strominger System via Stable Bundles on Calabi–Yau Threefolds," *Comm. Math. Phys.* **315** (2012) 153–168.
L.B. Anderson, J. Gray and E. Sharpe: "Algebroids, Heterotic Moduli Spaces and the Strominger System," *J. High E. Phys.* **07** (2014) 037.
X.C. de la Ossa and E.E. Svanes: "Holomorphic Bundles and the Moduli Space of $N=1$ Supersymmetric Heterotic Compactifications," *J. High E. Phys.* **10** (2014) 123.
M. Garcia-Fernandez: "Ricci Flow, Killing Spinors, and T-Duality in Generalized Geometry," *Adv. Math.* **350** (2019) 1059–1108.
T. Fei, Z. Huang and S. Picard: "A Construction of Infinitely Many Solutions to the Strominger System," *J. Diff. Geom.* **117** (1) (2021) 23–39.

A. Fino, G. Grantcharov and L. Vezzoni: "Solutions to the Hull–Strominger System with Torus Symmetry," *Comm. Math. Phys.* **388** (2) (2021) 947–967.

[301] M. Artebani, P. Comparin and R. Guilbot: "Families of Calabi–Yau Hypersurfaces in **Q**-Fano Toric Varieties," *J. Math. Pure & Appl.* **106** (2) (2016) 319–341.

[302] A. Ashmore, Y.-H. He, E. Heyes and B.A. Ovrut: "Numerical Spectra of the Laplacian for Line Bundles on Calabi–Yau Hypersurfaces," *J. High E. Phys.* **07** (2023) 164.

[303] P.S. Aspinwall and R.Y. Donagi: "The Heterotic String, the Tangent Bundle, and Derived Categories," *Adv. Th. Math. Phys.* **2** (1998) 1041–1074; P.S. Aspinwall and M.R. Douglas: "D-Brane Stability and Monodromy," *J. High E. Phys.* **05** (2002) 031.
E. Sharpe: "Lectures on D-Branes and Sheaves," `arXiv:hep-th/0307245`.
P.S. Aspinwall: "D-Branes on Calabi–Yau Manifolds," pp. 1–152 in *Proceedings,* **TASI** *(2003): Recent Trends in String Theory*, `arXiv:hep-th/0403166`.

[304] P.S. Aspinwall, B.R. Greene and D.R. Morrison: "Multiple Mirror Manifolds and Topology Change in String Theory," *Phys. Lett.* **B303** (1993) 249–259; "Calabi–Yau Moduli Space, Mirror Manifolds and Space-Time Topology Change in String Theory," *Nucl. Phys.* **B416** (1994) 414–480.

[305] A.C. Avram, P. Candelas, D. Jančić and M. Mandelberg: "On the Connectedness of Moduli Spaces of Calabi–Yau Manifolds," *Nucl. Phys.* **B465** (1996) 458–472.
A.C. Avram, M. Kreuzer, M. Mandelberg and H. Skarke: "The Web of Calabi–Yau Hypersurfaces in Toric Varieties," *Nucl. Phys.* **B505** (1997) 625–640.

[306] S. Banerjee, U. Danielsson, G. Dibitetto, S. Giri and M. Schillo: "Emergent de Sitter Cosmology From Decaying Anti-de Sitter Space," *Phys. Rev. Lett.* **121** (26) (2018) 261301 (6p); "de Sitter Cosmology on an Expanding Bubble," *J. High E. Phys.* **10** (2019) 164.
G. Dibitetto, N. Petri and M. Schillo: "Nothing Really Matters," *J. High E. Phys.* **08** (2020) 040.

[307] I.A. Batalin and I. Tyutin: "BRST-Invariant Algebra of Constraints in Terms of Commutators and Quantum Antibrackets," *Theor. Math. Phys.* **138** (1) (2004) 1–17.

[308] I.A. Batalin and G.A. Vilkovisky: "Gauge Algebra and Quantization," *Phys. Lett.* **B102** (1981) 27–31; "Quantization of Gauge Theories With Linearly Dependent Generators," *Phys. Rev.* **D28** (1983) 2567–2582.

[309] V.V. Batyrev and L.A. Borisov: "On Calabi–Yau Complete Intersections in Toric Varieties," in *Higher-Dimensional Complex Varieties (Trento, 1994)*, pp. 39–65 (de Gruyter, Berlin, 1996).

[310] V.V. Batyrev, I. Ciocan-Fontanine, B. Kim and D. van Straten: "Conifold Transitions and Mirror Symmetry for Calabi–Yau Complete Intersections in Grassmannians," *Nucl. Phys.* **B514** (1998) 640–666.

[311] V.V. Batyrev and M. Kreuzer: "Constructing New Calabi–Yau 3-Folds and Their Mirrors via Conifold Transitions," arXiv:0802.3376.
K. Fredrickson: "Extremal Transitions From Nested Reflexive Polytopes," *Comm. Math. Phys.* **335** (3) (2014) 1381–1395.

[312] K. Becker, M. Becker, K. Dasgupta and P.S. Green: "Compactifications of Heterotic Theory on Non-Kähler Complex Manifolds. 1.," *J. High E. Phys.* **04** (2003) 007.
K. Becker, M. Becker, P. S. Green, K. Dasgupta and E. Sharpe: "Compactifications of Heterotic Strings on Non-Kähler Complex Manifolds. 2.," *Nucl. Phys.* **B678** (2004) 19–100.

[313] F. Benini and S. Cremonesi: "Partition Functions of $\mathcal{N}=(2,2)$ Gauge Theories on S^2 and Vortices," *Comm. Math. Phys.* **334** (3) (2015) 1483–1527.
N. Doroud, J. Gomis, B. Le Floch and S. Lee: "Exact Results in $D=2$ Supersymmetric Gauge Theories," *J. High E. Phys.* **05** (2013) 093.
H. Jockers, V. Kumar, J.M. Lapan, D.R. Morrison and M. Romo: "Two-Sphere Partition Functions and Gromov-Witten Invariants," *Comm. Math. Phys.* **325** (2014) 1139–1170.
J. Gomis and S. Lee: "Exact Kahler Potential From Gauge Theory and Mirror Symmetry," *J. High E. Phys.* **04** (2013) 019.
F. Benini, R. Eager, K. Hori and Y. Tachikawa: "Elliptic Genera of Two-Dimensional $\mathcal{N}=2$ Gauge Theories With Rank-One Gauge Groups," *Lett. Math. Phys.* **104** (2014) 465–493.
K. Hori and M. Romo: "Exact Results In Two-Dimensional (2, 2)-Supersymmetric Gauge Theories With Boundary," arXiv:1308.2438.
N. Doroud and J. Gomis: "Gauge Theory Dynamics and Kähler Potential for Calabi–Yau Complex Moduli," *J. High E. Phys.* **12** (2013) 099.
D.R. Morrison: "Gromov–Witten Invariants and Localization," *J. Phys.* **A50** (44) (2017) 443004.
E. Kessler, A. Sheshmani and S.-T. Yau: "Super Gromov–Witten Invariants via Torus Localization," arXiv:2311.09074.

[314] B.V. Bento, D. Chakraborty, S.L. Parameswaran and I. Zavala: "A New de Sitter Solution With a Weakly Warped Deformed Conifold," *J. High E. Phys.* **12** (2021)124.

[315] P. Berglund, G. Butbaia, T. Hübsch, V. Jejjala, D. Mayorga Peña, C. Mishra and J. Tan: "Machine Learned Calabi–Yau Metrics and Curvature," *Adv. Th. Math. Phys.* **27** (3) (2023) awaiting publication; arXiv:2211.09801.

[316] P. Berglund, Y.-H. He, E. Heyes, E. Hirst, V. Jejjala and A. Lukas: "New Calabi–Yau Manifolds From Genetic Algorithms," *Phys. Lett.* **B850** (2024) 138504.

[317] P. Berglund and M. Henningson "Landau–Ginzburg Orbifolds, Mirror Symmetry and the Elliptic Genus," *Nucl. Phys.* **B433** (1995) 311–332.
M. Krawitz: *FJRW Rings and Landau–Ginzburg Mirror Symmetry*, PhD thesis (University of Michigan, Ann Arbor, MI, 2010).

M. Krawitz, N. Priddis, P. Acosta, N. Bergin and H. Rathnakumara: "FJRW-Rings and Mirror Symmetry," *Comm. Math. Phys.* **296** (1) (2009) 145–174.

[318] P. Berglund and T. Hübsch: "On a Residue Representation of Deformation, Koszul and Chiral Rings," *Int. J. Mod. Phys.* **A10** (1995) 3381–3430.

[319] P. Berglund and T. Hübsch: "On Calabi–Yau Generalized Complete Intersections From Hirzebruch Varieties and Novel K3-Fibrations," *Adv. Th. Math. Phys.* **22** (2) (2018) 261–303.

[320] P. Berglund and T. Hübsch: "A Generalized Construction of Calabi–Yau Models and Mirror Symmetry," *SciPost.* **4** (2) (2018) 009 (1–30).

[321] P. Berglund and T. Hübsch: "Hirzebruch Surfaces, Tyurin Degenerations and Toric Mirrors: Bridging Generalized Calabi–Yau Constructions," *Adv. Th. Math. Phys.* **26** (8) (2022) 2541–2598.

[322] P. Berglund and T. Hübsch: "Chern Characteristics and Todd–Hirzebruch Identities for Transpolar Pairs of Toric Spaces," `arXiv:2403.07139`.

[323] P. Berglund, T. Hübsch and D. Minic: "On Stringy de Sitter Spacetimes," *J. High E. Phys.* **2019** (2019) 166; "String Theory, the Dark Sector and the Hierarchy Problem," *LHEP* **2021** (2021) 186; "String Theory Bounds on the Cosmological Constant, the Higgs Mass, and the Quark and Lepton Masses," *Symmetry* **15** (9) (2023) 1660.

[324] P. Berglund, T. Hübsch and D. Minic: "Stringy Bubbles Solve de Sitter Troubles," *Universe* **7** (10) (2021) 363.

[325] P. Berglund, T. Hübsch and D. Minic: "Mirror Symmetry, Born Geometry and String Theory," in *Nankai Symposium on Mathematical Dialogues: Celebrating the 110th Anniversary of the Birth of Prof. S.-S. Chern*, Y.-H. He, M.-L. Ge, C.-M. Bai, J. Bao, and E. Hirst, eds. (Springer, Singapore, Sept., 2022).

[326] P. Berglund, T. Hübsch and D. Minic: "On de Sitter Spacetime and String Theory," *Int. J. Mod. Phys.* **D32** (9) (2023) 2330002 (111 pp).

[327] P. Berglund, S.H. Katz and A. Klemm: "Mirror Symmetry and the Moduli Space for Generic Hypersurfaces in Toric Varieties," *Nucl. Phys.* **B456** (1995) 153–204.

[328] P. Berglund and M. Lathwood: "Tropical Periods for Calabi–Yau Hypersurfaces in non-Fano Toric Varieties," `arXiv:2212.11906`.

[329] M. Bershadsky, K. Intriligator, S. Kachru, D.R. Morrison, V. Sadov and C. Vafa: "Geometric Singularities and Enhanced Gauge Symmetries," *Nucl. Phys.* **B481** (1996) 215–252.

[330] R. Blumenhagen, G. Honecker and T. Weigand: "Loop-corrected compactifications of the heterotic string with line bundles," *J. High E. Phys.* **06** (2005) 020.

[331] R. Blumenhagen, B. Jurke, T. Rahn and H. Roschy: "Cohomology of Line Bundles: A Computational Algorithm," *J. Math. Phys.* **51** (2010) 103525.
T. Rahn and H. Roschy: "Cohomology of Line Bundles: Proof of the Algorithm," *J. Math. Phys.* **51** (2010) 103520.

R. Blumenhagen, B. Jurke, T. Rahn and H. Roschy: "Cohomology of Line Bundles: Applications," *J. Math. Phys.* **53** (2012) 012302.

[332] R. Bousso and J. Polchinski: "Quantization of four-form fluxes and dynamical neutralization of the cosmological constant," *J. High E. Phys.* **0006** (2000) 006.

B.S. Acharya: "A moduli fixing mechanism in M theory," `arXiv:hep-th/0212294`.

F. Denef, M.R. Douglas, B. Florea, A. Grassi and S. Kachru: "Fixing all moduli in a simple F-theory compactification," *Adv. Th. Math. Phys.* **9** (2005) 861–929.

F. Denef "Lectures on Constructing String Vacua," pp. 483–610 in *Les Houches Summer School in Theoretical Physics: Session 87: String Theory and the Real World: From Particle Physics to Astrophysics*, vol. 87, C. Bachas, *et al.*, eds. (Elsevier, Amsterdam, 2008).

[333] V. Braun: "On Free Quotients of Complete Intersection Calabi–Yau Manifolds," *J. High E. Phys.* **04** (2011) 005.

[334] V. Braun, P. Candelas and R. Davies: "A Three-Generation Calabi–Yau Manifold With Small Hodge Numbers," *Fortsch. Phys.* **58** (2010) 467–502.

[335] V. Braun, Y.-H. He, B.A. Overut and T. Pantev: "A Heterotic Standard Model," *Phys. Lett.* **B618** (2005) 252–258; "A Standard Model From the $E(8)\times E(8)$ Heterotic Superstring," *J. High E. Phys.* **06** (2005)039; "Vector Bundle Extensions, Sheaf Cohomology, and the Heterotic Standard Model," *Adv. Th. Math. Phys.* **10** (4) (2006) 525–589.

[336] C.R. Brodie, A. Constantin, R. Deen and A. Lukas: "Machine Learning Line Bundle Cohomology," *Fortsch. Phys.* **68** (1) (2020) 1900087.

[337] C.R. Brodie, A. Constantin, R. Deen and A. Lukas: "Index Formulae for Line Bundle Cohomology on Complex Surfaces," *Fortsch. Phys.* **68** (2)(2020) 1900086.

[338] C. Brodie, A. Constantin, J. Gray, A. Lukas and F. Ruehle: "Recent Developments in Line Bundle Cohomology and Applications to String Phenomenology," in *Nankai Symposium on Mathematical Dialogues: Celebrating the 110th Anniversary of the Birth of Prof. S.-S. Chern*, Y.-H. He, M.-L. Ge, C.-M. Bai, J. Bao and E. Hirst, eds. (Springer, Singapore, Sept., 2022).

[339] F. Brown and O. Schnetz: "A K3 in ϕ^4," *Duke Math. J.* **161** (10) (2012) 1817–1862.

J.L. Bourjaily, Y.-H. He, A.J. Mcleod, M. von Hippel and M. Wilhelm: "Traintracks Through Calabi–Yau Manifolds: Scattering Amplitudes Beyond Elliptic Polylogarithms," *Phys. Rev. Lett.* **121** (7) (2018) 071603.

J.L. Bourjaily, A.J. McLeod, M. von Hippel and M. Wilhelm: "Bounded Collection of Feynman Integral Calabi–Yau Geometries," *Phys. Rev. Lett.* **122** (3) (2019) 031601; also as "A (Bounded) Bestiary of Feynman Integral Calabi–Yau Geometries," `arXiv:1810.07689`.

K. Bönisch, C. Duhr, F. Fischbach, A. Klemm and C. Nega: "Feynman Integrals in Dimensional Regularization and Extensions of Calabi–Yau Motives," *J. High E. Phys.* **09** (2022) 156.

J.L. Bourjaily, J. Broedel, E. Chaubey, C. Duhr and H. Frellesvig: "Functions Beyond Multiple Polylogarithms for Precision Collider Physics," in *Snowmass 2021* (2022), `arXiv:2203.07088`.

C. Duhr, A. Klemm, F. Loebbert, C. Nega and F. Porkert: "Geometry From Integrability: Multi-Leg Fishnet Integrals in Two Dimensions," `arXiv:2402.19034`.

[340] G. Brown: "A Database of Polarized K3 Surfaces," *Exp. Math.* **16** (1) (2007) 7 (20pp).

[341] G. Brown and A. Kasprzyk: "Four-Dimensional Projective Orbifold Hypersurfaces," *Exp. Math.* **25** (2) (2015) 176–193.

G. Brown, A. Kasprzyk and L. Zhu: "Gorenstein Formats, Canonical and Calabi–Yau Threefolds" *Exp. Math.* **31** (1) (2022) 146–164.

[342] G. Brown and A. Kasprzyk: "Kawamata Boundedness for Fano Threefolds and the Graded Ring Database," `arXiv:2201.07178`.

[343] D.A. Buchsbaum and D. Eisenbud: "Algebra Structures for Finite Free Resolutions, and Some Structure Theorems for Ideals of Codimension 3," *Am. J. Math.* **99** (3) (1977) 447–485.

[344] K. Bull, Y.-H. He, V. Jejjala and C. Mishra: "Getting CICY High," *Phys. Lett.* **B795** (2019) 700–706.

A. Ashmore, Y.-H. He and B.A. Ovrut: "Machine Learning Calabi–Yau Metrics," *Fortsch. Phys.* **68** (9) (2020) 2000068.

M.R. Douglas, S. Lakshminarasimhan and Y. Qi: "Numerical Calabi–Yau Metrics From Holomorphic Networks," *Proc. ML Res.* **107** (2021) 1–30, *2nd Annual Conference on Mathematical and Scientific Machine Learning,* J. Bruna, J.S. Hesthaven and L. Zdeborova, eds.

V. Jejjala, D.K. Mayorga Pena and C. Mishra: "Neural Network Approximations for Calabi–Yau Metrics," *J. High E. Phys.* **08** (2022) 105.

[345] G. Butbaia, D. Mayorga Peña, J. Tan, P. Berglund, T. Hübsch, V. Jejjala and C. Mishra: "Physical Yukawa Couplings in Heterotic String Compactifications," `arXiv:2401.15078`.

[346] A.C. Cadavid and S. Ferrara: "Picard–Fuchs Equations and the Moduli Space of Superconformal Field Theories," *Phys. Lett.* **B267** (1991) 193–199.

[347] P. Candelas, A. Constantin and C. Mishra: "Calabi–Yau Threefolds With Small Hodge Numbers," *Fortsch. Phys.* **66** (6) (2018) 1800029.

[348] P. Candelas and R. Davies: "New Calabi–Yau Manifolds With Small Hodge Numbers," *Fortsch. Phys.* **58** (2010) 383–466.

[349] P. Candelas, X.C. de la Ossa, Y.-H. He and B. Szendrői: "Triadophilia: A Special Corner in the Landscape," *Adv. Th. Math. Phys.* **12** (2) (2008) 429–473.

[350] P. Candelas, X.C. de la Ossa and S.H. Katz: "Mirror Symmetry for Calabi–Yau Hypersurfaces in Weighted $\mathbb{P}^4$ and Extensions of Landau–Ginzburg Theory," *Nucl. Phys.* **B450** (1995) 267–292.

[351] P. Candelas, X.C. de la Ossa and J. McOrist: "A Metric for Heterotic Moduli," *Comm. Math. Phys.* **356** (2) (2017) 567–612.

P. Candelas, X.C. de la Ossa, J. McOrist and R. Sisca "The Universal Geometry of Heterotic Vacua," *J. High E. Phys.* **02** (2019) 038.

A. Ashmore, C. Strickland-Constable, D. Tennyson and D. Waldram "Heterotic Backgrounds via Generalised Geometry: Moment Maps and Moduli," *J. High E. Phys.* **11** (2020) 071.

J. McOrist and E.E. Svanes: "Heterotic Quantum Cohomology," *J. High E. Phys.* **11** (2022) 096.

[352] P. Candelas, C.A. Lütken and R. Schimmrigk: "Complete Intersection Calabi–Yau Manifolds, 2: Three Generation Manifolds," *Nucl. Phys.* **B306** (1988) 113.

[353] P. Candelas, M. Lynker and R. Schimmrigk: "Calabi–Yau Manifolds in Weighted $P(4)$," *Nucl. Phys.* **B341** (1990) 383–402.

[354] J. Carifio, J. Halverson, D. Krioukov and B.D. Nelson: "ML in the String Landscape," *J. High E. Phys.* **1709** (2017) 157.

[355] F. Carta, J. Moritz and A. Westphal: "A Landscape of Orientifold Cacua," *J. High E. Phys.* **05** (2020) 107.

J. Moritz: "Orientifolding Kreuzer–Skarke," `arXiv:2305.06363`.

[356] L. Castellani, R. D'Auria and S. Ferrara: "Special Kähler Geometry: An Intrinsic Formulation From $N=2$ Space-Time Supersymmetry," *Phys. Lett.* **B241** (1990) 57–62.

B. de Wit, F. Vanderseypen and A. van Proeyen: "Symmetry Structure of Special Geometries," *Nucl. Phys.* **B400** (1993) 463–524.

A. Ceresole, R. D'Auria and S. Ferrara: "The Symplectic Structure of $N=2$ Supergravity and its Central Extension," *Nucl. Phys.* **B** *(Proc. Suppl.)* **46** (1996) 67–74.

A. Klemm, W. Lerche, P. Mayr, C. Vafa and N.P. Warner: "Selfdual Strings and $N=2$ Supersymmetric Field Theory," *Nucl. Phys.* **B477** (1996) 746–766.

L. Andrianopoli, M. Bertolini, A. Ceresole, R. D'Auria, S. Ferrara, P. Fre and T. Magri: "$N = 2$ Supergravity and $N = 2$ Super Yang–Mills Theory on General Scalar Manifolds: Symplectic Covariance, Gaugings and the Momentum Map," *J. Geom. Phys.* **23** (1997) 111–189.

B. Craps, F. Roose, W. Troost and A. van Proeyen: "What is Special Kähler Geometry?," *Nucl. Phys.* **B503** (1997) 565–613.

R. Dijkgraaf and C. Vafa: "Matrix Models, Topological Strings, and Supersymmetric Gauge Theories," *Nucl. Phys.* **B644** (2002) 3–20.

W. Lerche, P. Mayr and N. Warner: "Holomorphic $N=1$ Special Geometry of Open–Closed Type II Strings," `arXiv:hep-th/0207259`; "$N=1$ Special Geometry, Mixed Hodge Variations and Toric Geometry," `arXiv:hep-th/0208039`.

H. Jockers and J. Louis: "The Effective Action of $D7$-Branes in $N=1$ Calabi–Yau Orientifolds," *Nucl. Phys.* **B705** (2005) 167–211.
A. Belavin, B. Eremin and S. Parkhomenko: "Review on Special Geometry and Mirror Symmetry for Calabi–Yau Manifolds (Brief Review)," *JETP Lett.* **118** (2023) 701–709.
[357] E. Clader and Y. Ruan: "Mirror Symmetry Constructions," pp. 1–78 in *B-Model Gomov–Witten Theory*, E. Clader and Y. Ruan, eds., Trends in Mathematics (Birkhäuser, Cham, Switzerland, 2018).
[358] S. Cecotti, L. Girardello and A. Pasquinucci: "Singularity Theory and $N=2$ Supersymmetry," *Int. J. Mod. Phys.* **A6** (1991) 2427–2496.
[359] A. Ceresole, R. D'Auria, S. Ferrara, W. Lerche and J. Louis: "Picard-Fuchs Equations and Special Geometry," *Int. J. Mod. Phys.* **A8** (1993) 79–114.
[360] K. Chan, S.C. Lau and N.C. Leung: "SYZ Mirror Symmetry for Toric Calabi–Yau Manifolds," *J. Diff. Geom.* **90** (2) (2012) 177–250.
K. Chan, N.C. Leung and Z.N. Ma: "Fukaya's Conjecture on Witten's Twisted A_∞ Structure," *J. Diff. Geom.* **118** (3) (2021) 399–455.
Z.N. Ma: "SYZ Mirror Symmetry From Witten–Morse Theory," `arXiv:1811.09042`.
H. Nakanishi: "Homological Mirror Symmetry of Toric Fano Surfaces via Morse Homotopy," `arXiv:2303.07851`; "SYZ Mirror of Hirzebruch Surface $\mathbb{F}_k$ and Morse Homotopy," `arXiv:2312.01329`.
[361] C. Closset: "Toric Geometry and Local Calabi–Yau Varieties: An Introduction to Toric Geometry (for Physicists)," `arXiv:0901.3695`, based on lectures given at the Modave Summer School in Mathematical Physics, 2008.
[362] A. Constantin: "Generating Functions for Line Bundle Cohomology Dimensions on Complex Projective Varieties," `arXiv:2401.14463`.
[363] A. Constantin, C.S. Fraser-Taliente, T.R. Harvey, A. Lukas and B. Ovrut: "Computation of Quark Masses From String Theory," `arXiv:2402.01615`.
[364] A. Constantin, J. Gray and A. Lukas: "Hodge Numbers for all CICY Quotients," *J. High E. Phys.* **01** (2017) 001.
[365] A. Constantin, Y.-H. He and A. Lukas: "Counting String Theory Standard Models," *Phys. Lett.* **B792** (2019) 258–262.
[366] T.J. Courant: "Dirac Manifolds," *Trans. AMS* **319** (2) (1990) 631–661.
[367] J. Davenport, B. Poonen, J. Maynard, H. Helfgott, P.H. Tiep, L. Cruz-Filipe: "Machine-Assisted Proofs (ICM 2018 Panel)," `arXiv:1809.08062`.
[368] R. Davies: "Hyperconifold Transitions, Mirror Symmetry, and String Theory," *Nucl. Phys.* **B850** (2011) 214–231; "Classification and Properties of Hyperconifold Singularities and Transitions," `arXiv:1309.6778`.
[369] G.B. De Luca, E. Silverstein and G. Torroba: "Hyperbolic Compactification of M-Theory and de Sitter Quantum Gravity," *SciPost Phys.* **12** (2022) 083.
[370] R. Deen, Y.-H. He, S.-J. Lee and A. Lukas: "Machine Learning String Standard Models," *Phys. Rev.* **D105** (4) (2022) 046001.

[371] J. Distler and S. Kachru: "(0, 2) Landau–Ginzburg Theory," *Nucl. Phys.* **B413** (1994) 213–243.

[372] L.J. Dixon, V. Kaplunovsky and J. Louis: "On Effective Field Theories Describing (2, 2) Vacua of the Heterotic String," *Nucl. Phys.* **B329** (1990)27–82.

[373] R. Donagi, Y.-H. He, B.A. Ovrut and R. Reinbacher: "Higgs Doublets, Split Multiplets and Heterotic $SU(3)_C \times SU(2)_L \times U(1)_Y$ Spectra," *Phys. Lett.* **B618** (2005) 259–264; "The Spectra of Heterotic Standard Model Vacua," *J. High E. Phys.* **06** (2005) 070.
V. Bouchard and R. Donagi: "An $SU(5)$ Heterotic Standard Model," *Phys. Lett.* **B633** (2006) 783–791.
V. Braun, Y.-H. He, B.A. Ovrut, T. Pantev: "The Exact MSSM Spectrum From String Theory," *J. High E. Phys.* **05** (2006) 043.
L.B. Anderson, J. Gray, Y.-H. He and A. Lukas: "Exploring Positive Monad Bundles and a New Heterotic Standard Model," *J. High E. Phys.* **02** (2010) 054.
L.B. Anderson, J. Gray, A. Lukas and E. Palti: "Two Hundred Heterotic Standard Models on Smooth Calabi–Yau Threefolds," *Phys. Rev.* **D84** (2011) 106005; "Heterotic Line Bundle Standard Models," *J. High E. Phys.* **06** (2012) 113.
L.B. Anderson, A. Constantin, J. Gray, A. Lukas and E. Palti: "A Comprehensive Scan for Heterotic $SU(5)$ GUT Models," *J. High E. Phys.* **01** (2014) 047.
S. Mizoguchi and T. Sakaguchi: "Note on Three-Generation Models in Heterotic String and F-Theory on Elliptic Calabi–Yau Manifolds Over Hirzebruch Varieties," *Prog. Th. Exp. Phys.* **2016** (10) (2016) 101B01.
M. Cvetič, L. Lin, M. Liu and P.-K. Oehlmann: "An F-Theory Realization of the Chiral MSSM With $\mathbb{Z}_2$-Parity," *J. High E. Phys.* **09** (2018) 089.

[374] M.R. Douglas: "The statistics of string / M-theory vacua," *J. High E. Phys.* **05** (2003) 046

[375] K. Hori and D. Tong: "Aspects of Non-Abelian Gauge Dynamics in Two-Dimensional $N=(2,2)$ Theories," *J. High E. Phys.* **05** (2007) 079.
R. Donagi and E. Sharpe: "GLSM's for Partial Flag Manifolds," *J. Geom. Phys.* **58** (2008) 1662–1692.
H. Jockers, V. Kumar, J.M. Lapan, D.R. Morrison and M. Romo "Nonabelian 2D Gauge Theories for Determinantal Calabi–Yau Varieties," *J. High E. Phys.* **11** (2012) 166.
F. Benini, R. Eager, K. Hori and Y. Tachikawa: "Elliptic Genera of 2D $\mathcal{N}=2$ Gauge Theories," *Comm. Math. Phys.* **333** (3) (2015) 1241–1286.
C. Closset, S. Cremonesi and D.S. Park: "The Equivariant A-Twist and Gauged Linear Sigma Models on the Two-Sphere," *J. High E. Phys.* **06** (2015) 076.
H. Fan, T. Jarvis and Y. Ruan: "A Mathematical Theory of the Gauged Linear Sigma Model," *Geom. Topol.* **22** (2018) 235–303.

[376] A.R. Fletcher: "Working With Complete Intersections," *Max Planck Institute for Mathematics, Bonn* preprint MPI/89-35, 1989.

[377] S. Franco and A. Uranga: "Bipartite Field Theories From D-Branes," *J. High E. Phys.* **1404** (2014) 161.

[378] D.S. Freed: "Special Kähler Manifolds," *Comm. Math. Phys.* **203** (1999) 31–52.

[379] L. Freidel, R.G. Leigh and D. Minic "Metastring Theory and Modular Space-Time," *J. High E. Phys.* **06** (2015) 006.

[380] L. Freidel, F.J. Rudolph and D. Svoboda: "A Unique Connection for Born Geometry," *Comm. Math. Phys.* **372** (1) (2019) 119–150.

[381] B. Friedman, S. Picard and C. Suan: "Gromov–Hausdorff continuity of non-Kähler Calabi–Yau conifold transitions," arXiv:2404.11840.

[382] R. Friedman: "Simultaneous resolution of threefold double points," *Math. Ann.* **274** (4) (1986) 671–689.

[383] R. Friedman: "On threefolds with trivial canonical bundle," pp. 103–134 in *Complex Geometry and Lie Theory*, J.A. Carlson, C.H. Clemens, and D.R. Morrison, eds., vol. 53 (AMS, Providence, RI, 1991).

[384] A. Gadde, L. Rastelli, S.S. Razamat and W. Yan: "Gauge Theories and Macdonald Polynomials," *Comm. Math. Phys.* **319** (2013) 147–193.
A. Kapustin and N. Seiberg: "Coupling a QFT to a TQFT and Duality," *J. High E. Phys.* **04** (2014) 001.
M. Barkeshli, P. Bonderson, M. Cheng and Z. Wang: "Symmetry Fractionalization, Defects, and Gauging of Topological Phases," *Phys. Rev.* **B100** (11) (2019) 115147.
C.-M. Chang, Y.-H. Lin, S.-H. Shao, Y. Wang and X. Yin: "Topological Defect Lines and Renormalization Group Flows in Two Dimensions," *J. High E. Phys.* **01** (2019) 026.
I. Brunner, N. Carqueville, P. Fragkos and D. Roggenkamp "Truncated Affine Rozansky–Witten Models as Extended Defect TQFTs," arXiv:2307.06284.

[385] M. Gagnon and Q. Ho-Kim: "An Exhaustive List of Complete Intersection Calabi–Yau Manifolds," *Mod. Phys. Lett.* **A9** (1994) 2235.

[386] A. Garbagnati and B. van Geemen: "A Remark on Generalized Complete Intersections," *Nucl. Phys.* **B925** (2017) 135–143.

[387] S.J. Gates, Jr. and T. Hübsch: "Unidexterous Locally Supersymmetric Actions for Calabi–Yau Compactifications," *Phys. Lett.* **B226** (1989) 100; "Calabi–Yau Heterotic Strings and Unidexterous Sigma Models," *Nucl. Phys.* **B343** (1990) 741–774.

[388] S.J. Gates, Jr. and W. Siegel: "Variant Superfield Representations," *Nucl. Phys.* **B187** (1981) 389–396.
S.J. Gates, Jr., C.M. Hull and M. Roček: "Twisted Multiplets and New Supersymmetric Nonlinear Sigma Models," *Nucl. Phys.* **B248** (1984) 157.
S.J. Gates, Jr.: "Vector Multiplets and the Phases of $N=2$ Theories in 2-D: Through the Looking Glass," *Phys. Lett.* **B352** (1995) 43–49.

[389] F. Gmeiner, R. Blumenhagen, G. Honecker, D. Lust and T. Weigand: "One in a Billion: MSSM-Like D-Brane Statistics," *J. High E. Phys.* **0601** (2006) 004.

[390] B.R. Greene: "String Theory on Calabi–Yau Manifolds," pp. 543–726 in *Proceedings,* **TASI** *(1996): Fields, Strings, and Duality*, arXiv:hep-th/9702155.

A. Klemm, J. Manschot and T. Wotschke: "Quantum Geometry of Elliptic Calabi–Yau Manifolds," *Comm. Number Th. & Phys.* **6** (4) (2012) 849–917.

M.-X. Huang, A. Klemm, J. Reuter and M. Schiereck: "Quantum Geometry of del Pezzo Surfaces in the Nekrasov-Shatashvili Limit," *J. High E. Phys.* **02** (2015) 031.

S. Alexandrov, S. Feyzbakhsh, A. Klemm, B. Pioline and T. Schimannek: "Quantum Geometry, Stability and Modularity," arXiv:2301.08066.

[391] W. Gu, E. Sharpe and H. Zou: "GLSMs for Exotic Grassmannians," *J. High E. Phys.* **10** (2020) 200.

[392] P. Hacking and A. Keating: "Homological Mirror Symmetry for log Calabi–Yau Surfaces," *Geom. Topol.* **26** (8) (2022) 3747–3833.

[393] J. Halverson, V. Kumar and D. R. Morrison: "New Methods for Characterizing Phases of 2D Supersymmetric Gauge Theories," *J. High E. Phys.* **09** (2013) 143.

[394] A. Hanany and Y.-H. He: "Non-Abelian Finite Gauge Theories," *J. High E. Phys.* **9902** (1999) 013.

[395] A. Hanlon: "Monodromy of Monomially Admissible Fukaya–Seidel Categories Mirror to Toric Varieties," *Adv. Th. Math. Phys.* **350** (2019) 662–746.

Z.-C. Ong and M.-C. Tan: "Vafa–Witten Theory: Invariants, Floer Homologies, Higgs Bundles, a Geometric Langlands Correspondence, and Categorification," arXiv:2203.17115.

[396] T. Hausel and M. Thaddeus: "Mirror Symmetry, Langlands Duality, and the Hitchin System," *Invent. Math.* **153** (2003) 197.

[397] Y.-H. He: "On Algebraic Singularities, Finite Graphs and D-Brane Gauge Theories: A String Theoretic Perspective," arXiv:hep-th/0209230; "Lectures on D-Branes, Gauge Theories and Calabi–Yau Singularities," arXiv:hep-th/0408142.

J. Bao, Y.-H. He, E. Hirst and S. Pietromonaco "Lectures on the Calabi–Yau Landscape," arXiv:2001.01212.

[398] Y.-H. He: "Deep-Learning the Landscape," arXiv:1706.02714; q.v. *Science*, vol. 365, issue 6452, Aug 2019; "Machine-Learning the String Landscape," *Phys. Lett.* **B774** (2017) 564.

[399] Y.-H. He: "Machine-Learning Mathematical Structures," *Int. J. Data Sci. in Math. Sci.* **1** (01) (2023) 23–47.

[400] Y.-H. He, E. Heyes and E. Hirst: "Machine Learning in Physics and Geometry," arXiv:2303.12626.

[401] Y.-H. He, V. Jejjala, C. Matti, B.D. Nelson and M. Stillman: "The Geometry of Generations," *Comm. Math. Phys.* **339** (1) (2015) 149.

[402] Y.-H. He: "Prolegomena to the Bestiary," `arXiv:2405.05720`.

[403] E. Hirst and T. S. Gherardini: "Calabi–Yau Four/Five/Six-Folds as $\mathbb{P}^n_{\mathbf{w}}$ Hypersurfaces: Machine Learning, Approximation, and Generation," `arXiv:2311.17146`.

[404] N. Hitchin: "Generalized Calabi–Yau Manifolds," *Quart. J. Math.* **54** (3) (2003) 281–308; "Lectures on Generalized Geometry," *Surv. Diff. Geom.* **16** (2010) 79–124.
H. Iritani: "Quantum D-modules and Generalized Mirror Transformations," *Topology* **47** (4) (2008) 225–276.
M. Gualtieri: "Generalized Complex Geometry," *Ann. Math.* **174** (2011) 75–123; "Generalized Kähler Geometry," *Comm. Math. Phys.* **331** (1) (2014) 297–331.

[405] K. Hori, K. Hosomichi, D.C. Page, R. Rabadan and J. Walcher: "Non-Perturbative Orientifold Transitions at the Conifold," *J. High E. Phys.* **10** (2005) 026.

[406] K. Hori, H. Ooguri and C. Vafa: "Non-Abelian Conifold Transitions and $N=4$ Dualities in Three Dimensions," *Nucl. Phys.* **B504** (1997) 147–174.

[407] K. Hori and C. Vafa: "Mirror Symmetry," `arXiv:hep-th/0002222`.

[408] J. Horn: "Über die Konvergenz hypergeometrischer Reihen zweier und dreier Veränderlichen," *Math. Ann.* **34** (1990) 544–600.

[409] C.M. Hull: "σ Model Beta Functions and String Compactifications," *Nucl. Phys.* **B267** (1986) 266–276; "Compactifications of the Heterotic Superstring," *Phys. Lett.* **B178** (1986) 357–364.
A. Strominger: "Superstrings With Torsion," *Nucl. Phys.* **B274** (1986) 253.

[410] T. Hübsch: "On a Stringy Singular Cohomology," *Mod. Phys. Lett.* **A12** (1997) 521–533.

[411] T. Hübsch: "A Hitchhiker's Guide to Superstring Jump Gates and Other Worlds," *Nucl. Phys.* **B** *(Proc. Suppl.)* **52** (1997) 347–351.

[412] T. Hübsch: "Haploid $(2,2)$-Superfields in 2-Dimensional Space-Time," *Nucl. Phys.* **B555** (3) (1999) 567–628; "Weaving Worldsheet Supermultiplets From the Worldlines Within," *Adv. Th. Math. Phys.* **17** (5) (2013) 1–72.

[413] T. Hübsch and A. Rahman: "On the Geometry and Homology of Certain Simple Stratified Varieties," *J. Geom. Phys.* **53** (2005) 31–48.
A. Rahman: "A Perverse Sheaf Approach Toward a Cohomology Theory for String Theory," *Adv. Th. Math. Phys.* **13** (3) (2009) 667–693.

[414] A. Imparato: "About Homological Mirror Symmetry," `arXiv:2306.13589`, MSci Thesis, ETH, Zürich, 2022.

[415] K. Intriligator, H. Jockers, P. Mayr, D.R. Morrison and M.R. Plesser: "Conifold Transitions in M-theory on Calabi–Yau Fourfolds With Background Fluxes," *Adv. Th. Math. Phys.* **17** (3) (2013) 601–699.

[416] D. Joyce: "Lectures on Calabi–Yau and Special Lagrangian Geometry," `arXiv:math/0108088`.

[417] S. Kachru, R. Kallosh, A.D. Linde, J.M. Maldacena, L.P. McAllister and S.P. Trivedi: "Towards Inflation in String Theory," *J. Cosm. Astroparticle. Phys.* **10** (2003) 013.
V. Balasubramanian, P. Berglund, J.P. Conlon and F. Quevedo: "Systematics of Moduli Stabilisation in Calabi–Yau Flux Compactifications," *J. High E. Phys.* **03** (2005) 007.
M. Grana: "Flux Compactifications in String Theory: A Comprehensive Review," *Phys. Rep.* **423** (2006) 91–158.
M.R. Douglas and S. Kachru: "Flux Compactification," *Rev. Mod. Phys.* **79** (2007) 733–796.
R. Blumenhagen, D. Herschmann and F. Wolf: "String Moduli Stabilization at the Conifold," *J. High E. Phys.* **08** (2016) 110.
D. Prieto: "Moduli Stabilization and Stability in Type II/F-theory Flux Compactifications," PhD Dissertation, Departamento de Física Teórica Universidad Autónoma de Madrid, 2024.

[418] S. Kachru, R. Kallosh, A.D. Linde and S.P. Trivedi: "de Sitter Vacua in String Theory," *Phys. Rev.* **D68** (2003) 046005.

[419] S. Kachru and E. Silverstein: "Chirality Changing Phase Transitions in 4-D String Vacua," *Nucl. Phys.* **B504** (1997) 272–284.

[420] M.M. Kapranov: "A Characterization of A-Discriminantal Hypersurfaces in Terms of the Logarithmic Gauss Map," *Math. Ann.* **290** (2) (1991) 277–285.

[421] A. Kapustin and E. Witten: "Electric-Magnetic Duality and the Geometric Langlands Program," *Commun. Num. Theor. Phys.* **1** (2007) 1–236.

[422] S.H. Katz, A. Klemm and C. Vafa: "Geometric Engineering of Quantum Field Theories," *Nucl. Phys.* **B497** (1997) 173–195.
W. Lerche: "Fayet–Iliopoulos Potentials From Four Folds," *J. High E. Phys.* **11** (1997) 004.
N.C. Leung and C. Vafa: "Branes and Toric Geometry," *Adv. Th. Math. Phys.* **2** (1998) 91–118.
A. Klemm and E. Zaslow: "Local Mirror Symmetry at Higher Genus," *AMS/IP Stud. Adv. Math.* **23** (2001) 183–207.
T.M. Chiang, A. Klemm, S.-T. Yau and E. Zaslow: "Local Mirror Symmetry: Calculations and Interpretations," *Adv. Th. Math. Phys.* **3** (1999) 495–565.
M. Aganagic, A. Karch, D. Lust and A. Miemiec: "Mirror Symmetries for Brane Configurations and Branes at Singularities," *Nucl. Phys.* **B569** (2000) 277–302.
F. Cachazo, B. Fiol, K.A. Intriligator, S. Katz and C. Vafa "A Geometric Unification of Dualities," *Nucl. Phys.* **B628** (2002) 3–78.
M. Marino: "Open String Amplitudes and Large Order Behavior in Topological String Theory," *J. High E. Phys.* **03** (2008) 060.

[423] S. Katz and T. Schimannek: "New Non-Commutative Resolutions of Determinantal Calabi–Yau Threefolds From Hybrid GLSM," arXiv: 2307.00047.

[424] A. Khovanskii and A. Pukhlikov: "A Riemann-Roch Theorem for Integrals and Sums of Quasipolynomials Over Virtual Polytopes," *St. Petersburg Math. J.* **4** (4) (1993) 789–812, Transl. from *Algebra and Analysis* **4** (4) (1992) 188–216.
Y. Karshon and S. Tolman: "The Moment Map and Line Bundles Over Presymplectic Toric Manifolds," *J. Diff. Geom.* **38** (3) (1993) 465–484.
A.C. da Silva, V. Guillemin and A.R. Pires: "Symplectic Origami," *Int. Math. Res. Not.* **2011** (18) (2010) 4252–4293.

[425] J.K. Kim, C.J. Park and Y. Yoon: "Calabi–Yau Manifolds From Complete Intersections in Products of Weighted Complex Projective Spaces," *Phys. Lett.* **B224** (1989) 108–114.

[426] I.R. Klebanov and E. Witten: "Superconformal Field Theory on Three-Branes at a Calabi–Yau Singularity," *Nucl. Phys.* **B536** (1998) 199.

[427] M. Kontsevich and Y. Manin: "Gromov–Witten Classes, Quantum Cohomology, and Enumerative Geometry," *Comm. Math. Phys.* **164** (1994) 525–562.
R. Gopakumar and C. Vafa: "M Theory and Topological Strings. 1.," `arXiv:hep-th/9809187`; "*ditto*, 2.," `arXiv:hep-th/9812127`; "On the Gauge Theory/Geometry Correspondence," *Adv. Th. Math. Phys.* **3** (1999) 1415–1443.
D. Maulik, N. Nekrasov, A. Okounkov and R. Pandharipande: "Gromov–Witten Theory and Donaldson–Thomas Theory, I," *Compos. Math.* **142** (05) (2006) 1263–1285.
R. Pandharipande and R.P. Thomas: "Curve Counting via Stable Pairs in the Derived Category," *Invent. Math.* **178** (2009) 407–447; "13/2 Ways of Counting Curves," pp. 282–333 in *London Math. Soc. Lect. Note Ser.* no. 411 (Cambridge University Press, Cambridge, 2014).

[428] D. Krefl and R. K. Seong: "Machine Learning of Calabi–Yau Volumes," *Phys. Rev.* **D96** (6) (2017) 066014

[429] M. Kreuzer and H. Skarke: "On the Classification of Reflexive Polyhedra," *Comm. Math. Phys.* **185** (1997) 495; "Reflexive Polyhedra, Weights and Toric Calabi–Yau Fibrations," *Rev. Mod. Phys.* **14** (2002) 343.

[430] M. Kreuzer and H. Skarke: "Complete Classification of Reflexive Polyhedra in Four Dimensions," *Adv. Th. Math. Phys.* **4** (2000) 1209–1230.

[431] M. Kreuzer and H. Skarke: "Calabi–Yau Data," http://hep.itp.tuwien.ac.at/~kreuzer/CY/.

[432] M. Larfors, A. Lukas and F. Ruehle: "Calabi–Yau Manifolds and $SU(3)$ Structure," *J. High E. Phys.* **01** (2019) 171.

[433] S.-C. Lau, L.-S. Tseng and S.-T. Yau: "Non-Kähler SYZ Mirror Symmetry," *Comm. Math. Phys.* **340** (1) (2015) 145–170.

[434] Y. Lekili and T. Perutz: "Arithmetic Mirror Symmetry for the 2-Torus," `arXiv:1211.4632` (being revised for *Memoirs of the AMS*).

[435] A. Libgober and J. Teitelbaum: "Lines on Calabi–Yau Complete Intersections, Mirror Symmetry, and Picard–Fuchs Equations," *Int. Math. Res. Notices* **1993** (1) (1993) 29–39.

[436] A. Lukas, B.A. Ovrut and D. Waldram: "Nonstandard embedding and five-branes in heterotic M theory," *Phys. Rev.* **D59** (1999) 106005.

[437] A. Maharana and E. Palti: "Models of Particle Physics From Type IIB String Theory and F-theory: A Review," *Int. J. Mod. Phys.* **A28** (2013) 1330005.

[438] J. M. Maldacena: "The Large N Limit of Superconformal Field Theories and Supergravity," *Adv. Th. Math. Phys.* **2** (1998) 231, rep.: *Int. J. Th. Phys.* **38** (1999) 1113.

[439] M. Masuda: "Unitary Toric Manifolds, Multi-Fans and Equivariant Index," *Tohoku Math. J. (2)* **51** (2) (1999) 237–265; "From Convex Polytopes to Multi-Polytopes," *Notes Res. Inst. Math. Analysis* **1175** (2000) 1–15.

A. Hattori and M. Masuda: "Theory of Multi-Fans," *Osaka J. Math.* **40** (2003) 1–68.

M. Masuda and T. Panov: "On the Cohomology of Torus Manifolds," *Osaka J. Math.* **43** (3) (2006) 711–746.

Y. Nishimura: "Multipolytopes and Convex Chains," *Proc. Steklov Inst. Math.* **252** (2006) 212–224.

H. Ishida, Y. Fukukawa and M. Masuda: "Topological Toric Manifolds," *Moscow Math. J.* **13** (1) (2013) 57–98.

A. Higashitani and M. Masuda: "Lattice Multipolygons," *Kyoto J. Math.* **57** (4) (2017) 807–828.

[440] J. McKay: "Graphs, Singularities, and Finite Groups," *Proc. Symp. Pure Math.* **37** (1980) 183–186.

[441] J. McNamara and C. Vafa: "Cobordism Classes and the Swampland," `arXiv:1909.10355`.

[442] D. Morrison: "Mirror Symmetry and Rational Curves on Quintic Threefolds: A Guide for Mathematicians," *J. Amer. Math. Soc.* **6** (1) (1993) 223–247.

[443] D.R. Morrison and M.R. Plesser: "Summing the Instantons: Quantum Cohomology and Mirror Symmetry in Toric Varieties," *Nucl. Phys.* **B440** (1995) 279–354.

[444] D.R. Morrison and M.R. Plesser: "Towards Mirror Symmetry as Duality for two Dimensional Abelian Gauge Theories," *Nucl. Phys.* **B** *(Proc. Suppl.)* **46** (1996) 177–18.

[445] D.R. Morrison and C. Vafa: "Compactifications of F Theory on Calabi–Yau Threefolds. 1.," *Nucl. Phys.* **B473** (1996) 74–92; "*ditto*, 2.," *ibid.* **B476** (1996) 437–469.

[446] S. Müller-Stach, S. Weinzierl and R. Zayadeh: "Picard–Fuchs Equations for Feynman Integrals," *Comm. Math. Phys.* **326** (2014) 237–249.

[447] H. Ooguri and C. Vafa: "On the Geometry of the String Landscape and the Swampland," *Nucl. Phys.* **B766** (2007) 21–33.

[448] P. Orlik and R. Randell: "The Structure of Weighted Homogeneous Polynomials," in *Proc. Sympos. Pure Math.*, vol. 30, pp. 57–64, R.O. Wells, Jr., ed. (AMS, Providence, RI, 1977).

[449] H. Otsuka and K. Takemoto: "$SO(32)$ Heterotic Standard Model Vacua in General Calabi–Yau Compactifications," *J. High E. Phys.* **11** (2018) 034; "Deep Learning and k-Means Clustering in Heterotic String Vacua With Line Bundles," *J. High E. Phys.* **05** (2020) 047.

[450] H. Partouche and B. Pioline: "Rolling Among $G(2)$ Vacua," *J. High E. Phys.* **03** (2001) 005.

[451] S. Picard "The Strominger System and Flows by the Ricci Tensor," `arXiv:2402.17770`.

[452] M.I. Qureshi and B. Szendroi: "Calabi–Yau Threefolds in Weighted Flag Varieties," *Adv. High E. Phys.* **2012** (2012) 547317.

[453] M.A. Reid: "Gorenstein in Codimension 4: the General Structure Theory," *Adv. Stud. Pure Math.* **2015** (2015) 201–227.

[454] A.G. Riess *et al.*, Supernova Search Team Collaboration: "Observational Evidence From Supernovae for an Accelerating Universe and a Cosmological Constant," *Astron. J.* **116** (1998) 1009–1038.
S. Perlmutter *et al.*, Supernova Cosmology Project Collaboration: "Measurements of Omega and Lambda from 42 High Redshift Supernovae," *Astrophys. J.* **517** (1999) 565–586.

[455] Y. Ruan and G. Tian: "A Mathematical Theory of Quantum Cohomology," *J. Diff. Geom.* **42** (2) (1995) 259–367.
Y. Ruan: "Topological Sigma Model and Donaldson-Type Invariants in Gromov Theory," *Duke Math. J.* **83** (2) (1996) 461–500.

[456] F. Ruehle: "Evolving NNs With Genetic Algorithms to Study the String Landscape," *J. High E. Phys.* **08** (2017) 038.

[457] F. Ruehle: "Data Science Applications to String Theory," *Phys. Rev.* **839** (2020) 1–117.

[458] A. Sagnotti "Open strings and their symmetry groups," in *NATO Advanced Summer Institute on Nonperturbative Quantum Field Theory (Cargese Summer Institute)*, G. 't Hooft, A. Jaffe, G. Mack, P. Mitter, and R. Stora, eds., p. 521 (Pergamon Press, New York, 1987.
J. Dai, R. G. Leigh and J. Polchinski: "New connections between string theories," *Mod. Phys. Lett.* **A4** (1989) 2073–2083.
P. Horava: "Strings on world sheet orbifolds," *Nucl. Phys.* **B327** (1989) 461–484.
M. Bianchi and A. Sagnotti "Twist symmetry and open string Wilson lines," *Nucl. Phys.* **B361** (1991) 519–538.
A. Sen: "F-theory and orientifolds," *Nucl. Phys.* **B475** (1996) 562.

[459] S. Schäfer-Nameki and T. Weigand: "F-Theory and 2d $(0,2)$ Theories," *J. High E. Phys.* **05** (2016) 059.

[460] H. Schenck, M. Stillman and B. Yuan: "Calabi–Yau Threefolds in $\mathbb{P}^n$ and Gorenstein Rings," *Adv. Th. Math. Phys.* **26** (3) (2022) 764–792.
S. Moshin, S. Nazir and M.I. Qureshi: "Constructions and Deformations of Calabi–Yau 3-Folds in Codimension 4," `arXiv:2312.17341`.

[461] P. Seidel: "Homological Mirror Symmetry for the Quartic Surface," *Mem. AMS* **236** (1116) (AMS, Providence, RI, 2015).

[462] A. SEN and B. ZWIEBACH: "String field theory: A review," arXiv:2405.19421.

[463] E. SHARPE: "A Survey of Recent Developments in GLSMs," arXiv:2401.11637.

[464] J. SMITH: "Superfiltered A_∞-Deformations of the Exterior Algebra, and Local Mirror Symmetry," *J. London Math. Soc.* **105** (4) (2022) 2203–2248.

L. AMORIM and C.-H. CHO: "Ungraded Matrix Factorizations as Mirrors of Non-Orientable Lagrangians," arXiv:2205.01046.

S. GANATRA, A. HANLON, J. HICKS, D. POMERLEANO and N. SHERIDAN: "Integrality of Mirror Maps and Arithmetic Homological Mirror Symmetry for Greene–Plesser Mirrors," arXiv:2312.01949.

[465] I. SMITH, R.P. THOMAS and S.-T. YAU: "Symplectic Conifold Transitions," *J. Diff. Geom.* **62** (2002)209–242.

M. ROSSI: "Geometric Transitions," *J. Geom. Phys.* **56** (9) (2006) 1940–1983; "Deforming Geometric Transitions," *Int. J. Geom. Meth. Mod. Phys.* **12** (05) (2015) 1550059.

J. FU, J. LI and S.-T. YAU: "Balanced metrics on non-Kähler Calabi–Yau threefolds," *J. Diff. Geom.* **90** (1) (2012) 81–129.

T.C. COLLINS, S. GUKOV, S. PICARD and S.-T. YAU: "Special Lagrangian Cycles and Calabi–Yau Transitions," *Comm. Math. Phys.* **401** (1) (2023) 769–802.

T.C. COLLINS, S. PICARD and S.-T. YAU: "Stability of the Tangent Bundle Through Conifold Transitions," *Comm. Pure App. Math.* **77** (1) (2024) 284–371.

[466] S. STROGATZ: "What Makes for 'Good' Mathematics?" *Quanta Magazine*, Feb. 1, 2024.

[467] A. STROMINGER: "Massless Black Holes and Conifolds in String Theory," *Nucl. Phys.* **B451** (1995) 96–108.

B.R. GREENE, D.R. MORRISON and A. STROMINGER: "Black Hole Condensation and the Unification of String Vacua," *Nucl. Phys.* **B451** (1995) 109–120.

[468] A. STROMINGER, S.T. YAU and E. ZASLOW: "Mirror Symmetry is T-Duality," *Nucl. Phys.* **B479** (1996) 243–259.

[469] S. TANABÉ: "On Horn–Kapranov Uniformisation of the Discriminantal Loci," *Adv. Stud. Pure Math.* **46** (2007) 223–249.

J. HUH: "Varieties With Maximum Likelihood Degree One," *J. Alg. Stat.* **5** (1) (2014) 1–17.

[470] G. TIAN: "Gauge Theory and Calibrated Geometry, 1," *Ann. Math.* **151** (2000) 193–268.

[471] A.A. TSEYTLIN: "Duality Symmetric Formulation of String World Sheet Dynamics," *Phys. Lett.* **B242** (1990) 163; "Duality Symmetric Closed String Theory and Interacting Chiral Scalars," *Nucl. Phys.* **B350** (1991) 395.

T.W. GRIMM: "The Effective Action of Type II Calabi–Yau Orientifolds," *Fortsch. Phys.* **53** (2005) 1179–1271.

W. MERRELL, L.A. PANDO ZAYAS and D. VAMAN: "Gauged (2, 2)-Sigma Models and Generalized Kähler Geometry," *J. High E. Phys.* **0712** (2007) 039.

M. GRANA, R. MINASIAN, M. PETRINI and D. WALDRAM: "T-Duality, Generalized Geometry and Non-Geometric Backgrounds," *J. High E. Phys.* **04** (2009) 075.

C.M. HULL and B. ZWIEBACH: "Double Field Theory," *J. High E. Phys.* **09** (2009) 099; "The Gauge Algebra of Double Field Theory and Courant Brackets," *J. High E. Phys.* **09** (2009) 090.

P. KOERBER: "Lectures on Generalized Complex Geometry for Physicists," *Fortsch. Phys.* **59** (2011) 169–242.

O. HOHM, C. HULL and B. ZWIEBACH: "Generalized Metric Formulation of Double Field Theory," *J. High E. Phys.* **08** (2010) 008.

D.S. BERMAN and M.J. PERRY: "Generalized Geometry and M Theory," *J. High E. Phys.* **06** (2011) 074.

C.M. HULL, U. LINDSTROM, M. ROČEK, R. VON UNGE and M. ZABZINE: "Generalized Kähler Geometry in (2, 1) Superspace," *J. High E. Phys.* **06** (2012) 013.

[472] A. VERMEERSCH: "Deep Learning for K3 Fibrations in Heterotic/Type IIA String Duality," *Nucl. Phys.* **B993** (2023) 116279.

[473] J. WALCHER: "Opening Mirror Symmetry on the Quintic," *Comm. Math. Phys.* **276** (2007) 671–689.

V. BOUCHARD, A. KLEMM, M. MARINO and S. PASQUETTI: "Remodeling the B-model," *Comm. Math. Phys.* **287** (2009) 117–178.

S. BLOCH, M. KERR and P. VANHOVE: "Local Mirror Symmetry and the Sunset Feynman Integral," *Adv. Th. Math. Phys.* **21** (2017) 1373–1453.

B. EYNARD and N. ORANTIN: "Computation of Open Gromov–Witten Invariants for Toric Calabi–Yau 3-Folds by Topological Recursion, a Proof of the BKMP Conjecture," *Comm. Math. Phys.* **337** (2) (2015) 483–567.

[474] S.-S. WANG: "On the Connectedness of the Standard Web of Calabi–Yau 3-Folds and Small Transitions," *Asian J. Math.* **22** (6) (2018) 981–1004.

[475] T. WEIGAND: "F-Theory," pp. 1–131 in *Proceedings,* **TASI 2017** (2018) 016, `arXiv:1806.01854`.

[476] E. WITTEN: "New Issues in Manifolds of $SU(3)$ Holonomy," *Nucl. Phys.* **B268** (1986) 79.

[477] E. WITTEN: "Phase Transitions in M Theory and F Theory," *Nucl. Phys.* **B471** (1996) 195–216.

[478] E. WITTEN: "Topological Sigma Models," *Comm. Math. Phys.* **118** (1988) 411; "Topological Quantum Field Theory," *Comm. Math. Phys.* **117** (1988) 353–386; "Mirror Manifolds and Topological Field Theory," pp. 121–160 in *Mirror Symmetry I*, S.-T. Yau, ed., vol. 9 of *Studies in Advanced Mathematics* (AMS, Providence, RI and Intl. Press, Cambridge, MA, 1998).

[479] E. Witten "Small instantons in string theory," *Nucl. Phys.* **B460** (1996) 541–559.

O.J. Ganor and A. Hanany: "Small E(8) instantons and tensionless non-critical strings," *Nucl. Phys.* **B474** (1996) 122–140.
B.A. Ovrut, T. Pantev and J. Park: "Small instanton transitions in heterotic M theory," *J. High E. Phys.* **05** (2000) 045.
E. Buchbinder, R. Donagi and B.A. Ovrut: "Vector bundle moduli and small instanton transitions," *J. High E. Phys.* **06** (2002) 054.

[480] M. Yamazaki: "Brane Tilings and Their Applications," *Fortsch. Phys.* **56** (2008) 555.

Index

D

www.ingramcontent.com/pod-product-compliance
Ingram Content Group UK Ltd.
Pitfield, Milton Keynes, MK11 3LW, UK
UKHW021826091026
9241IPUK00004B/40

9 781800 616424